OPTICAL PROPERTIES
OF SOLIDS

Optical Physics and Engineering

Series Editor: William L. Wolfe
Honeywell Inc. Radiation Center
Boston, Mass.

1968:

M. A. Bramson
Infrared Radiation: A Handbook for Applications

1969:
Sol Nudelman and S. S. Mitra, Editors
Optical Properties of Solids

OPTICAL PROPERTIES OF SOLIDS

Edited by
Sol Nudelman and S. S. Mitra
Department of Electrical Engineering
University of Rhode Island
Kingston, Rhode Island

Papers from the
NATO Advanced Study Institute on Optical Properties of Solids
Held August 7-20, 1966, at Freiburg, Germany.

℗ PLENUM PRESS · NEW YORK · 1969

Library of Congress Catalog Card Number 68-15008

To
FRANK L. MATOSSI
1902-1968

Preface

This book is an account of the manner in which the optical phenomena observed from solids relate to their fundamental properties. Written at the graduate level, it attempts a threefold purpose: an indication of the breadth of the subject, an in-depth examination of important areas, and a text for a two-semester course.

The first two chapters present introductory theory as a foundation for subsequent reading. The following ten chapters broadly concern electronic properties associated with semiconductors ranging from narrow to wide energy gap materials. Lattice properties are examined in the remaining chapters, in which effects governed by phonons in perfect crystals, point defects, their vibrational and electronic spectra, and electron–phonon interactions are stressed.

Fun and hard work, both in considerable measure, have gone into the preparation of this volume. At the University of Freiburg, W. Germany, from August 7–20, 1966, the occasion of a NATO Advanced Study Institute on "The Optical Properties of Solids," the authors of these various chapters lectured for the Institute; this volume provides essentially the "Proceedings" of that meeting. Many major revisions of original lectures (contractions and enlargements) were required for better organization and presentation of the subject matter. Several abbreviated chapters appear mainly to indicate the importance of their contents in optical properties research and to indicate recently published books that provide ample coverage.

We are indebted to many people: the authors for their efforts and patience; our host at the University of Freiburg, the late Professor Dr. F. Matossi, and his assistants (particularly Dr. H. Teitge); and Professor C. Polk of the University of Rhode Island, who provided encouragement and support throughout the preparation of the manuscript. Proofreading and index preparation were helped by Mr. K. V. Namjoshi and Mr. R. S. Singh, and secretarial assistance beyond the call of duty was performed by Mrs. M. Barney.

Most of all, heartfelt thanks go to NATO and to the co-sponsors of the Institute (the Office of Naval Research and the Advanced Research Projects Agency). Without the assistance of Dr. H. Arnth-Jensen (NATO), Mr. F. Isakson (ONR), and Dr. R. Zirkind (ARPA), this Institute could not have taken place.

S. Nudelman and S. S. Mitra
University of Rhode Island

July 1968

Contents

CHAPTER 1

Group Theory in Crystal Physics

H. Jones

Department of Mathematics
Imperial College
London, England

Group theory is used in crystal physics to obtain a classification of electronic states and of normal modes of lattice vibrations. In one important application it is used to determine all possible symmetries of the wave functions of a single electron in a given space-periodic field. Another application consists of the determination of the symmetries of the states of an impurity atom subject to a field imposed by the neighboring atoms of crystal. The same theory applies in a very similar form to the analogous situation of lattice vibrations, either in a perfect lattice or in a lattice containing impurity atoms.

The most important part of the theory of groups for solid-state physics is that part which deals with the representation of the abstract group by matrices, and this chapter is concerned almost exclusively with this aspect of the theory. The groups of interest are those whose operations leave invariant some differential equation or system of equations. To gain precision of statement and without loss of essential generality, the argument is restricted to the case of the Schrödinger equation for a single particle in a space-periodic field. The operations of the group to be considered are those which leave the Hamiltonian operator $H(r)$ unchanged, and since these are linear transformations of the coordinates which are orthogonal transformations, or linear displacements, the Laplacian operator is invariant and the group is defined by the symmetry of the potential energy $V(r)$.

The crystal is considered to be of infinite extent but divided into finite regions of equal size and shape. The condition that every property of the crystal at every point and instant repeats exactly over these "periodicity regions" is imposed. If ψ is the dependent variable of any differential equation relating to physical properties, then $\psi(r + R_N) = \psi(r)$, where R_N is a displacement of any point in one region to the corresponding point of another. Such a displacement represents the identity operator of the group to be considered. The operations which leave $V(r)$ unchanged are now finite

1

in number and satisfy the following conditions: (1) they contain the identity, (2) the product of any pair is a member of the set, and (3) the inverse of every operator is included in the set. These operations, therefore, form a group of finite order.

I. THE SPACE GROUP

Let $\mathbf{a}_1$, $\mathbf{a}_2$, $\mathbf{a}_3$ be the primitive vectors which define the crystal lattice. Hence, if $\mathbf{n} = (n_1, n_2, n_3)$ is a triple of integers, a lattice point is specified by the vector

$$\mathbf{R_n} = n_1 \, \mathbf{a}_1 + n_2 \, \mathbf{a}_2 + n_3 \, \mathbf{a}_3 \tag{1.1}$$

and, as $\mathbf{n}$ takes all integral values, $\mathbf{R_n}$ maps out of the translation lattice. A sum of two vectors may be written as follows:

$$\mathbf{R_n} + \mathbf{R_m} = (n_1 + m_1) \, \mathbf{a}_1 + (n_2 + m_2) \, \mathbf{a}_2 + (n_3 + m_3) \, \mathbf{a}_3 = \mathbf{R_{(n+m)}} \tag{1.2}$$

Let $T(\mathbf{R_n})$ denote the operator which displaces a point in space specified by the vector $\mathbf{r}$ to a point specified by $\mathbf{r} + \mathbf{R_n}$. The potential energy $V(\mathbf{r})$ is invariant under all such operations as the following equations show:

$$T(\mathbf{R_n}) \, V(\mathbf{r}) = V(\mathbf{r} + \mathbf{R_n}) = V(\mathbf{r}) \tag{1.3}$$

Let the periodicity region of the crystal be defined by the vectors $N_1 \mathbf{a}_1$, $N_1 \mathbf{a}_2$, $N_3 \mathbf{a}_3$, where N_1, N_2, N_3 are large integers. All properties of the crystal, and in particular all wave functions, repeat exactly with respect to the displacements $\mathbf{R_N}$, that is, for all states $\psi(\mathbf{r} + \mathbf{R_N}) \equiv \psi(\mathbf{r})$. Hence, the $N_1 N_2 N_3$ operators $T(\mathbf{R_n})$ form a translation group of order $\gamma = N_1 N_2 N_3$. This follows because (1) the set contains the identity $E = T(\mathbf{R_N})$, (2) the product of any two operators is a third of the set $T(\mathbf{R_n}) \, T(\mathbf{R_m}) = T(\mathbf{R_{n+m}})$, and (3) the inverse of each operation is contained in the set

$$T^{-1}(\mathbf{R_n}) = T(-\mathbf{R_n}) = T(\mathbf{R_{N-n}})$$

This translation group, denoted by Γ, is a subgroup of the crystallographic space group. It is Abelian because all operators commute with each other:

$$T(\mathbf{R_n}) \, T(\mathbf{R_m}) = T(\mathbf{R_{n+m}}) = T(\mathbf{R_m}) \, T(\mathbf{R_n}) \tag{1.4}$$

In addition to the pure translations, there are other operations which leave $V(\mathbf{r})$ invariant. These operations depend on the nature of the primitive vectors $\mathbf{a}_1, \mathbf{a}_2, \mathbf{a}_3$. For example, if $\mathbf{a}_1, \mathbf{a}_2, \mathbf{a}_3$ are equal vectors at right angles to each other, they define a cubic lattice, and a linear transformation of the coordinates which corresponds to a rotation through π or $\pi/2$, for instance, will leave $V(\mathbf{r})$ unchanged. Clearly also a reflection in a lattice plane would be another such operation. A different situation arises if a_1 and a_2 are inclined at 120° to each other and a_3 is perpendicular to both vectors. In this case, the lattice would be hexagonal and operations which rotate the crystal through 60° about the direction of a_3 would leave $V(\mathbf{r})$ unaltered.

In general, let these operations be denoted by $A, B, C, \ldots$. We then speak of $\{E, A, B, C, \ldots\}$ as a complex. This complex does not always form a group by itself, although in some important cases it does do so. If we multiply all operations of Γ by A, for example, we get the set $A\Gamma$ which is called a coset. If the subgroup Γ is such that $X\Gamma = \Gamma X$, or equivalently $\Gamma = X^{-1}\Gamma X$, where X is any member of the group, then we say that Γ is an invariant subgroup or normal divisor. The translation group Γ is such an invariant subgroup, because it is clear that although $X^{-1}T(\mathbf{R_n})X$ will not, in general, be equal to $T(\mathbf{R_n})$, yet it must be another translation, and the operators $X^{-1}\Gamma X$ will consist therefore of the same set of translations as Γ. Hence, the whole space group, denoted by G, may be written as

$$G = \Gamma + A\Gamma + B\Gamma + \cdots \qquad (1.5)$$

where no difference exists between the right and left cosets.

II. GROUP REPRESENTATIONS

A group is characterized by the way its operators multiply together. This property is exemplified by the group table. It is not necessary in general to specify the precise nature of the operators. Two different sets of operators which multiply together in the same way, i.e., have the same group table, are said to be isomorphic. A set of square matrices which have the same multiplication properties as the operators of the group is said to form a representation of the group. The application of group theory to crystal physics is almost entirely concerned with the representation of groups by matrices.

III. BASIS FUNCTIONS

The construction of matrices to form a representation is quite simple. Let X be an operator of the group and $f(\mathbf{r})$ a solution of the Schrödinger equation

$$H(\mathbf{r})f(\mathbf{r}) = \varepsilon f(\mathbf{r}) \qquad (1.6)$$

not necessarily satisfying boundary conditions and therefore not necessarily a wave function. The operator X transforms $\mathbf{r}$ to $\mathbf{r}'$, i.e., $X\mathbf{r} = \mathbf{r}'$, and therefore application of this operator to equation (1.6) gives

$$H(X\mathbf{r})f(X\mathbf{r}) = \varepsilon f(X\mathbf{r}) \qquad (1.7)$$

and since $H(X\mathbf{r}) = H(\mathbf{r})$ it follows that $f(X\mathbf{r})$, which may be written $f_X(\mathbf{r})$, is also a solution of the Schrödinger equation. If the order of the group is n, there are n such functions—$f_E(\mathbf{r})$, $f_A(\mathbf{r})$, $f_B(\mathbf{r})$ etc.,—each satisfying the Schrödinger equation; since the operators form a group, any operator applied to any f produces another function of the set.

It is convenient to regard the basis functions as the components of an n-dimensional column vector sometimes written $\{f_E, f_A, \ldots\}$ for convenience of printing. If we now operate on this vector with any operator P of the group, this merely changes the order of the components as follows:

$$P \begin{pmatrix} f_E \\ f_A \\ \vdots \end{pmatrix} = \begin{pmatrix} 0 & 0 & 1 & 0 & \ldots \\ 1 & 0 & 0 & \ldots \\ & & \text{etc.} \end{pmatrix} \begin{pmatrix} f_E \\ f_A \\ \vdots \end{pmatrix} \tag{1.8}$$

where the matrices have zeros everywhere except at one point in each row where there is the numeral one. The traces are zero except for the identity operator E which may be expressed as follows:

$$T_r P = \begin{cases} 0 & P \neq E \\ n & P = E \end{cases} \tag{1.9}$$

The mathematical problem to be solved in the wave-mechanical application is the reduction of this representation. This means that the matrices must be brought as near as possible to diagonal form, and this in turn implies the choice of solutions which are linear combinations of the basis functions which have specified symmetries.

IV. THE TRANSLATION GROUP Γ

We begin by considering the representations of the subgroup Γ, and, because this group is Abelian, the matrices which represent the operators must commute with each other. Any particular matrix A can always be brought to diagonal form by a similarity transformation, i.e., an S exists such that $S^{-1}AS$ is diagonal. Now since all matrices commute it follows that the same S will diagonalize all of them. Hence, the representation of an Abelian group can always be brought into diagonal form. If the operators do not commute, this is not possible.

The reduction of the matrices representing the operators means that a set of linear combinations of the f's can always be found, say, ψ_1, ψ_2, etc., such that with this new basis all matrices are diagonal. Thus, we write

$$\psi_i = \sum_A \alpha_{iA} f_A(\mathbf{r}) \tag{1.10}$$

Since the ψ_i's lead to diagonal matrices, it follows that $P\psi_i = \lambda\psi_i$, where P is an operator of the group and λ is a number. Our first problem, therefore, is to find these numbers for the translation group.

Since the operators of Γ are the translations $T(\mathbf{R_n})$, we can specify the basis functions as follows: $T(\mathbf{R_n})f(\mathbf{r}) = f_{\mathbf{n}}(\mathbf{r})$. Thus, the $\gamma = N_1 N_2 N_3$ functions are specified by γ triples of integers $\mathbf{n}$. Consider the linear combinations

$$\psi_{\mathbf{k}}(\mathbf{r}) = \sum_{\mathbf{n}} e^{-i\mathbf{k}\cdot\mathbf{R_n}} f_{\mathbf{n}}(\mathbf{r}) \qquad (1.11)$$

where the sum is taken over all γ values of $\mathbf{n}$ and where $\mathbf{k}$ is some vector to be detemined.

Operating with $T(\mathbf{R_s})$, we get

$$\psi_{\mathbf{k}}(\mathbf{r}+\mathbf{R_s}) = \sum_{\mathbf{n}} e^{-i\mathbf{k}\cdot\mathbf{R_n}} f_{\mathbf{n}+\mathbf{s}}(\mathbf{r})$$

$$= e^{i\mathbf{k}\cdot\mathbf{R_s}} \sum e^{-i\mathbf{k}\cdot\mathbf{R_m}} f_{\mathbf{m}}(\mathbf{r}) = e^{i\mathbf{k}\cdot\mathbf{R_s}} \psi_{\mathbf{k}}(\mathbf{r}) \qquad (1.12)$$

where in the second line we have made the substitution $\mathbf{n}+\mathbf{s} = \mathbf{m}$. The coefficients of $\psi_{\mathbf{k}}(\mathbf{r})$ are the required numbers and will therefore appear on the diagonals of the reduced representation. Now because $T(\mathbf{R_N})$ is the identity

$$\mathbf{k}\cdot\mathbf{R_N} = 2\pi(\text{integer}) \qquad (1.13)$$

To interpret the vector $\mathbf{k}$, we introduce the reciprocal lattice. Let

$$\mathbf{K_m} = m_1\mathbf{b}_1 + m_2\mathbf{b}_2 + m_3\mathbf{b}_3 \qquad (1.14)$$

where the $\mathbf{b}_i$'s are to be defined so that

$$\mathbf{K_m}\cdot\mathbf{R_n} = 2\pi\mathbf{m}\cdot\mathbf{n} = 2\pi(m_1 n_1 + m_2 n_2 + m_3 n_3) \qquad (1.15)$$

This requires that

$$\mathbf{b}_i\cdot\mathbf{a}_j = 2\pi\delta_{ij} \qquad (1.16)$$

and therefore defines the three $\mathbf{b}_i$'s. These vectors are called the primitive vectors of the reciprocal lattice. If we write $\mathbf{k} = h_1\mathbf{b}_1 + h_2\mathbf{b}_2 + h_3\mathbf{b}_3$, then $\mathbf{k}\cdot\mathbf{R_N} = 2\pi(h_1 N_1 + h_2 N_2 + h_3 N_3)$, and equation (1.13) requires that $(h_1 N_1 + h_2 N_2 + h_3 N_3)$ shall be an integer. This requirement is satisfied if $h_1 = l_1/N_1$, etc., where l_1, l_2, l_3 are integers less than N_1, N_2, N_3, respectively. Thus, $\mathbf{k}$ is a vector in the reciprocal lattice space and takes γ distinct values, each one of which specifies a position along the diagonal of the reduced representation.

It has been seen that the traces of all matrices must be zero except that of the matrix which represents the identity. Thus, since a similarity transformation does not change the trace, it follows that

$$T_r \, T(\mathbf{R_s}) = \sum_{l_1, l_2, l_3} \exp\left[2\pi i\left(\frac{l_1}{N_1} s_1 + \frac{l_2}{N_2} s_2 + \frac{l_3}{N_3} s_3 \right) \right] \qquad (1.17)$$

and this is zero unless $s = 0$ as required by equation (1.9). Thus, we have obtained the representation of the operators of Γ in reduced, i.e., diagonal, form. This reduction of Γ constitutes Bloch's theorem, which states that

$$T(\mathbf{R_n}) \, \psi_\mathbf{k}(\mathbf{r}) = e^{i\mathbf{k} \cdot \mathbf{R_n}} \, \psi_\mathbf{k}(\mathbf{r}) \qquad (1.18)$$

This condition is satisfied for all $T(\mathbf{R_n})$'s if the wave function is written in the form

$$\psi_\mathbf{k}(\mathbf{r}) = e^{i\mathbf{k} \cdot \mathbf{r}} u_\mathbf{k}(\mathbf{r}) \qquad (1.19)$$

where $u_\mathbf{k}(\mathbf{r} + \mathbf{R_n}) = u_\mathbf{k}(\mathbf{r})$.

V. THE POINT GROUP

When the complex of operations $\{E, A, B, ...\}$, which together with Γ constitutes the space group $G = \Gamma + A\Gamma + \cdots$, itself forms a group, it is known as the point group. It is made up of rotations, reflections, and combinations of these called improper reflections. One object of the group-theoretical analysis is the classification of wave functions according to their symmetry with respect to the operations of these point groups. The reduction of the representation of Γ to diagonal form enabled us to define a wave vector $\mathbf{k}$ and to express all crystal wave functions in the Bloch form. We now examine the symmetry of $\psi_\mathbf{k}(\mathbf{r})$ with respect to the operations of the point group at a particular point in k-space. The first step is clearly to construct the group multiplication table. For the cubic crystal of highest symmetry, there are 48 operations in the point group and a systematic approach is therefore necessary.

Let S denote a rotation about a particular axis. For example, in a cubic crystal a rotation about the z axis (coincident with a fourfold axis) is given by

$$\begin{pmatrix} x' \\ y' \\ z' \end{pmatrix} = \begin{pmatrix} \cos\gamma & \sin\gamma & 0 \\ -\sin\gamma & \cos\gamma & 0 \\ 0 & 0 & 1 \end{pmatrix} \begin{pmatrix} x \\ y \\ z \end{pmatrix} \qquad (1.20)$$

A rotation through the same angle γ about any other axis can be written as $P^{-1}SP$, where P denotes a transformation to another system of axes. The result of such an operation is shown by the following equation:

$$(P^{-1} SP) \, \mathbf{R_n} = \mathbf{R_{n'}} \qquad (1.21)$$

i.e., the vector $\mathbf{R_n}$ suffers a rotation to become $\mathbf{R_{n'}}$. If $\mathbf{n'}$ is also a triple of integers, it means that each lattice point $\mathbf{n}$ is transferred to another point $\mathbf{n'}$ and therefore $P^{-1}SP$ is an operation of the point group.

In any matrix representation of the group, the operations $P^{-1}SP$ and S must have the same trace or diagonal sum. Thus, rotations of a given amount about different axes will have the same trace in any representation. For example, rotations of $\pi/2$ about any of the coordinate axes (the $\langle 100 \rangle$ axes) will have the same trace.

A reflection in the xy plane would be represented by

$$\begin{pmatrix} x' \\ y' \\ z' \end{pmatrix} = \begin{pmatrix} 1 & 0 & 0 \\ 0 & 1 & 0 \\ 0 & 0 & -1 \end{pmatrix} \begin{pmatrix} x \\ y \\ z \end{pmatrix} \tag{1.22}$$

Let this be denoted by M. Then $P^{-1}MP$ would be a reflection in some other plane determined by the transformation matrix P. Again the traces of all matrices denoting reflections will be the same.

VI. STEREOGRAPHIC PROJECTION

The stereographic projection provides a convenient means of visualizing the operations of the point group. Figure 1 shows an example of a stereogram for the cubic system. The great circles which pass through the pole project as straight lines. Those great circles whose normals are inclined to the vertical project as sectors of circles. The intersections denote the directions of the symmetry axes of the cube: fourfold axes are indicated by small squares, diad axes by elongated dots, and trigonal axes by triangles. Operations of the point group can be represented by displacements in the stereogram, for example, the reflection in the xz plane implies $A \to A'$ and a rotation about a trigonal axis through $120°$ implies $B \to B'$. In order to represent directions which make an angle greater than $\pi/2$ with the upward vertical, the projection is taken from the opposite pole and denoted by a small circle

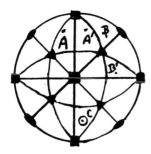

Fig. 1. A stereogram for the cubic system.

as shown at C, and thus two directions with direction cosines (l, m, n) and $(l, m, -n)$ are indicated by $\odot$ in the appropriate place in the unit circle. The order of the point group is equal to the total number of dots and small circles in the stereogram.

VII. CLASSES

The first step toward discovering the structural properties of a group is the arrangement of the operators into classes. A class is defined in the following way. Let A and X be two operations of the group; then, by definition, $X^{-1}AX$ belongs to the same class as A. The g elements $X^{-1}AX$ which result from allowing X to stand in turn for each operation of the group of order g consist only of the elements of a single class. Each element of the class is repeated g/h_i times, where h_i is the number of distinct operations in the class denoted by the symbol $\mathscr{C}_i$. This may be expressed as follows

$$\mathscr{C}_i = \frac{h_i}{g} \sum_X X^{-1}AX \tag{1.23}$$

where the summation is over each operation of the group.

In any representation of the group by a set of matrices, the matrices of one class can be transformed into each other by a similarity transformation and therefore all have the same trace. This, it may be recalled, was shown to be the property of rotations of a given angle about different axes. Such rotations might, therefore, be expected to form one class of the point group. Similarly, reflections in a plane of any orientation are represented by matrices with the same trace. It does not follow, however, that all reflections with the same trace must belong to one class, since the similarity transformation matrix has to be a matrix of the representation and cannot be chosen arbitrarily. An example makes this clear: The symmetry operations which can be performed on a two-dimensional square lattice are indicated in Fig. 2, which is analogous to a stereogram, and the formation of two classes is shown in Table I. It will be observed that both lines satisfy equation (1.23) where $g = 8$ and $h_1 = h_2 = 2$. Note that the reflections m_1, m_2 and m'_1, m'_2 belong to difference classes although they would have the same trace.

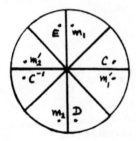

Fig. 2. Diagram showing the symmetry operations that can be performed on a two-dimensional square lattice.

Table I

X	E	m_1	C	m_1'	D	m_2	C^{-1}	m_2'
$X^{-1} m_1 X$	m_1	m_1	m_2	m_2	m_1	m_1	m_2	m_2
$X^{-1} m_1' X$	m_1'	m_2'	m_2'	m_1'	m_1'	m_2'	m_2'	m_1'

One important feature of the classes is that they always commute with each other. Thus, if we define a class symbol $\mathscr{C}_i$ by

$$\mathscr{C}_i = A + B + C + \cdots + P \qquad (1.24)$$

where A, B, etc., belong to one class, then

$$\mathscr{C}_i \mathscr{C}_j = \mathscr{C}_j \mathscr{C}_i \qquad (1.25)$$

For example, if $A + B$ and $C + D$ are two classes,

$$(A + B)(C + D) = (A + B)C + (A + B)D \qquad (1.26)$$

and, since $X^{-1}(A + B)X = A + B$ for any X in G, it follows that

$$(A + B)X = X(A + B)$$

and therefore that the right-hand side of equation (1.26) is equal to

$$C(A + B) + D(A + B)$$

As a result of equation (1.25), it follows that the matrices which represent the classes can always be brought into diagonal form. Again it follows that

$$\mathscr{C}_i \mathscr{C}_j = \sum_s c_{ij,s} \mathscr{C}_s \qquad (1.27)$$

where the coefficients $c_{ij,s}$ are positive integers. This can be seen as follows. The product of $\mathscr{C}_i$ and $\mathscr{C}_j$ will be a sum of operations, for example,

$$\mathscr{C}_i \mathscr{C}_j = A + B + \cdots \qquad (1.28)$$

and hence

$$X^{-1} \mathscr{C}_i X X^{-1} \mathscr{C}_j X = X^{-1}(A + B + \cdots)X \qquad (1.29)$$

By the basic property of classes, the left-hand side of equation (1.28) is unchanged by the similarity transformation for any X in the group. Hence, if $(A + B + \cdots)$ does not consist of complete classes, the extra operators must be transformed into others of their class by some X. Hence, the equality of the right-hand sides of equations (1.28) and (1.29) would be violated.

VIII. REDUCTION OF THE REPRESENTATION
OF THE POINT GROUP G_P

Let y_E, y_A, y_B, etc., be solutions generated by the operators E, A, B, ... of the point group. The column vector $\{y_E, y_A, ...\}$ is transformed by an element of G_P in a manner similar to equation (1.8), namely,

$$P\begin{pmatrix} y_E \\ y_A \\ \vdots \end{pmatrix} = \begin{pmatrix} 0 & 0 & 1 & - - - \\ 1 & 0 & & - - - \\ & & \cdot & \\ & & & 0 \end{pmatrix}\begin{pmatrix} y_E \\ y_A \\ \vdots \end{pmatrix} \qquad (1.30)$$

where, however, the matrix now has only g rows and columns, not $n = \gamma g$ as previously. These matrices will be referred to as the regular representation. Clearly again

$$T_r P = \begin{cases} 0 & P \neq E \\ g & P = E \end{cases} \qquad (1.31)$$

Since the operations of this point group do not commute, there is no similarity transformation which will bring all the matrices onto diagonal form as was the case with the translation group Γ. When the matrices of equation (1.30) are brought as near to diagonal form as possible, for all operations of the group, they are said to be completely reduced and the submatrices constitute the irreducible representations. Each matrix of the regular representation will be of the following form:

$$\begin{pmatrix} (\Gamma_1) & & & \\ & (\Gamma_2) & & \\ & & (\Gamma_3) & \\ & & & \cdot \end{pmatrix} \qquad (1.32)$$

Each set of submatrices, for example, $\Gamma_i(A)$ for all A, forms by itself a group isomorphic with the point group and therefore constitutes a representation. When the Γ_i's are made as small as possible, they are said to be the irreducible representations of G_P. Γ_1 is called the identity representation and consists of the numeral 1 at the same position along the diagonal for all matrices of the group. It always exists because one basis function can be chosen as the sum of all y_A. Clearly Σy_A is a linear invariant, i.e., it transforms into itself under every operation of G_P.

The degrees of these representations, i.e., the number of rows or columns of the submatrices, vary from 1 to 3 in the crystallographic point groups. A submatrix of given degree may occur several times along the diagonal. Some of these can be transformed into each other by a single similarity transformation for all operations of the group. Such submatrices (representations) are said to be equivalent. If two representations cannot be transformed into each other in this way, they are said to be inequivalent. Equivalent representations have the same trace; inequivalent ones have different traces for at least some of the operations of the group.

An example of an inequivalent representation is provided by the representation of degree 2 in the point group of the tetragonal system. In one representation the basis functions transform as x and y, and in the other, as xz and yz. The inversion J is one operation of the group. Hence, in the first representation,

$$J \begin{pmatrix} x \\ y \end{pmatrix} = \begin{pmatrix} -x \\ -y \end{pmatrix} = \begin{pmatrix} -1 & 0 \\ 0 & -1 \end{pmatrix} \begin{pmatrix} x \\ y \end{pmatrix} \tag{1.33}$$

and in the second

$$J \begin{pmatrix} xz \\ yz \end{pmatrix} = \begin{pmatrix} xz \\ yz \end{pmatrix} = \begin{pmatrix} 1 & 0 \\ 0 & 1 \end{pmatrix} \begin{pmatrix} xz \\ yz \end{pmatrix} \tag{1.34}$$

When a matrix is transformed into another by a similarity transformation, the trace remains unchanged. Hence, equivalent representations have the same trace, or character, and inequivalent representations unequal characters. In the above example, one character is -2, the other $+2$.

A. Theorem: The Number of Inequivalent Irreducible Representations is Equal to the Number of Classes

To prove this theorem, regard the class symbols $\mathscr{C}_1, \mathscr{C}_2, \ldots, \mathscr{C}_r$ as the components of a column vector and operate on this vector with $\mathscr{C}_i$. We obtain

$$\mathscr{C}_i \begin{pmatrix} \mathscr{C}_1 \\ \mathscr{C}_2 \\ \vdots \\ \mathscr{C}_r \end{pmatrix} = \begin{pmatrix} \mathscr{C}_i \mathscr{C}_1 \\ \mathscr{C}_i \mathscr{C}_2 \\ \vdots \\ \mathscr{C}_i \mathscr{C}_r \end{pmatrix} = \begin{pmatrix} \sum c_{i1s} \mathscr{C}_s \\ \sum c_{i2s} \mathscr{C}_s \\ \vdots \\ \sum c_{irs} \mathscr{C}_s \end{pmatrix} = \begin{pmatrix} c_{i11} & c_{i12} & \cdots & c_{i1r} \\ c_{i21} & & & \vdots \\ \vdots & & & \\ c_{ir1} & \cdots & & c_{irr} \end{pmatrix} \begin{pmatrix} \mathscr{C}_1 \\ \mathscr{C}_2 \\ \vdots \\ \mathscr{C}_r \end{pmatrix} \tag{1.35}$$

Thus, each $\mathscr{C}_i$ may be represented by a matrix similar to the one shown above where the elements are the coefficients of equation (1.21) which are determined directly from the group multiplication table. The r classes $\mathscr{C}_i$ form an Abelian group of order r so that the matrices $(c_{ij, s})$ may be brought simultaneously onto diagonal form. Let this diagonalized form of the matrix representing $\mathscr{C}_i$ be

$$
\begin{pmatrix}
x_i^{(1)} & & & \\
& x_i^{(2)} & & \\
& & \ddots & \\
& & & x_i^{(r)}
\end{pmatrix}
\tag{1.36}
$$

Now consider what the matrix representing $\mathscr{C}_i$ will be in the completely reduced form of the regular representation. The character (i.e., trace) of the νth irreducible representation of the operator A is $\chi^{(\nu)}(A)$ and since $\mathscr{C}_i$ consists of the sum of the following h_i operations

$$
\mathscr{C}_i = A + B + \cdots
\tag{1.37}
$$

and since the character of each of these is the same, it follows that the trace of the submatrix of $\mathscr{C}_i$ is $h_i \chi_i^{(\nu)}$, where $\chi_i^{(\nu)} = \chi^{(\nu)}(A) = \chi^{(\nu)}(B)$, etc. The degree of the representation must be $\chi_E^{(\nu)}$ because the matrix representing E is always the unit matrix. Now the matrix of $\mathscr{C}_i$ in any given irreducible representation commutes with the matrices of all operators A, B, etc., of the same irreducible representation and these in general are not diagonal. Hence, the matrix representing $\mathscr{C}_i$ in the irreducible representation must be the unit matrix multiplied by some numerical factor. The matrix of $\mathscr{C}_i$ in the νth irreducible representation must therefore be a diagonal matrix in which each diagonal element is $h_i \chi_i^{(\nu)}/\chi_E^{(\nu)}$. If there are n_ν equivalent matrices in the νth irreducible representation, then in the reduced form of the regular representation $h_i \chi_i^{(\nu)}/\chi_E^{(\nu)}$ will occur just $n_\nu \chi_E^{(\nu)}$ times along the diagonal. Hence, there will be as many different terms along the diagonal as there are irreducible representations. However, we have seen that the diagonal matrices (1.36), when suitably extended by repetition along the diagonal to degree g, must be identical with the reduced regular representation. As there are just r different numbers $x_i^{(s)}$ in matrices (1.36), we conclude that there are just r inequivalent submatrices, i.e., r irreducible representations.

B.   Theorem: The Number of Times an Irreducible Representation Occurs Along the Diagonal of the Reduced Regular Representation is Equal to its Degree

What is implied by this theorem is the following. The submatrix $\Gamma_\nu(A)$ occurs $\chi_E^{(\nu)}$ times along the diagonal and hence the diagonal sum, for the identity operator, of these submatrices is $[\chi_E^{(\nu)}]^2$. Since we have shown that ν varies from 1 to r and the trace of the total matrix of the identity operator is g, we have

$$
[\chi_E^{(1)}]^2 + [\chi_E^{(2)}]^2 + \cdots + [\chi_E^{(r)}]^2 = g
\tag{1.38}
$$

This relation enables us to obtain the degree of the irreducible representations immediately. The square lattice is an example. Here $g = 8$ and there are five classes. Hence,

$$1^2 + 1^2 + 1^2 + 1^2 + 2^2 = g \qquad (1.39)$$

and we see that there are four single irreducible representations and one double one. This means that the wave functions, classified by such a group, fall into four nondegenerate types and one doubly degenerate type.

IX. OUTLINE OF PROOF

An outline of the argument which leads to the above theorem is as follows. Let Γ_ν and Γ_μ be two irreducible representations of degrees n and m. These matrices will therefore operate on column vectors of n and m components, respectively. Let these vectors be denoted by $\{x_1, x_2, ..., x_n\}$ and $\{y_1, y_2, ..., y_m\}$. Now form products of all x's and y's and take these to be the nm components of a column vector

$$\{x_1 y_1, x_1 y_2, ..., x_1 y_m; x_2 y_1, ...; ..., x_n y_m\}$$

The matrices Γ_ν and Γ_μ acting together produce a linear transformation of the components of this vector. Such a transformation can also be brought about by a matrix of nm rows and columns. This matrix is said to be the direct product of Γ_ν and Γ_μ and is written $\Gamma_\nu \times \Gamma_\mu$. These direct-product matrices also form a representation of the group, for the product of direct products obeys the following rule:

$$[\Gamma_\nu(A) \times \Gamma_\mu(A)][\Gamma_\nu(B) \times \Gamma_\mu(B)] = \Gamma_\nu(A)\,\Gamma_\nu(B) \times \Gamma_\mu(A)\,\Gamma_\mu(B) \qquad (1.40)$$

For example, if $AB = C$ then since Γ_ν and Γ_μ are representations $\Gamma_\nu(A)\,\Gamma_\nu(B) = \Gamma_\nu(C)$ and $\Gamma_\mu(A)\,\Gamma_\mu(B) = \Gamma_\mu(C)$, and, hence, the right-hand side of equation (1.40) is $\Gamma_\nu(C) \times \Gamma_\mu(C)$.

The representation $\Gamma_\nu \times \Gamma_\mu$ of degree nm will in general be reducible, and since there is only one set of irreducible representations of a given group it follows that, by a similarity transformation, direct-product matrices can be brought to a form in which, for every operation of the group, just the submatrices Γ_s occur along the diagonals and zeros occur everywhere else. This may be expressed by the equation

$$\Gamma_\nu(A) \times \Gamma_\mu(A) = \sum_s g_{\nu\mu s}\,\Gamma_s(A) \qquad (1.41)$$

where the sum on the right-hand side has the above meaning and hence the $g_{\nu\mu s}$'s are positive integers.

In the regular representation by matrices of degree g, the identity representation always occurs in the reduced form. This follows, as we have seen, from the obvious fact that Σy_A is invariant under all operations of the group. The question now arises whether the identity representation Γ_1 appears on the right-hand side of equation (1.41) in the reduction of the direct-product representation. The answer depends on whether a bilinear invariant $\sum_{i,j} a_{ij} x_i y_j$ exists or not.

First it will be shown that if $n \neq m$ no bilinear invariant can exist. Let Γ and Γ' be any two matrix representations of the group of degrees n and m, operating on the n x's and m y's, respectively. We now show that, if $n \neq m$ and if a bilinear invariant exists, then Γ or Γ' must be reducible, and thus, if Γ and Γ' are stated to be irreducible, no such invariant can exist. By introducing new variables, the invariant can always be written as

$$I = \sum_{i=1}^{n} \sum_{j=1}^{m} a_{ij} x_i y_j = \xi_1 \eta_1 + \xi_2 \eta_2 + \cdots + \xi_r \eta_r \qquad (1.42)$$

where r is less than or equal to either n or m. The change of variables replaces the n x's by n ξ's and the m y's by m η's, but ξ_{r+1} to ξ_n and η_{r+1} to η_m do not occur in the invariant. The matrices Γ and Γ' which operate on the column vectors $\{x_1, x_2, \ldots, x_n\}$ and $\{y_1, y_2, \ldots, y_m\}$, respectively, can be transformed for all operations of the group so as to apply to the column vectors

$$\{\xi_1, \xi_2, \ldots, \xi_n\} \quad \text{and} \quad \{\eta_1, \eta_2, \ldots, \eta_m\}$$

It is now clear that if I is an invariant Γ can mix together only the first r ξ's not all n, and Γ' mixes only the first r η's not all m. Hence, if either $r < n$ or $r < m$, Γ or Γ' is reducible or both are reducible. Only if $r = n = m$ can both be irreducible. In this case, Γ' must be the inverse, or the transposed, matrix of Γ, i.e., they must have the same trace. Hence, if Γ is the irreducible representation Γ_v, then no invariant exists unless Γ' is an equivalent matrix of the same representation; i.e., if $v \neq \mu$, then $g_{v\mu,1} = 0$, but if $v = \mu$ an invariant exists which means that Γ_1 occurs on the right-hand side of equation (1.41) and therefore that $g_{vv,1} = 1$.

If we now take the trace of equation (1.41) we get

$$\chi^{(v)}(A) \chi^{(\mu)}(A) = \sum_{s} g_{v\mu,s} \chi^{(s)}(A) \qquad (1.43)$$

because the trace of a direct product is equal to the product of the traces of the two matrices.

To make further progress, the following result is required

$$\sum_{A} \chi^{(s)}(A) = \begin{cases} 0 & \text{if } s \neq \text{identity representation} \\ g & \text{if } s = \text{identity representation} \end{cases} \qquad (1.44)$$

and this will now be proved.

From the original basis functions we form the following functions:

$$u = \sum_A y_A$$

$$z_A = y_A - y_E \tag{1.45}$$

$$z_B = y_B - y_E, \quad \text{etc.}$$

Altogether there are $g-1$ of these z variables. Now $(\Sigma A)z_X = 0$ for all X since there are just as many positive as negative terms and each occurs just once. Also $\chi^{(1)}(A) = 1$ for all A and therefore

$$\sum_A \chi^{(1)}(A) = g \tag{1.46}$$

Since (ΣA) operating on any linear combination of z's is zero, this operator will always lead to a zero on the diagonal and therefore

$$\sum_A \chi^{(s)}(A) = 0 \qquad s \neq \text{identity representation} \tag{1.47}$$

and thus we have obtained equation (1.44).

Now by summing equation (1.43) with respect to A we get

$$\sum_A \chi^{(v)}(A)\chi^{(\mu)}(A) = \sum_s g_{v\mu,s} \sum_A \chi^{(s)}(A) \tag{1.48}$$

and, hence, by equation (1.44)

$$\sum_A \chi^{(v)}(A)\chi^{(\mu)}(A) = \begin{cases} g & \text{if } \mu = v \\ 0 & \text{if } \mu \neq v \end{cases} \tag{1.49}$$

If $\Pi(A)$ denotes a $g \times g$ matrix of the regular representation, we have seen that this can be written as

$$\Pi = \sum_{v=1}^{r} n_v \Gamma_v \tag{1.50}$$

where r is the number of classes, but n_v has not yet been determined. Since

$$T_r \Pi(A) = \begin{cases} 0 & A \neq E \\ g & A = E \end{cases} \tag{1.51}$$

it follows from equation (1.50) that

$$\sum_{v=1}^{r} n_v \chi^{(v)}(A) = \begin{cases} 0 & A \neq E \\ g & A = E \end{cases} \tag{1.52}$$

Multiply equation (1.52) by $\chi^{(\mu)}(A)$ and sum over A. Thus,

$$\sum_{\nu=1}^{r} n_\nu \sum_A \chi^{(\nu)}(A)\chi^{(\mu)}(A) = \sum_A \chi^{(\mu)}(A) \begin{cases} 0 & A \neq E \\ g & A = E \end{cases} \tag{1.53}$$

Equation (1.49) now shows that only the term $\nu = \mu$ survives on the left and only the term $A = E$ on the right and therefore

$$n_\mu g = \chi^{(\mu)}(E)g \quad \text{or} \quad n_\mu = \chi^{(\mu)}(E) \tag{1.54}$$

Thus, we have proved that the number of equivalent matrices of the νth irreducible representation in the regular representation is $\chi_E^{(\nu)}$ and thus these numbers are determined by the equation

$$\sum_{\nu=1}^{r} [\chi_E^{(\nu)}]^2 = g \tag{1.55}$$

It is now a comparatively simple matter to determine the characters of the other classes. We have seen that the eigenvalues of the $r \times r$ matrix $(c_{ij,s})$, where i is constant, are the quantities

$$x_i^{(\nu)} = h_i \chi_i^{(\nu)} / \chi_E^{(\nu)} \tag{1.56}$$

The values of the $c_{ij,s}$'s follow from the group table, the h_i's from the analysis into classes, the $\chi_E^{(\nu)}$'s from the above theorem, and, hence, the $\chi_i^{(\nu)}$'s are determined by relations (1.36) and (1.56).

In order to determine which eigenvalues $x_i^{(\nu)}$ belong to the same irreducible representation, it is necessary to use orthogonality relations (1.52) and (1.49) which may be written as

$$\sum_{\nu=1}^{r} \chi_E^{(\nu)} \chi_i^{(\nu)} = \begin{cases} g & i = E \\ 0 & i \neq E \end{cases} \tag{1.57}$$

$$\sum_{i=1}^{r} h_i \chi_i^{(\nu)} \chi_i^{(\mu)} = \begin{cases} g & \nu = \mu \\ 0 & \nu \neq \mu \end{cases} \tag{1.58}$$

Equation (1.57) is derived from equation (1.52) by using equation (1.54), and equation (1.58) is derived from equation (1.49) by using the fact that $X^{(\nu)}(A)$ is the same for all members of the same class.

The zinc-blende structure provides a convenient illustration of the use of the group character table. There are five classes in this group which may be written as follows

$$E, \, 3C_4^2, \, 8C_3, \, 6JC_4, \, 6JC_2$$

where the subscripts denote the nature of the axes—2 for diad axes, 3 for trigonal axes, and 4 for fourfold axes— and where the superscript 2 indicates

Table II

	E xyz	$3\,C_4^2$ $\bar{x}\bar{y}z$	$8\,C^3$ zxy	$6\,JC_4$ $y\bar{x}\bar{z}$	$6\,JC_2$ $\bar{y}\bar{x}z$	
Γ_1	1	1	1	1	1	$1 + xyz + \cdots$
Γ_2	1	1	1	-1	-1	$x^4(y^2-z^2) + y^4(z^2-x^2) + z^4(x^2-y^2)$
Γ_3	2	2	-1	0	0	$[z^2-\frac{1}{2}(x^2+y^2)]$, $[x^2-y^2]$
Γ_4	3	-1	0	-1	1	x, y, z
Γ_5	3	-1	0	1	-1	$z(x^2-y^2)$, $x(y^2-z^2)$, $y(z^2-x^2)$

a rotation through two right angles. Typical operations of each class are shown as coordinate transformations in the second row of Table II. The degrees of the irreducible representations are given by the equation

$$1^2 + 1^2 + 2^2 + 3^2 + 3^2 = 24 \tag{1.59}$$

The character table (Table II) may be found by the process already described. Inspection shows that both the rows and columns are orthogonal in accordance with equations (1.57) and (1.58). The basis functions are given in the last column and from these the matrices may be readily constructed and the character table checked.

This analysis shows that the wave functions of electrons in a field with $\bar{4}3\,m$ symmetry can be classified into states such as Γ_1, which has the complete symmetry of the field and is nondegenerate, and Γ_2, which is nondegenerate and of lower symmetry. Because this last wave function has many nodal planes, this state usually occurs with high energy. Another state is that specified by Γ_5 which is threefold degenerate. Sometimes Γ_1 is said to be s-like, Γ_4 p-like, and Γ_3 d-like, in analogy with atomic states. These analogies have some qualitative value.

X. THE WAVE-VECTOR GROUP

Let f_k denote one of the basis functions of the reduced form of Γ, i.e., it has the translational symmetry of a Bloch function at k, and let A be an operator of the crystallographic point group; then

$$AT(\mathbf{R}_n)f_k = Ae^{i\mathbf{k}\cdot\mathbf{R}_n}f_k \tag{1.60}$$

If $\mathbf{k}\cdot\mathbf{R}_n$ is invariant under A, the right-hand side of equation (1.60) may be written as $e^{i\mathbf{k}\cdot\mathbf{R}_n}Af_k$, where Af_k is a function distinct from f_k. Thus, those operations of G_P which leave $\mathbf{k}\cdot\mathbf{R}_n$ unchanged constitute the operations of a group called the wave-vector group G_k. As A takes in turn all values in G_k, the functions Af_k constitute the basis for the representation of G_k. When a point group exists, the order of G_k will always be a factor of the order of G_P. Examples of the operations of G_k are clearly rotations belonging to G_P which do not change k. Thus, if k lies on an axis of the reciprocal lattice, all rotations about this axis which belong to G_P will form part of G_k.

Other elements of the wave-vector group can arise as follows. The wave vector k by its definition is limited to the region of the reciprocal lattice space known as the Brillouin zone. If k is a vector on the surface of this zone and K_n is a reciprocal lattice vector, then $k + K_n$, also on the surface, specifies the same state as k. The condition which determines whether an operation Q belongs to the wave-vector group G_k is

$$k \cdot Qr = (k + K_n) \cdot r \qquad (1.61)$$

By definition of the reciprocal lattice space, when r undergoes a transformation Q, k undergoes $\tilde{Q}$, the transposed matrix, and because the transformations are orthogonal, $\tilde{Q} = Q^{-1}$, it follows that $k \cdot Qr = kQ \cdot r$. Hence, Q^{-1} will belong to G_k if

$$kQ = k + K_n \qquad (1.62)$$

For rotations about an axis on which k lies, $K_n = 0$, but if the point k lies on a Brillouin zone boundary a rotation about another axis can transform k to its equivalent point separated from k by a reciprocal lattice vector. The simple cubic lattice shows this effect. The operators of the wave-vector group at Δ are shown in Fig. 3. The operators of the wave-vector group X are illustrated by the same stereogram with the addition of small circles around each dot, since the operation of reflection in the k_x, k_y plane is contained in this group. In G_Δ, there are eight operations which fall into the following five classes: E, C_4^2, JC_2, JC_4^2, C_4. Hence, since

$$1^2 + 1^2 + 1^2 + 1^2 + 2^2 = 8$$

there are four nondegenerate Δ states and one doubly degenerate Δ state. For G_X, there are 16 operations and 10 classes with eight nondegenerate states and two pairs of degenerate states.

Fig. 3. A stereogram showing the operators of the wave-vector group at Δ.

XI. SPACE GROUPS WHICH CONTAIN GLIDES
AND SCREW DISPLACEMENTS

Let $T(0, 0, \frac{1}{2})$ denote a displacement of half a lattice distance along the $\mathbf{a}_3$ axis, and m a reflection of G_P in a plane which contains $\mathbf{a}_3$. The combined operation $mT(0, 0, \frac{1}{2})$ is called a glide reflection. Again, let Q denote some rotation of G_P about the $\mathbf{a}_3$ axis; then $QT(0, 0, \frac{1}{2})$ is called a screw displacement. More generally, glides and screws can be of the form mT and QT, where $T = T(s_1, s_2, s_3)$ and where the s's are proper fractions of the primitive vectors.

When the space group G contains operations of this kind, a complex $\{E, A, B, ...\}$ defined by $G = \Gamma + A\Gamma + B\Gamma + \cdots$ does not itself form a subgroup of G. For instance, if Q is a rotation through π, $\{QT(0, 0, \frac{1}{2})\}^2$ is equal to $T(0, 0, 1)$ which is an operation of Γ not of the complex. However, the cosets $(1/\gamma)\Gamma$, $(1/\gamma)A\Gamma$, $(1/\gamma)B\Gamma$, etc., regarded as elements, do form a group. The factor $(1/\gamma)$, i.e., the reciprocal of the order of Γ, has been included for convenience. Consider, for example, $[(1/\gamma)A\Gamma]^2$, where A denotes a screw $QT(0, 0, \frac{1}{2})$ or a glide $mT(0, 0, \frac{1}{2})$; this quantity is then equal to $(1/\gamma)\Gamma$, which is the identity element of the factor group. Hence, the factor group contains as many elements as the number of elements in the complex. Its multiplication table may be constructed and the irreducible representations determined in exactly the same way as was done for the point group.

The significance of the irreducible representations of the factor group in relation to the electronic states in crystals may be seen as follows. Let $u(\mathbf{r})$ be a solution of the Schrödinger equation which has the translational symmetry of the space lattice. Then $(1/\gamma)\Gamma u = u_E$ and $(1/\gamma)A\Gamma u = u_A$, etc., and the set of functions $u_E, u_A, ...$, equal in number to the number of operations of the complex, forms a basis for the representation of the factor group. The irreducible representations, therefore, determine the symmetries and degeneracies of the states at the center of the Brillouin zone. Hence, as far as the states at the origin are concerned, no new feature is introduced by the presence of screws and glides.

The factor group is in general isomorphic with one of the point groups and therefore its irreducible representations are the same as those for that point group. For example, the space group of the close-packed hexagonal structure is $P(6_3/m)mc$ and the factor group is isomorphic with $(6/m)mm$, which is the point group obtained by replacing screws by simple rotations and glides by ordinary reflections. The determination of all possible symmetries of wave functions at the origin of the Brillouin zone for crystals which contain glides and screws is not, therefore, essentially more difficult than for crystals with a simple point group. However, for wave vectors other than $(0, 0, 0)$, the situation is more complicated and is best explained by an example. We shall consider the case of the wave vector at the center of the top or bottom hexagonal face of the Brillouin zone.

Let

$$\psi = e^{i(\pi/c)z} f(\mathbf{r}) \tag{1.63}$$

where $f(\mathbf{r})$ has the translational symmetry of the lattice but in general no other symmetry. We now determine the set of operations of G which will transform ψ into other functions of this type, i.e., associated with the points $[0, 0, \pm (\pi/c)]$. If $G = \Gamma + A\Gamma + B\Gamma + \cdots$, where Γ is the translation group of the simple hexagonal lattice, then one member of the complex $\{E, A, B, ...\}$ will be a glide reflection mT, where m denotes a reflection and T a displacement through $\frac{1}{2} c$ along the z axis. The square of mT is the operation which produces a simple translation through the lattice distance c and therefore belongs to Γ. However, $(mT)^2$ applied to ψ gives $-\psi$ and this must be included in the set of basis functions which leads to the regular representation of the wave-vector group of $[0, 0, \pm (\pi/c)]$. Thus although there are only 24 operations of the type $E, A, B, ...$, we have to include $(mT)^2$ and all products of this operator with $A, B, ...$, making altogether a set of 48 operators. These 48 operators applied to ψ produce 48 basis functions which can be regarded as the components of a column vector. The associated matrices form the regular representation of the wave-vector group.

The same 48 operations also transform the functions associated with the center of the Brillouin zone into each other so that the wave-vector group at $[0, 0, \pm (\pi/c)]$ of order 48 includes also the wave-vector group of the origin $(0, 0, 0)$. Hence, on reduction, some of the irreducible representation will refer to the states at the center and some to the states at $[0, 0, \pm (\pi/c)]$. The criterion which determines which representations refer to $[0, 0, \pm (\pi/c)]$ and which to $(0, 0, 0)$ is their behavior with respect to the operator $(mT)^2$. Thus, representations which leave the basis functions unchanged under $(mT)^2$ refer to $(0, 0, 0)$ and those which change the sign of the basis functions refer to $[0, 0, \pm (\pi/c)]$. This distinction is shown in the character table by the sign of the character of the class which contains $(mT)^2$.

XII. TIME-REVERSAL SYMMETRY

An example which illustrates the simultaneous reduction of two wave-vector groups and has a conveniently small character table is the case of the wave vectors $(k_x, k_y, 0)$ and $[k_x, k_y, \pm (\pi/c)]$. These are any points which lie on the basal plane or the two hexagonal faces of the Brillouin zone and on a common vertical line. This example also illustrates the effect of what is known, in this case rather inappropriately, as time-reversal symmetry. What is really involved in this symmetry is that the Hamiltonian is a real operator

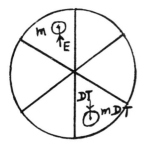

Fig. 4. Stereogram related to the wave vector $(k_x, k_y, 0)$.

and therefore that the Schrödinger equation is satisfied by the complex conjugate of any solution.
The wave function is

$$\psi = e^{i(k_x x + k_y y) + i(\pi/c)z}\, u(\mathbf{r}) \tag{1.64}$$

where $u(\mathbf{r})$ is invariant under all integral translations. Consider the four operations indicated in the stereogram shown in Fig. 4, where m is a reflection in the basal plane, D is a rotation through π about the principal axis, and T is a displacement of $\tfrac{1}{2}c$ along this axis. Clearly E and m applied to ψ produce two functions belonging to the same Bloch state, i.e., k_x and k_y are unchanged and π/c is transferred to $-(\pi/c)$, i.e., $(k_x, k_y, \pi/c)$ is transferred to the equivalent point. DT and mDT, which leave the lattice unchanged, do change ψ into functions belonging to a different Bloch state. However, if $\mathscr{J}$ denotes the operation of taking the complex conjugate, then $DT\mathscr{J}$ and $mDT\mathscr{J}$ are operations of the group which transforms ψ into functions referring to the same Bloch state. By multiplying these operations together, we find a group of order 8 which includes $S = (DT)^2$ and which has five classes as follows:

$\mathscr{C}_1$	$\mathscr{C}_2$	$\mathscr{C}_3$	$\mathscr{C}_4$	$\mathscr{C}_5$
E	S	$m+mS$	$DT\mathscr{J}+DT^{-1}\mathscr{J}$	$mDT\mathscr{J}+mDT^{-1}\mathscr{J}$

where
$$E = x, y, z \qquad S = x, y, z+c \qquad m = x, y, -z$$

$$DT = \left[-x, -y, \left(z + \frac{c}{2}\right)\right] \qquad mDT = \left[-x, -y, -\left(z + \frac{c}{2}\right)\right]$$

The character table (Table III) is readily deduced.
In the first four irreducible representations $S\psi = \psi$ and these must therefore refer to points on the $k_z = 0$ plane, i.e., $(k_x, k_y, 0)$. The fifth representation (shown in the fifth row) gives $S\psi = -\psi$ and therefore belongs to $[k_x, k_y, \pm(\pi/c)]$. The fact that this is the only irreducible represen-

Table III

$\mathscr{C}_1$	$\mathscr{C}_2$	$\mathscr{C}_3$	$\mathscr{C}_4$	$\mathscr{C}_5$
1	1	1	1	1
1	1	-1	1	-1
1	1	-1	-1	1
1	1	1	-1	-1
2	-2	0	0	0

tation for the wave vector on the upper and lower faces and that it is of the second degree shows that only doubly degenerate states exist on these faces. Hence, energy bands always touch over the whole of the hexagonal faces of the Brillouin zone of the close-packed hexagonal structure in the absence of electron-spin effects.

XIII. REFERENCES*

1. A. Speiser, *Theorie der Gruppen von Endlicher Ordnung*, Springer-Verlag (Berlin), 1937.
2. Allen Nussbaum, *Solid State Phys.* **18**: 165 (1966).
3. H. Jones, *Theory of Brillouin Zones and Elecronic States in Crystals*, North-Holland Publishing Co. (Amsterdam), 1960.

---

* The material used in this chapter is standard and therefore only general references need be given. Many books concerned with the theory of the solid state have chapters on group theory. Three references closely related to the subject matter of this chapter which the reader might find useful are given in this section.

CHAPTER 2

Principles and Methods of Band Theory

L. Pincherle

University of London
London, England

I. FOUNDATIONS AND APPROXIMATIONS OF BAND THEORY

Band theory is based on the one-electron approximation and relies essentially on the Hartree–Fock equations. Modern many-body theory has shown when and how we can use these equations which, for electrons in solids, are in most cases reliable and useful.

The wave function is expressed in a completely antisymmetric form

$$\Psi(\mathbf{r}_1 \ldots \mathbf{r}_N) = \frac{1}{\sqrt{N!}} \begin{vmatrix} \psi_1(\mathbf{r}_1) & \ldots & \psi_1(\mathbf{r}_N) \\ \psi_2(\mathbf{r}_1) & \ldots & \psi_2(\mathbf{r}_N) \\ \vdots & & \vdots \\ \psi_N(\mathbf{r}_1) & \ldots & \psi_N(\mathbf{r}_N) \end{vmatrix} \tag{2.1}$$

where the ψ_i's are one-particle wave functions and the coordinate $\mathbf{r}_j$ denotes both space and spin variables of one particle. The Hamiltonian, with obvious notation, is

$$H = \sum_{i=1}^{N} \left[-\frac{\hbar^2}{2m} \nabla_i^2 + V(\mathbf{r}_i) + \tfrac{1}{2} \sum_j \frac{e^2}{r_{ij}} \right] \tag{2.2}$$

The best Ψ which can be constructed is the one that minimizes the expectation value of the energy, subject to orthonormality of the one-electron functions. Application of the variational principle leads to the Hartree–Fock equations for the one-electron functions, e.g., for $\psi_i(\mathbf{r}_1)$:

$$\left[-\frac{\hbar^2}{2m} \nabla_1^2 + V(\mathbf{r}_1) + \sum_j e^2 \int \frac{|\psi_j(\mathbf{r}_2)|^2 \, d\mathbf{r}_2}{|\mathbf{r}_1 - \mathbf{r}_2|} \right] \psi_i(\mathbf{r}_1)$$

$$- \sum_j e^2 \left[\int \frac{\psi_j^*(\mathbf{r}_2) \psi_i(\mathbf{r}_2) \, d\mathbf{r}_2}{|\mathbf{r}_1 - \mathbf{r}_2|} \right] \psi_j(\mathbf{r}_1) = E_i \psi_i(\mathbf{r}_1) \tag{2.3}$$

The two sums express, respectively, the average electrostatic potential energy of all the electrons and the exchange interaction. The energy E_i is the negative of the energy required to remove an electron in state i from the system (Koopman's theorem[†]). Generally the space parts of the wave functions for opposite spin are taken as identical. If this is not possible (magnetic problems), we have the "unrestricted" Hartree–Fock approximation.

Although the concept of energy bands does not depend on the Hartree–Fock approximation, equations (2.3) are the basis of all practical calculations. They are a complicated system of nonlinear integrodifferential equations. The exchange term is the most awkward as it couples to ψ_i all the other ψ's. A simplification generally adopted is to introduce Slater's exchange potential

$$V_{\text{exch.}}(\mathbf{r}_1) = \frac{1}{\psi_i(\mathbf{r}_1)} \sum_j \int \psi_j^*(\mathbf{r}_2)\, \psi_j(\mathbf{r}_1)\, \frac{e^2\, \psi_i(\mathbf{r}_2)}{|\mathbf{r}_1 - \mathbf{r}_2|}\, d\mathbf{r}_2 \qquad (2.4)$$

so that the exchange term becomes $- V_{\text{exch.}}(\mathbf{r}_1)\, \psi_i(\mathbf{r}_1)$ and one is left with a separate equation for each $\psi_i(\mathbf{r}_1)$.

Particular solutions of equations (2.3) are obtained by the method of self-consistent fields. The coulomb and exchange integrals are calculated with an assumed set of wave functions and the equations solved. The two integrals are recalculated with the new functions and the process is repeated until the successive iterations agree within some assigned limit. In solids there is the complication that one must sum the charge density over a very large number of states.

The single-electron approximation reflects the experimental fact that electrons and holes in crystals behave very much as free particles. The necessary modifications are that the free electron mass must be replaced by an effective mass m^*, generally a tensor, and all coulomb interactions must be divided by the static dielectric constant.

Within the one-electron approximation, two approaches are possible —the band approach and the valence-bond approach. In the first, the electrons are assumed to move through the whole crystal, and any electron has the same probability of being at equivalent points in the crystal. In the second, one starts by considering each electron as bound to a given nucleus. The latter approach is appropriate for electrons of deep shells, and sometimes d electrons in transition metals. The localized functions that should in principle be used are called Wannier functions (see section IV). We shall deal mainly with the band model.

[†] As always with interacting particles, the total energy is not just the sum of the one-particle energies E_i, but depends on the distribution of the particles.

As is well known, the name "band theory" arises because in a solid the allowed electronic energy levels group themselves into bands, often separated by forbidden energy gaps. The broadening of atomic levels into bands can be understood on the basis of the uncertainty principle, since the electron spends only a short time orbiting around each atom. In a perfect crystal, the electronic energy bands are a particular case of the frequency bands arising in every wave propagation in a periodic structure. In such cases, the dispersion relations have the following main characteristics: (1) v (or $E = hv$) is a periodic function of the wave vector $\mathbf{k}$, which indicates that it must be possible to restrict consideration to one period only; (2) for any $\mathbf{k}$ there are generally a number of possible frequencies; and (3) values of v within certain frequency intervals are not found for any value of $\mathbf{k}$.

One can also consider what happens when the atoms gradually come together in a crystal. With two atoms, each atomic level splits into two molecular levels. Then the addition of every new atom adds one more energy level, and, in the limit for N atoms, one gets practically continuous bands, each containing N levels (N is very large), all deriving in a sense from an atomic level. This approach indicates that bands are not restricted to periodic structures, though here we shall confine attention to these structures.

In dealing with band theory, one should always remember its main approximations: (1) the action of other electrons upon the electron under study can be replaced by a periodic potential; (2) anything in the nature of a multiplet structure is neglected, treating the problem as, in the case of atoms, that of one electron outside closed shells, and, thus, the theory is not applicable when interactions between particles are essential, as in an "exciton"; (3) the energy bands are calculated assuming fixed nuclei, and the interaction with the lattice vibrations is introduced later as a perturbation. This is not justified when the interaction is very strong, as in ionic crystals. The appropriate formalism then becomes that of "polaron" theory.

II. FORM OF THE ONE-ELECTRON WAVE FUNCTIONS

We must solve Schrödinger's equation

$$\nabla^2 \psi(\mathbf{r}) + \frac{2m}{\hbar^2}[E - V(\mathbf{r})]\,\psi(\mathbf{r}) = 0 \qquad (2.5)$$

where the potential V, including the effect of all the other electrons, has the periodicity of the lattice; V is invariant with respect to all the translations T_i, where T_i is the operation of translating by a fundamental translation $\mathbf{R}_i$. Thus, the Hamiltonian commutes with all the T_i's, and ψ must be a simultaneous eigenfunction of all these operators, that is, for every T_i,

$$T_i \psi(\mathbf{r}) \equiv \psi(\mathbf{r} + \mathbf{R}_i) = C_i \psi(\mathbf{r}) \qquad (2.6)$$

Consider two translation operators T_i and T_j; then,

$$T_j T_i \psi(\mathbf{r}) = C_j C_i \psi(\mathbf{r}) \tag{2.7}$$

but $\mathbf{R}_i + \mathbf{R}_j = \mathbf{R}_m$, another fundamental translation; thus, the product $T_j T_i$ is itself a translation T_m (necessary since the T_i's form a group) and

$$T_m \psi(\mathbf{r}) = C_m \psi(\mathbf{r}) \tag{2.8}$$

It follows from equations (2.7) and (2.8) that $C_m = C_j C_i$, for any two lattice vectors $\mathbf{R}_j$ and $\mathbf{R}_i$. Therefore, the numbers C_i must depend exponentially on $\mathbf{R}$, i.e., $C(\mathbf{R}) = \exp(\mathbf{b} \cdot \mathbf{R})$, where the vector $\mathbf{b}$ characterizes the particular ψ. However, if $\mathbf{b}$ is real, ψ tends exponentially to infinity as $\mathbf{R}$ increases. This behavior is not admissible, and, thus, $\mathbf{b}$ must have the form $i\mathbf{k}$ ($\mathbf{k}$ is real) and

$$T_i \psi_{\mathbf{k}}(\mathbf{r}) \equiv \psi_{\mathbf{k}}(\mathbf{r} + \mathbf{R}_i) = \exp(i\mathbf{k} \cdot \mathbf{R}_i) \psi_{\mathbf{k}}(\mathbf{r}) \tag{2.9}$$

This is known as the Bloch condition and is the fundamental periodicity condition for the one-electron wave functions in a perfect lattice.

Writing now $u_{\mathbf{k}}(\mathbf{r}) = \exp(-i\mathbf{k} \cdot \mathbf{r}) \psi_{\mathbf{k}}(\mathbf{r})$, we have

$$u_{\mathbf{k}}(\mathbf{r} + \mathbf{R}_i) = \exp\left[-i\mathbf{k} \cdot (\mathbf{r} + \mathbf{R}_i)\right] \psi_{\mathbf{k}}(\mathbf{r} + \mathbf{R}_i) = e^{-i\mathbf{k} \cdot \mathbf{r}} \psi_{\mathbf{k}}(\mathbf{r}) = u_{\mathbf{k}}(\mathbf{r})$$

That is, $u_{\mathbf{k}}(\mathbf{r})$ has the periodicity of the lattice. Thus,

$$\psi_{\mathbf{k}}(\mathbf{r}) = e^{i\mathbf{k} \cdot \mathbf{r}} u_{\mathbf{k}}(\mathbf{r}) \tag{2.10}$$

i.e., ψ is a plane wave modulated by a function with the periodicity of the lattice. Such a wave function is called a Bloch function. The vector $\mathbf{k}$ plays the role of a set of three quantum numbers. As for its physical meaning, if V is constant, $\hbar\mathbf{k}$ is exactly the momentum of the electron. It may be proved that also in a crystal $\hbar\mathbf{k}$ has the properties of momentum, as far as external forces are concerned [e.g., in an applied electric field $\mathbf{F}$, $\hbar(d\mathbf{k}/dt) = e\mathbf{F}$], and may be called "crystal momentum." It is not, however, an eigenvalue of the operator $-i\hbar$ grad and is not equal to the average value of the momentum of the electron in state $\mathbf{k}$. In any case, $\mathbf{k}$ is a constant of the motion. That is, an electron in a state $\mathbf{k}$ is not scattered at all into other states by a perfect lattice. Incidentally, in a disordered structure $\mathbf{k}$ is complex ($= \mathbf{k}_1 + i\mathbf{k}_2$), which corresponds to a wave decaying as $\exp(-k_2 r)$ in the direction of propagation. The reciprocal of the modulus of $\mathbf{k}_2$ is thus essentially the mean free path. The disorder scatters the electrons into states with other $\mathbf{k}$ values allowed by conservation of energy.

Substituting equation (2.10) into equation (2.5), we obtain the equation satisfied by $u_{\mathbf{k}}(\mathbf{r})$

$$\nabla^2 u + 2\, i\mathbf{k} \cdot \operatorname{grad} u - k^2 u + (2\,m/\hbar^2) [E - V(\mathbf{r})] u = 0$$

or, since $-i\hbar$ grad is the operator $\mathbf{p}$,

$$\nabla^2 u - (2/\hbar)\,\mathbf{k}\cdot\mathbf{p}u + (2\,m/\hbar^2)\left[\left(E - \frac{\hbar^2 k^2}{2\,m}\right) - V(\mathbf{r})\right]u = 0 \qquad (2.11)$$

This equation is used to calculate u corresponding to a certain $\mathbf{k}$ when E and u are known at a nearby $\mathbf{k}_0$; the term containing $\mathbf{k}\cdot\mathbf{p}$ is treated as a perturbation ($\mathbf{k}\cdot\mathbf{p}$ approximation).

III. RECIPROCAL SPACE—BRILLOUIN ZONES

A crystal lattice has several periods. Choose periods $\mathbf{a}_1$, $\mathbf{a}_2$, $\mathbf{a}_3$ such that (1) $\mathbf{a}_1$ is the shortest period in the lattice (or one of several equally short ones), (2) $\mathbf{a}_2$ is the shortest not parallel to $\mathbf{a}_1$, and (3) $\mathbf{a}_3$ is the shortest not coplanar to $\mathbf{a}_1$ and $\mathbf{a}_2$. Then any period $\mathbf{R}$ of the lattice is given by

$$\mathbf{R} = l\mathbf{a}_1 + m\mathbf{a}_2 + n\mathbf{a}_3 \qquad (2.12)$$

where l, m, and n are integers. The lattice so defined is called the Bravais lattice of the crystal. The crystal is entirely described if the content of one parallelepiped of sides $\mathbf{a}_1$, $\mathbf{a}_2$, $\mathbf{a}_3$ is known. This unit is called a primitive cell of the crystal (Fig. 1) and may contain any number of atoms. It was proved in crystallography that there are just 14 different Bravais lattices, e.g., the hexagonal, the trigonal, the simple cubic, the face-centered cubic, and the body-centered cubic. In each case, the primitive cell may be filled in a number of different ways having different symmetry, and in this way the 14 Bravais lattices become diversified into 230 complete space groups. Each of these groups can be the symmetry type of several different crystal structures. For instance, the NaCl structure (Fig. 2), the diamond structure, and the zinc-blende structure (Fig. 3) all have the face-centered cubic Bravais lattice (but different space groups). In these structures, there are two atoms per primitive cell; the crystals are made up of two interpenetrating face-centered cubic lattices. In all cases, the basic cube of the face-centered cubic structure contains four primitive cells. Wigner and Seitz proposed

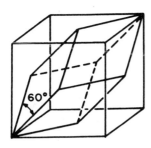

Fig. 1. Primitive cell of face-centered cubic lattice.

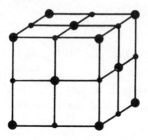

Fig. 2.   Sodium chloride structure.

a cell which is primitive and yet has the symmetry of the lattice (the primitive cell formed by $\mathbf{a}_1$, $\mathbf{a}_2$, $\mathbf{a}_3$ may miss some symmetry elements, e.g., the cell for the face-centered cubic structure misses the axes of fourfold symmetry). The Wigner–Seitz cell about a lattice point $\mathbf{R}$ is the locus of all points that are closer to $\mathbf{R}$ than to any other lattice point. Thus, for the face-centered cubic structure, the Wigner–Seitz cell is a rhombododecahedron. In polyatomic crystals, there may be several atoms in the Wigner–Seitz cell, or on its surface.

Some of the rotational symmetries of the Bravais lattice may not be present in a crystal, or retained in a modified form; for instance, a rotation about a certain axis may have to be combined with a translation by a fraction of a lattice displacement. Thus, in the diamond structure, the cubic fourfold symmetry becomes a screw symmetry: the screw axes are perpendicular and pass through the centers of the faces of the cube. The presence of screw axes, or glide planes, has important effects on the band structure; it is more important than, say, the number of atoms per primitive cell.

Given a lattice with basis vectors $\mathbf{a}_1$, $\mathbf{a}_2$, $\mathbf{a}_3$, one defines as its reciprocal lattice that generated by three basis vectors $\mathbf{b}_1$, $\mathbf{b}_2$, $\mathbf{b}_3$, which satisfy the relation

$$\mathbf{a}_i \cdot \mathbf{b}_j = 2\pi\delta_{ij} \tag{2.13}$$

The reciprocal lattice is said to be in "reciprocal space." If the term 2π in the right-hand side of equation (2.13) is omitted, we have the so-called "k-space," or "wave-number space." If the right-hand side of the equation is multiplied by $\hbar$, we have "momentum space." However, often the three

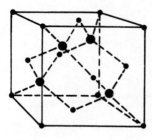

Fig. 3.   Zinc-blende structure showing tetrahedral bonds. The two kinds of atoms are identical in the diamond structure.

terms are interchangeable. Solving the equation (2.13), we have

$$\mathbf{b}_1 = \frac{2\pi\,\mathbf{a}_2 \times \mathbf{a}_3}{\mathbf{a}_1 \cdot \mathbf{a}_2 \times \mathbf{a}_3} \qquad (2.14)$$

and similar expressions for $\mathbf{b}_2$ and $\mathbf{b}_3$. Any fundamental vector $\mathbf{K}_n$ of the reciprocal lattice is given by

$$\mathbf{K}_n = p\mathbf{b}_1 + q\mathbf{b}_2 + r\mathbf{b}_3 \qquad (2.15)$$

where p, q, and r are integers. These vectors are the "frequencies" for Fourier expansion of the periodic potential

$$V(\mathbf{r}) = \sum_{\mathbf{K}_n} V(\mathbf{K}_n) \exp(i\mathbf{K}_n \cdot \mathbf{r}) \qquad (2.16)$$

If $\mathbf{R}_i$ is any direct lattice vector, we have from equation (2.13)

$$\mathbf{K}_n \cdot \mathbf{R}_i = 2\pi(pl + qm + rn) = 2\pi g$$

where g is an integer, and thus

$$\exp(i\mathbf{K}_n \cdot \mathbf{R}_i) = 1 \qquad (2.17)$$

The volume of the primitive cell of the reciprocal lattice is $8\pi^3$ divided by the volume Ω of the primitive cell of the direct lattice. Each vector $\mathbf{K}_n$ is perpendicular to a set of planes of atoms in the crystal, and the spacing of these planes is $2\pi/\mathbf{K}_n$. Each set of reticular planes can be labelled by a vector $\mathbf{K}_n$. From equation (2.14), one recognizes immediately that the reciprocal lattice of a face-centered cubic lattice is body-centered, and *vice versa*.

Consider now two states, defined by vectors $\mathbf{k}$ and $\mathbf{k}'$, differing by a fundamental vector $\mathbf{K}_n$ of the reciprocal lattice—$\mathbf{k} = \mathbf{k}' + \mathbf{K}_n$; we have

$$\psi_\mathbf{k}(\mathbf{r} + \mathbf{R}_i) = \exp[i(\mathbf{k}' + \mathbf{K}_n) \cdot \mathbf{R}_i]\,\psi_\mathbf{k}(\mathbf{r})$$

and using equation (2.17), we have

$$\psi_\mathbf{k}(\mathbf{r} + \mathbf{R}_i) = e^{i\mathbf{k}' \cdot \mathbf{R}_i}\psi_\mathbf{k}(\mathbf{r}) \qquad (2.18)$$

That is, the state $\psi_\mathbf{k}$ satisfies Bloch's theorem as if it had the wave vector $\mathbf{k}'$. The original label $\mathbf{k}$ is not unique: every state has an infinity of possible wave vectors, differing from one another by the fundamental vectors of the reciprocal lattice. This may be expressed by saying that the one-electron wave functions of a perfect solid are periodic in reciprocal space; thus, $E(\mathbf{k})$ also must have the periodicity of the reciprocal lattice.

It is thus unnecessary to consider more than one of all the equivalent vectors $\mathbf{k} + \mathbf{K}_n$. The most convenient procedure is to reduce all the equivalent vectors to their smallest possible value by substracting vectors of the reciprocal lattice. The end points of all the reduced vectors fill a region around the origin of **k**-space which is called the first, or central, Brillouin zone: it contains all nonequivalent **k** vectors. We shall call it simply and somewhat improperly just the Brillouin zone (BZ). It is in effect the Wigner–Seitz cell of the reciprocal lattice. Its volume is the same as that of the primitive cell in **k**-space, that is, $8\pi^3/\Omega$. One can construct a second zone including all the second smallest nonequivalent vectors possible, and so on. The construction is carried out by considering all the planes that bisect perpendicularly the lines joining the origin of **k**-space to all the reciprocal lattice points. All the zones so constructed have the same volume and each can be reassembled so as to fit exactly the central zone.

The BZ for the direct face-centered cubic lattice is a truncated octahedron (Fig. 4). The octahedric faces bisect perpendicularly the lines joining the point $k = 0$ to its eight nearest neighbors along the $(1, 1, 1)$ directions; the cubic faces bisect the lines joining it to its six second nearest neighbors along the $(1, 0, 0)$ directions. The BZ for the body-centered cubic lattice is a rhombododecahedron.

Note that the shape of the BZ is determined only by the type of Bravais lattice, not by the particular crystal structure. There are thus just 14 BZ's, and, for instance, the BZ's of Al, NaCl, Ge, and InSb are all the above-mentioned truncated octahedron. In all cases, it is necessary to study the dispersion relations only within the first BZ.

This scheme in which all **k** vectors are considered to lie in the BZ is called the "reduced" zone scheme. If the totality of **k**-space in considered, we have the "extended" zone scheme. Also, having found the E–**k** relations, or the constant energy surfaces, in the first BZ, it is sometimes useful to consider the same surfaces, or relations, in all cells of reciprocal space. This is called the "repeated" zone scheme.

To illustrate, consider a one-dimensional lattice with period a. The period of the reciprocal lattice is $2\pi/a$. The first BZ extends from $k = -\pi/a$ to $k = +\pi/a$. The second zone extends from $-2\pi/a$ to $-\pi/a$ and

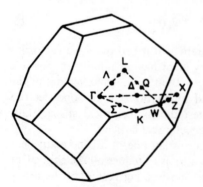

Fig. 4. Brillouin zone for a face-centered cubic lattice, showing symmetry points and axes.

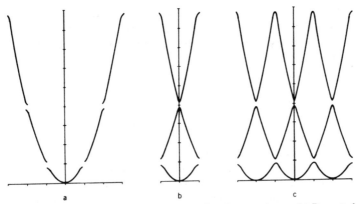

Fig. 5.   (a) Extended zone scheme. (b) Reduced zone scheme. (c) Repeated zone scheme.

from $+\pi/a$ to $2\pi/a$, and so on. Suppose that the periodic potential is weak, so that the $E(k)$ relation deviates little from $E = \hbar^2 k^2/2m$. As we shall see, the deviations occur at the zone boundaries where there are energy gaps and where the E–k curves must have horizontal tangents. Plotting the (slightly modified) parabola for all values of k constitutes the extended zone scheme [Fig. 5(a)].

Now reduce all k vectors to the central zone by adding or subtracting an appropriate number of periods $2\pi/a$. We obtain then the reduced zone scheme [Fig. 5(b)]. In this scheme, E is a multivalued function of k. Every curve is a piece of the original parabola, displaced by $2\pi n/a$. Thus, for vanishing potential (empty lattice)

$$E_n(k) = (\hbar^2/2m) \left(k + \frac{2\pi n}{a} \right)^2 \qquad n = 0, \pm 1, \pm 2, \ldots$$

and in general in three dimensions

$$E_n(\mathbf{k}) = (\hbar^2/2m)\, |\mathbf{k} + \mathbf{K}_n|^2 \tag{2.19}$$

Repeating the curves throughout $\mathbf{k}$-space, we obtain the repeated zone scheme [Fig. 5(c)], which brings out the periodicity of E with k. Which scheme to use is purely a matter of convenience.

In the simplest cases, the energies of a zone form a band, but in general there are complications. Since E (in the reduced zone scheme) is a multi-valued function of $\mathbf{k}$, a fourth quantum number, generally called the band index n, denoting the particular band, is necessary to identify completely a given state. When the electrons are tightly bound, n may be taken to coincide with the usual notation for atomic levels, e.g., $3s, 4d, \ldots$. When the electrons are almost free, equation (2.19) indicates that n is represented by a fundamental vector of the reciprocal lattice. In general, at **k**-points of prominent symmetry, n is represented by symbols derived from a group-theoretical study. At points of no symmetry and no degeneracy, the band index is often conveniently assigned by using the energy, namely,

$$n = 1, 2, 3, \ldots$$

and

$$n' > n \text{ if } E_{n'}(\mathbf{k}) > E_n(\mathbf{k})$$

Difficulties arise concerning the topology of the bands because the prescription breaks down at degeneracy points (Fig. 6). However, bands can cross only at symmetric points, so that these points of degeneracy form at most lines in **k**-space. When two bands cross, the prescription interchanges the subscripts n and $n+1$. It follows that the curves $E(\mathbf{k})$ are not differentiable at these points. E is, however, a continuous function of **k**. The set of E_i values, for all possible values of **k** within the BZ, the band index i being assigned as stated, constitutes one band of energy levels. In one dimension, the various sets are always separated by forbidden energy gaps; in three dimensions, overlapping is very common, so that $E_n(\mathbf{k}_1)$ may be larger than $E_{n+1}(\mathbf{k}_2)$. Generally, while energy gaps exist between all bands at the boundary of the BZ, there are exceptions and two bands may meet at points or over an area of the boundary. This occurs if the crystal contains screw axes (diamond) or glide planes. When the discontinuity is present, the derivative of E in the direction normal to the a face of the BZ must vanish, for symmetry reasons, if the face in question is parallel to a plane of reflection symmetry. As is clear from its derivation from an atomic level, each non-degenerate band can accommodate two electrons with opposite spin for each fundamental cell of the crystal. There are exactly as many allowed wave vectors in the BZ as there are unit cells in the block of the crystal.

If Ω is the volume of the unit cell of the crystal, the volume of the BZ is $8\pi^3/\Omega$. If there are N cells in the volume V, then $\Omega = V/N$. The volume of **k**-space per allowed **k** vector is thus $8\pi^3/V$. If we introduce the factor 2 due to spin, there are $V/4\pi^3$ states per unit volume of reciprocal space, a well-

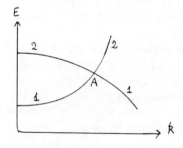

Fig. 6.   Interchange of band index at a point of degeneracy.

known result of statistical mechanics. Per unit volume of the crystal, the number of states is

$$n_s = (4\,\pi^3)^{-1} \tag{2.20}$$

The number of states per unit volume per unit energy interval, dn_s/dE, is called the density of states. In a volume A of k-space, the number of states per unit volume of the crystal is

$$n_s = (4\,\pi^3)^{-1} \int_A d\mathbf{k}$$

Choosing $d\mathbf{k}$ to be bounded on two faces by surfaces of constant energy, we have

$$d\mathbf{k} = dS\,dE/|\mathrm{grad}_\mathbf{k}\,E|$$

where dS is an element of area of the constant energy surface. Then the density of states is

$$dn_s/dE = (4\,\pi^3)^{-1} \int_A dS/|\mathrm{grad}_\mathbf{k}\,E| \tag{2.21}$$

the integral extending over the constant energy surface $E(\mathbf{k}) = E$ in k-space. As is well known, $\mathrm{grad}_\mathbf{k}\,E$ ("group velocity") is proportional to the velocity of the electron in state $\mathbf{k}$. It is evident that dn_s/dE may have singularities if the integration includes points at which $\mathrm{grad}_\mathbf{k}\,E = 0$. Such points are called critical points. It has been proved that the minimum set of critical points in any band is one maximum, one minimum, and six saddle points and that the density of states is continuous at a critical point, but has a discontinuity in its first derivative.

IV. MOMENTUM EIGENFUNCTIONS AND WANNIER FUNCTIONS

We have seen that the periodicity of the crystal enables us to expand the potential in a Fourier series [equation (2.16)] in which each Fourier coefficient is associated with a point in the reciprocal lattice. If there is more than one atom in the unit cell, some of the $V(\mathbf{K}_n)$'s may vanish or be connected by simple relations. Also, the coefficients are the same for equivalent vectors, e.g., for the eight vectors $(\pm\mathbf{K}, \pm\mathbf{K}, \pm\mathbf{K})$. The coefficients are real if the crystal has a center of inversion at $k = 0$; otherwise, $V(-\mathbf{K}_n) = V^*(\mathbf{K}_n)$. The coefficients $V(\mathbf{K}_n)$ are found by the usual integral expression, and there is the advantage that the waves $\exp(i\mathbf{K}_n \cdot \mathbf{r})$ form a complete orthogonal system within the unit cell. The unit cell may be either the primitive cell of the Bravais lattice, or any equivalent cell, such as the Wigner–Seitz cell. Thus,

$$V(\mathbf{K}_n) = (\Omega)^{-1} \int_{\substack{\text{one} \\ \text{cell}}} V(\mathbf{r}) \exp(-i\mathbf{K}_n \cdot \mathbf{r}) \, d\mathbf{r} \qquad (2.22)$$

Similarly, we can expand the periodic part u of the Bloch function (omitting the band index)

$$u_{\mathbf{k}}(\mathbf{r}) = \sum_{\mathbf{K}_n} v(\mathbf{K}_n) \exp(i\mathbf{K}_n \cdot \mathbf{r}) \qquad (2.23)$$

Multiplying by $\exp(i\mathbf{k} \cdot \mathbf{r})$ and relabelling the v's, we get

$$\psi_{\mathbf{k}}(\mathbf{r}) = \sum_{\mathbf{K}_n} v(\mathbf{k} - \mathbf{K}_n) \exp[i(\mathbf{k} + \mathbf{K}_n) \cdot \mathbf{r}] \qquad (2.24)$$

with

$$v(\mathbf{k} - \mathbf{K}_n) = (\Omega)^{-1} \int_{\substack{\text{unit} \\ \text{cell}}} \psi_{\mathbf{k}}(\mathbf{r}) \exp[-i(\mathbf{k} + \mathbf{K}_n) \cdot \mathbf{r}] \, d\mathbf{r} \qquad (2.25)$$

The $v(\mathbf{k} - \mathbf{K}_n)$'s are called momentum eigenfunctions, since, when the bands are well separated so that all the nondiagonal matrix elements of momentum with respect to band index vanish, they are proportional to the probability amplitudes for the possible values of momentum, $\hbar(\mathbf{k} - \mathbf{K}_n)$. This follows immediately from the general principles of quantum mechanics, since the plane waves are the eigenfunctions of the momentum operator. Only a discrete set of values of momentum is possible for an electron in a state $\mathbf{k}$. If ψ does not deviate much from a plane wave, only one coefficient is large. The higher coefficients are due to the oscillations of the periodic part of ψ, e.g., to the wiggles of the atomic-like orbitals near the nucleus.

If we let $\mathbf{k}$ take all values inside the BZ, all the coefficients vary and are in fact continuous functions of $\mathbf{k}$. The important point is that, because of the periodicity in $\mathbf{k}$-space, they are all the same function of $\mathbf{k}$, only displaced by fundamental vectors of the reciprocal lattice. This point is illustrated in Fig. 7, which shows schematically the momentum eigenfunctions for one band in a one-dimensional case. The intercepts of any vertical line $k = k_s$ with the various curves give the various coefficients. Each v is a localized function centered on one of the reciprocal lattice sites [which is why they are denoted by $v(\mathbf{k} - \mathbf{K}_n)$]. They are all identical, so that we speak of one function $v(\mathbf{k})$ which determines the probability distribution of momentum for the whole band. Each band is represented by a different v function, provided that the bands are still nondegenerate. If the band is narrow and deviates little from an atomic level, v approaches the probability amplitude of momentum for the corresponding atomic state. For almost free electrons, only one of the coefficients is different from zero in each zone and is almost constant through the zone.

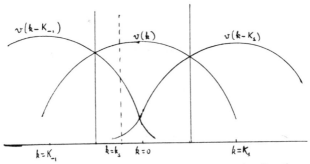

Fig. 7.   Schematic representation of momentum eigenfunctions.

The coefficients $v(\mathbf{k} - \mathbf{K}_n)$ satisfy a system of equations, equivalent to Schrödinger's equation in p-space, that is important because most band structure calculations are based on it. Consider Schrödinger's equation in $\mathbf{p}$-space

$$(2\,m)^{-1}\,(p_x^2 + p_y^2 + p_z^2)\,\psi(\mathbf{p}) + V(\mathbf{p})\,\psi(\mathbf{p}) = E\psi(\mathbf{p}) \qquad (2.26)$$

where V must be expressed as an operator in $\mathbf{p}$-space. Now

$$V(\mathbf{r}) = \sum_{\mathbf{K}_m} V(\mathbf{K}_m)\exp(i\mathbf{K}_m \cdot \mathbf{r})$$

and according to the principles of quantum mechanics, x must be replaced by $i\hbar\partial/\partial p_x$, etc. We also remember that

$$\exp(iK_{mx}\,i\hbar\,\partial/\partial p_x)\,\psi(p_x) = \psi(p_x - \hbar K_{mx}) \qquad (2.27)$$

so that equation (2.26) becomes

$$[(\hbar^2 k^2/2\,m) - E]\,\psi(\mathbf{p}) + \sum_{\mathbf{K}_m} V(\mathbf{K}_m)\,\psi(\mathbf{p} - \hbar\mathbf{K}_m) = 0 \qquad (2.28)$$

Having fixed $\mathbf{k}$, this is a system of linear algebraic equations for the coefficients $\psi(\mathbf{k} + \mathbf{K}_n)$ which are exactly those previously denoted by $v(\mathbf{k} - \mathbf{K}_n)$. Thus, in atomic units, the required system of equations is, for any $v(\mathbf{k} - \mathbf{K}_n)$,

$$(|\mathbf{k} + \mathbf{K}_n|^2 - E)\,v(\mathbf{k} - \mathbf{K}_n) + \sum_{\mathbf{K}_m} V(\mathbf{K}_m)\,v(\mathbf{k} - \mathbf{K}_n - \mathbf{K}_m) = 0 \qquad (2.29)$$

Alternatively, this system may be obtained by substituting equations (2.16) and (2.24) in Schrödinger's equation $-\nabla^2\psi + V\psi = E\psi$, multiplying by $\exp[-i(\mathbf{k} + \mathbf{K}_n) \cdot \mathbf{r}]$, and integrating over the unit cell. The system (2.29) has nonzero solutions only if the determinant of the coefficients of the v's vanishes, and this condition is a secular equation for E, determining the various E_n's for the chosen value of $\mathbf{k}$. If $\mathbf{k}$ is a point of high symmetry, many of the vectors $\mathbf{k} - \mathbf{K}_n$ are equivalent, and one must then construct, according to the rules of group theory, combinations of plane waves of the correct symmetry; the number of equations is thereby reduced. In any case, the system has an infinite number of equations and the solution can be obtained only by approximate methods.

Consider now the periodicity in k-space. We can write expansions in terms of "plane waves in k-space," whose "wave vectors" are the fundamental translation $\mathbf{R}_i$ of the direct lattice. Thus,

$$E(\mathbf{k}) = \sum_{\mathbf{R}_i} E(\mathbf{R}_i) \exp (i\mathbf{R}_i \cdot \mathbf{k}) \tag{2.30}$$

with

$$E(\mathbf{R}_i) = (\Omega_B)^{-1} \int_{\substack{\text{unit cell} \\ \text{of k-space}}} E(\mathbf{k}) \exp (-i\mathbf{R}_i \cdot \mathbf{k}) \, d\mathbf{k} \tag{2.31}$$

where Ω_B, the volume of the BZ, is $8\pi^3/\Omega$. When ψ is developed in k-space, the coefficients $a(\mathbf{R}_i)$ depend on the point $\mathbf{r}$ at which the expansion is performed. By analogy with equation (2.24), denote them $a(\mathbf{r} - \mathbf{R}_i)$; then,

$$\psi(\mathbf{k}, \mathbf{r}) = \sum_{\mathbf{R}_i} a(\mathbf{r} - \mathbf{R}_i) \exp (i\mathbf{R}_i \cdot \mathbf{k}) \tag{2.32}$$

A formula of this type was suggested by Bloch in his first paper on the quantum theory of solids. He assumed that the functions $a(\mathbf{r} - \mathbf{R}_i)$ could be taken as atomic orbitals centered at $\mathbf{R}_i$. Bloch's expression was an approximation; equation (2.32) is exact. It can be inverted in the usual way giving

$$a(\mathbf{r} - \mathbf{R}_i) = (\Omega_B)^{-1} \int \psi(\mathbf{r}, \mathbf{k}) \exp (-i\mathbf{R}_i \cdot \mathbf{k}) \, d\mathbf{k} \tag{2.33}$$

where the integration is over the unit cell of k-space, namely, over all the states of one band. Complications may arise because of the topology of the bands. The $a(\mathbf{r} - \mathbf{R}_i)$'s are called Wannier functions.

There is a simple relation between $a(\mathbf{r})$ and $v(\mathbf{k})$. If in equation (2.33) we express ψ by means of equation (2.24), it follows that

$$a(\mathbf{r} - \mathbf{R}_i) = (\Omega_B)^{-1} \int_{\substack{\text{unit cell} \\ \text{of k-space}}} \exp(-i\mathbf{R}_i \cdot \mathbf{k}) \, d\mathbf{k} \sum_{\mathbf{K}_n} v(\mathbf{k} - \mathbf{K}_n) \exp [i(\mathbf{k} + \mathbf{K}_n) \cdot \mathbf{r}]$$

$$= (\Omega_B)^{-1} \int_{\text{unit cell}} d\mathbf{k} \sum_{\mathbf{K}_n} v(\mathbf{k} - \mathbf{K}_n) \exp [i(\mathbf{k} + \mathbf{K}_n) \cdot (\mathbf{r} - \mathbf{R}_i)]$$

since $\mathbf{K}_n \cdot \mathbf{R}_i = 2\pi n$. The summation over all cells of k-space coupled to the integration over one of these cells is equivalent to integration over the whole of k-space. Thus, the argument of v can be denoted just by $\mathbf{k}$ and, accordingly,

$$a(\mathbf{r} - \mathbf{R}_i) = (\Omega/8\pi^3) \int_{\substack{\text{whole} \\ \text{k-space}}} v(\mathbf{k}) \exp\left[i\mathbf{k} \cdot (\mathbf{r} - \mathbf{R}_i)\right] d\mathbf{k} \qquad (2.34)$$

In particular,

$$a(\mathbf{r}) = (\Omega/8\pi^3) \int^{\infty} v(\mathbf{k}) e^{i\mathbf{k} \cdot \mathbf{r}} d\mathbf{k} \qquad (2.35)$$

Thus, $v(\mathbf{k})$ and $a(\mathbf{r})$ form a pair of integral transforms.

The Wannier functions can be considered also in two ways, that is, either as a single function $a(\mathbf{r})$ extending to the whole of space, or as a set of functions $a(\mathbf{r} - \mathbf{R}_i)$ defined in one cell. The diagram (Fig. 7) used to illustrate the momentum eigenfunctions applies also to the Wannier functions, with x replacing k. The Wannier functions for one band are thus identical functions localized and centered at each lattice site. There is one Wannier function for each band; for inner-core electrons, the Wannier function coincides almost exactly with the atomic orbital. If the bands are well separated, the expression $\Sigma_{\mathbf{R}_i} a^*(\mathbf{r}_0 - \mathbf{R}_i) a(\mathbf{r}_0 - \mathbf{R}_i)$ gives, at each point $\mathbf{r}_0$, the probability that an electron is to be found there, averaged over all the $\mathbf{k}$ states of the band.

We state without proof two further properties of the Wannier functions: (1) the Wannier functions centered on different lattice points are orthogonal, which is an advantage over the atomic orbitals which generally are orthogonal only for strongly localized electrons; and (2) the matrix element of the one-particle Hamiltonian between two Wannier functions located at different lattice points, $\mathbf{R}_i$ and $\mathbf{R}_j$, is equal to the corresponding Fourier coefficient of the energy, $E(\mathbf{R}_j - \mathbf{R}_i)$. If the atoms are far apart, there is little overlap between the Wannier functions and only the diagonal element $E(0)$ is appreciably different from zero; it gives the average energy of the states of the (narrow) band.

A straightforward calculation shows that the system of differential equations satisfied by the Wannier functions is

$$Ha(\mathbf{r} - \mathbf{R}_j) = \sum_{\mathbf{R}_s} a(\mathbf{r} - \mathbf{R}_j + \mathbf{R}_s) E(\mathbf{R}_s) \qquad (2.36)$$

where H is the one-particle Hamiltonian and the $E(\mathbf{R}_s)$'s are the Fourier coefficients of the energy, given by equation (2.31). The system (2.36) may in principle be used instead of the Schrödinger equation or instead of equations (2.29) as a starting point for band structure calculations. It is, however, a complicated system and can be dealt with only by variational methods. It has not found much application. Only the Wannier function of the band under consideration appears in equation (2.36). There are no matrix elements of H between Wannier functions of different bands except where degeneracies occur.

V. REPRESENTATION THEORY IN THE BAND MODEL

If the system is described by vectors $\mathbf{k}$ and their probability amplitudes $v(\mathbf{k})$, we have what is called the "crystal momentum representation" in which $\mathbf{k}$ is diagonal and which corresponds to the momentum representation of ordinary continuum mechanics. If instead the system is described by Wannier functions, we have the analog of the coordinate representation.

Now, as is well known, the ordinary momentum operator $\mathbf{p}$ has the role of a generator of infinitesimal translations, that is, the operator

$$\exp(i\varepsilon p_x) = 1 + i\varepsilon p_x = 1 + \hbar\varepsilon \frac{d}{dx}$$

(where ε is infinitesimal) applied to $\psi(x)$ changes it into

$$\psi(x) + \hbar\varepsilon \frac{d\psi}{dx} = \psi(x + \hbar\varepsilon)$$

That is, the operator applied to any state shifts its position by the infinitesimal amount $\varepsilon\hbar$. In three dimensions, $1 + i\varepsilon\mathbf{n}\cdot\mathbf{p}$ shifts any state by $\varepsilon\hbar$ in the direction of $\mathbf{n}$. Since H is invariant with respect to all translations, $\mathbf{p}$ is a constant of motion. In a crystal, however, it is not $\mathbf{p}$ which is a constant of motion, but the crystal momentum $\hbar\mathbf{k}$. The corresponding operator is called $\mathbf{P}$ and

$$\mathbf{P}\psi_{n\mathbf{k}}(\mathbf{r}) = \hbar\mathbf{k}\psi_{n\mathbf{k}}(\mathbf{r}) \tag{2.37}$$

What translations does it generate? If the operator $\exp(i\mathbf{P}\cdot\mathbf{R}/\hbar)$ (where for the moment $\mathbf{R}$ is any vector) is applied to

$$\psi_{n\mathbf{k}}(\mathbf{r}) = u_{n\mathbf{k}}(\mathbf{r}) e^{i\mathbf{k}\cdot\mathbf{r}}$$

there results

$$u_{n\mathbf{k}}(\mathbf{r}) \exp[i\mathbf{k}\cdot(\mathbf{r}+R)]$$

since for any function of $\mathbf{P}$, $f(\mathbf{P})\psi = F(\hbar\mathbf{k})\psi$. The exponential part of the Bloch function has been shifted by $\mathbf{R}$, while the periodic part has remained the same. The new function bears no simple relation to the original one, except when $\mathbf{R}$ is a lattice translation $\mathbf{R}_i$, because then

$$u_{n\mathbf{k}}(\mathbf{r}) = u_{n\mathbf{k}}(\mathbf{r}+\mathbf{R}_i)$$

so that,

$$\exp(i\mathbf{P}\cdot\mathbf{R}_i/\hbar)\,\psi(\mathbf{r}) = \psi(\mathbf{r}+\mathbf{R}_i)$$

The crystal momentum operator $\mathbf{P}$ generates the lattice translations.

The operator $\mathbf{Q}$, canonically conjugated to $\mathbf{p}$, must be somewhat analogous to the position operator $\mathbf{r}$ of continuous media. The eigenfunctions of $\mathbf{r}$ are δ functions; of course, we could construct perfectly localized functions using Bloch functions from all bands, since they form a complete set. But, since $\mathbf{P}$ commutes with the band index n, it is convenient to develop a formalism in which $\mathbf{Q}$ also does the same, so that the various bands are all treated separately. Here the most appropriate definition of band index may not be that based on the energy values. The results of the analysis will be described. $\mathbf{Q}$ turns out to have discrete eigenvalues (as might well have been expected, since the reduced zone scheme is used here), and these eigenvalues are the lattice translation vectors. Its eigenfunctions are localized to the approximate extension of a primitive cell; they are found to have all the properties of the Wannier functions, so that they are identified with these functions. In particular, the elements of the transformation matrix $\langle n\mathbf{k}|n\mathbf{R}_i\rangle$ from the Bloch states to the new states are found to be of constant modulus, so that each Wannier state has all Bloch states represented with equal "intensity." These elements are given by equation (2.33). However, a phase factor should be considered, so that the more complete expression for the Wannier functions is

$$a_n(\mathbf{r}-\mathbf{R}_i) = (\Omega_B)^{-1} \int_{\text{unit cell}} d\mathbf{k}\, u_{n\mathbf{k}}(\mathbf{r}) \exp\left[i\mathbf{k}\cdot(\mathbf{r}-\mathbf{R}_i) - i\varphi_n(\mathbf{k})\right] \quad (2.38)$$

The phases are arbitrary (of course, all results that can be verified experimentally are independent of their choice). It is found that the most general admissible $\varphi_n(\mathbf{k})$ consists of a periodic function of $\mathbf{k}$ plus the scalar product of $\mathbf{k}$ with some lattice vector $\mathbf{R}_m$. The effect of the last term is merely to shift the whole Wannier function by $\mathbf{R}_m$ ($\mathbf{R}_m$ is then chosen so that the mean value of $\mathbf{r}$ in a Wannier state lies in the cell belonging to that $\mathbf{R}_i$ that labels the given Wannier function). Then the periodic function in the phase can be conveniently chosen so as to minimize the mean value of $|\mathbf{r}-\mathbf{R}_i|^2$. With this choice of the phases, the Wannier function of a nondegenerate band vanishes at infinity faster than any finite r^{-n} (like an atomic orbital). If the band is degenerate, the corresponding Wannier function vanishes at ∞ much more slowly. This can be easily verified, for instance, by considering an empty simple cubic lattice.

With all the above conventions, the eigenvalue equation for $\mathbf{Q}$ is thus

$$\mathbf{Q} a_n(\mathbf{r}-\mathbf{R}_i) = \mathbf{R}_i a_n(\mathbf{r}-\mathbf{R}_i) \quad (2.39)$$

VI. APPROACHES TO BAND CALCULATIONS

Band structure calculations can be approached from two main points of view: (1) starting from the assumption of free electrons, and (2) starting from the assumption of tightly bound electrons. This refers to the choice of the type of wave function to use in the self-consistent procedure.

The approximation of completely free electrons consists of assuming that the potential V is constant (and then it may be taken as zero). Then

$$\psi_k(\mathbf{r}) = e^{i\mathbf{k} \cdot \mathbf{r}} \equiv |\mathbf{k}\rangle$$

with energy $E(\mathbf{k}) = \hbar^2 k^2/2m$. In fact, the potential is supposed to be the limit of a very weak periodic potential, so that there is, for any $\mathbf{k}$, an infinite number of possible states with energies given by equation (2.19).

One of the most useful applications of the model is to the discussion of the Fermi surfaces of metals. It is found that often this very simple model gives surprisingly accurate results. The surfaces of constant energy are spheres in $\mathbf{k}$-space. As is well known, if there are n electrons per unit volume, the radius k_F of the Fermi sphere (FS) is $k_F = (3\pi^2 n)^{\frac{1}{3}}$. At 0°K all states within this sphere are occupied, and all states outside it are unoccupied. The Fermi energy, or Fermi level, is the energy corresponding to the top of this distribution $E_F = \hbar^2 k_F^2/2m$.

The FS is, in fact, a sphere for univalent metals. When there is more than one electron per atom, the BZ structure must be taken into account. Thus, according to the free electron model, the FS of, for instance, aluminum (three free electrons per atom) would be as in Fig. 8, i.e., bulging out of the faces of the BZ. If we wish to consider only the central BZ, we must shift into this zone all the parts of the FS situated outside it. In this way the FS is, as it were, reflected inward from the boundary. The reflected portions are those that would be included in the higher zones (Fig. 9). It is clear that on adding more electrons the first zone continues to fill, while states in the higher zones are already occupied.

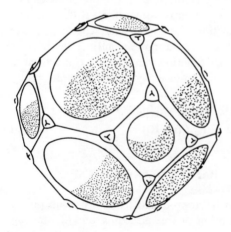

Fig. 8.   Fermi surface of aluminum.

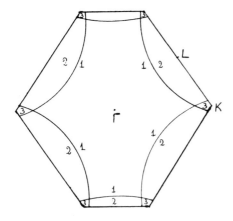

Fig. 9. Reduced free electron Fermi surface of aluminum, showing the main 1-2 Fermi surface and pockets of electrons in the third zone (section with plane $k_x = k_y$).

The above considerations suggest a simple way for constructing the FS. Draw about each reciprocal lattice point a sphere whose volume corresponds to the appropriate number of free electrons in the metal (equal to the volume of the BZ if there are two free electrons per Bravais cell). These spheres intercept when the number of electrons is sufficiently large; the inter-sections occur at the surfaces of the (repeated) BZ. Considering any point **k**, if it is contained within a single sphere, one free electron state at that point in the first zone is occupied. If it lies within two spheres, states belonging to the first and second zones are occupied, and so on. The various parts of the FS thus separate regions of **k**-space where n zones are occupied from regions where $n+1$ are occupied. In Fig. 9 the main FS is a 1-2 Fermi surface, while the pockets at the corners form the 2-3 surface.

Sometimes the FS is not closed, but multiply connected, and it extends all through reciprocal space from one cell to the other. This has important consequences for many properties, e.g., magnetoresistance.

We now examine, in the example of the face-centered cubic lattice, how the band structure is constructed in the free electron approximation. The BZ is the truncated octahedron of Fig. 4. The center point, $k = 0$, is conventionally denoted by Γ. Other prominent points are shown: points along the $(1, 0, 0)$ directions are denoted by Δ, the end points by X, points along the $(1, 1, 1)$ directions by Λ, the end point by L, Σ is any point along the $(1, 1, 0)$ directions, and W is the end point of the $(0, 2, 1)$ directions. Assume for convenience that the side of the elementary cube in the direct lattice is 2π; then the distance $\Gamma - X$ is 1. The energy is given by equation (2.14) in atomic units as follows:

$$E(k_x, k_y, k_z, n) = (k_x + K_{nx})^2 + (k_y + K_{ny})^2 + (k_z + K_{nz})^2 \qquad (2.40)$$

where $\mathbf{K}_n$ is any fundamental vector of the reciprocal lattice. The constant energy surfaces are spheres centered at all reciprocal lattice points. Since the reciprocal lattice is body-centered, the first reciprocal lattice vectors are

$$\mathbf{K}_0 = 0$$
$$\mathbf{K}_1 = (\pm 1, \pm 1, \pm 1) \qquad \text{8 vectors of length } 3^{\frac{1}{2}}$$
$$\mathbf{K}_2 = (\pm 2, \quad 0, \quad 0) \qquad \text{6 vectors of length } 2$$
$$\mathbf{K}_3 = (\pm 2, \pm 2, \pm 2) \qquad \text{12 vectors of length } 8^{\frac{1}{2}}$$
$$\mathbf{K}_4 = (\pm 3, \pm 1, \pm 1) \qquad \text{24 vectors of length } 11^{\frac{1}{2}}$$

etc.

Thus, the successive energy levels at $k = 0$ are $E_0 = 0$, $E_1 = 3$, $E_2 = 4$, $E_3 = 8$, ...; E_1 is eight times degenerate, E_2 six times, etc.

Consider now the Δ direction, for instance, $(0, 1, 0)$; along this direction, $k_x = k_z = 0$ while $0 \leqslant k_y \leqslant 1$. It is not necessary to consider negative values since $E(\mathbf{k}) = E(-\mathbf{k})$. Starting with $\mathbf{K}_0$ in equation (2.40), we have the nondegenerate band $E = k_y^2$. Taking $\mathbf{K}_1$, we have the possibilities $k_x + K_x = \pm 1$, $k_y + K_y = k_y \pm 1$, and $k_z + K_z = \pm 1$, with two possible bands

$$E = 2 + (k_y - 1)^2 \qquad \text{four times degenerate}$$
$$E = 2 + (k_y + 1)^2 \qquad \text{four times degenerate}$$

At $k = 0$, both bands start at the eight-times degenerate level $E_1 = 3$; for one band the energy decreases as k_y increases; for the other, it increases. Now taking $\mathbf{K}_2$, we have the following three possibilities:

$$k_x + K_x = \pm 2 \qquad k_y + K_y = k_y \qquad k_z + K_z = \quad 0$$
$$k_x + K_x = \quad 0 \qquad k_y + K_y = k_y \pm 2 \qquad k_z + K_z = \quad 0$$
$$k_x + K_x = \quad 0 \qquad k_y + K_y = k_y \qquad k_z + K_z = \pm 2$$

and, thus, the three bands

$$E = (k_y - 2)^2 \qquad \text{nondegenerate}$$
$$E = 4 + k_y^2 \qquad \text{four times degenerate}$$
$$E = (k_y + 2)^2 \qquad \text{nondegenerate}$$

The curves along other directions are found in a similar way and are illustrated in Fig. 10. The values of K_x, K_y, and K_z are indicated on these curves. Also indicated are the degeneracies of the various states and the position of the Fermi level $E_F = k_F^2 = (3 Z/2 \pi)^{2/3}$ since we have taken the number of cells per unit volume to be $(2 \pi^3)^{-1}$. This simple picture is very useful in making qualitative predictions on the band structure, particularly of metals.

The next approximation, the nearly free electron (NFE) approximation, consists of introducing a small periodic potential $V(\mathbf{r})$ that may be treated as a perturbation. The unperturbed states are the free electron states $E_k^0 = \hbar^2 k^2/2 m$. To second order

$$E(\mathbf{k}) = E^0(\mathbf{k}) + \langle \mathbf{k} | V | \mathbf{k} \rangle + \sum_{k'} \frac{|\langle \mathbf{k} | V | \mathbf{k}' \rangle|^2}{E_k^0 - E_{k'}^0} \qquad (2.41)$$

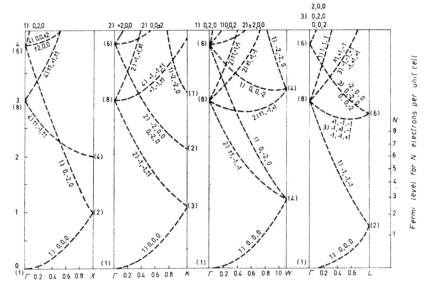

Fig. 10.   Free electron bands along prominent directions of a face-centered cubic lattice.

However,

$$\langle k \,|\, V \,|\, k' \rangle \equiv \int_{\substack{\text{unit} \\ \text{cell}}} \exp \left[i(k' - k) \cdot r \right] V(r) \, dr$$

vanishes unless $(k - k')$ is a vector of the reciprocal lattice K_i, when it is the Fourier coefficient $V(K_i)$; thus,

$$E(k) = E^0(k) + V_0 + \sum_{K_i \neq 0} \frac{|V(K_i)|^2}{E_k^0 - E_{k'}^0} \qquad (2.42)$$

This result is satisfactory if (1) the $V(K_i)$'s tend to zero rapidly as K_i increases (discussed below), and (2) there is no degeneracy of the form $E^0(k) = E^0(k')$. The condition $E^0(k) = E^0(k')$ means that k lies on the perpendicular bisector of the reciprocal lattice vector K_i (Fig. 11), the condition for Bragg reflection. Thus the expansion is not valid when k lies on or near a zone boundary.

To find the energy in such cases, the perturbation equations must be considered explicitly. Accordingly, this means using equations (2.19), but considering only those Fourier coefficients of the potential that couple the degenerate bands. To illustrate, consider the two free electron bands degenerate at the point X, of energy $E = 2$, in the example previously discussed referring to a face-centered cubic lattice (Fig. 10). The relevant

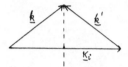

Fig. 11.   Condition for Bragg reflection.

Fourier coefficient of the potential is $V(\mathbf{K}_2)$, abbreviated as V_2, and the only equations of the system (2.29) taken into consideration are

$$(k^2 - E)v_0 + V_2 v_2 = 0$$
$$[(k-2)^2 - E]v_2 + V_2 v_0 = 0$$

where v_0 denotes $v(\mathbf{k})$ and v_2 denotes $v(\mathbf{k} - \mathbf{K}_2)$. These equations are compatible only if the determinant of the coefficients of v_0 and v_2 vanishes, or if

$$(k^2 - E)\,[(k-2)^2 - E] - V_2{}^2 = 0 \tag{2.43}$$

At the point $X(k = 1)$, $E = 1 \pm V_2$, and the two first-order wave functions (symmetrized wave functions) corresponding to the two levels are the symmetric and antisymmetric combinations of the two unperturbed plane waves e^{ikx} and $e^{i(k-2)x}$ (standing waves—Bragg reflection).

Thus, the perturbation introduces an energy gap which is twice the magnitude of the relevant Fourier coefficient of the potential. This result is general in the NFE model. The Fourier coefficients of the potential can be estimated from X-ray diffraction data, and thus one can readily form an idea of the planes in **k**-space across which the main energy gaps are situated in metals and alloys.

Differentiating equation (2.43) with respect to k and letting $k = 1$, we verify that $dE/dk(k = 1) = 0$.

For the validity of the NFE model, the other condition is that the $V(\mathbf{K}_i)$'s should tend to zero rapidly as $|\mathbf{K}_i|$ becomes large. This does not seem likely, since the field near any ion is strong, which means that it has Fourier components of short wavelength. Despite this, the NFE method can be made formally valid by the introduction of pseudopotentials. This concept is better discussed after a description of the method of orthogonalized plane waves (OPW). The main disadvantage of the use of simple plane waves is that the lowest eigenvalues correspond to core states, in which the electron is strongly localized, so that a series of plane waves is a poor approximation to its wave function. To avoid this, one makes the plane waves, or the symmetrized plane waves, orthogonal to the core states, so that the core-state eigenvalues are automatically excluded from the roots of the secular equation for the energy.

Let the localized core wave functions be $b_t(\mathbf{r} - \mathbf{R}_i)$. We construct the following Bloch-like function:

$$b_{t,\,\mathbf{k}}\,(\mathbf{r}) = \sum_{\mathbf{R}_i} \exp\,(i\mathbf{k}\cdot\mathbf{R}_i)\,b_t(\mathbf{r} - \mathbf{R}_i) \tag{2.44}$$

This is a solution of the Schrödinger equation of the whole crystal corresponding to one of the core levels of energy E_t (for narrow bands the Wannier functions and the atomic orbitals coincide). The higher states ψ_k must be orthogonal to $b_{t,k}$:

$$\langle b_{t,k}, \psi_k \rangle = 0 \tag{2.45}$$

Since ψ_k must behave as a free electron wave in the regions between the atoms, we take as basis functions for the higher states

$$\chi_k = e^{ik \cdot r} - \sum_{t'} Z_{t'k} b_{t'k}(r) \tag{2.46}$$

For states of high symmetry, one would have symmetrized combinations of plane waves. If the $b_{t,k}$'s form an orthogonal system, χ is orthogonal to the core states if the coefficients $Z_{t,k}$ are taken as

$$Z_{t,k} = \langle b_{t,k}, e^{ik \cdot r} \rangle \tag{2.47}$$

The function

$$\chi_k = e^{ik \cdot r} - \sum_{t} \langle b_{t,k}, e^{ik \cdot r} \rangle b_{t,k} \tag{2.48}$$

is called an OPW. This function approaches a plane wave in the interatomic region, where the b's are small, while within the core it is orthogonal to the core states. These OPW's are used as basis states for the wave function

$$\psi_k(r) = \sum_{K_i} A_{k-K_i} \chi_{k-K_i}(r) \tag{2.49}$$

and the variational principle is applied to minimize the expectation value of the energy; this gives the eigenvalues and determines the coefficients A_{k-K_i}. One finds in practice that the method often converges rapidly on adding more OPW's, so that the modified NFE method gives better results than we should expect. An argument, first given by Phillips and Kleinman, explains this. Suppose that ψ_k is the exact wave function with some definite values of the A_{k-K_i}'s. Consider the function

$$\varphi_k(r) = \sum_{K_j} A_{k-K_j} \exp[i(k-K_j) \cdot r] \tag{2.50}$$

where simple plane waves replace the OPW's. Then, from equation (2.48),

$$\psi = \varphi - \sum_{t} \langle b_t, \varphi \rangle b_t \tag{2.51}$$

Substitution of equation (2.51) into $H\psi = E\psi$ yields

$$H\varphi - \sum_{t} \langle b_t, \varphi \rangle Hb_t = E\varphi - E \sum_{t} \langle b_t, \varphi \rangle b_t$$

However, $Hb_t = E_t b_t$ and thus

$$H\varphi + \sum_t (E-E_t) \langle b_t, \varphi \rangle b_t = E\varphi \qquad (2.52)$$

We may look upon this as a new Schrödinger equation

$$(H+V_R)\varphi = E\varphi \qquad (2.53)$$

where the operator V_R means formally

$$V_R \varphi = \sum_t (E-E_t) b_t \langle b_t, \varphi \rangle \qquad (2.54)$$

The "smoothed wave function" or "pseudo wave function" φ thus satisfies an equation of which the Hamiltonian is

$$H+V_R = -(\hbar^2/2m) \nabla^2 + V + V_R$$

thus providing an effective potential ("pseudopotential")

$$V_{eff} = V + V_R \qquad (2.55)$$

which contains the unknown function φ. The important point is that the eigenvalues E of the valence states are unaltered.

Since φ is a series of plane waves, the coefficients $A_{k-\kappa_j}$ (momentum eigenfunctions) satisfy equation (2.29), except that the Fourier components of V_{eff} appear instead of the Fourier components of the true potential V. Now there are plausible arguments which show that V_{eff} is small, so that the NFE model must be satisfactory. First of all, there must be some cancellation between V, which is negative, and V_R, which is positive, since it contains $E-E_t$ and the square of an atomic orbital. Second, this cancellation can be improved by suitable choice of the core functions. In fact, the expression (2.54) for V is not unique. We can replace $(E-E_t) \langle b_t, \varphi \rangle$ by $\langle F_t, \varphi \rangle$, where the F_t's are arbitrary functions. This is because the operator V_R expresses φ as a linear combination of core states, and the noncore wave functions are orthogonal to these states, so that they are not affected by changes in the coefficients of the linear combination. A possible choice of the F_t that will achieve good cancellation is

$$F_t = -Vb_t$$

and then

$$V_{eff} \varphi = V\varphi + \sum_t \langle -Vb_t, \varphi \rangle b_t = V\varphi - \sum_t \langle b_t, V\varphi \rangle b_t \qquad (2.56)$$

That is, we subtract from V any part that can be expanded as a sum of core functions. If these formed a complete system, the difference would vanish; that is, with this choice of the F_t, it depends on the degree to which the occupied core states form a complete system. It is found in practice that the cancellation is often remarkable.

Note that the argument that V_{eff} may be treated as a weak localized potential is only approximate. V_R is a nonlocalized pseudopotential operator; from equation (2.54),

$$V_R = \sum_t (E - E_t) b_t(\mathbf{r}) \int b_t^*(\mathbf{r}') \varphi(\mathbf{r}') d\mathbf{r}' = \int \sum_t (E - E_t) b_t(\mathbf{r}) b_t^*(\mathbf{r}') \varphi(\mathbf{r}') d\mathbf{r}'$$

and with

$$V_R(\mathbf{r}, \mathbf{r}') = \sum_t (E - E_t) b_t(\mathbf{r}) b_t^*(\mathbf{r}')$$

then

$$V_R \varphi = \int V_R(\mathbf{r}, \mathbf{r}') \varphi(\mathbf{r}') d\mathbf{r}' \tag{2.57}$$

At the other extreme of the approach based on free electrons is the "tight binding" approach, based on Bloch's original suggestion of approximating the wave functions in crystals by linear combinations of atomic orbitals (LCAO) of the form

$$\psi_{\mathbf{k},t}(\mathbf{r}) = N \sum_{\mathbf{R}_i} \exp(i\mathbf{k} \cdot \mathbf{R}_i) b_t(\mathbf{r} - \mathbf{R}_i) \tag{2.58}$$

where N is a normalizing factor. The factor $\exp(i\mathbf{k} \cdot \mathbf{R}_i)$ ensures that equation(2.9) is satisfied. The b_t's are atomic orbitals, e.g., normalized $1s$, $2s$, $2p$, ... eigenfunctions, centered at the various lattice sites. With these $\psi_{\mathbf{k},t}$'s a standard calculation gives for the average value of the energy

$$\bar{E}_t(\mathbf{k}) \equiv \int \psi_{\mathbf{k},t}^*(\mathbf{r}) H \psi_{\mathbf{k},t}(\mathbf{r}) d\mathbf{r} = \frac{\sum_{\mathbf{R}_m} \exp(i\mathbf{k} \cdot \mathbf{R}_m) H_t(\mathbf{R}_m)}{\sum_{\mathbf{R}_m} \exp(i\mathbf{k} \cdot \mathbf{R}_m) S_t(\mathbf{R}_m)} \tag{2.59}$$

where the overlap integral is

$$S_t(\mathbf{R}_m) = \int_{\substack{whole \\ space}} b_t^*(\mathbf{r}) b_t(\mathbf{r} - \mathbf{R}_m) d\mathbf{r} \tag{2.60}$$

and the matrix element of the Hamiltonian between two atomic orbitals located on atoms separated by $\mathbf{R}_m$ is

$$H_t(\mathbf{R}_m) = \int b_t^*(\mathbf{r}) H b_t(\mathbf{r} - \mathbf{R}_m) d\mathbf{r} \tag{2.61}$$

The integrals $H_t(\mathbf{R}_m)$ are complicated because they are three center integrals:

$b(\mathbf{r})$ is centered on one atom, $b(\mathbf{r} - \mathbf{R}_m)$ is centered about another atom, and H involves potentials centered about third atoms.

Having evaluated the $E(\mathbf{k})$ relation with only one type of atomic orbitals, the next step is to consider wave functions made up of several b_i's and to use them in a variational procedure. These calculations are cumbersome and do not generally give good results; the method is particularly unsuitable for excited states. At present, it is chiefly used to gain a qualitative picture of a band structure. Essentially, one simply uses equation (2.54) for nondegenerate levels and considers as few orbitals as possible for degenerate cases. Also, one sums only over very few $\mathbf{R}_m$'s; often all terms are neglected for which $\mathbf{R}_m$ is not the shortest fundamental vector (only "interaction with nearest neighbors" is considered). The tight binding formulas may be used as interpolation formulas; the values of the integrals are derived from experimental data or from the results of other calculations which are generally carried out at prominent symmetry points. Then the tight binding formulas give the energy at all $\mathbf{k}$-points.

VII. METHODS FOR BAND STRUCTURE CALCULATION

The problem divides into two parts: the setting up of the best one-electron Hamiltonian and the solution of the corresponding Schrödinger equation with the appropriate boundary conditions. However, the first part involves the second, because the Hartree–Fock equations must be solved in a self-consistent way. Self-consistent calculations have been carried out only in the last few years, partly following the introduction of pseudo-potentials, and partly due to the increased use of larger computers. We shall not examine the problem of the choice of the starting potential for the self-consistent procedure and shall discuss only the methods of solution of the Schrödinger equation with an assumed Hamiltonian.

The general procedure is to expand ψ as

$$\psi = \sum_{i=1}^{n} a_i f_i(\mathbf{r}, \mathbf{k}) \tag{2.62}$$

where the f_i's have the correct symmetry and are chosen out of an orthonormal set and where n depends on the computational facilities. The f_i's may be OPW's, atomic orbitals, etc. If

$$H_{ij} = \int f_i^* H f_j \, d\mathbf{r} \tag{2.63}$$

the standard variational procedure leads to the following secular equation for the energy:

$$
\begin{vmatrix}
H_{11}-E & H_{12} & \cdots & H_{1n} \\
H_{21} & H_{22}-E & \cdots & H_{2n} \\
\cdot\cdot\cdot\cdot\cdot\cdot\cdot\cdot\cdot\cdot\cdot\cdot\cdot \\
H_{n1} & H_{n2} & \cdots & H_{nn}-E
\end{vmatrix} = 0 \qquad (2.64)
$$

The energy may appear also implicitly, e.g., when the f_i's have been obtained by integration of Schrödinger's equation with a trial energy.

Most of the methods in current use are based on plane wave expansions. We have examined the OPW method; another is the method of augmented plane waves (APW). Alternatively, one may use ψ's obtained by integration of Schrödinger's equation around each atom and devise methods for satisfying as accurately as possible the continuity and periodicity conditions (cellular method). Another interesting method goes under the name of the "S-matrix," or "integral equation," method.

A. Cellular Method

The crystal is divided into cells by drawing all planes bisecting and perpendicular to the lines joining each atom in the lattice to its neighbors. In monatomic lattices, these cells are the Wigner–Seitz cells. In each cell the potential is assumed to be spherically symmetric (deviations from spherical symmetry may be considered later as a perturbation). Then ψ may be separated in each cell in spherical polar coordinates, the angular part being expressed by means of associated spherical harmonics. From group theory it is determined which spherical harmonics (or linear combinations) are present at each $\mathbf{k}$-point. The radial part of ψ is determined by integration of the radial equation with the assumed potential and use of trial values of the energy. One has then an expansion in each cell, and the problem is to "match" these solutions, that is, to satisfy the appropriate boundary conditions at the surfaces of the polyhedral cells. This is more complicated for polyatomic substances. If we take a monatomic substance and two opposite faces of a cell, the boundary conditions at two equivalent points (A, B) separated by a lattice vector $\mathbf{R}$ are

$$
\psi(A) = \psi(B)\, e^{i\mathbf{k}\cdot\mathbf{R}}
$$

$$
\frac{d\psi(A)}{dn} = -\frac{d\psi(B)}{dn}\, e^{i\mathbf{k}\cdot\mathbf{R}}
$$

One method is to satisfy these conditions at a sufficient number of chosen points; least-squares or variational procedures may be used. In any case, some compatibility relation is obtained, which is satisfied only for certain values of the trial energy. The cellular method has almost been abandoned since it has become clear that it can give reliable results only if a very large number of terms is included in the expansion of ψ. Moreover, the type of potential employed is incorrect in the interatomic regions.

B. Augmented Plane Waves

Here one makes the more realistic assumption that the potential is constant between the ion cores (though this assumption is not essential for the method). This so-called "muffin-tin" potential is spherically symmetric within some radius R_s about each atom (consider here for simplicity monatomic substances—the extension to polyatomic substances is not difficult) and constant in the interstitial regions. The spheres are not large enough to intersect. Within each sphere the solution is written as an expansion in spherical harmonics; in the region between the spheres, it is expressed as a plane wave $\psi_0 = \exp(i\mathbf{k}\cdot\mathbf{r})$. To match these solutions on the surface of the spheres, we use the expansion of a plane wave in terms of Legendre functions

$$\psi_0 = \sum_l i^l (2l+1) P_l(\cos\theta) j_l(kr) \tag{2.65}$$

where j_l is a spherical Bessel function [e.g., $j_0(r) = (\sin r/r)$], and the angle θ is measured relative to $\mathbf{k}$. Choosing a trial value of the energy E_t, one finds the radial solution of Schrödinger's equation within the sphere for each l, and one constructs a linear combination of these solutions so as to join continuously to equation (2.65) on the surface. There will be a discontinuity in grad ψ. The expectation value of the energy $\bar{E}$ is calculated for the APW so constructed. There are contributions to $\bar{E}$ from the regions inside and outside the sphere and from the discontinuity in grad ψ at the surface. $\bar{E}$ is a function of E_t, and one varies E_t to make $\bar{E}$ stationary, assuming that the stationary value is the best approximation to the correct energy. One finds that the value of E_t that makes $\bar{E}$ stationary is equal to the stationary value $\bar{E}$ itself. Thus, an APW yields an $\bar{E}$ such that the function within each sphere is an eigenfunction of Schrödinger's equation belonging to the eigenvalue $\bar{E}$. It is not an exact solution outside the sphere because $\bar{E}$ is not generally equal to k^2, but the error due to this is compensated by the contribution from the discontinuity in grad ψ. The relation between $\bar{E}$ and $\mathbf{k}$ can be expressed analytically: If Ω is the volume of the Wigner–Seitz cell minus the volume of the sphere of radius R_s, then

$$\Omega(\bar{E}-k^2) = 4\pi \sum_l R_s^2(2l+1)j_l^2(kR_s)\left[\frac{d}{dr}\ln R_l(\bar{E}, r)\right]_{r=R_s} \tag{2.66}$$

Note that, as far as the radial functions are concerned, only their logarithmic derivatives at $r = R_s$ are required; these derivatives can be obtained from experimental spectroscopic data (quantum defect method). Equation (2.66) gives a series of $\bar{E}$–$\mathbf{k}$ curves: One chooses the value of $\bar{E}$ that is nearest to k^2. The next step is to take the APW's as basis functions for the variational procedure, and the number of APW's is increased until the required energy values have converged satisfactorily. The convergence is found to be good. In the variational calculation, the restriction to a muffin-tin potential can be removed.

C. Scattering Matrix Approach

In the original S-matrix method, taking again a muffin-tin potential, every atomic sphere is considered to be the source of an outgoing wave of wave vector $\mathbf{k}$, corresponding to electrons of energy E propagating in the region of zero potential. Let $k_0^2 = E$. The waves emitted by two spheres $\mathbf{R}_i$ apart must satisfy the relation

$$\varphi(\mathbf{r} + \mathbf{R}_i) = \exp(i\mathbf{k} \cdot \mathbf{R}_i)\, \varphi(\mathbf{r}) \qquad (2.67)$$

Now, at the surface of each sphere, there is the outgoing wave from the sphere (φ_{out}) and the waves coming from all the other spheres (φ_{in}), and the two must be the same. However, φ_{in} can be expressed in terms of φ_{out} by means of equation (2.67). Expand both in spherical harmonics, the coefficients being a_{lm} for φ_{out} and b_{lm} for φ_{in}. The condition $\varphi_{\text{in}} = \varphi_{\text{out}}$ gives equations for a_{lm} and b_{lm}, in which the energy enters as an unknown parameter. On the other hand, φ_{out} must be related to φ_{in}, as scattered and incident waves, through a scattering matrix. If the potential is spherically symmetric, the S-matrix is diagonal in lm and

$$a_{lm} = \tfrac{1}{2} b_{lm} [1 - \exp(-2 i\eta_l)]$$

Here η_l is a phase shift, a function of E, determined from the condition that the solution $K(r)$ of

$$\frac{d^2 K}{dr^2} + \left[k_0^2 - V(r) - \frac{l(l+1)}{r^2} \right] K = 0$$

outside the sphere must have the form

$$K(r) = C \sin(k_0 r - \tfrac{1}{2} \pi l + \eta_l)$$

In any case, there is a second set of equations for a_{lm} and b_{lm}, and the two can be satisfied simultaneously only for certain values of the energy: $\mathbf{k}$ enters into the first set of equations and thus the dispersion relation is determined. The first part of the calculation, that is, the determination of φ_{in}

as the superposition of the waves coming from all lattice points, requires
the calculation of lattice sums for all values of **k**. It does not depend on the
potential and, thus, apart from scaling factors, is the same for all crystals
with the same structure and can be carried out once and for all for any given
structure. The second part depends on the particular potential, but is
independent of **k**.

The integral equation approach is based on the integral form of
Schrödinger's equation

$$\psi(\mathbf{r}) = \int_{\substack{\text{unit} \\ \text{cell}}} G(\mathbf{r}, \mathbf{r}') V(\mathbf{r}') \psi(\mathbf{r}') \, d\mathbf{r}' \qquad (2.68)$$

where the subscript **k** is omitted and where $G(\mathbf{r}, \mathbf{r}')$ is the Green function:

$$G(\mathbf{r}, \mathbf{r}') = (\Omega)^{-1} \sum_{\mathbf{K}_n} \frac{\exp\left[i(\mathbf{k}+\mathbf{K}_n)\cdot(\mathbf{r}-\mathbf{r}')\right]}{E - |\mathbf{k}+\mathbf{K}_n|^2}$$

Note that if, instead of $\psi(\mathbf{r})$, one uses the probability amplitude of momentum $\psi(\mathbf{p})$, equation (2.68) may be written in the equivalent form

$$(p^2 - E)\psi(\mathbf{p}) = \int V(\mathbf{p}-\mathbf{p}')\psi(\mathbf{p}') \, d\mathbf{p}' \qquad (2.69)$$

where $V(\mathbf{p}-\mathbf{p}')$ is the Fourier transform of the potential, reducing in the
periodic problem to a sum of δ functions with coefficients $V(\mathbf{K}_n)$, and thus
to the equations (2.29). Formally, equations (2.68) and (2.69) may be solved
by successive iterations. A succession of iterations and variations leads most
rapidly to a satisfactory approximation of the eigenvalue. However, the
integral equation is difficult to solve without restrictions on the potential,
but things are simpler with a muffin-tin potential, because then the integration
has to be carried out only within the sphere. The procedure is then to replace
the integral equation by a variational principle using, as trial functions,
expansions in spherical harmonics. Also the Green function is expanded in
spherical harmonics and then the formalism is found to become identical
to that of the S-matrix approach.

VIII. CLASSIFICATION OF BANDS

In the free electron model, the band index n can be identified with
the fundamental vectors of the reciprocal lattice; in the tight binding scheme,
n coincides with a set of atomic quantum numbers. For the general case, the
classification must be based on symmetry properties. We saw how the
symmetry under translations leads to a classification of the various states

by means of **k** vectors. Now, at every **k**-point we must consider the reflection and rotation symmetry properties of that point. In fact, the types of bands possible at any **k**-point are entirely determined by the corresponding point group, that is, by the group of operations that leave the point in the original position or, if it is on the boundary of the BZ, that move it to an equivalent **k**-point. This applies provided that these operations are symmetry operations of the crystal as a whole (e.g., for the point group at $k = 0$ in the zinc-blende structure, only half the operations must be considered than for the diamond structure, because the zinc-blende structure does not possess a center of inversion—this despite the fact that the BZ is the same for the two structures). At a general **k**-point of no symmetry, the group of **k** consists of the single identity operation. Note that since all the operations of any point group commute with the one-electron Hamiltonian of the crystal, the energy has the full symmetry of the BZ. This has the practical consequence, for instance, that in a cubic structure it is necessary to determine the E–**k** relations only in $1/48$ of the BZ.

The definitions of representations of a group, irreducible representations, characters of a representation, etc., are given elsewhere. For reference, however, Table I gives the character table of the full cubic group. It applies at $k = 0$ in the BZ of any cubic structure. The group has ten classes, corresponding to the different types of rotation (e.g., the class $C_4{}^2$ includes the three rotations by π about the three axes of fourfold symmetry), coupled or not to the inversion J. The reason for denoting the representations with the subscripts $s, p, d, \ldots$ is that they denote the spherical harmonics of smallest l appearing in the development, in terms of these functions, of the wave functions associated with the given representation.

Table I. Character Table of the Γ Group (Full Cubic Group)

Irreducible representations	Classes									
	E	$C_4{}^2$	C_4	C_2	C_3	J	$JC_4{}^2$	JC_4	JC_2	JC_2
Γ_s	1	1	1	1	1	1	1	1	1	1
Γ_p	3	-1	1	-1	0	-3	1	-1	1	0
Γ_d	2	2	0	0	-1	2	2	0	0	-1
$\Gamma_d{}'$	3	-1	-1	1	0	3	-1	-1	1	0
Γ_f	1	1	-1	-1	1	-1	-1	1	1	-1
$\Gamma_f{}'$	3	-1	-1	1	0	-3	1	1	-1	0
Γ_g	3	-1	1	-1	0	3	-1	1	-1	0
Γ_h	2	2	0	0	-1	-2	-2	0	0	1
Γ_i	1	1	-1	-1	1	1	1	-1	-1	1
Γ_j	1	1	1	1	1	-1	-1	-1	-1	-1

We recall the two main principles for the application of group theory to quantum mechanics.

1. The possible states of a system are those whose wave functions transform, under the operations of the group under consideration, according to irreducible representations of this group; that is, the eigenfunctions belonging to any one energy level form a basis for one of the irreducible representations of the group. For instance, at $k = 0$ in a cubic crystal, there must be states Γ_s whose wave function is left unchanged by all the operations of the full cubic group. Each state is thus labelled by an irreducible representation, which is therefore the most general type of band index. Of course, there is an infinite number of bands of each type. At a point of no symmetry, all the bands belong to the identity representation. Wave functions belonging to different irreducible representations are orthogonal, and all matrix elements of H between states belonging to different irreducible representations vanish.

2. The degeneracy of each state is equal to the dimension of the corresponding irreducible representation (the dimension is equal to the character of the class E).

For the proper identification of the various levels, it is necessary to relate the crystal states to the free electron and to the atomic states. One may wonder why we treat in detail the points of high symmetry, since they constitute the exception. The reasons are that energy extrema are normally found at these points, so that they are very important in semiconductors, and that band structure calculations are simpler at these points.

A. Free Electron Correspondence

How this is established will be better understood by examining a simple illustration. Consider the Γ states of a face-centered cubic lattice. In the free electron model there is one nondegenerate state of energy $E_0 = 0$, one eight-times degenerate state of energy $E_1 = (4\pi^2/a^2)(1^2+1^2+1^2)$, one six-times degenerate state of energy $E_2 = (4\pi^2/a^2)(2^2+0^2+0^2)$, etc. We wish to find how these free electron states are related to the "natural" Γ states and, in particular, how the degeneracy is partly removed by the introduction of the periodic potential.

Consider E_1. The eight plane waves $\exp[(2\pi i/a)(\pm x \pm y \pm z)]$ can be taken as basis for a representation of the 48 operations of the group, since the vector $\mathbf{K}_1$ is $(\pm 1, \pm 1, \pm 1)$. Take the waves in the order—with obvious symbols—$(+++)$, $(++-)$, $(+-+)$, $(-++)$, $(+--)$, $(-+-)$, $(--+)$, $(---)$. Each symmetry operation merely interchanges these functions and is represented by a square matrix whose elements are all zero except for one element in each row or column. For instance, the three operations of class $C_4{}^2$, which change (xyz), respectively, into $(x, -y, -z)$, $(-x, y, -z)$, $(-x, -y, z)$ are represented by

$$
\begin{vmatrix}
0 & 0 & 0 & 0 & 1 & 0 & 0 & 0 \\
0 & 0 & 1 & 0 & 0 & 0 & 0 & 0 \\
0 & 1 & 0 & 0 & 0 & 0 & 0 & 0 \\
0 & 0 & 0 & 0 & 0 & 0 & 0 & 1 \\
1 & 0 & 0 & 0 & 0 & 0 & 0 & 0 \\
0 & 0 & 0 & 0 & 0 & 0 & 1 & 0 \\
0 & 0 & 0 & 0 & 0 & 1 & 0 & 0 \\
0 & 0 & 0 & 1 & 0 & 0 & 0 & 0
\end{vmatrix}
\quad
\begin{vmatrix}
0 & 0 & 0 & 0 & 0 & 1 & 0 & 0 \\
0 & 0 & 0 & 1 & 0 & 0 & 0 & 0 \\
0 & 0 & 0 & 0 & 0 & 0 & 0 & 1 \\
0 & 1 & 0 & 0 & 0 & 0 & 0 & 0 \\
0 & 0 & 0 & 0 & 0 & 0 & 1 & 0 \\
1 & 0 & 0 & 0 & 0 & 0 & 0 & 0 \\
0 & 0 & 0 & 0 & 1 & 0 & 0 & 0 \\
0 & 0 & 1 & 0 & 0 & 0 & 0 & 0
\end{vmatrix}
\quad
\begin{vmatrix}
0 & 0 & 0 & 0 & 0 & 0 & 1 & 0 \\
0 & 0 & 0 & 0 & 0 & 0 & 0 & 1 \\
0 & 0 & 0 & 1 & 0 & 0 & 0 & 0 \\
0 & 0 & 1 & 0 & 0 & 0 & 0 & 0 \\
0 & 0 & 0 & 0 & 0 & 1 & 0 & 0 \\
0 & 0 & 0 & 0 & 1 & 0 & 0 & 0 \\
1 & 0 & 0 & 0 & 0 & 0 & 0 & 0 \\
0 & 1 & 0 & 0 & 0 & 0 & 0 & 0
\end{vmatrix}
$$

The whole set of 48 8×8 matrices is easily obtained and they form a representation of the Γ group. This representation is reducible. We could have started with suitable linear combinations of the eight plane waves and would have obtained the matrices in the form of a direct sum of irreducible representations Γ_n. To determine which Γ's occur in the direct sum, we make use of the theorem stating that the character of a reducible representation (for each class) is equal to the sum of the characters of the irreducible representations of which it is the direct sum. The characters are the same for all the operations belonging to any one class and they are found immediately by writing down one matrix for each class. They are given in Table II.

One then picks up from the character table of the Γ group those irreducible representations whose characters add up for each class to the values given in Table II. In this case, by inspection of the characters for the classes E and JC_2, it is found immediately that the irreducible representations are $\Gamma_s + \Gamma_p + \Gamma_{d'} + \Gamma_f$. The direct sum is always unique. Thus in this case the original eight-times degenerate free electron level splits up into two nondegenerate and two triply degenerate levels. The leading term of the expansion of the corresponding eigenfunctions in spherical harmonics is, respectively, 1; x, y, z; xy, yz, zx; xyz. Similarly, the six-times degenerate level E_2 is found to split into $\Gamma_s + \Gamma_p + \Gamma_d$. The angular dependence of the leading term of the expansion of $\psi(\Gamma_d)$ is of type $x^2 - y^2$.

Table II. Characters of Reducible Representation of the Γ Group Using exp $(\pm ix \pm iy \pm iz)$ as Basis Functions

Class	Character	Class	Character
E	8	J	0
$C_4{}^2$	0	$JC_4{}^2$	0
C_4	0	JC_4	0
C_2	0	JC_2	4
C_3	2	JC_3	0

The next problem is the determination of the linear combination of plane waves corresponding to each irreducible representation. The prescription for finding it is as follows: Consider an irreducible representation r, for instance, a p-like representation, with basis functions x, y, z. Each operation R is represented by a matrix with diagonal elements $[D^{(r)}(R)]_{ii}$ (e.g., i is x). Then the i partner of the symmetrized combination is given by

$$\psi_i^{(r)} = \sum_R [D^{(r)}(R)]_{ii} \, R\psi$$

where

$$\psi = \exp[i(\mathbf{k}+\mathbf{K}_n)\cdot\mathbf{r}]$$

Although the combinations corresponding to different i are generally different, they belong, of course, to the same eigenvalue, so that only one need be considered in energy determinations. The symbol operating on ψ may be considered as a projection operator, which projects out of any arbitrary function the portion that transforms according to the irreducible representation r. If r is one-dimensional, $[D^{(r)}(R)]_{ii}$ reduces to the character for the class to which R belongs and for the irreducible representation r. For instance, suppose we want the linear combination of the eight waves $\exp[i(\pm x \pm y \pm z)]$ that transforms according to the representation Γ_f. Take the wave $(+++)$. By applying the operations of the cubic group, we find that it is brought into itself by some operations of classes E, C_3, JC_2; that it is brought into $(++-)$, or $(+-+)$, or $(-++)$ by some operations of classes C_4, C_2, JC_4^2, JC_3, etc. By inspecting the table of characters, the required combination is

$$(+++) + (+--) + (-+-) + (--+)$$
$$-(++-) - (+-+) - (-++) - (---)$$

B. Atomic Correspondence

When atoms come together in a lattice, the quantum numbers l, etc., appropriate to atoms cease to be "good" quantum numbers, since the corresponding observables are no longer constants of the motion. Instead of $s, p, d, \ldots$ levels we have, for each value of $\mathbf{k}$, states belonging to one of the irreducible representations of the group that leaves $\mathbf{k}$ invariant. For instance, at $k = 0$ each atomic level splits generally into a number of Γ levels. To investigate this splitting (atomic correspondence), reducible representations of the Γ group are found using as basis functions the wave functions suitable for a free atom, namely, the spherical harmonics. Then the reducible representations are expressed as direct sums of irreducible representations found again by means of the characters.

To find the character of every class of the Γ group using the spherical harmonics as basis functions, note that every operation of the cubic group (apart from the inversion which simply brings in a factor -1 for harmonics of odd l) is a rotation of an angle θ about an axis. Since there are no preferred directions for an atom, this can always be taken to be the axis of the spherical harmonics. Consider then the $2l+1$ functions belonging to a given l, for which $m = l, l-1, \ldots, -l$. After the rotation, the function $P_l^m (\cos \theta)$ $\exp(im\theta)$ becomes $P_l^m(\cos \theta) \exp[im(\theta + \varphi)]$, and thus the $(2l+1)$-dimensional representation is diagonal

$$
\begin{vmatrix}
e^{il\varphi} & 0 & \ldots & 0 \\
0 & e^{i(l-1)\varphi} & \ldots & 0 \\
\vdots & & & \vdots \\
0 & 0 & \ldots & e^{-il\varphi}
\end{vmatrix}
$$

with character

$$
\sum_{m=-l}^{l} e^{im\varphi} = e^{-il\varphi}(1 + e^{i\varphi} + \cdots + e^{2\,il\varphi}) = \frac{\sin (l+\frac{1}{2})\,\varphi}{\sin \frac{1}{2}\varphi}
$$

For the class $C_4{}^2$, $\varphi = \pi$; for C_4, $\varphi = \frac{1}{2}\pi$; for C_3, $\varphi = 2\pi/3$; for C_2, $\varphi = \pi$. Thus, the character table for the reducible representation is as shown in Table III.

Comparing with the character table for the irreducible representations Γ, we recognize that an s level becomes Γ_s, a p level becomes Γ_p, a d level splits into $\Gamma_d + \Gamma_{d'}$, an f level splits into $\Gamma_p + \Gamma_f + \Gamma_{f'}$, a g level splits into $\Gamma_s + \Gamma_d + \Gamma_{d'} + \Gamma_g$, etc. Clearly the subscripts given to the levels denote the atomic level of the smallest l to which each level is related in the atomic correspondence. If the wave function of a Γ level is expanded in spherical harmonics, we recognize immediately from the table which orders will be

Table III. Characters of Reducible Representations of the Γ Group Using the Spherical Harmonics as Basis Functions

Representation	Classes									
	E	$C_4{}^2$	C_4	C_2	C_3	J	$JC_4{}^2$	JC_4	JC_2	JC_3
$l = 0$	1	1	1	1	1	1	1	1	1	1
$l = 1$	3	-1	1	-1	0	-3	1	-1	1	0
$l = 2$	5	1	-1	1	-1	5	1	-1	1	-1
$l = 3$	7	-1	-1	-1	1	-7	1	1	1	-1
$l = 4$	9	1	1	1	0	9	1	1	1	0
$\ldots$	.	.	.	.	.	.	.	.	.	.

present in the development. For instance, the development of the Γ_s wave function will contain one s, one g function, etc. A more complete study is necessary to determine which combination of the $2l+1$ spherical harmonics of order l appears. The spherical harmonics of the lowest l can be used as basis functions for the corresponding irreducible representations. Thus, the basic functions appropriate to Γ_p are x, y, z, etc.

For **k**-points of lower symmetry, the splitting is different: there are generally more levels for a given l, and the wave functions of the various irreducible representations include more orders of spherical harmonics. At a point of no symmetry, there is only one type of band, so that a d level, for instance, splits into five bands, all of the same symmetry. The wave functions of these states include all the spherical harmonics (for instance, if the expansion is limited to $l = 4$, there are $1+3+5+7+9 = 25$ terms). Compare the case of Γ_s, for which there are only two terms up to $l = 4$.

The atomic correspondence that has been sketched applies strictly only to atoms with one electron outside closed shells. In general, things are more complicated. Thus, for instance, the states of the valence band in Si or Ge correspond to an atomic sp^3 hybrid.

Another problem that can be solved by comparing the relevant tables of characters is the determination of which irreducible representations along a certain direction in **k**-space [e.g., the $(1, 0, 0)$ direction] are compatible with the irreducible representations of the end point (e.g., $\mathbf{k} = 0$). The sum of the characters of the compatible representations along the axis must be equal to the character of the representation at the end point.

To complete the study of band structure by group theory, one has still to consider the effect of spin-orbit coupling. Consideration of spin doubles the degeneracy of all levels, but the spin-orbit coupling may remove some degeneracies. If the structure has a center of inversion, there is twofold degeneracy at each point and $E(-\mathbf{k}) = E(\mathbf{k})$; if the structure has no center of inversion, the degeneracy is removed, but still $(E(-\mathbf{k}) = E(\mathbf{k})$ (Fig. 12). From the point of view of group theory, consideration of spin has the effect of changing every **k**-point group into a "double group." The irreducible representations and corresponding basis functions found for the single group hold also for the double one, but there are other irreducible representations, called the extra representations, whose basis functions must be written in a form involving the spin functions. These extra representations are the ones

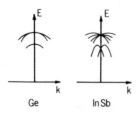

Fig. 12. Splitting of levels, due to spin-orbit coupling, at the top of the valence band in Ge and InSb.

that must be considered when studying the effect of spin-orbit interaction. The wave functions must be basis functions for one of the extra representations.

IX. REFERENCES*

1. P. W. Anderson, *Concepts in Solids*, Benjamin (New York), 1963.
2. E. I. Blount, "Formalism of Band Theory," in: *Solid-State Physics, Vol. 13*, Academic Press (New York), 1962.
3. L. Brillouin, *Wave Propagation in Periodic Structures*, McGraw-Hill (New York), 1946.
4. J. Callaway, *Energy Band Theory*, Academic Press (New York), 1964.
5. W. A. Harrison, *Pseudopotentials in the Theory of Metals*, Benjamin (New York), 1965.
6. V. Heine, *Group Theory in Quantum Mechanics*, Pergamon Press (Oxford), 1960.
7. H. Jones, *The Theory of Brillouin Zones and Electronic States in Crystals*, North-Holland Publishing Co. (Amsterdam), 1960.
8. C. Kittel, *Quantum Theory of Solids*, John Wiley & Sons (New York), 1963.
9. G. F. Koster, "Group Theory," in: *Solid-State Physics, Vol. 5*, Academic Press (New York), 1957.
10. N. F. Mott and H. Jones, *The Theory of the Properties of Metals and Alloys*, Clarendon Press (Oxford), 1936.
11. S. Raimes, *The Wave Mechanics of Electrons in Metals*, North-Holland Publishing Co. (Amsterdam), 1961.
12. F. Seitz, *Modern Theory of Solids*, McGraw-Hill (New York), 1940.
13. J. C. Slater, "Electronic Structure of Solids," in: *Handbuch der Physik, Vol. 19*, Springer-Verlag (Berlin), 1957.
14. J. C. Slater, *Quantum Theory of Molecules and Solids, Vol. 11*, McGraw-Hill (New York), 1965.
15. R. A. Smith, *Wave Mechanics of Crystalline Solids*, Chapman and Hall (London), 1961.
16. G. Weinreich, *Solids: Elementary Theory for Advanced Students*, John Wiley & Sons (New York), 1965.
17. A. H. Wilson, *The Theory of Metals*, Cambridge University Press (Cambridge), 1953.
18. J. M. Ziman, *Principles of the Theory of Solids*, Cambridge University Press (Cambridge), 1964.

* The entries in this section represent books or review articles covering various aspects of the subject. Most contain references to the original literature.

CHAPTER 3

Electric-Susceptibility Mass of Free Carriers in Semiconductors

Jack R. Dixon

U. S. Naval Ordnance Laboratory
White Oak, Silver Spring, Maryland

I. INTRODUCTION

The relationship between the electric-susceptibility mass m_s of free carriers in a semiconductor and the optical properties of the material in the infrared region of the spectrum was first pointed out and applied by Spitzer and Fan [1]. As part of their general treatment of this subject, they showed that reliable values of m_s could often be obtained from simple measurements of the normal reflectivity as a function of wavelength. Since that time, this method has been used widely as an experimental tool for studying the nature of charge carriers in semiconductors [2-23].

Such measurements are of interest primarily for two reasons. First, m_s is simply related to electronic band structure, and a knowledge of its value often serves as a stringent consistency test of band models. Second, m_s is equal to the conductivity mass m_c, which plays a dominant role in determining the electrical transport properties of a semiconductor. Such an independent measurement of m_c is often useful in detailed studies of electrical conductivity. The fact that measurements of m_s can normally be carried out over an unusually large range of carrier concentration and temperature makes this tool particularly useful.

This chapter will describe briefly the fundamentals on which these measurements are based, present various methods of data analysis, and indicate the types of complications which can arise. In addition, the relationship of m_s to band structure is discussed.

II. ELEMENTARY DESCRIPTION OF PHENOMENA

The experimental determination of m_s involves the measurement of reflectivity at normal incidence as a function of wavelength. As illustrated

61

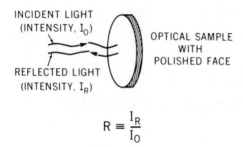

INCIDENT LIGHT
(INTENSITY, I_0)

REFLECTED LIGHT
(INTENSITY, I_R)

OPTICAL SAMPLE
WITH
POLISHED FACE

$$R \equiv \frac{I_R}{I_0}$$

Fig. 1. Features of reflectivity measurements at normal incidence used to determine the free-carrier electric-susceptibility mass in semiconductors.

in Fig. 1, this requires the determination of the ratio

$$R = I_R/I_0 \tag{3.1}$$

where I_R and I_0 are the reflected and incident intensities of a monochromatic beam, respectively. That there should be a relationship between R and m_s can be understood easily if one considers the elementary phenomena illustrated in Fig. 2. When light is shone upon the surface of a semiconductor, the associated alternating electric field will cause the free carriers to move. This motion involves acceleration and, consequently, reradiation of the light. In the classical picture, it is this reradiation which gives rise to the reflected beam. One would expect, therefore, that R should depend upon the physical characteristics of the carriers, such as their concentration N and effective mass m_s. In addition, a dependence upon the carrier-scattering mechanisms would be expected, since the scattering will, in general, influence the nature of the motion. However, simple reasoning suggests that the influence of carrier scattering should vary with the period P (or frequency) of the incident light, as illustrated in Fig. 2. The situation shown in this

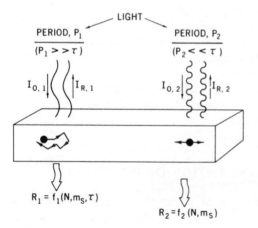

LIGHT

PERIOD, P_1
$(P_1 >> \tau)$

$I_{0,1}$ $I_{R,1}$

$R_1 = f_1(N, m_s, \tau)$

PERIOD, P_2
$(P_2 << \tau)$

$I_{0,2}$ $I_{R,2}$

$R_2 = f_2(N, m_s)$

Fig. 2. Elementary phenomena involved in the interaction of light with free carriers. The dependence of the reflectivity R upon the carrier concentration N, the effective mass m_s, and the average scattering time τ is expected to vary with the period or frequency of the light.

figure at the left applies to radiation having a period P_1 which is long in comparison with the average scattering time τ. For this case, the characteristic motion of a carrier during a period P_1 is strongly influenced by the numerous scattering events which occur. Thus, the expected functional dependence is

$$R_1 = f_1(N, m_s, \tau) \tag{3.2}$$

The influence of scattering is considerably different for $P_2 \ll \tau$, as shown at the right in Fig. 2. For this case, scattering events occur only rarely during a single period, and one would expect that R_2 would be relatively independent of τ, or

$$R_2 = f_2(N, m_s) \tag{3.3}$$

This functional dependence is of particular interest, since it represents a means for determining m_s independently of the scattering mechanism. It is this characteristic which makes the method being described here especially appealing and useful.

The general ideas which have been presented above can be formalized in terms of Maxwell's equations, optical constants, and dispersion relations. The major features of the formalization will be outlined in the next two sections.

III. MACROSCOPIC DESCRIPTION OF PHENOMENA

A. Maxwell's Equations

The macroscopic behavior of an electromagnetic wave in a solid is described by Maxwell's equations. For a nonmagnetic cubic crystal of dielectric constant ε and conductivity σ, these relations in the cgs system of units are

$$\nabla \cdot \mathbf{E} = 0 \qquad \nabla \cdot \mathbf{H} = 0$$

$$\nabla \times \mathbf{E} = -\frac{1}{c}\frac{\partial \mathbf{H}}{\partial t} \qquad \nabla \times \mathbf{H} = \frac{1}{c}\left(\varepsilon \frac{\partial \mathbf{E}}{\partial t} + 4\pi\sigma \mathbf{E}\right) \tag{3.4}$$

in which the symbols have the usual meanings. Simple manipulation of these equations yields the wave relation

$$\nabla^2 \mathbf{E} = \frac{1}{c^2}\left(\varepsilon \frac{\partial^2 \mathbf{E}}{\partial t^2} + 4\pi\sigma \frac{\partial \mathbf{E}}{\partial t}\right) \tag{3.5}$$

A solution of this differential equation is

$$\tilde{\mathbf{E}} = \tilde{\mathbf{E}}_0\, e^{-i\omega[t-(x/v)]} \tag{3.6}$$

which represents a plane wave traveling in the $+x$ direction. It is characterized by an amplitude $\tilde{E}_0$, angular frequency ω, and phase velocity v. The condition that equation (3.6) be a solution of equation (3.5) is

$$\frac{c^2}{v^2} = \varepsilon + i\,\frac{2\sigma}{v} \tag{3.7}$$

where v is the frequency of the radiation ($\omega = 2\pi v$). Equation (3.6) is a good representation of the electromagnetic waves normally employed in experimental measurements of the type being discussed here. It follows that if ε and σ are known for a material, then the nature of the electromagnetic wave passing through it is specified by equations (3.6) and (3.7).

B. Optical Constants

The nature of an electromagnetic wave in a solid is usually described in terms of a complex dielectric constant $\tilde{\varepsilon}$ or index of refraction $\tilde{N}$, rather than in terms of ε and σ. The defining relations for these parameters are based upon the condition expressed by equation (3.7). They are

$$\tilde{\varepsilon} \equiv \varepsilon_1 + i\varepsilon_2 \equiv \left(\frac{c}{v}\right)^2 \tag{3.8}$$

$$\tilde{N} \equiv n + ik \equiv \frac{c}{v} \tag{3.9}$$

where ε_1 and ε_2 are real and imaginary parts of $\tilde{\varepsilon}$, n is the index of refraction, and k is the extinction coefficient. Several useful relations between these quantities, obtained by manipulation of equations (3.7), (3,8), and (3.9), are

$$\varepsilon_1 = n^2 - k^2 = \varepsilon \tag{3.10}$$

$$\varepsilon_2 = 2\,nk = \frac{2\sigma}{v} \tag{3.11}$$

These latter relations can be solved to give

$$n^2 = \frac{\varepsilon_1 + (\varepsilon_1{}^2 + \varepsilon_2{}^2)^{\frac{1}{2}}}{2}$$

$$k^2 = \frac{-\varepsilon_1 + (\varepsilon_1{}^2 + \varepsilon_2{}^2)^{\frac{1}{2}}}{2} \tag{3.12}$$

It follows from the last four equations that if ε and σ are known, then the pairs $(\varepsilon_1, \varepsilon_2)$ and (n, k) are specified. Thus, either of these pairs of parameters can be used to characterize the electromagnetic wave given by equation (3.6). A common and convenient way of demonstrating this fact is to write equation (3.6) in terms of n and k. This can be done by employing equation (3.9). The result is

$$\tilde{\mathbf{E}} = \tilde{\mathbf{E}}_0 \, e^{-i\omega(t - nx/c)} \, e^{-\omega kx/c} \tag{3.13}$$

which represents an attenuated plane wave of phase velocity c/n and amplitude attenuation coefficient $\omega k/c$.

C. Reflectivity

As indicated above, this method of determining m_s is based upon measurements of the reflectivity R at normal incidence. An expression for R in terms of the optical constants can be derived by rewriting equation (3.1) in terms of the time-averaged Poynting vectors $\bar{\mathbf{S}}_I$ and $\bar{\mathbf{S}}_R$ associated with the incident and reflected beams, as follows:

$$R = \frac{|\bar{\mathbf{S}}_R|}{|\bar{\mathbf{S}}_I|} \tag{3.14}$$

The Poynting vectors are related to the electric fields of the radiation by

$$|\bar{\mathbf{S}}_I| = \frac{c}{8\pi} \, \tilde{\mathbf{E}}_I \cdot \tilde{\mathbf{E}}_I^* $$

$$|\bar{\mathbf{S}}_R| = \frac{c}{8\pi} \, \tilde{\mathbf{E}}_R \cdot \tilde{\mathbf{E}}_R^* \tag{3.15}$$

The substitution of these relations into equation (3.14) and the use of equation (3.13) leads to

$$R = \left(\frac{\tilde{E}_{0,R}}{\tilde{E}_{0,I}}\right) \cdot \left(\frac{\tilde{E}_{0,R}}{\tilde{E}_{0,I}}\right)^* \tag{3.16}$$

The ratio of amplitudes $\tilde{E}_{0,R}/\tilde{E}_{0,I}$ can be related to the optical constants by applying the condition that the tangential components of $\mathbf{E}$ and $\mathbf{H}$ be continuous across the reflecting interface. The result[†] is

$$\frac{\tilde{E}_{0,R}}{\tilde{E}_{0,I}} = \frac{1 - (n + ik)}{1 + (n + ik)} \tag{3.17}$$

[†] The general relations between components of $\mathbf{E}$ and $\mathbf{H}$ associated with the incident, reflected, and transmitted electromagnetic radiation at a boundary were first derived by Fresnel. They are referred to as the Fresnel relations. Equation (3.17) follows directly from these for the special case of normal incidence.

The substitution of this relation into equation (3.16) gives the familiar expression

$$R = \frac{(n-1)^2 + k^2}{(n+1)^2 + k^2} \tag{3.18}$$

Thus, a knowledge of any one of the pairs (ε, σ), (n, k), and $(\varepsilon_1, \varepsilon_2)$ specifies the reflectivity of a material.

IV. MICROSCOPIC DESCRIPTION OF PHENOMENA

The analysis outlined in the preceding section did not take into account the microscopic phenomena involved in the interaction of light with a semiconductor. The relations presented thus far serve only to describe the situation macroscopically. Therefore, an experimental determination of the dielectric constants by itself cannot be expected to yield information concerning the microscopic nature of a solid. In order to obtain such information from studies of the optical properties, it is necessary to develop a theory which takes into account the details of the interaction of light with matter. Such a subject is referred to as dispersion theory because the results are often expressed in terms of the energy or wavelength dependence of the dielectric constants. A comparison of such theoretical relations with experimentally determined dielectric constants often yields useful information about the microscopic nature of solids. The topic of this chapter is an example of such a case.

A. Dispersion Mechanisms

Electromagnetic radiation can interact with a solid in many ways. The physical process by which such an interaction takes place is referred to as a dispersion mechanism. Three dispersion mechanisms which are of major significance in semiconductors involve electromagnetic interaction with: (1) bound electrons, electrons in a crystal which do not contribute to electrical conduction; (2) free carriers, electrons or holes in a crystal which contribute to electrical conduction; and (3) ions, atoms of the crystal which interact with an electromagnetic field by way of their ionic charge. The dispersion associated with ions is normally referred to as lattice dispersion.

A theoretical relation for the $\tilde{\varepsilon}$ associated with a given dispersion mechanism is obtained by deriving an expression for the corresponding electric polarization $\tilde{P}$. The polarization is defined as the electric dipole per unit volume and is given by

$$\tilde{P} = Nq\tilde{x} \tag{3.19}$$

where N is the concentration of interacting entities, q is their charge, and $\tilde{x}$ is their displacement under the influence of the electric field. The dielectric constant is related to $\tilde{P}$ by

$$\tilde{\varepsilon} = 1 + 4\pi\tilde{\chi}$$

where

$$\tilde{\chi} \equiv \frac{\tilde{P}}{\tilde{E}} \tag{3.20}$$

$\tilde{\chi}$ is referred to as the electric susceptibility or polarizability. If all three mechanisms described above contribute simultaneously but independently, their polarizations are additive and the total dispersion is described by

$$\tilde{\varepsilon} = 1 + 4\pi(\tilde{\chi}_{BE} + \tilde{\chi}_{FC} + \tilde{\chi}_{L}) \tag{3.21}$$

The subscripts refer to bound electron, free carrier, and lattice dispersion, respectively.

The dispersion in a particular spectral region is often dominated by a single mechanism. Typical regions of dominance for the case of a semiconductor are indicated in Fig. 3. Thus, by selecting the appropriate spectral region, a single dispersion mechanism can be studied somewhat independently of the others. This appealing characteristic is a major reason for the extensive use of optics in basic studies of semiconductors.

B. Classical Free-Carrier Dispersion

The method of determining m_s being described here is based upon an experimental study of optical dispersion in the infrared. In this spectral

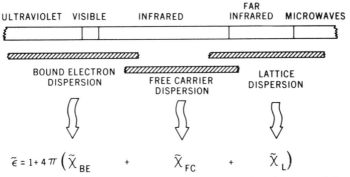

Fig. 3. Major optical dispersion mechanisms in semiconductors and the spectral regions in which they typically dominate. The corresponding contributions to the dielectric constant are indicated in terms of their electric susceptibilities $\tilde{\chi}$.

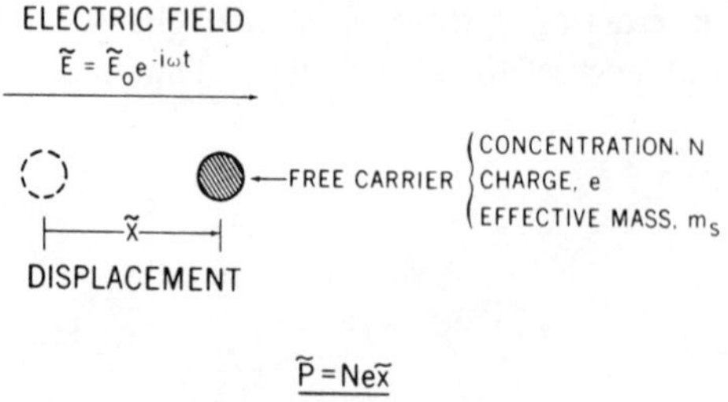

ELECTRIC FIELD
$$\tilde{E} = \tilde{E}_0 e^{-i\omega t}$$

FREE CARRIER
$\left\{\begin{array}{l}\text{CONCENTRATION. N} \\ \text{CHARGE, e} \\ \text{EFFECTIVE MASS. } m_s\end{array}\right.$

DISPLACEMENT

$$\tilde{P} = N e \tilde{x}$$

Fig. 4.   Model on which the classical theory of free-carrier dispersion is based.

region, the free-carrier dispersion mechanism often dominates. A classical treatment of free-carrier dispersion can be carried out by deriving an expression for the corresponding polarization $\tilde{P}$, based upon the model illustrated in Fig. 4. According to this model, a free carrier of mass m_s and charge e is displaced by an amount $\tilde{x}$ as a result of interaction with the electromagnetic field $\tilde{E}$. Associated with this motion is a damping force which is proportional to the velocity of the carrier, the coefficient being $m_s \gamma$. The dynamical equation describing the motion* is

$$m_s \frac{d^2 \tilde{x}}{dt^2} + m_s \gamma \frac{d\tilde{x}}{dt} = e\tilde{E}_0 \, e^{-i\omega t} \tag{3.22}$$

which can be solved to give

$$\tilde{x} = -\frac{e\tilde{E}}{m_s \omega(\omega + i\gamma)} \tag{3.23}$$

Using this result and equation (3.19), one obtains

$$\tilde{P}_{FC} = -\frac{Ne^2}{m_s} \frac{\tilde{E}}{\omega(\omega + i\gamma)} \tag{3.24}$$

The corresponding expression for $\tilde{\chi}_{FC}$ can be substituted into equation (3.21) to obtain the general relation for $\tilde{\varepsilon}$. For many semiconductors this general relation can be simplified when applied to the spectral region of interest for m_s measurements. The simplification arises from the facts that (1) the lattice polarization is negligible, and (2) the bound-electron contribution can be represented by

$$1 + 4\pi\tilde{\chi}_{BE} \equiv \varepsilon_\infty \tag{3.25}$$

---

* In writing this relation, one assumes that the effective field acting upon free carriers in a semiconductor is $\tilde{E}$. The validity of this assumption is discussed on p. 343 of the work by Stern [24].

The optical dielectric constant ε_∞ defined by this relation is real and energy-independent. Thus, for such situations the total dispersion is described by

$$\tilde{\varepsilon} = \varepsilon_\infty - \frac{4\pi Ne^2}{m_s} \frac{1}{\omega(\omega + i\gamma)} \tag{3.26}$$

This is the dispersion relation upon which the experimental determination of m_s is often based. Some examples of its application for this purpose will be given in the next section.

V. TYPICAL DATA AND METHODS OF ANALYSIS

A. Analysis of Reflectivity Spectrum

The conditions applying to the bound-electron and lattice dispersions upon which equation (3.26) is based are satisfied for many semiconductors. A typical reflectivity spectrum for such a semiconductor is shown in Fig. 5. The variations in R are the result of the free-carrier contribution to $\tilde{\varepsilon}$. It is sometimes possible to describe these variations rather accurately in terms of equation (3.26). For such cases, the physical parameters ε_∞, m_s, and γ of equation (3.26) can be established by straightforward curve-fitting procedures. The calculations involved are normally carried out on a computer. Good fits are unusual, however, since equation (3.26) is based upon the classical assumption that carrier scattering can be represented by an energy-independent damping coefficient. Such an assumption is generally not valid for semiconductors. Therefore, the analysis is most useful when applied to spectral regions in which the reflectivity is independent of carrier scattering, as discussed in section II. Such a region is indicated in Fig. 5 and is characterized by the conditions:

$$\omega^2 \gg \gamma^2 \qquad \text{condition 1}$$

$$\varepsilon_1 \gg \varepsilon_2 \qquad \text{condition 2}$$

When condition 1 is applied to equation (3.26), one obtains

$$\varepsilon_1 = \varepsilon_\infty - \frac{e^2}{\pi c^2}\left(\frac{N}{m_s}\right)\lambda^2 \tag{3.27}$$

$$\varepsilon_2 = \frac{e^2}{2\pi^2 c^3}\left(\frac{N}{m_s}\right)\gamma\lambda^3 \tag{3.28}$$

In addition, it follows from condition 2 and equations (3.12) and (3.18) that

$$n^2 \gg k^2 \tag{3.29}$$

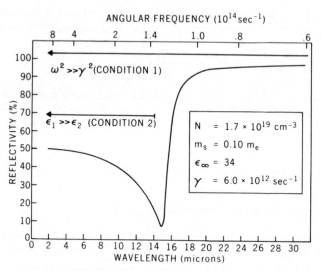

Fig. 5. Typical reflectivity spectrum arising from classical free-carrier dispersion. Conditions 1 and 2 discussed in the text are satisfied over the regions indicated.

$$\varepsilon_1 \approx n^2 \tag{3.30}$$

$$R \approx \frac{(n-1)^2}{(n+1)^2} \tag{3.31}$$

Equations (3.30) and (3.31) can be manipulated algebraically to give

$$\varepsilon_1 \approx \frac{(1+\sqrt{R})^2}{(1-\sqrt{R})^2} \tag{3.32}$$

This last equation indicates that an experimental determination of R in a spectral region in which conditions 1 and 2 are satisfied determines the corresponding values of ε_1. Such experimental values can be used in conjunction with the theoretical expression (3.27) to obtain values of ε_∞ and m_s. Data which have been analyzed in this way are presented in Fig. 6, as an example. They apply to p-type lead telluride having hole concentrations ranging from 3.5×10^{18} cm^{-3} (curve A) to 4.8×10^{19} cm^{-3} (curve D). Values of ε_1 calculated from reflectivities on the short-wavelength side of the minimum of curve B are plotted in Fig. 7. It can be seen that they vary linearly with λ^2, as would be expected on the basis of equation (3.27). It follows that the slope of the experimental line shown in Fig. 7 yields a

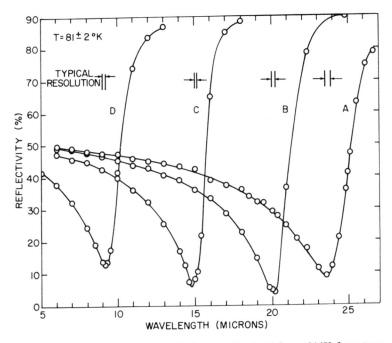

Fig. 6. Wavelength dependence of the normal reflectivity at 81 °K for p-type lead telluride samples having carrier concentrations ranging from 3.5×10^{18} cm^{-3} (spectrum A) to 4.8×10^{19} cm^{-3} (spectrum D). Taken from Fig. 1 of Dixon and Riedl [22].

value of m_s if all the other quantities in the coefficient of λ^2 in equation (3.27) are known. In addition, the $\lambda^2 = 0$ intercept determines ε_∞, as shown.

B. Analysis of Reflectivity Minimum

Another method of analysis based upon equation (3.26) requires the experimental determination of only the wavelength (λ_{min}) at which the reflectivity minimum (R_{min}) occurs [3, 20, 24]. The analysis is particularly simple when R_{min} is small (e.g., less than 5%), as it often is for semiconductors with large carrier mobilities, such as indium antimonide and indium arsenide [1, 9]. For such cases, conditions 1 and 2 are satisfied at λ_{min} and the following approximations are valid:

$$R_{min} \approx \frac{(n-1)^2}{(n+1)^2} \approx 0$$

$$\varepsilon_1 \approx n^2 \approx 1$$

(3.33)

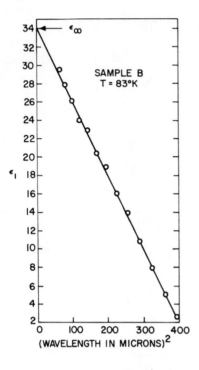

Fig. 7. Real part of the dielectric constant at 83 °K for p-type lead telluride having a carrier concentration of 5.7×10^{18} cm^{-3}. The values were determined by applying equation (3.32) to the reflectivity spectrum B of Fig. 6. As discussed in the text, these data can be analyzed in terms of equation (3.27) to obtain values of the effective mass m_s and the optical dielectric constant ε_∞. Taken from Fig. 3 of Dixon and Riedl [22].

By applying this last relation to equation (3.27), one obtains

$$m_s \approx \left(\frac{1}{\varepsilon_\infty - 1}\right) \frac{Ne^2}{\pi c^2} \lambda_{\min}^2 \qquad (3.34)$$

Thus, a simple measurement of $\lambda_{\min}$ serves to determine m_s if N and ε_∞ are known.

A more complicated analysis is required when $R_{\min}$ is not small. For this case, Lyden has shown that equations (3.18) and (3.26) can be manipulated to form a cubic equation in m_s [20]. With the use of this relation, m_s can again be determined from a measurement $\lambda_{\min}$. However, the analysis requires an independent determination of γ as well as N and ε_∞. Lyden has shown that good results are obtained in this way for the case of germanium [20] and lead telluride [21].

Analyses based only upon an experimental determination of $\lambda_{\min}$ must be employed with care. The difficulty arises because the method requires that the parameters ε_∞ and γ be estimated from the results of other types of measurements or from theory. Such estimated values are sometimes considerably different from those required to describe the optical dispersion. For example, the damping coefficient γ estimated from electrical conductivity and Hall measurements has been found in some cases to be too small [22, 23]. In addition, there are semiconductors for which the carrier concentration and temperature dependences of ε_∞ are considerably different from those predicted by theory [23]. It follows that such estimates introduce uncertainties into the evaluation of m_s from λ_{min} data. This problem does not arise in connection with the analysis of the reflectivity spectrum described in section V-A. In that case, values of ε_∞ and γ are direct by-products of the curve-fitting procedures involved. It is for this reason that the latter method has been chosen as the basis for discussion in the following sections.

C. Complications Due to Lattice Dispersion

Equation (3.26), which serves as a basis for the analyses described in the preceding section, is based upon the assumption that lattice dispersion is negligible in the spectral region of interest. For some ionic semiconductors, however, lattice dispersion contributes significantly [3, 23, 25]. For such cases, equation (3.26) must be replaced by

$$\tilde{\varepsilon} = \varepsilon_\infty - \frac{4\pi Ne^2}{m_s} \frac{1}{\omega(\omega + i\gamma_{FC})} + \frac{(\varepsilon_0 - \varepsilon_\infty)\,\omega_{TO}^2}{\omega_{TO}^2 - \omega^2 - i\gamma_L\,\omega} \qquad (3.35)$$

The contribution of the lattice, described by the last term, is characterized by the static dielectric constant ε_0, optical dielectric constant ε_∞, transverse optical mode frequency ω_{TO}, and lattice damping coefficient γ_L. These lattice dispersion parameters must be known in order to separate the free-carrier and lattice contributions to the total dispersion. They can sometimes be determined from experimental studies of the optical dispersion using samples in which the concentration of free carriers is negligible [3]. For semiconductors in which the carrier concentration cannot be reduced sufficiently, useful information regarding the lattice dispersion parameters can often be obtained from curve-fitting procedures which take both dispersion mechanisms into account [23].

Particular care must be exercised when studying a semiconductor for which the lattice contribution to the total dispersion is unknown. Under certain rather common circumstances, the free carriers and lattice contribute to the dispersion in similar ways. When this occurs, the two mechanisms are difficult to identify and separate, and serious errors in analysis can result. For example, consider a portion of the spectrum in which

$$\omega^2 \gg \gamma_{FC}^2$$

$$\omega^2 \gg \omega_{TO}^2 \text{ and } \gamma_L^2$$

The real part of equation (3.35) then becomes

$$\varepsilon_1 = \varepsilon_\infty - \left[\frac{e^2 N}{\pi c^2 m_s} + \varepsilon_\infty \left(\frac{1}{\lambda_{LO}^2} - \frac{1}{\lambda_{TO}^2} \right) \right] \lambda^2 \tag{3.36}$$

in which λ_{LO} and λ_{TO} are the wavelengths corresponding to the longitudinal and transverse optical lattice modes. Under such conditions, ε_1 varies linearly with λ^2, just as in the case of equation (3.27) which applies to free-carrier dispersion acting alone. Thus, an unknown lattice contribution could easily be mistaken for a free-carrier effect. An analysis carried out on this basis would lead to errors in the evaluation of free-carrier characteristics such as m_s. Lead sulfide is a good example of a semiconductor for which this situation arose [23].

D. Complications Due to Bound-Electron Dispersion

Equation (3.26) is based also upon the assumption that the contribution of bound electrons to the total dispersion can be represented by an energy-independent, optical dielectric constant ε_∞. This assumption will be satisfied if bound-electron absorption occurs at wavelengths considerably shorter than the spectral region involved in the reflectivity measurements. This condition is satisfied for many semiconductors. For those cases in which bound-electron and free-carrier dispersion both contribute strongly in the same spectral region, equation (3.26) must be replaced by

$$\tilde{\varepsilon} = \tilde{\varepsilon}_{BE} - \frac{4\pi Ne^2}{m_s} \frac{1}{\omega(\omega + i\gamma)} \tag{3.37}$$

The dielectric constant $\tilde{\varepsilon}_{BE}$ represents the energy-dependent, bound-electron dispersion. Tin telluride is an example of a semiconductor to which this equation has been applied [26]. Optical data for a sample having a carrier concentration of 1.4×10^{20} cm^{-3} are presented in Fig. 8. It can be seen that the variation in R which is typical of free-carrier dispersion occurs in the same spectral region as the fundamental absorption edge. This edge indicates the onset of bound-electron absorption. The calculated reflectivity curve represents the best fit based upon equation (3.26). The discrepancy between calculation and experiment at short wavelengths is the result of overlapping bound-electron and free-carrier dispersion. Therefore, an analysis of these reflectivity data to obtain information about the free carriers must be done in terms of equation (3.37).

E. Complications Due to Interband, Free-Carrier Dispersion

The free-carrier dispersion described classically in section IV-B can be thought of as arising from intraband electronic transitions occurring

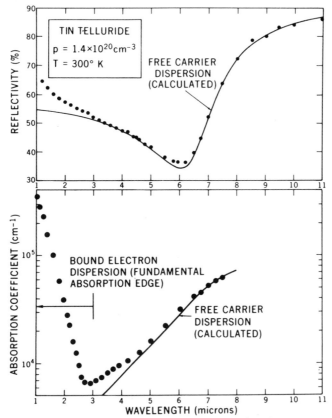

Fig. 8.   Reflectivity and absorption spectra of p-type tin telluride. The graphs illustrate the influence of overlapping bound-electron and free-carrier dispersion upon the reflectivity, as discussed in the text. These are unpublished data of H. R. Riedl and R. B. Schoolar [26].

within a given conduction or valence band, as shown at the left in Fig. 9. Free-carrier dispersion can result also from interband transitions within a conduction or valence band system, as indicated on the right side of the figure. Interband transitions of the types shown are common in semiconductors [1, 27, 28]. For such cases, equation (3.26) must be replaced by

$$\tilde{\varepsilon} = \varepsilon_\infty - \frac{4\pi N e^2}{m_s} \frac{1}{\omega(\omega + i\gamma)} + 4\pi \tilde{\chi}'_{FC} \tag{3.38}$$

where $\tilde{\chi}'_{FC}$ is the susceptibility associated with interband transitions. Normally, the theoretical description of interband free-carrier dispersion is complicated. For this reason, the evaluation of $\tilde{\chi}'_{FC}$ is usually done on the

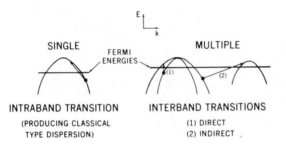

Fig. 9. Types of electronic transitions within valence band systems which can give rise to free-carrier dispersion. Analogous transitions occur within conduction bands.

basis of experiment. Both p-type germanium [1] and p-type lead telluride [22] are good examples of semiconductors for which equation (3.38) has been applied in analyzing reflectivity data to obtain values of m_s.

VI. RELATIONSHIP TO BAND STRUCTURE

The electric-susceptibility mass is directly related to the nature of the band structure near the Fermi energy. In general, m_s represents an average of the mass parameters characterizing the band at that level. Because of this, experimental studies of m_s ordinarily do not yield detailed information about band structures. Such measurements are, however, very useful as a tool for exploring the general nature of an unknown band structure. In addition, they often serve as a stringent consistency check of band models proposed on the basis of other, more detailed experiments. Since these measurements can be carried out over a large range of temperature and carrier concentration, they often augment other experimental studies of the band structure which are limited to low temperatures and carrier concentrations.

A.   Relation to Band Parameters

Relationships between m_s and the parameters describing various band models have been developed in detail by several authors [1, 5, 22]. Only the general approach and major results of their analyses will be presented here.

Spitzer and Fan have applied the Boltzmann equation to analyze the motion of a free carrier under the influence of an alternating electric field [1]. They have used these results to obtain expressions for the m_s associated with several simple band structure systems. Extensions of their work for the case of cubic crystals have led to more general relations between m_s and the band structure [5, 22]. A particularly useful expression of this type [22] is

$$\frac{N}{m_s} = -\frac{\Gamma}{3}\frac{2}{(2\pi)^3}\frac{1}{\hbar}\int_0^\infty \frac{\partial f_0}{\partial E}\left(\int_{S_k} v\,dS_k\right)_E dE \qquad (3.39)$$

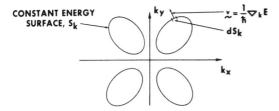

CONSTANT ENERGY
SURFACE, S_k

$v = \dfrac{1}{\hbar} \nabla_k E$

dS_k

Fig. 10. Graphical representation of the symbols used in equation (3.39) for the case of ellipsoidal constant energy surfaces in two-dimensional k-space.

This equation is valid when either of the following conditions pertaining to the average scattering time τ is satisfied: (1) τ is independent of energy or (2) τ and the frequency of excitation ω are such that $(\omega\tau)^2 \gg 1$. The meanings of the symbols used in equation (3.39) are illustrated in Fig. 10 for the case of two-dimensional ellipsoidal energy surfaces. Γ is the number of equivalent surfaces; f_0, the Fermi–Dirac distribution function; E, the energy; S_k, the area of a constant energy surface in wave number k-space; and v, the carrier velocity. The carrier velocity is related to the band structure by

$$v = \frac{1}{\hbar} \nabla_k E \qquad (3.40)$$

This relation is evaluated using the $E = f(k)$ relation describing the band model being considered. The magnitude of the velocity is integrated over a constant energy surface, as indicated by the term in parentheses in equation (3.39). The factor $\partial f_0/\partial E$ involved in the integration over energy is also of particular significance. This arises from the fact that it has nonzero values only at energies which are within a few kT of the Fermi energy. Thus, m_s is normally dependent only upon the nature of the energy bands near the Fermi energy, as stated previously.

Equation (3.39) is relatively easy to solve for the cases of parabolic spherical and ellipsoidal energy bands. Using the expressions

$$E = \frac{\hbar^2 k^2}{2 m^*} \qquad \text{spherical} \qquad (3.41)$$

$$E = \frac{\hbar^2}{2} \left(\frac{k_t^2}{m_t} + \frac{k_l^2}{m_l} \right) \qquad \text{ellipsoidal} \qquad (3.42)$$

which describe such bands, one obtains

$$m_s = m^* \qquad \text{spherical} \qquad (3.43)$$

$$\frac{1}{m_s} = \frac{1}{3} \left(\frac{2}{m_t} + \frac{1}{m_l} \right) \qquad \text{ellipsoidal} \qquad (3.44)$$

In these expressions m^*, m_t, and m_l are band masses, and the subscripts t and l refer to the transverse and longitudinal axes of the ellipsoidal constant energy surfaces.

B. Influence of Equivalent Energy Surfaces

A useful characteristic of m_s, indicated by equations (3.43) and (3.44), is that it does not depend upon the number of equivalent energy surfaces Γ associated with the band. This is in contrast to the density-of-states mass m_d which for the case of ellipsoids is given by

$$m_d = \Gamma^{\frac{2}{3}}(m_t^2\, m_l)^{\frac{1}{3}} \tag{3.45}$$

This latter mass is involved in the theoretical description of many optical, electrical, and thermal properties of semiconductors. It is commonly determined from measurements of the thermoelectric power or the Burstein shift of the fundamental absorption edge.

A consequence of the facts presented above is that experimental values of m_s and m_d can be used to obtain a useful estimate of Γ. For example, when parabolic bands are involved the analysis is based upon the relation

$$\Gamma^{\frac{2}{3}} = (m_d/m_s)\,[3\,K^{\frac{2}{3}}/(2\,K+1)] \tag{3.46}$$

which is derived from equations (3.44) and (3.45). The symbol K represents the ratio m_l/m_t and is referred to as the anisotropy of the ellipsoids. Equation (3.46) is a slowly varying function of K over the usual range of this number. Thus, for most cases the anisotropy need not be known accurately in order to obtain a reliable value of Γ in this way. The analysis required for nonparabolic bands is considerably more complicated. Nevertheless, the method ordinarily yields useful estimates of Γ for such bands.

C. Multiple-Band Effects

When free carriers occupy more than one band, their contribution to $\tilde{\varepsilon}$ as expressed in equation (3.21) is represented by the sum

$$\tilde{\chi}_{FC} = \sum_n \tilde{\chi}_{FC,n} \tag{3.47}$$

where $\tilde{\chi}_{FC,n}$ is the susceptibility arising from carriers in the nth band. It follows that, under the conditions which led to equation (3.27) for the single-band case, the dielectric constant for a multiple-band system is given by

$$\varepsilon_1 = \varepsilon_\infty - \frac{e^2}{\pi c^2}\left(\sum_n \frac{N_n}{m_{s,n}}\right)\lambda^2 \tag{3.48}$$

A total susceptibility mass $m_{s,T}$ arising from carriers in multiple bands is defined under these conditions by

$$\frac{N_T}{m_{s,T}} = \sum_n \frac{N_n}{m_{s,n}} \tag{3.49}$$

where

$$N_T = \sum_n N_n$$

This is the mass determined experimentally when data applying to a multiple-band semiconductor are analyzed in terms of the single-band relation, equation (3.27).

Multiple-band effects can be important in determining the magnitude and carrier-concentration dependence of m_s. This point is illustrated by the calculated results presented in Fig. 11. They apply to the band systems shown at the top of the figure. For simplicity, the temperature was assumed to be absolute zero, and all the individual bands were taken to be parabolic and spherical. For such bands, m_s is equal to the band mass m^* as indicated by equation (3.43). Thus for system A

$$m_{s,T} = m_s \tag{3.50}$$

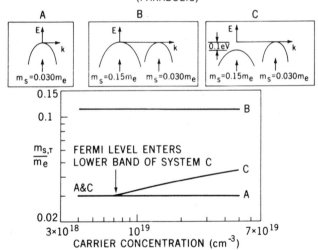

Fig. 11. Influence of multiple bands upon the magnitude and the carrier-concentration dependence of the electric-susceptibility mass. The calculated results given in the graph apply to the band structure systems illustrated at the top and to a temperature of absolute zero.

and there is no carrier-concentration dependence, as shown by line A in the figure. When two bands are involved, as in systems B and C, $m_{s,T}$ is given by

$$\frac{N_T}{m_{s,T}} = \frac{N_1}{m_{s,1}} + \frac{N_2}{m_{s,2}} \tag{3.51}$$

The carrier concentrations N_1 and N_2 corresponding to a particular N_T were determined from the relations

$$N_T = N_1 + N_2$$

$$N_1 = \frac{8\,\pi}{3\,h^3}\,(2\,m_1{}^* E_{F,1})^{\frac{3}{2}} \tag{3.52}$$

$$N_2 = \frac{8\,\pi}{3\,h^3}\,(2\,m_2{}^* E_{F,2})^{\frac{3}{2}}$$

The Fermi energies $E_{F,1}$ and $E_{F,2}$ are measured from their respective band extrema. Calculations based upon these relations for system B led to the carrier-concentration-independent result shown by line B. In this case, however, the magnitude of $m_{s,T}$ is affected, being different from either of the masses associated with the individual bands. System C represents the more general situation of two bands having extrema at different energies. The calculated results show that a sharp break occurs in the carrier-concentration dependence when the Fermi energy enters the lower, heavier-mass band. Beyond that point, $m_{s,T}$ increases with carrier concentration.

Since the variations in $m_{s,T}$ discussed above are characteristic of those observed in more complicated multiband semiconductors, experimental studies of $m_{s,T}$ serve as a useful exploratory tool for investigating complex band structures [8, 9].

D. Nonparabolic Band Effects

As discussed in section VI-A, m_s depends upon the nature of the energy bands near the Fermi level. This assertion is expressed analytically by equation (3.39). Thus, measurements of m_s as a function of N serve to probe valence and conduction bands at various energies. This point is illustrated in Fig. 12 for a spherical band. There it is shown that as N is increased the Fermi level moves deeper into the band. It follows directly from equation (3.39) that for such a case m_s is related to the band parameters by

$$\frac{1}{m_s} = \frac{1}{\hbar^2 k_F}\left|\frac{\partial E}{\partial k}\right|_F \tag{3.53}$$

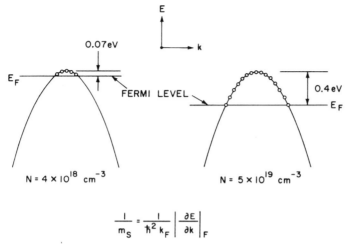

$$\frac{1}{m_S} = \frac{1}{\hbar^2 k_F} \left| \frac{\partial E}{\partial k} \right|_F$$

Fig. 12.   Influence of nonparabolicity upon the carrier-concentration dependence of the electric-susceptibility mass. A spherical nonparabolic band is shown for two different carrier concentrations. The corresponding change in the Fermi level leads to a variation in m_s as indicated by equation (3.53).

at low temperatures. The subscript F indicates that the quantity is to be taken at the Fermi level. For a parabolic band, described by equation (3.41), the above relation yields the result given by equation (3.43). Thus, m_s is independent of E_F and, consequently, of the carrier concentration. The situation is considerably different for the case of nonparabolic spherical bands for which

$$E = \frac{\hbar^2 k^2}{2 m^* (k)} \tag{3.54}$$

In this case the band mass $m^*(k)$ is a function of the wave number k. Because of this, m_s will vary with the Fermi energy, as can be seen by evaluating equation (3.53). Thus, nonparabolicity is associated with a carrier-concentration-dependent m_s.

The relationships described above between the carrier-concentration dependence of m_s and band parabolicity are general characteristics; they are not limited to the simple spherical bands used here for illustration. Thus, measurements of m_s are a generally useful method of determining whether or not an unknown band is parabolic. In addition, such measurements can be used as a test of proposed band models [5, 9, 22, 23]. A verification requires agreement between the experimental results and those calculated on the basis of the model and equation (3.39). Such an application is illustrated by the data and calculated curve presented in Fig. 13. They apply to p-type

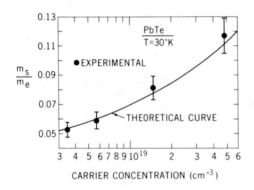

Fig. 13. Observed carrier-concentration dependence of the electric-susceptibility mass of free carriers in p-type lead telluride at 30°K. The large variation has been shown to be the result of nonparabolic ellipsoidal energy bands, as discussed in the text. These results were taken from Fig. 6 of Dixon and Riedl [22].

lead telluride at 30°K. The large variation of m_s with N is the result of nonparabolicity. The theoretical curve was based upon a model involving nonparabolic ellipsoids. The agreement between experiment and calculation served to establish the appropriateness of this model over a large range of carrier concentrations [22].

VII. SUMMARY

The electric-susceptibility mass m_s of free carriers in semiconductors can be determined from relatively simple measurements of normal reflectivity in the infrared region. Results are usually obtainable over large ranges of carrier concentration and temperature. They characteristically yield useful information concerning the conduction and valence bands. In addition, experimental values of m_s obtained in this way are often of considerable value in studies of electrical transport phenomena. This arises from the fact that m_s is identical to the conductivity mass m_c.

VIII. ACKNOWLEDGMENTS

I am indebted to my associates E. J. Alexander, R. F. Bis, H. R. Riedl, and R. B. Schoolar for their numerous and significant contributions to the manuscript.

IX. REFERENCES

1. W. G. Spitzer and H. Y. Fan, *Phys. Rev.* **106**: 882 (1957).
2. L. C. Barcus, A. Perlmutter, and J. Callaway, *Phys. Rev.* **111**: 167 (1958).
3. R. J. Collins and D. A. Kleinman, *J. Phys. Chem. Solids* **11**: 190 (1959).
4. W. G. Spitzer and J. M. Whelan, *Phys. Rev.* **114**: 59 (1959).
5. M. Cardona, W. Paul, and H. Brooks, *Helv. Phys. Acta* **33**: 329 (1960).

6. W. Albers, C. Haas, H. J. Vink, and J. D. Wasscher, *J. Appl. Phys.* **32**: 2220 (1961).
7. M. Cardona, *Proceedings of the International Conference on the Physics of Semiconductors, Prague*, Publishing House of the Czechoslovak Academy of Sciences (Prague), 1961, p. 388.
8. M. Cardona, *J. Phys. Chem. Solids* **17**: 336 (1961).
9. M. Cardona, *Phys. Rev.* **121**: 752 (1961).
10. C. Haas and M. M. G. Corbey, *J. Phys. Chem. Solids* **20**: 197 (1961).
11. T. C. Harman, A. J. Strauss, D. H. Dickey, M. S. Dresselhaus, G. B. Wright, and J. G. Mavroides, *Phys. Rev. Letters* **7**: 403 (1961).
12. W. W. Piper and D. T. F. Marple, *J. Appl. Phys.* **32**: 2237 (1961).
13. W. J. Turner, A. S. Fischler, and W. E. Reese, *J. Appl. Phys.* **32**: 2241 (1961).
14. J. R. Dixon and H. R. Riedl, *Proceedings of the International Conference on the Physics of Semiconductors, Exeter*, The Institute of Physics and the Physical Society (London), 1962, p. 179.
15. R. Sehr and L. R. Testardi, *J. Phys. Chem. Solids* **23**: 1219 (1962).
16. G. B. Wright, A. J. Strauss, and T. C. Harman, *Phys. Rev.* **125**: 1534 (1962).
17. D. McWilliams and D. W. Lynch, *Phys. Rev.* **130**: 2248 (1963).
18. R. Sehr and L. R. Testardi, *J. Appl. Phys.* **34**: 2754 (1963).
19. M. D. Blue, *Phys. Rev.* **134**: A226 (1964).
20. H. A. Lyden, *Phys. Rev.* **134**: A1106 (1964).
21. H. A. Lyden, *Phys. Rev.* **135**: A514 (1964).
22. J. R. Dixon and H. R. Riedl, *Phys. Rev.* **138**: A873 (1965).
23. J. R. Dixon and H. R. Riedl, *Phys. Rev.* **140**: A1283 (1965).
24. F. Stern, in: *Solid State Physics, Vol. 15*, F. Seitz and D. Turnbull (eds.), Academic Press Inc. (New York), 1963, p. 350.
25. T. S. Moss, *Optical Properties of Semiconductors*, Academic Press Inc. (New York), 1959, p. 235.
26. H. R. Riedl and R. B. Schoolar, U. S. Naval Ordnance Laboratory, White Oak, Silver Spring, Maryland, unpublished data.
27. F. Matossi and F. Stern, *Phys. Rev.* **111**: 472 (1958).
28. H. R. Riedl, *Phys. Rev.* **127**: 162 (1962).

CHAPTER 4

Magneto-optics

S. D. Smith

J. J. Thomson Physical Laboratory
University of Reading
Reading, England

I. SIMPLE QUANTUM THEORY

The basis of magneto-optical effects in solids is, of course, transitions between the magnetically quantized electronic states. For free electrons, the nature of these states is well known from Landau's [1] solution of the Schrödinger equation in a magnetic field. We have

$$H = \frac{1}{2m} (\mathbf{p} + e\mathbf{A})^2 \tag{4.1}$$

and we may take for a magnetic field represented by the vector potential $\mathbf{A}$

$$A_x = -By \qquad A_y = A_z = 0$$

If we substitute for $\mathbf{A}$ and write

$$p_x = \frac{\hbar}{i} \frac{\partial}{\partial x}, \text{ etc.}$$

the Schrödinger equation becomes

$$-\frac{\hbar^2}{2m} \left[\left(\frac{\partial}{\partial x} - \frac{ieBy}{\hbar} \right)^2 + \frac{\partial^2}{\partial y^2} + \frac{\partial^2}{\partial z^2} \right] \psi(r) = E\psi(r) \tag{4.2}$$

in Cartesian coordinates. This equation is separable if we put

$$\psi(r) = e^{i(k_x x + k_z z)} g(y)$$

which on substitution into equation (4.2) yields

$$\frac{d^2 g}{dy^2} + \frac{2m}{\hbar^2} \left[E - \frac{\hbar^2 k_z^2}{2m} - \frac{1}{2m} (\hbar k_x - eBy)^2 \right] g(y) = 0 \tag{4.3}$$

If we now introduce the classical cyclotron resonance frequency $\omega_c = eB/m$ and define a quantity $y_0 = \hbar k_x/eB$, we may rewrite equation (4.3) as

$$\frac{d^2 g}{dy^2} + \frac{2m}{\hbar^2}\left[\left(E - \frac{\hbar^2 k_z^2}{2m}\right) - \tfrac{1}{2} m\omega_c^2 (y - y_0)^2\right] g(y) = 0 \qquad (4.4)$$

This is the equation of a simple harmonic oscillator of frequency ω_c, centered at y_0 which in quantum mechanics has eigenvalues of $[E - (\hbar^2 k_z^2/2m)]$ given by

$$\left(E - \frac{\hbar^2 k_z^2}{2m}\right)_n = (n + \tfrac{1}{2})\,\hbar\omega_c$$

where n is any positive integer. Thus the energy levels in the field B are

$$E_{n,k_z} = \frac{\hbar^2 k_z^2}{2m} + (n + \tfrac{1}{2})\,\hbar\omega_c \qquad (4.5)$$

and are known as Landau levels [1, 2] (Fig. 1).

For the motion of electrons and holes near band extrema in semiconductors it is, of course, well known that the free-electron result is valid with the substitution of an appropriate effective mass. Equation (4.5) therefore predicts that the three-dimensional parabolic band structure is split up into a series of lines, the oscillator levels $(n + \tfrac{1}{2})\,\hbar\omega_c$. These may be associated with classical cyclotron resonance motion of the carriers, i.e., circles in a plane *perpendicular* to the field together with a one-dimensional motion in

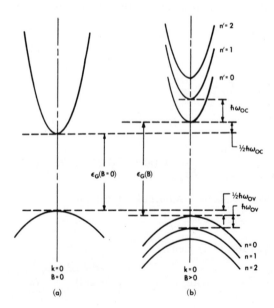

Fig. 1. Landau levels in the field B.

a direction *parallel* to the field unchanged from the zero-field case. However, the energy of the lowest state is raised to $\frac{1}{2}\hbar\omega_c$.

The Landau levels are highly degenerate with respect to the choice of "orbit center" y_0, and the number of states with a particular quantum number n increases with B. We can estimate this degeneracy by considering the system to be contained in a box with sides L_x, L_y, L_z. The number of possible values of k_x in a small interval Δk_x is

$$L_x \frac{\Delta k_x}{2\pi} \tag{4.6}$$

and a similar result is obtained for k_y and k_z. The maximum value of k_x is determined from $y_0 = \hbar k_x/eB$ by the requirement that the orbit center y_0 lie inside the box, i.e.,

$$-\tfrac{1}{2}L_y \leqslant y_0 \leqslant \tfrac{1}{2}L_y$$

Substitution for y_0 yields

$$\frac{-eBL_y}{2\hbar} \leqslant k_x \leqslant \frac{eBL_y}{2\hbar}$$

The number of states in a single Landau level is then, with use of equation (4.6),

$$L_x \frac{\Delta k_x}{2\pi} = \frac{eBL_xL_y}{2\pi\hbar}$$

With $L_z(\Delta k_z/2\pi)$ states in an interval Δk_z, the number of Landau levels associated with n and Δk_z is then

$$\left(\frac{eBV}{4\pi^2\hbar}\right)\Delta k_z$$

where $V = L_xL_yL_z$. The degeneracy is thus directly proportional to the applied field B, and this is illustrated in comparison with the zero-field case in Fig. 2.

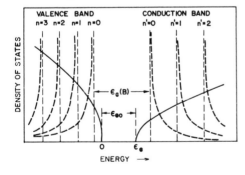

Fig. 2. The degeneracy is directly proportional to the applied field, as illustrated in comparison with the zero-field case.

These results provide all that is required to understand the magneto-optic experiments in a *qualitative* manner since we may apply the results to semi-conductor energy bands with relatively minor modifications. Assume that the above results hold if we introduce the effective mass m^* in place of the free-electron mass m. The average energy of two neighboring Landau states, n and $n+1$, is then (if $k_z = 0$)

$$\bar{E} = \frac{E_n + E_{n+1}}{2} = (n+1)\,\hbar\omega_c$$

Since we can write this kinetic energy semiclassically [3] as

$$\bar{E} = \tfrac{1}{2}\,m^*\,v^2$$

where v is the velocity, then

$$v = \sqrt{\frac{2(n+1)\,\hbar\omega_c}{m^*}}$$

The radius of the orbit is then

$$r = \frac{v}{\omega_c} = \frac{2(n+1)\,\hbar}{m^*\,\omega_c}$$

with $\omega_c = eB/m^*$. For B equal to 1 Wb/m^2 (10,000 G), and when m^*/m is ~ 1, i.e., many times greater than the lattice constants, r is ~ 200 Å. These are, therefore, circumstances in which the use of an effective mass is valid. A further useful formula is

$$\hbar\omega_c = \hbar\left(\frac{eB}{m^*}\right) = 1.1577 \times 10^{-4}\left[\frac{B}{(m^*/m)}\right]\text{eV} \qquad (4.7)$$

where B is again in Wb/m^2, m is the mass of the free electron, and one works with the effective mass ratio.

It is therefore permissible to use effective masses for both electrons m_c^* and holes m_v^* to describe orbital effects in conduction and valence bands in a simple model. Electron spin (not included since our theory is nonrelativistic) may now be readily introduced by simply including an extra term

$$g\left(\frac{e\hbar}{2m}\right)BM_J$$

where g is an effective g-factor, which may vary considerably from the free-electron value 2.00, $e\hbar/2m$ is equal to β, the gyromagnetic ratio or Bohr magneton, and M_J is the spin quantum number which may take values $\pm\tfrac{1}{2}$.

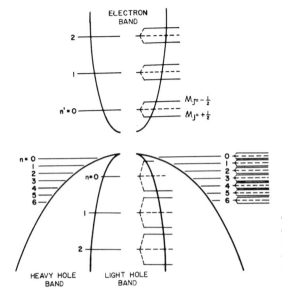

Fig. 3. The energy bands for electrons and holes for the case of degenerate valence bands. At high quantum numbers we can consider the states as derived from the light-hole and heavy-hole bands.

The energy bands for electrons and holes then become

$$E_c = E_{c0} + \frac{\hbar^2 k_z^2}{2m_c^*} + (n + \tfrac{1}{2})\hbar\omega_{cc} + g_c \beta B M_J \tag{4.8}$$

$$E_v = E_{v0} - \frac{\hbar^2 k_z^2}{2m_v^*} - (n + \tfrac{1}{2})\hbar\omega_{cv} + g_v \beta B M_J \tag{4.9}$$

These are shown in Fig. 3 which gives the case for degenerate valence bands (i.e., group IV and groups III–V semiconductors) which have two types of holes, light and heavy. These equations describe satisfactorily the behavior in a magnetic field of simple bands with extrema at **k** = 0 in the region in which they are *parabolic*, i.e., $E \propto k^2$ and m^* is a constant. Since the effective g-factors are dependent on effective mass, equations (4.8) and (4.9) also imply that g_c and g_v are constants.

II. TRANSITIONS BETWEEN MAGNETIC LEVELS

In order that optical effects due to transitions between the magnetically quantized levels be observable, it is necessary that the matrix elements for either electric or magnetic dipole transitions between the states in question be nonzero. We are mainly concerned in magneto-optics with *electric* dipole transitions. The probability of transition is proportional to the square of the quantity

$$\langle f | \alpha \mathbf{p} | i \rangle = i\hbar \int \psi_f{}^*(\mathbf{r}) \, \boldsymbol{\alpha} \cdot \nabla \psi_i(\mathbf{r}) \, d\mathbf{r}$$

where ψ_f and ψ_i are final and initial wave functions, respectively, and $\boldsymbol{\alpha}$ is a polarization vector. Applying this for the case of Landau-like wave functions, we obtain selection rules which can be classified as follows.

A. Intraband (Free-Carrier) Transitions

These occur *within* one band, such as the conduction or valence band, and require the presence of a finite number of *free carriers* for observation.

Allowed transition	Effects
(1) $\Delta k_z = 0$, $\Delta M_J = 0$	Cyclotron resonance, Free-carrier Faraday effect
but $\Delta n = \pm 1$	Voigt effect, etc.
(2) $\Delta k_z = 0$, $\Delta M_J = 1$ and $\Delta n = 0$	Transition between the spin split levels of *one* Landau state

The first is an electric dipole effect, a transition *between* Landau levels, and the fundamental effect of magneto-optics in semiconductors. The second is a *magnetic dipole* effect, i.e., spin resonance. Effects (1) and (2) can be *mixed* (Yafet [4]), and thus two further effects exist — combination resonances.

B. Interband Transitions

These transitions govern *interband magnetoabsorption* and a variety of related effects. They are usually between valence (full) and conduction band (empty); then

$\Delta k_z = 0$, $\Delta n = 0$
$\Delta M_J = +1$ for left-circularly polarized radiation
$\Delta M_J = -1$ for right-circularly polarized radiation
$\Delta M_J = 0$ for plane-polarized radiation when the magnetic
          field **B** is perpendicular to the light beam.

III. CLASSICAL ANALOGIES

Many of the magneto-optical effects can be understood by considering the classical theory of cyclotron resonance and comparing it with the results of quantum-mechanical calculations of the energy states of an electron in a magnetic field.

The basic analogy is that of the transition between two Landau levels $\Delta n = +1$ and a classical oscillator of natural frequency $\omega_c = eB/m^*$. With the magnetic induction B along the z axis, the classical equation of motion of an electron driven by applied fields $E_0 e^{\pm i\omega t}$ is

$$m^* \frac{d^2 s}{dt^2} + gm^* \frac{ds}{dt} - iBe \frac{ds}{dt} = eE_0 e^{\pm i\omega t}$$

where $s = (x+iy)$ is a complex displacement of the electron and the $\pm i\omega t$ exponents describe two contrarotating, circularly polarized waves. The useful solution is a complex amplitude

$$s = \frac{e/m^* E_0 e^{\pm i\omega t}}{(-\omega \pm \omega_c) \pm ig}$$

where $\omega_c = eB/m^*$ is the cyclotron resonance frequency. Hence from the polarization $P = Nes$ and the dielectric constants $\varepsilon_\pm = 1 + (P/E\varepsilon_0)$ we obtain

$$\varepsilon^\pm = (n_\pm - i\varkappa_\pm)^2 = 1 - \frac{Ne^2/m^* \omega\varepsilon_0}{(-\omega \pm \omega_c \pm ig)}$$

The absorbing part is

$$2n_\pm \varkappa_\pm = \frac{Ne^2 g/m^* \omega\varepsilon_0}{(\omega \pm \omega_c)^2 + g^2}$$

where g is the line width. If the minus sign is used in the term $(\omega - \omega_c)$ in the denominator, this absorption resonates at $\omega = \omega_c$ and is called *cyclotron resonance* (CR) absorption. If the positive sign is used, there is no resonance. It is necessary to fulfill the condition $\omega_c \tau > 1$ for a well-defined resonance. Consequently, either a high ω_c (i.e., high B since $\omega_c = eB/m^*$) or long τ (i.e., low temperature and pure material) is required.

The dispersive part

$$n_\pm^2 - \varkappa_\pm^2 = 1 - \frac{Ne^2/m^* \omega\varepsilon_0}{(\omega \pm \omega_c)^2 + g^2}$$

has a less powerful dependence on $g(\equiv 1/\tau)$, and it is thus possible to calculate the *Faraday rotation* θ from the relation

$$\theta = \frac{\omega l}{2c} (n_- - n_+)$$

where l is the length of the specimen

$$\theta = \frac{-\omega l}{4nc} (n_+^2 - n_-^2)$$

If $\varkappa^2 \ll n^2$ and $\omega^2 \gg \omega_c{}^2$ and g^2, we obtain

$$\theta = \frac{BNe^3 l}{2nc\varepsilon_0 m^{*2} \omega^2}$$

which is independent of g. Thus the condition on $\omega_c \tau$ required above need not be fulfilled for observation of Faraday rotation.

Thus classical theory enables us to understand the relation between *absorptive effects* (depending on $\varkappa$) and *dispersive effects* (depending on n). To discuss dispersive effects in *quantum theory*, we must work through absorptive effects using the Kramers–Kronig relations between real and imaginary parts of the dielectric constant (see Section IV). With this background, a list of most of the effects discussed thus far is presented in Table I.

Table I.   Magneto-Optical Effects*

Effect	Orientation	$n, \varkappa$	Resonant or nonresonant	Reference to observation
		1. Free-carrier Effects		
		In transmission		
Cyclotron resonance	F or V	$\varkappa_+$	Res.	Burstein et al., 1956[23] Zwerdling et al., 1956[31]
Faraday rotation	F	$n_+ - n_-$	Nonres.	Smith and Moss, 1958[25]
Resonant Faraday rotation		n_+	Res.	Palik, 1963(see[7])
Faraday ellipticity	F	$\varkappa_+ - \varkappa_-$	Nonres.	Smith and Pidgeon, 1960[30]
Voigt effect	V	$n_\perp - n_{\parallel}$	Nonres.	Teitler and Palik, 1960(see[7,31])
Interference fringe shift	F or V	$n_\pm,$ $n_{\parallel}$ or $n_\perp$	Nonres.	Palik, 1963(see[7])
Oscillatory free-carrier absorption		$\varkappa_+$		Palik and Wallis, 1963(see[7])
		In reflection		
Magnetoplasma reflection	F or V	$n_\pm$ $n_{\parallel}, n_\perp$	Shift of resonance	Lax and Wright, 1960[31]

* F and V refer to orientation of the Poynting vector of the radiation beam with respect to the magnetic field, Faraday configuration (parallel to B) and the Voigt configuration (perpendicular to B) respectively; n and $\varkappa$ refer, respectively, to the optical constants for right- and left-circularly polarized radiation associated with F; $n_\perp$, $n_{\parallel}$, $\varkappa_\perp$, and $\varkappa_{\parallel}$ refer to parallel and perpendicularly polarized radiation associated with V.

Table I. Magneto-Optical Effects (*continued*)

Effect	Orientation	$n, \varkappa$	Resonant or nonresonant	Reference to observation
Magnetoplasma rotation (Kerr effect)	F	$\varkappa_\pm$	—	Lax, 1960 [31], Palik *et al.*, 1962 (see [7])
Magnetoplasma ellipticity	F	$n_\pm$	—	Lax, 1960 [31], Palik *et al.*, 1962 (see [7])

2. Interband Effect

In transmission

Interband magneto-absorption (I.M.O.)	F or V	$\varkappa_+, \varkappa_-$ $\varkappa_{||}, \varkappa_\perp$	Res.	Burstein *et al.*, 1956 [23] Zwerdling *et al.*, 1956 [31]
Faraday rotation	F	$n_+ - n_-$	Nonres.	Kimmel, 1957 [31]
Resonant Faraday rotation	F	$n_+ - n_-$	Res.	Nishina *et al.*, 1962 [3] Smith *et al.*, 1962 [28] Mitchell and Wallis, 1963 [12]
Faraday ellipticity	F	$\varkappa_+ - \varkappa_-$	Nonres.	Smith and Pidgeon, 1960 [7]
Resonant Faraday ellipticity	F	$\varkappa_+ - \varkappa_-$	Res.	Mitchell and Wallis, 1963 [12]
Voigt effect	V	$n_{||} - n_\perp$	Nonres.	Not reported
Resonant Voigt effect	V	$n_{||} - n_\perp$	Res.	Nishina, Kolodziejczak and Lax, 1962 [3]
Zeeman effect for impurities	F or V	$\varkappa_\pm,$ $\varkappa_{||}, \varkappa_\perp$	Resonance shift	Fan and Fisher, 1958 (see [7]) Boyle, 1958 (see [7])
Magnetoabsorption of the impurity photoionization spectrum	F or V	$\varkappa_+, \varkappa_-$ $\varkappa_{||}, \varkappa_\perp$	Resonance shift	Boyle, 1958 (see [7])
Zeeman effect of excitons	F or V	$\varkappa_+, \varkappa_-$ $\varkappa_{||}, \varkappa_\perp$	Resonance shift	Gross and Zaharcenja, 1957 (see [7])
Cross-field magneto-absorption			Res.	Vrehen and Lax, 1964 (see [7])

In reflection

Interband magneto-reflection	F or V	$n_\pm, \varkappa_\pm$ $n_{||} n_\perp, \varkappa_\perp \varkappa_{||}$	Res.	Wright and Lax, 1961 [31]
Interband Kerr rotation	F	$\varkappa_\pm$		Lax and Nishina, 1961 [31]
Interband Kerr ellipticity	F	$n_\pm$		

IV. MACROSCOPIC THEORY

A. General

The theory of magneto-optical effects is semiclassical in the sense that the electromagnetic field is treated classically by Maxwell's equations but the magnetic states of the electrons are treated by quantum theory. Consequently, one can discuss magneto-optic effects at the level of the macroscopic concepts of dielectric constant and conductivity. This approach has a number of uses. (1) The dielectric constant or conductivity is a tensor in the presence of magnetic fields and the symmetry properties of effects may therefore be discussed. (2) In some cases, such as the nonresonant low-field Faraday effect, the quantum treatment of the states may be accurately explained by less rigorous theory. Thus one may use, for example, the Boltzmann transport equation generalized to high frequencies. This approach also enables us to make use of the results of transport theory and to see analogies between optical and electrical effects. (3) Consideration of the dielectric constant tensor is a convenient method of treating dispersive effects. This is possible since the Kramers–Kronig dispersion relations between the real and imaginary parts of the tensor components of this quantity (or the conductivity tensor) hold and *absorption* is correlated with transitions.

The classical analogy between the harmonic oscillator of natural frequency ω_c and the transitions between Landau levels readily shows that the fundamental motion of the electron is the circulating cyclotron motion. This may be excited by *circularly polarized radiation* of the appropriate sense. Such modes can be written

$$E(\omega) = (E_{0x} \pm iE_{0y})\, e^{i(\omega t - kz)} \tag{4.10}$$

for a wave propagating in the z direction and the $\pm$ signs refer to right- and left-circularly polarized modes looking *along* the direction of propagation (other conventions also exist). This wave is a solution of the wave equation

$$\nabla^2 \mathbf{E} + \mu_0 \frac{\partial j}{\partial t} - \mu_0 \frac{\partial^2 \mathbf{D}}{\partial t^2} = 0 \tag{4.11}$$

The complex conductivity $\boldsymbol{\sigma}$ and complex permittivity ε can be introduced from

$$\mathbf{j} = \boldsymbol{\sigma}\mathbf{E} \quad \text{or} \quad \mathbf{D} = \varepsilon\mathbf{E}$$

These can give alternative descriptions related by

$$i\omega\varepsilon = \boldsymbol{\sigma} \tag{4.12}$$

For a cubic crystal in the presence of a magnetic field along the z axis, ε has the form

$$\varepsilon = \begin{pmatrix} \varepsilon_{xx} & \varepsilon_{xy} & 0 \\ -\varepsilon_{xy} & \varepsilon_{yy} & 0 \\ 0 & 0 & \varepsilon_{zz} \end{pmatrix} \tag{4.13}$$

The complex displacement

$$\mathbf{D}(\omega) = \varepsilon(\omega) \cdot \mathbf{E}(\omega) \tag{4.14}$$

is then understood in the sense that

$$\mathbf{E} = \int_{-\infty}^{+\infty} \mathbf{E}(\omega)\, d\omega$$

Then

$$\mathbf{D} = \int_{-\infty}^{+\infty} \varepsilon(\omega) \cdot \mathbf{E}(\omega)\, d\omega$$

and a similar result obtains for $\mathbf{j}$ and $\boldsymbol{\sigma}$.

That the field and displacement be real then requires

$$\varepsilon_{ij}(-\omega) = \varepsilon_{ij}^{*}(\omega) \tag{4.15}$$

for all components of the tensor ε.

1. The Propagation Constant

Substituting from equations (4.10) and (4.14) in the wave equation (4.11), we obtain the equation defining k, the propagation constant, in terms of ε and the frequency ω

$$-\mathbf{k}(\mathbf{E}_0 \cdot \mathbf{k}) + k^2 \mathbf{E}_0 = \omega^2 \mu_0 \varepsilon \mathbf{E}_0$$

and we obtain a similar equation in terms of $\boldsymbol{\sigma}$ using equation (4.12).

2. Longitudinal Propagation—Faraday Configuration

For propagation along the z axis and with B along the same axis, we have

$$k^2 \mathbf{E}_0 = \omega^2 \mu_0 \varepsilon \mathbf{E}_0$$

from which, using equation (4.13), we can write the component equations

$$k^2 E_{0x} = \omega^2 \mu_0 (\varepsilon_{xx} E_{0x} + \varepsilon_{xy} E_{0y})$$

$$k^2 E_{0y} = \omega^2 \mu_0 (-\varepsilon_{xy} E_{0x} + \varepsilon_{xx} E_{0y})$$

Multiplying the latter equation by $\pm i$ and adding we obtain

$$-k^2 E_{0\pm} = \omega^2 \mu_0 \varepsilon_{\pm} E_{0\pm}$$

where

$$E_{0\pm} = E_{0x} \pm i E_{0y}$$

and

$$\varepsilon_{\pm} = \varepsilon_{xx} \mp i\varepsilon_{xy} \tag{4.16}$$

Thus we have formed circularly polarized modes, the upper sign corresponding to the propagation along $+z$ of waves which appear right-circularly polarized to an observer looking in the same direction, and *vice versa* for the lower signs. Accordingly, we are led to define the permittivity $\varepsilon_{\pm}$ governing the propagation of right- and left-circularly polarized modes in terms of the components of the tensor [equation (4.16)].

Similarly $\sigma_{\pm}$ is defined by

$$\sigma_{\pm} = \sigma_{xx} \mp i\sigma_{xy}$$

3. Transverse Propagation—Voigt Configuration

The magnetic field is again chosen along the z axis, but we now consider propagation in the x–y plane. Since $\varepsilon_{xx} = \varepsilon_{yy}$, there is no loss in generality in choosing k to be in the y direction. Once again writing out components we obtain

$$-k^2 E_{0x} = \omega^2 \mu_0 (\varepsilon_{xx} E_{0x} + \varepsilon_{xy} E_{0y}) \tag{4.17}$$

as before, but for the y direction

$$-k^2 E_{0y} + k^2 E_{0y} = 0 = \omega^2 \mu_0 (-\varepsilon_{xy} E_{0x} + \varepsilon_{xx} E_{0y}) \tag{4.18}$$

and

$$-k^2 E_{0z} = \omega^2 \mu_0 \varepsilon_{zz} E_{0z}$$

Note that by virtue of the polarization of the medium there is a component of $\mathbf{E}$ along the direction of propagation given from equation (4.17) as

$$E_{0y} = \frac{\varepsilon_{xy} E_{0x}}{\varepsilon_{xx}} \tag{4.19}$$

There are two cases for transverse propagation—the incident beam may be polarized so that $\mathbf{E}$ is either *perpendicular* or *parallel* to $\mathbf{B}$ (or a polarized beam in some other direction can be resolved into parallel or perpendicular components). For the *perpendicular* case $E_{0z} = 0$, and substituting for E_{0y} from equations (4.19) and (4.17), we find

$$-k^2 E_{0x} = \omega^2 \mu_0 \left(\varepsilon_{xx} + \frac{\varepsilon_{xy}^2}{\varepsilon_{xx}} \right) E_{0x}$$

The quantity

$$\varepsilon_\perp = \varepsilon_{xx} + \frac{\varepsilon_{xy}^2}{\varepsilon_{xx}} \qquad (4.20)$$

is defined as the permittivity for "perpendicular" radiation. For the *parallel* case $E_{0x} = 0$, the propagation constant is given from equation (4.18) so that

$$\varepsilon_\parallel = \varepsilon_{zz} \qquad (4.21)$$

where ε_{zz} is just the value of the permittivity in the absence of the magnetic field. This case corresponds therefore to motion of the electron parallel to the magnetic field, and the Lorentz force expression $\mathbf{F} = e\mathbf{v} \times \mathbf{B}$ clearly states that no interaction can take place. Conversely, the perpendicular component would cause a free particle to undergo cyclotron resonance if driven at the correct frequency ω_c.

4. *The Complex Refractive Index*

From the result

$$k^2 \mathbf{E} = \omega^2 \mu_0 \varepsilon \mathbf{E} \qquad (4.22)$$

note that when ε is complex, $\mathbf{k}$ is also complex and the wave is attenuated. The equivalent symmetry of the real and imaginary parts of ε ensures that the directions defined by the real and imaginary parts of $\mathbf{k}$ coincide. Accordingly the optical constants may be introduced by writing

$$\mathbf{k} = \frac{\omega}{c}(n - i\varkappa)\hat{\mathbf{k}} \qquad (4.23)$$

where $\hat{\mathbf{k}}$ is real unit vector in the direction of $\mathbf{k}$. The quantity $(n - i\varkappa)$ is the *complex* refractive index. The real part n is the refractive index and controls the phase velocity in the medium from the relation

$$v_p = c/n$$

The imaginary part $\varkappa$ is called the *absorption index* and this gives the attenuation of the intensity (i.e., the time average of the Poynting vector) from the absorption coefficient, where

$$\alpha = \left(\frac{2\omega}{c}\right)\varkappa$$

and has dimensions of meters^{-1} in mks units. Magneto-optical constants appropriate to longitudinal and transverse propagation may now be defined from the appropriate tensor components.

5. *Longitudinal Propagation*

From equations (4.23), (4.22), and (4.16), we have

$$(n_\pm - i\varkappa_\pm)^2 = \varepsilon_\pm = \varepsilon_{xx} \mp i\varepsilon_{xy} \qquad (4.24)$$

6. *Transverse Propagation*

Similarly from equations (4.20)–(4.23)

$$(n_\perp - i\varkappa_\perp)^2 = \varepsilon_\perp = \varepsilon_{xx} + \varepsilon_{xy}^2/\varepsilon_{xx} \qquad (4.25)$$

and

$$(n_\parallel - i\varkappa_\parallel)^2 = \varepsilon_\parallel = \varepsilon_{zz}$$

In terms of conductivity, there are the equivalent expressions

$$(n_\pm - i\varkappa_\pm)^2 = \sigma_\pm/i\omega = (i\sigma_{xx} \mp \sigma_{xy})/\omega$$

$$(n_\perp - i\varkappa_\perp)^2 = \sigma_\perp/i\omega = (\sigma_{xx} + \sigma_{xy}^2/\sigma_{xx})i\omega$$

The magneto-optical constants $n_\pm$, $\varkappa_\pm$, $n_\perp$, $n_\parallel$, $\varkappa_\perp$, and $\varkappa_\parallel$ are the quantities appearing in Table I.

B. Magneto-Optical Effects Expressed in Terms of Magneto-Optical Constants and the Tensor Components

1. *Magnetoabsorption*

a. *Cyclotron Resonance Absorption.* For free carriers the resonance is given by either $\varkappa_+$ or $\varkappa_-$ depending upon the sign of the charge, which selects the circularly polarized modes that resonate.

(1) *Faraday Configuration.* For *electrons* only right-circularly polarized radiation is absorbed, so cyclotron resonance absorption is given by $\alpha_+ = (2\omega/c)\varkappa_+$ which from equation (4.24) can be written

$$\alpha_+ = -\frac{\omega}{cn_+}\varepsilon_+^I$$

where ε_+^I is the imaginary part of ε_+. If the free-carrier concentration and magnetic fields are small, the magnet contribution to n will be small, even near resonance, and $n_\pm$ is approximately equal to n, the zero-field value. Then

$$\alpha_+ \approx \frac{\omega}{cn}(\varepsilon_{xx}^I + \varepsilon_{xy}^R)$$

or more appropriately for an absorption effect

$$\alpha_+ = (\sigma_{xx}^R + \sigma_{xy}^I)/cn$$

in terms of conductivity.

(2) *Voigt Configuration and Plane-Polarized Radiation.* The above result is for circularly polarized radiation. In the Voigt configuration we consider plane-polarized radiation and obtain $\alpha = (\omega/c)\varkappa_+$, i.e., half the value for circularly polarized radiation must be resolved into two circular components, only one of which is absorbed. This result also applies to plane-polarized radiation in the Faraday configuration.

b. *I.M.O. Absorption.* The above results also apply to interband effects except that absorption takes place for both circular components.

c. *Dispersive Effects*

(1) *The Faraday Effect.* The rotation of the plane of polarization arises from the different phase velocities c/n_+ and c/n_- of the two circular components making up the plane-polarized beam incident on the crystal. At the end of a crystal of length l the two components recombine to give a rotation according to

$$\theta = \frac{\omega l}{2c}(n_- - n_+) \tag{4.26}$$

The rotation is such that positive θ is a clockwise rotation for a wave propagating along the magnetic field for an observer looking in that direction. This is the situation as it is observed for free electrons. From equation (4.24)

$$\varepsilon_\pm{}^R = n_\pm{}^2 - \varkappa_\pm{}^2 = \varepsilon_{xx}^R \pm \varepsilon_{xy}^I \tag{4.27}$$

and we may rewrite equation (4.26)

$$\theta = -\frac{\omega l}{4nc}(n_+{}^2 - n_-{}^2)$$

where $n = (n_+ + n_-)/2$ and may be taken as the refractive index at zero field. Thus we obtain from equation (4.26) if $\varkappa^2 \ll n^2$

$$\theta = \frac{-\omega l}{4nc}(\varepsilon_+{}^R - \varepsilon_-{}^R)$$

$$= \frac{\omega l}{2nc}\varepsilon_{xy}^I \tag{4.28}$$

$$= \frac{\sigma_{xy}^R l}{2nc} \tag{4.29}$$

The nonresonant Faraday effect described by equation (4.28) or (4.29) is an isotropic effect in a cubic crystal.

(2) *Faraday Ellipticity.* In addition to the differential magnetodispersion responsible for rotation there is the corresponding absorption effect depending upon $\varkappa_{\pm}$. This attenuates one of the circular modes more than the other and makes the emergent beam elliptically polarized. The attenuation of amplitude is given by $E = E_0 e^{-w\varkappa/c} \pm z$ and for small absorption this can be expanded to $(1 - \omega \varkappa z/c) E_0$. In this approximation the ellipticity Δ, the amplitude ratio of the axes of the ellipse, is given by

$$\Delta = \frac{\omega l}{2c} (\varkappa_- - \varkappa_+)$$

$$\approx \frac{\omega l}{2cn} (n_- \varkappa_- - n_+ \varkappa_+)$$

$$= \frac{\omega l}{cn} \varepsilon_{xy}^R$$

$$= \frac{l}{cn} \sigma_{xy}^I$$

(3) *Voigt Effect.* The Voigt effect, which also has been called the Cotton–Mouton effect, is magnetic birefringence and arises from the difference between $n_{\parallel}$ and $n_{\perp}$ in the transverse orientation. It is usually observed by inclining incident plane-polarized radiation with the electric vector at 45° to the direction of B. The components resolved parallel and perpendicular to B then have different phase velocities ($c/n_{\parallel}$ and $c/n_{\perp}$) and recombine at the end of the crystal to give emergent radiation which is elliptically polarized. The measure of this ellipticity then determines the magnitude of the Voigt effect.

The Voigt phase shift δ is given by

$$\delta = \frac{\omega l}{c} (n_{\parallel} - n_{\perp})$$

In the absence of any absorption, this phase shift is related to the resultant ellipticity Δ by

$$\Delta^2 = \frac{1 - \cos \delta}{1 + \cos \delta} = \tan^2 \frac{\delta}{2}$$

and the ellipse is oriented at 45° to B.

In the presence of absorption the ellipse is *rotated* by differential attenuation caused by $k_\parallel$ and $k_\perp$ and the angle of orientation is determined from

$$\tan 2\phi = \frac{2f\cos\delta}{f^2-1}$$

where $f = e^{(\varkappa_\parallel - \varkappa_\perp)}$. ϕ becomes $45°$ when $f = 0$. Rewriting as

$$\delta = \frac{\omega l}{2cn}(n_\parallel{}^2 - n_\perp{}^2)$$

we can find δ in terms of the tensor components from equation (4.25) where $\varkappa \ll n$ as

$$\delta = \frac{\omega l}{2cn}(\varepsilon_{zz} - \varepsilon_{xy} + \varepsilon_{xy}^2/\varepsilon_{xx}) \tag{4.30}$$

From symmetry relations of the tensor components and equation (4.30), it can be shown that the Voigt effect is second-order in the magnetic field and, unlike the nonresonant Faraday effect described by equations (4.28) or (4.29), it can be *anisotropic* in a cubic crystal.

2. *Magnetoreflection*

The direct effects in reflectivity can be derived by simply substituting the magneto-optical constant $n_\pm \varkappa_\pm$, $n_\parallel n_\perp$, $\varkappa_\parallel \varkappa_\perp$ in the Fresnel formulas for the reflection coefficient, i.e.,

$$R_\pm = \frac{(n_\pm - 1)^2 + \varkappa_\pm{}^2}{(n_\pm + 1)^2 + \varkappa_\pm{}^2}$$

for the Faraday configuration and

$$R_{\parallel,\perp} = \frac{(n_{\parallel,\perp} - 1)^2 - \varkappa_{\parallel,\perp}^2}{(n_{\parallel,\perp} + 1)^2 - \varkappa_{\parallel,\perp}^2}$$

for the Voigt configuration. Usually $\varkappa \ll n$ so that the absorption terms can be neglected.

The rotation effect in reflection arises from the differential phase shift δ, for the Faraday configuration, given by

$$\tan \delta_\pm = 2k_\pm/(1 - n_\pm{}^2 - \varkappa_\pm)^2$$

which leads to a rotation

$$\theta = \tfrac{1}{2}(\delta_- - \delta_+)$$

The differential reflectivity, on the other hand, gives ellipticity

$$\Delta = \frac{(|r_+|-|r_-|)}{(|r_+|+|r_-|)}$$

where $r_\pm$ are Fresnel amplitude coefficients:

$$r_\pm = (1-n_\pm - i\varkappa_\pm)/(1+n_\pm - i\varkappa_\pm)$$

The role of absorption and dispersion is thus reversed compared with the Faraday rotation and ellipticity in transmission.

Corresponding effects also exist for the Voigt configuration, but the separation of causes is not so complete.

Using the methods of the previous section and equations (4.14) and (4.15), we can express the magnetoreflection effects in terms of the tensor components.

For some magneto-optical experiments where only the *frequency* of resonances is concerned, the direction relation between two energy levels and the absorption makes it possible to formulate quantum-mechanical theory of the effect without requiring use of the macroscopic theory. When dispersion effects, nonresonant effects, or the *magnitude* of effect in general is concerned, we usually are required to use the results of this section.

3. Dispersion Relations

No simple picture exists for dispersion and it is necessary to use the relation between absorption and dispersion to treat the dispersive effects. The appropriate formulas are variously known as the Kramers–Kronig or Kramers–Heisenberg dispersion relations. The real and imaginary parts of the complex permittivity (i.e., dielectric constant) or conductivity tensors are related merely as a consequence of causality and the properties of a complex variable. The causality condition is that if $\mathbf{E}(t) = 0$ for all $t < 0$ then $\mathbf{D}(t) = \varepsilon\mathbf{E}$ is also zero for $t < 0$. The resultant dispersion relations are quite general and may be applied to any system, classical or quantum-mechanical. The dispersion relations may be written, after Boswarva, Howard, and Lidiard [5], as

$$\varepsilon_{ij}^R(\omega) - \delta_{ij} = \frac{1}{\pi} P \int_{-\infty}^{+\infty} \frac{\varepsilon_{ij}^I(\omega')\,d\omega'}{\omega'^2 - \omega^2}$$

and

$$\varepsilon_{ij}^I(\omega) = \frac{1}{\pi} P \int_{-\infty}^{\infty} \frac{[\varepsilon_{ij}^R(\omega') - \delta_{ij}]}{\omega'^2 - \omega^2} d\omega'$$

where P denotes the principal value of the integral. These forms do not hold between the real and imaginary parts of ε_- [defined in equation (4.16)], and this has led to error in certain theoretical treatments:

$$\sigma_{ij}^R(\omega) = -\frac{2}{\pi} P \int_0^\infty \frac{\omega' \sigma_{ij}^I(\omega') \, d\omega'}{\omega'^2 - \omega^2}$$

$$\sigma_{ij}^I(\omega) = \frac{2}{\pi} P \int_0^\infty \frac{\omega' \sigma_{ij}^R(\omega') \, d\omega'}{\omega'^2 - \omega^2}$$

The dispersion relations can be transformed into other forms, e.g., the relation between Faraday rotation and magnetoabsorption or Faraday rotation and ellipticity. For the former we have

$$\theta = \frac{\omega^2}{2\pi n} \int_0^\infty \frac{(n_- \alpha_- - n_+ \alpha_+)}{\omega'(\omega'^2 - \omega^2)} \, d\omega' \qquad (4.31)$$

Macroscopic theory is also discussed by Lax [3] and Smith [7].

V. BANDS IN REAL SEMICONDUCTORS—EXTENSION OF k · p THEORY TO THE CASE OF MAGNETIC LEVELS

The simplified theory quoted thus far, together with the macroscopic theory, gives us a basic understanding of magneto-optic effects. However, in order to extract band-structure parameters in real cases—the primary aim of most magneto-optical experiments—a realistic theory of the energy states in a high magnetic field is required for many of the effects, in particular, interband magnetoabsorption. (It should also be noted that some effects, notably free-carrier effects at low fields, do not require a full treatment of the magnetic states in order to extract fairly explicit parameters; see Section II.)

A suitable theory must take into account anisotropy of energy surfaces, effects of band degeneracy, and must include spin. In this respect the semiempirical k · p theory has been outstandingly successful, giving as its basic result the shape of energy bands with respect to extrema whose position is assumed known. The theory makes use of a small number of experimentally determined parameters; useful experiments are those in which the required parameters are most explicitly determined.

For zero fields we have the Schrödinger one-electron equation

$$\left[\frac{p^2}{2m} + V(r) \right] \psi_K = E_K \psi_K$$

where $\psi_K = U_K e^{i\mathbf{k} \cdot \mathbf{r}}$ is the Bloch function. The operator $p^2 (p = \hbar/i \, \nabla)$

acting on $e^{i\mathbf{k}\cdot\mathbf{r}}$ then gives the basic equation of $\mathbf{k}\cdot\mathbf{p}$ theory

$$\left[\frac{p^2}{2m} + V(r) + \frac{\hbar}{m}\mathbf{k}\cdot\mathbf{p} + \frac{\hbar^2 k^2}{2m}\right]U_K = E_K U_K$$

The term in $\mathbf{k}\cdot\mathbf{p}$ is then considered as a perturbation and separated from the unperturbed Hamiltonian,

$$H_0 = \frac{p^2}{2m} + V(r)$$

The equation is then summed over a group of bands within which the mutual interaction appearing through the $\mathbf{k}\cdot\mathbf{p}$ term is significant. An example might be, for group IV and groups III–V semiconductors, the doubly (spin) degenerate conduction bands together with the six valence bands at $K = 0$. Then we have

$$\psi_K = \sum_1^8 a_i u_{0i}$$

where the set of wave functions at $K = 0$ are taken as known, i.e., $u_{0i}(i = 1, ..., 8)$. Thus

$$(H_0 + H_{\mathbf{k}\cdot\mathbf{p}})\sum_1^8 a_i u_{0i} = E_i'\sum_1^8 a_i u_{0i}$$

where

$$E' = \left(E - \frac{\hbar^2 K^2}{2m}\right)$$

We proceed by multiplying each term by u_{0i}^* and integrating over all space giving a set of simultaneous equations; the condition for their solution is that

$$|E_0 + (H_{\mathbf{k}\cdot\mathbf{p}}) + E'(I)| = 0$$

in which $E_{0i} = u_{0i}^* H_0 u_{0i}$ so that E_{0i} are the assumed energy gaps. This leads to a relation between E', E_{0i} and the interaction terms which contain powers of $\mathbf{k}$ and matrix elements of p between the coupled states, i.e., constants to be determined empirically from the experiments. If the number of interacting bands is small and the symmetry sufficient, the 8×8 matrix may reduce sufficiently to give explicit expressions; usually numerical methods are used.

    The extension of this method to magnetic states was pioneered by Luttinger and Kohn [8] who used an expansion in terms of Bloch functions at $\mathbf{k} = 0$ (or band extremum if not at $\mathbf{k} = 0$).

    A simpler method of treating magnetic states of electrons in crystals has been given recently by Harper [9] which leads easily to the Kohn–

Luttinger equations by using an extension of the Bloch theorem. Harper [10] has shown quite generally that crystalline magnetic states have the form

$$\psi(\mathbf{r}) = e^{i\mathbf{k}x} U(\mathbf{x}, \mathbf{r})$$

where $\mathbf{x} = \mathbf{k} - (e/\hbar)\mathbf{A}(\mathbf{r})$, $\mathbf{B} = \nabla_x \mathbf{A}(\mathbf{r})$, and $\mathbf{x}$ is not the same quantity as $\varkappa$ in $n - i\varkappa$. The effect of the dependence of $\mathbf{x}$ on $\mathbf{r}$, through the operation of $\mathbf{p}(=\hbar/i\,\nabla_r)$, is to introduce the terms $(1/m)\pi\cdot\mathbf{p}+(1/2m)\,\pi^2$ into the Hamiltonian where

$$\pi = \hbar\mathbf{x}+e\mathbf{A}(-i\nabla_x)$$

The Schrödinger equation now becomes

$$\left[\frac{1}{2m}\mathbf{p}^2 + V(r) + \frac{1}{m}\,\pi\cdot\mathbf{p} + \frac{1}{2m}\,\pi^2 - E\right] U(\mathbf{x}, \mathbf{r}) = 0$$

which is six-dimensional. The fourth term taken alone gives the Landau levels of *free* electrons; the first two terms are the usual crystal Hamiltonian in the absence of a field. The *third* term, *directly analogous to* $\mathbf{x}\cdot\mathbf{p}$ at zero field, couples the magnetic and crystal terms together and so expresses interaction between bands in the presence of a magnetic field. The $\pi\cdot\mathbf{p}$ term is, in fact, just the interaction treated in the conventional Luttinger–Kohn method. The $\mathbf{x}\cdot\mathbf{p}$ interaction can be removed to first order for distant bands and groups of adjacent bands treated exactly as in the Luttinger–Kohn case. Such calculations for groups III–V compounds (treating conduction bands together with the degenerate valence band set exactly and higher bands to order k^2) have been made by Pidgeon and Brown [11] and the $\mathbf{k}\cdot\mathbf{p}$ method has been extended to the energy levels of the lead salts in a magnetic field by Mitchell and Wallis [12].

The results for degenerate valence bands are of interest in that at high quantum numbers we can consider the states as derived from the light-hole and heavy-hole bands as shown in Fig. 3. However, at low quantum numbers the mixing gives rise to *irregularly* spaced levels, sometimes referred to as "quantum effects."

For the simple band picture described by equations (4.8) and (4.9), the spin splitting of the electron and hole states was described by an effective g-factor. The splitting in practical cases is anomalous with respect to a free-electron g-factor of 2.00.

To explain this effect we consider a conduction band edge at $\mathbf{k} = 0$, having only time reversal (spin) degeneracy. In the absence of a magnetic field, the energy can be written [13] in terms of coefficients $D_{\alpha\beta}$

$$E(k) = \varepsilon_{\alpha\beta} D_{\alpha\beta} k_\alpha k_\beta$$

where $\alpha, \beta = x, y, z$ and this includes symmetric terms of the form

$$(D_{\alpha\beta} + D_{\beta\alpha})\, k_\alpha k_\beta$$

In the presence of a field we can replace $\mathbf{k}$ by the operator $(\mathbf{p} - e\mathbf{A})$ expressed as $\bar{k}$. The energy then becomes

$$E(\bar{k}) = \varepsilon_{\alpha\beta}\, D_{\alpha\beta}\, \bar{k}_\alpha \bar{k}_\beta \tag{4.32}$$

$D_{\alpha\beta}$ can now be written in the form

$$
\begin{aligned}
D_{\alpha\beta} &= \tfrac{1}{2}\,(D_{\alpha\beta} + D_{\beta\alpha})\,[\bar{k}_\alpha \bar{k}_\beta + \bar{k}_\alpha \bar{k}_\beta] \\
&+ \tfrac{1}{2}\,(D_{\alpha\beta} - D_{\beta\alpha})\,[\bar{k}_\alpha \bar{k}_\beta - \bar{k}_\alpha \bar{k}_\beta]
\end{aligned} \tag{4.33}
$$

When $\mathbf{B} = 0$, the commutator $[\bar{k}_\alpha \bar{k}_\beta - \bar{k}_\beta \bar{k}_\alpha] \equiv [k_\alpha, k_\beta] = 0$, but for $\mathbf{B} \neq 0$, some components are nonzero; e.g., if $\mathbf{A} = \mathbf{B}(0, x, 0)$, $[k_x, k_z] = 0$ and $[k_y, k_z] = 0$ but $[k_x, k_y] = ieB_z$.

Using this result we can express equation (4.32) as the Hamiltonian

$$H = \sum_{\alpha\beta} D_{\alpha\beta}^s\,[\bar{k}_\alpha \bar{k}_\beta + \bar{k}_\beta \bar{k}_\alpha] + ieD_{xy}^A\, B_z + \mu_\beta\, \boldsymbol{\sigma} \cdot \mathbf{B} \tag{4.34}$$

in which the tensor $D_{\alpha\beta}$ has been separated into symmetric $(D_{\alpha\beta}^s)$ parts and antisymmetric $(D_{\alpha\beta}^A)$ parts according to equation (4.31) and a spin interaction term has been added $(\mu_\beta \boldsymbol{\sigma} \cdot \mathbf{B})$. The form of the antisymmetric part is

$$
D_{\alpha\beta}^A =
\begin{bmatrix}
0 & D_{xy} & -D_{xz} \\
-D_{xy} & 0 & D_{yz} \\
D_{xz} & -D_{yz} & 0
\end{bmatrix}
$$

and represents an axial vector.

Thus the *antisymmetric* part of the mass tensor $D_{\alpha\beta}$ together with the spin interaction term in equation (4.34) constitute the coefficient of B_z; by analogy with the form familiar in atomic theory, $\mathbf{L} \cdot \mathbf{B} + \mathbf{S} \cdot \mathbf{B}$, they give orbital $(\mathbf{L})$ and spin $(\mathbf{S})$ contributions to angular momentum and between them give the effective g-factor.

The *symmetric* part of the tensor $(D_{\alpha\beta}^s)$ gives the Landau levels in terms of the effective mass. The coefficients $D_{\alpha\beta}$ are obtained from $\mathbf{k} \cdot \mathbf{p}$ perturbation theory. Combining the orbital and spin terms provides

$$E(k) = \sum D_{\alpha\beta}^s\,[\bar{k}_\alpha \bar{k}_\beta + \bar{k}_\beta \bar{k}_\alpha] - \mu^* \boldsymbol{\sigma} \cdot B$$

where the anomalous magnetic moment μ^* is defined by

$$\frac{\mu^*}{\mu_B} = (\langle \gamma | L_z | \gamma \rangle + 1)$$

where

$$iD_{xy}^A = \frac{1}{2m} \langle \gamma | L_z | \gamma \rangle$$

and the effective g-factor is

$$g = \mu^* / \mu_B$$

Using the results of $\mathbf{k} \cdot \mathbf{p}$ perturbation theory, we get

$$\frac{\mu^*}{\mu_\beta} = 1 + \frac{1}{2im} I \left(\sum_\delta \frac{\langle \gamma | p_x | \delta \rangle \langle \delta | p_y | \gamma \rangle}{\varepsilon_\gamma - \varepsilon_\delta} \right)$$

where I denotes "imaginary part of." Using this treatment, Roth, Lax, and Zwerdling [14] have given an expression for g_c in terms of the effective mass m^* and the energy gap E_G of the spin-orbit splitting of the valence states Δ as follows:

$$g_c^* = 2 \left[1 - \left(\frac{m}{m^*} - 1 \right) \frac{\Delta}{3 E_G + 2 \Delta} \right]$$

This expression shows that $g_c \rightarrow 2$ as the spin-orbit splitting $\Delta \rightarrow 0$, so that the presence of spin-orbit interaction is essential for anomalous g-factors. Equation (4.34) applies to cases where we have p-like valence states and s-like conduction states as in Ge, Si, and InSb.

The effective g-factor may be either positive or negative and as large as -50 (e.g., InSb) or as small as $+0.2$ (e.g., GaAs) according to the effective mass, the energy gap, and spin-orbit splitting. Thus the spin splitting can be quite significant and occasionally comparable with the orbital splittings between Landau levels. In a nonparabolic band the g-factor will, like the effective mass, vary with energy.

In the degenerate valence bands the arrangement of the ladders does not lend itself to the definition of effective g-factors, as the first few levels are irregularly spaced and the levels arising by the mixing of wave functions are considered as labeled by their quantum numbers.

A detailed discussion of g-factors is given by Yafet [4].

VI. FREE-CARRIER EFFECTS

A. General

The observation of free-carrier or intraband effects requires the presence of a finite number of free electrons or holes in the appropriate conduction or valence band. The Fermi distribution of these carriers then becomes important. We can distinguish two limiting cases which have practical application. The first occurs when $\hbar\omega_c$ is much less than the range over which the derivative of the Fermi function with respect to energy (df/dE) is finite. That is, many Landau levels occur over this range. The details of the magnetic states are then not important and the Boltzmann transport equation may be used to calculate the conductivity tensor (or imaginary part of the dielectric constant tensor). By using

$$\mathbf{j} = \sigma\mathbf{E} = \frac{2}{(2\pi)^3} \int e\mathbf{v}f(k)\,dk$$

substituting

$$\mathbf{v} = \frac{1}{\hbar}\frac{dE}{d\mathbf{k}}$$

and for $f(\mathbf{k})$, one then introduces the E–k relations, but at this stage one does not need to known the exact form of $E(\mathbf{k})$. For certain experiments in this approximation, e.g., microwave cyclotron resonance and low-field Faraday effect, *explicit* information on $E(\mathbf{k})$ is obtained *without recourse to band theory* for fitting to experimental results. The classical treatment of such phenomena together with their relation to the Boltzmann treatment is described by Houghton and Smith [15]. The second limiting case is when $\hbar\omega_c$ is much greater than the range of df/dE finite, and quantum effects of transitions between magnetic states became observable.

B.   Cyclotron Resonance

1.   *Microwave Cyclotron Resonance*

The first observation of cyclotron resonance of free carriers in semiconductors was made at microwave frequencies by Dresselhaus, Kip, and Kittel [16]. These experiments, which gave very explicit information about the energy surfaces near the extrema of the valence and conduction bands in very pure silicon and germanium, have been extensively reviewed by Lax and Mavroides [17].

Microwave observations typically employ a field of ~ 1000 G and the splitting of the Landau levels is relatively small. Even at liquid-helium temperatures this can be in the classical region that can be treated by Boltzmann theory. The experiments were carried out at low temperatures to satisfy the condition

$$\omega_c \tau \gg 1$$

required for the observation of resonance in which τ is the collision time for the carrier. In pure materials τ can be lengthened by cooling and suppressing the lattice vibrations. This criterion also restricted the early measurements to pure germanium and silicon, but it had more recently been possible to work at higher temperatures and with compounds such as InSb (see, for example, Bagguley, Stradling, and Whiting [18]).

Since the range of the band explored by CR experiments is restricted to near the band extrema, the dependence of energy upon **k** is accurately quadratic, i.e., the bands are parabolic. Under these circumstances the Boltzmann expression contains terms of the form $d^2 E/dk^2$ which are constant, i.e., the effective mass [given by $(1/m^*) = (1/\hbar^2)\,(\partial^2 E/\partial k^2)$] is a constant with energy. The Boltzmann expression then reduces to a single-particle analysis, independent of the statistics, and CR can be described by an equation with m^* in place of m and $\omega_0 = 0$. The anisotropy of the mass tensor is readily measured since for ellipsoidal surfaces, as for electrons in germanium, ω_c varies with direction and is given by

$$\omega_c = \frac{eB}{m^*} = eB\sqrt{\frac{m_1\,\alpha^2 + m_2\,\beta^2 + m_3\,\gamma^2}{m_1\,m_2\,m_3}}$$

where the α's are direction cosines to axes 1, 2, and 3.

From the results of resonance experiments at various orientations it is found experimentally that the conduction band in germanium can be expressed as

$$E = \frac{\hbar^2}{2}\left(\frac{k_1^2}{m_t} + \frac{k_2^2}{m_t} + \frac{k_3^2}{m_l}\right)$$

where

$$m_t = (0.0819 \pm 0.0003)m_0$$

and

$$m_l = (1.64 \pm 0.03)m_0$$

The effective masses and energy surfaces for holes may be obtained for p-type material. From $\mathbf{k}\cdot\mathbf{p}$ theory, the valence bands in diamond, silicon, and germanium are given near $\mathbf{k} = 0$ by

$$E = -\frac{\hbar^2}{2m}\left\{Ak^2 \pm [B^2 k^4 + C^2(k_x^2 k_y^2 + k_y^2 k_z^2 + k_x^2 k_z^2)]^{\frac{1}{2}}\right\}$$

for the light $(-)$ and heavy $(+)$ holes and

$$E = -\Delta - \frac{\hbar^2}{2m}Ak^2$$

for the split-off band. Some recent values of A, B, and C (from Stickler,

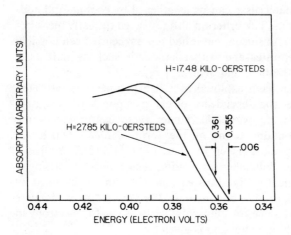

Fig. 4. A plot of CR absorption strength *versus* photon energy.

Zeiger, and Heller [19]) are

	A	B	C
Ge	− 13.12	8.2	13.3
Si	− 4.22	1.0	4.34

If $\hbar\omega_c$ is increased or the sample temperature reduced, or both, the quantum situation can be reached and the quantum effects predicted from the Luttinger–Kohn theory observed. The resonance spectrum becomes complex and presents interpretational problems since theory predicts many more lines than have been observed. Quantum effects in germanium were first noticed by Fletcher, Yager, and Merritt [20] and a recent extension was made by Hensel [21] working at 4 mm (50–60 Gcps).

A further important experiment on CR at 4 mm has extended CR technique by using *monochromatic* infrared radiation to excite the carriers. This experiment, carried out by Rauch [22] on type IIb (*p*-type) diamond, has determined the masses of the light-hole, heavy-hole, and split-off bands, giving $m^*/m = 0.76$, 2.18, and 1.06, respectively, and also determined the spin-orbit splitting Δ as 0.006 eV. This is achieved by setting the magnetic field for resonance in one of the degenerate bands and varying the photon energy of the radiation used to excite the holes into the valence band from the acceptor centers. A plot of CR absorption strength *versus* photon energy is shown in Fig. 4. The first energy value is that corresponding to the difference between the acceptor level and the valence band. The resonance is then tuned to that appropriate to the split-off band and the experiment repeated. No CR absorption is seen until a photon energy corresponding to the energy difference between the *split-off band and the acceptor* is reached. From the energy difference of these "edges" the spin-orbit splitting is determined. The spectrum of CR absorption *versus* exciting radiation energy showed a pronounced structure; this is due to preferential trapping of the carriers by optical phonon processes and is also observed in photoconductivity. There was some considerable experimental difficulty in obtaining a diamond with a sufficiently long relaxation time τ even at 1.2°K and the field required for resonance of heavy holes at 4 mm was 54 kG. Values for A, B, and C obtained were

$$A = 0.94 \qquad B^2 \sim 0.2 \qquad C^2 < 0.16$$

2. Infrared Cyclotron Resonance

At infrared frequencies the conditions for the observation of quantum effects, $\hbar\omega_c \gg kT$, and for observation of CR, $\omega_c \tau \to 1$, can be readily met at low temperatures. However, since high magnetic fields are required to raise $\omega_c = eB/m^*$ into this frequency range, observations have been restricted to compound semiconductors in which, at least in the conduction band, effective masses are relatively low.

The earliest infrared experiments were carried out at NRL [23] on InSb. For such conditions, i.e., 40 μ, 40 kG, 300°K, and $\omega_c \sim 2kT$, several Landau levels are populated, and both the magnetic states, including spin effects and the distribution of the electrons, must be considered for proper analysis. Later results at 77°K for several intermetallic compounds are shown in Fig. 5. These results show a change of mass with energy indicating how the increasing magnetic splitting is exploring the nonparabolic nature of the conduction bands.

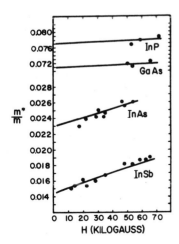

Fig. 5. Variation of low-temperature effective mass ratio m^*/m with magnetic field H (from cyclotron resonance data for several n-type III–V compound semiconductors).

A recent development of interest is the use of laser sources in the infrared for CR experiments. The quantum effects in the valence band of germanium have been studied at NML with use of the 300-μ cyanide laser.

C. Free-Carrier Dispersive Effects

1. Faraday Effect

This effect is caused by the dispersion associated with the cyclotron resonance absorption of free carriers as can be readily seen from the form of the Kramers–Kronig relation given in equation (4.31). The rotation is given in terms of conductivity by equation (4.29) as

$$\theta = \frac{\omega l}{2 c} (n_- - n_+) = \frac{\sigma_{xy}^R l}{2 n c}$$

It is therefore possible to obtain a relation between θ and the band structure parameters by using the Boltzmann expression to find this component of the conductivity tensor in the lowfield case.

For the case of spherical energy surfaces and a parabolic band, the Boltzmann expression reduces to a single-particle classical dynamics as in the case of CR. The indices n_- and n_+ may then be deduced directly from equation (4.29)

$$\theta = \frac{BNe^3 l}{2n \, c\varepsilon_0 \, m^{*2} \, \omega^2} \tag{4.35}$$

which is the expression obtained by Mitchell [24]. The dependence of θ upon the inverse square of the mass implies that Faraday rotation is an accurate method for determination of effective mass and, as noted by Mitchell, the relaxation time τ does not appear in the expression. This is a consequence of choosing the measuring frequency $\omega \gg \omega_c$, the cyclotron frequency. The infrared free-carrier effect for which this applies must be distinguished from the microwave effect which depends upon the square of the relaxation time and gives information about the mobility $(e\tau/m^*)$ rather than the mass of the carriers.

We may obtain further insight into the problem by noting that equation (4.35) may be obtained by inserting values of $\alpha_\pm$ in equation (4.31) where $\alpha_\pm$ are obtained from classical theory of cyclotron resonance, including the nonresonant component $(\omega + \omega_c)$. The absorption can be assumed to have a Lorentzian line shape with half-width $1/\tau$. In the approximation quoted, the integral of the absorption over this line is independent of τ. Thus τ can be chosen to be so short that the line width is large and the CR would be unobservable, i.e., $\omega_c \tau > 1$ does not have to be satisfied. The

Faraday rotation will still be observable and given by equation (4.35) under these conditions—a valuable property since high temperatures and large impurity content (factors which make τ small) can be tolerated.

The earliest infrared free-carrier measurements, made by Smith and Moss [25], showed that the rotation was accurately proportional to B and to λ^2 as predicted by equation (4.35) and that the rotation could be easily measured (see Fig. 6). The simplicity and flexibility of the Faraday effect experiment has subsequently resulted in the performance of many more such experiments than the more difficult cyclotron resonance. Some 20 materials had been examined by 1965.

Given that the rotation is accurately proportional to magnetic field and to the square of the wavelength, the effective mass deduced can be interpreted more generally in the case of nonspherical energy surfaces or nonparabolic bands, or both, by comparison between equation (4.35) and the appropriate Boltzmann theory for the rotation. The success of the early experiments stimulated Stephen and Lidiard [26] to calculate expressions for the rotation due to free carriers in this form and to give general results for cubic crystals:

$$\frac{n\theta}{Bl} = \frac{e^3}{8\pi^3\hbar^4\varepsilon_0 c\omega^2}\int\frac{\partial f_o}{\partial E}\frac{\partial E}{\partial k_x}\left(\frac{\partial E}{\partial k_y}\frac{\partial^2 E}{\partial k_x\partial k_y} - \frac{\partial E}{\partial k_x}\cdot\frac{\partial^2 E}{\partial k_y^2}\right)d^3\mathbf{k} \quad (4.36)$$

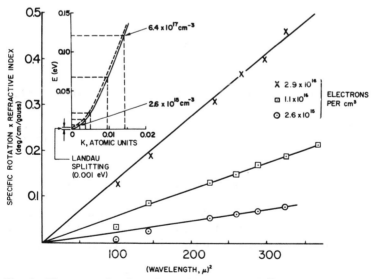

Fig. 6. The proportionality of rotation to B and λ^2 as predicted by equation (4.35).

The carrier density N can be included in equation (4.36) from the relation

$$N = \frac{2}{(2\pi)^3} \int_0^\infty f_0 \, d^3 \mathbf{k} \qquad (4.37)$$

so that equations (4.34) and (4.33) can be compared. This comparison gives an interpretation of the effective mass measured in the Faraday rotation experiments. We note that so far no assumption has been made about the form of the E–$\mathbf{k}$ relation; this relation is required to perform the integral in equation (4.36) and so evaluate the Faraday rotation. There are four particular cases of interest.

First, when E has a quadratic dependence on $\mathbf{k}$ and the energy surfaces are spherical. Comparison between equations (4.35) and (4.36) and including equation (4.37) then gives

$$\frac{1}{m^*} = \frac{1}{\hbar^2} \frac{d^2 E}{dk^2} \qquad (4.38)$$

and the result is independent of the distribution function f, i.e., electrons in all the populated states behave with the same effective mass since $d^2 E/dk^2$ is a constant in a parabolic band.

Second, with a quadratic dependence of E upon k, but with, e.g., ellipsoidal energy surfaces such as for electrons in germanium. A combination of transverse and longitudinal masses is then measured in the Faraday effect given by

$$\frac{1}{m^*} = \left[\frac{\kappa(\kappa + 2)^{\frac{1}{2}}}{3} \right]^{\frac{1}{2}} \frac{1}{m_l}$$

where $\kappa = m_l/m_t$.

Third, in a semiconductor with spherical energy surfaces in which the carrier concentration and temperature are such that df_0/dE is finite only at the Fermi level E_F and negligible elsewhere and that the Fermi level lies within one of the bands. Under these circumstances equation (4.36) yields

$$\frac{1}{m^*} = \frac{1}{\hbar k_F} \left(\frac{\partial E}{\partial k} \right)_{k_F}$$

The carrier distribution in such cases is said to be "degenerate." This result is very important since with the degenerate distribution the wave vector at the Fermi surface k_F can be found from the carrier density N by the relation

$$N = \frac{2}{(2\pi)^3} \cdot \frac{4\pi k_F^3}{3}$$

and so can be found from a Hall effect experiment. Thus values of k_F and $(\partial E/\partial k)_{k_F}$ (the slope of the band at k_F) can be found from the Faraday rotation. Hence the E–k relation can be determined directly over a range of k by varying the doping and progressively filling up the band since this result holds when the band is nonparabolic. In the parabolic case the result reduces to that in the first case noted [equation (4.38)].

Fourth, if the band is nonparabolic and the distribution nondegenerate, df/dE is finite over a significant range of the E–k curve. Thus electrons at different energies have different effective masses and an average is measured. Evaluation of the integrals in equation (4.36) is now needed and requires the insertion of a known relation between E and $\mathbf{k}$.

Interpretation of Faraday rotation under the conditions noted in the first case is straightforward; probably the most useful case is the third, as the band shape can be directly determined by numerical integration of the experimentally determined quantity dE/dk *without the use of a theory of the band shape*. The results can then be used as a direct test of such theory. Such a method is also important for comparison and checking with high-field experiments, such as CR in nonparabolic bands, the interpretation of which requires both use of a band shape theory and its extension to the case of high magnetic fields.

The most favorable material for such investigation is InSb since CR results have shown that the energy surfaces for electrons are spherical and the small effective mass ensures a low density of states. Such an experiment has been carried out by Smith, Moss, and Taylor [27] and further refined by Smith, Pidgeon, and Prosser [28]. The range of carrier and Fermi levels is shown in Fig. 6, and the experimental values of dE/dk *versus* k in Fig. 7. An interesting feature of this plot is that the slope of the dE/dk curve at $\mathbf{k} = 0$, which is of course $d^2 E/dk^2$, gives the mass at the band minimum,

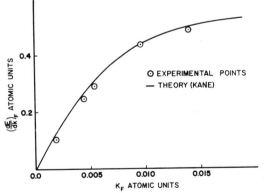

Fig. 7. Graph of experimentally determined dE/dk *versus* k.

where in the limit the band will be parabolic. This gives a useful technique for finding the zero-k mass.

The results are compared with the $\mathbf{k \cdot p}$ perturbation calculation of Kane [29] which yields the result

$$E'(E' - E_G)(E' + \Delta) - k^2 P^2(E' + 2\Delta/3) = 0 \qquad (4.39)$$

when the conduction–valence band interaction is treated exactly, and the interactions with higher bands are neglected, and

$$E' = E - \frac{\hbar^2 k^2}{2m}$$

where m is the free-electron mass. The parameters P, E_G, and Δ are determined from experiment. The energy gap E_G may be found from absorption edge analysis or, better, from I.M.O. measurements at low temperatures where the $\mathbf{k \cdot p}$ theory is appropriate. At higher temperatures, as will be seen later, some corrections to the optically measured energy gap are required. The expression is insensitive to Δ; the spin-orbit splitting and values derived from atomic data may be used without significant error. The parameter P is the momentum matrix element between the valence and conductor states and is related to the previously defined quantity p_{cv}

$$p_{cv} = -\sqrt{\tfrac{2}{3}}\, P$$

Determination of the band shape is therefore essentially a determination of the matrix element P. P may be found directly from the zero-k mass since for small k (parabolic region) equation (4.38) may be expounded in powers of k. Retaining only the term in k^2 yields

$$E = \left(\frac{\hbar^2}{m_0^{*2}} - \frac{\hbar^2}{m^2}\right) k^2 = \frac{2P^2}{3}\left(\frac{2}{E_G} + \frac{1}{E_G + \Delta}\right) k^2$$

from which neglecting $1/m$ in comparison with $1/m_0^*$, where m_0^* is the effective mass at $\mathbf{k} = 0$, we get

$$\frac{\hbar^2}{m_0^{*2}} = \frac{2P^2}{3}\left(\frac{2}{E_G} + \frac{1}{E_G + \Delta}\right)$$

Alternatively P may be found from a best fit over a range of the E–k curve. Such a fit is obtained from an analysis by Pidgeon [30]. A very good fit between experiment and theory is obtained yielding a value of $P^2 = 0.395$ atomic units and corresponding to a zero-k mass of 0.0145 m. It should be noted that Faraday effect cannot explicitly separate differing hole masses, which are better studied by CR. When difficulty is experienced in fulfilling the conditions for observation of resonance, $\omega_c \tau > 1$, Faraday rotation measurements are particularly useful in that a variety of temperatures and impurity concentrations can be tolerated in the specimens.

2. Faraday Ellipticity and Voigt Effect

Plane-polarized light undergoing Faraday rotation emerges elliptically polarized due to differential absorption of the two circular components. This ellipticity (amplitude ratio of axes) is given by

$$\Delta = \frac{\omega l}{2c}(\varkappa_+ - \varkappa_-)$$

for small absorption and hence classical theory gives

$$\Delta = \frac{BNe^3(1/\tau)}{nc\varepsilon_0 m^* \omega^3}$$

Thus $\theta/\Delta = \omega\tau/2$. Combining rotation and ellipticity therefore gives a method of finding the relaxation time τ (Fig. 8).

The Voigt effect, or magnetic double refraction, occurs for **B** perpendicular to the radiation. The phase lag δ in classical theory is

$$\delta = \frac{\omega l}{2c}(n_\parallel - n_\perp) = \frac{B^2 Ne^4}{2nc\varepsilon_0 m^{*3} \omega^3}$$

This effect, unlike Faraday effect, can detect anisotropy of energy surfaces in a cubic crystal.

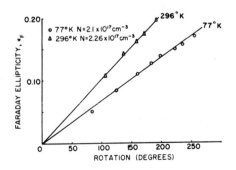

Fig. 8. A method of finding the relaxation time τ from the rotation and the ellipticity.

VII. INTERBAND MAGNETOABSORPTION (I.M.O.) EFFECTS

In the absence of a magnetic field the absorption by interband direct transitions is of the form

$$\alpha(0) = A(\hbar\omega - E_G)^{\frac{1}{2}}$$

for simple bands. In an analogous manner the absorption in the presence of a field can be shown to be

$$\alpha(H) = \frac{A\hbar\omega_c}{2} \sum_l (\hbar\omega - E_n)^{-\frac{1}{2}}$$

where

$$E_n = E_G + (n+\tfrac{1}{2})\,\hbar\omega_c + (g_c\,M_{Jc} - g_v\,M_{Jv})\,\beta H$$

and is obtained by finding the energy difference between the Landau levels. In this case, $\omega_c = eB/\overline{m}$, where $\overline{m} = m_c m_v/(m_c + m_v)$ is the reduced effective mass of electrons and holes. The density of states of the one-dimensional magnetic subbands is shown in Fig. 2. Since the absorption follows the density of states, it is readily seen that the absorption is oscillatory in character, sometimes above and sometimes below that of the zero-field case. Such effects have been extensively observed at NRL and MIT [31]; thin samples, high resolution, and low temperatures are required in addition to large magnetic fields. Materials studied include Ge, InSb, InAs, and GaSb [31].

Selection rules require that $\Delta n = 0$

$$\Delta M_J = 0 \qquad \text{for} \qquad E \parallel H$$

and

$$\Delta M_J = \pm \qquad \text{for} \qquad E \perp H$$

At the onset of absorption, the edge is shifted to higher photon energies, i.e., to $E_G + \tfrac{1}{2}\hbar\omega_c$.

Three important parameters can be extracted from the results of such experiments: (1) by plotting the positions of the minima against magnetic field, converging linear plots are obtained, which when extrapolated to zero field give accurate values for the energy gap E_G (Fig. 9); (2) the value of the cyclotron frequency corresponding to the reduced mass

$$\overline{m}^* = m_c m_v/(m_c + m_v)$$

can be obtained from the slopes; and (3) by fitting high-resolution results to calculated energy values, the splitting of lines for different states of polarization enables one to deduce effective g-values. Anomalous values are obtained, e.g., -2.5 for Ge and -48 for InSb.

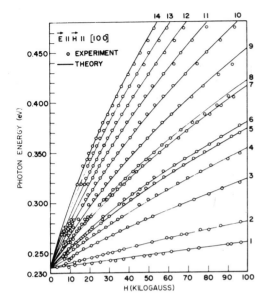

Fig. 9. Graph of mininum photon energy *versus* magnetic field strength. Extrapolating the converging linear plots to zero field gives accurate values of the energy gap. (Results obtained by computer and adjusted for best fit to the experiment.) From Pidgeon and Brown[11].

High-resolution results show a wealth of fine structure—to be expected theoretically due to the complicated nature of the valence band states in a magnetic field as discussed in Section V. I. M. O. experiments contain more information than other effects and require detailed fitting of the observed lines for comparison with the theoretical energy levels to extract the band

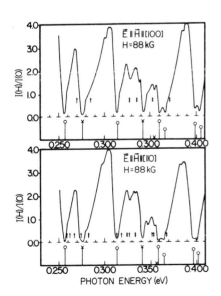

Fig. 10. Results of analysis of InSb by Pidgeon and Brown. Circular and plane-polarized radiation were employed in both Faraday (*B* parallel to radiation) and Voigt (*B* perpendicular to radiation) configurations in [100] and [110] crystal directions.

Table II. Summary of Experimental Parameters and Techniques for Their Measurement

Determination	Experiment	Comment
Effective mass at band extrema including anistropy	Cyclotron resonance (microwave)	Extends only $\sim 10^{-3}$ eV from extrema.
Effective mass and position of band extrema lying above or below the minimum gap	CR (microwave) with selective optical excitation of carriers from n- or p-type impurities	,,
Effective mass in materials with short relaxation time or in heavily doped materials or both	High-field infrared CR. Free-carrier Faraday effect (with reflectivity minimum)	No anisotropy observable.
Temperature dependence of effective mass	Free-carrier Faraday effect	,,
Exploration of band shape (nonparabolicity) away from extrema	Free-carrier Faraday effect with heavy doping CR (infrared with high fields) Interband magneto-absorption (I.M.O.)	Only low-field theory. No anisotropy. Need high-field theory. Involves both valence and conduction bands. Thin samples.
Energy gaps	I.M.O.	Convergence plot *versus B*.
Effective mass at direct gap minima when indirect gap is smaller	I.M.O.	Involves valence band.
Conduction band g-factors	I.M.O. Resonant interband Faraday effect	
Study of light-hole and heavy-hole valence states	I.M.O. Resonant interband Faraday effect	Using various orientations of B. Opposite sign of rotation.
Very deep levels, or where absorption is high	Interband magnetoreflection	Particularly for semimetals.
Relaxation times (1) Free carriers	Free-carrier ellipticity and Faraday rotation combined	
(2) Interband processes	Combination of resonant interband Faraday effect and Voigt effect	

structure parameters. A recent and instructive example is the analysis of InSb by Pidgeon and Brown. Some of their results are shown in Fig. 10. In this work, circular and plane-polarized radiation was employed in both Faraday (B parallel to radiation) and Voigt (B perpendicular to radiation) configurations in [100] and [110] crystal directions. Comparison of $E \perp B$ and $E \parallel B$ spectra aids assignment of the transitions since selection rules differ, favoring transitions from light-hole and heavy-hole states, respectively.

The theory, an extension of the Luttinger–Kohn method, treats the conduction–valence band interaction in a magnetic field exactly, treats higher bands to order K^2, and includes anisotropy and electron spin. The mixing of the valence states changes the selection rules (usually $\Delta n = 0$) to include $\Delta n = -2$ and working induced transitions with $\Delta n = \pm 4, -6$, all of which can give rise to fine structure. The results are obtained by computer calculation and adjusted for best fit to the experiment (Fig. 9) in terms of the following parameters (given with best-fit values):

$$P^2 = 0.403 \text{ atomic units}$$
$$E_G = 0.2355 \text{ eV}$$
$$\Delta = 0.9$$
$$\gamma_1{}^L = 32.5$$
$$\gamma_2{}^L = 14.3$$
$$\gamma_3{}^L = 15.4$$
$$L^L = 13.4$$

The momentum matrix element is the same quantity discussed earlier, as are E_G and Δ. The γ parameters determine valence-band anisotropy according to the formalism of Luttinger and Kohn [8]. This is the first determination of a complete set of band parameters from one type of magneto-optical experiment.

Table II presents a list of parameters which are possible, or derivable, to obtain together with what presently appear to be the most suitable techniques or combination of techniques for their measurement.

VIII. REFERENCES

1. L. Landau, *Z. Physik* **64**:629 (1930).
2. J. Callaway, *Energy Band Theory*, Academic Press (New York), 1964, p. 252.
3. B. Lax, *Proc. Intern. School of Phys. Enrico Fermi*, Course XXII, Academic Press (New York), 1963, pp. 240-340.
4. Y. Yafet, *Solid State Phys.* **14**:1 (1963).
5. I. M. Boswarva, R. E. Howard, and A. B. Lidiard, *Proc. Roy. Soc. (London)* **A269**: 125 (1962).
6. B. Lax, *Proc. Intern. School of Phys. Enrico Fermi*, Course XXII, Academic Press (New York), 1963, p. 312.

7. S. D. Smith, *Handbuch der Physik Band*, **XXV** /2a, pp. 234-318.
8. J. M. Luttinger and W. Kohn, *Phys. Rev.* **97**:869 (1955).
9. P. G. Harper, to be published in *J. Phys. Chem. Solids*.
10. P. G. Harper, *Proc. Phys. Soc. (London)* **A68**:879 (1955).
11. C. R. Pidgeon and R. N. Brown, *Phys. Rev.* **146**:575 (1966).
12. D. L. Mitchell and R. F. Wallis, *Phys. Rev.* **151**: 581 (1966).
13. C. Kittel, *Quantum Theory of Solids*, John Wiley and Sons (New York), 1963, p. 279.
14. L. M. Roth, B. Lax, and S. Zwerdling, *Phys. Rev.* **114**:90 (1959).
15. J. T. Houghton and S. D. Smith, *Infra-red Physics*, Oxford University Press (London), 1966, Chap. IV.
16. G. Dresselhaus, A. F. Kip, and C. Kittel, *Phys. Rev.* **93**:827 (1953); **98**:368 (1955).
17. B. Lax and J. G. Mavroides, *Solid State Phys.* **11**:161 (1960).
18. D. M. S. Bagguley, R. A. Stradling, and J. S. S. Whiting, *Proc. Roy. Soc. (London)* **A262**:342 (1961); *Phys. Letters* **6**:143 (1963).
19. J. J. Stickler, H. J. Zeiger, and G. S. Heller, *Phys. Rev.* **127**:1077 (1961).
20. R. C. Fletcher, W. A. Yager, and F. R. Merritt, *Phys. Rev.* **100**:747 (1955).
21. J. C. Hensel, *Proc. Intern. Conf. Phys. Semicond. Exeter*, 1962, p. 281.
22. C. J. Rauch, *Proc. Intern. Conf. Phys. Semicond. Exeter*, 1962, p. 276.
23. E. Burstein, G. S. Picus, and H. A. Gebbie, *Phys. Rev.* **103**:825 (1956).
24. E. W. J. Mitchell, *Proc. Phys. Soc. (London)* **B68**:973 (1965).
25. S. D. Smith and T. S. Moss, *Brussels Solid State Conference, Vol. II*, John Wiley and Sons (New York), 1958, p. 671.
26. M. J. Stephen and A. B. Lidiard, *J. Phys. Chem. Solids* **9**:43 (1958).
27. S. D. Smith, T. S. Moss, and K. W. Taylor, *J. Phys. Chem. Solids* **11**:131 (1959).
28. S. D. Smith, C. R. Pidgeon, and V. Prosser, *Proc. Intern. Conf. Phys. Semicond. Exeter*, 1962, p. 301.
29. E. Kane, *J. Phys. Chem. Solids* **1**:249 (1957).
30. C. R. Pidgeon, Ph.D. Thesis, University of Reading, 1962.
31. B. Lax and S. Zwerdling, *Progr. Semicond.* **5**:221 (1960).

CHAPTER 5

Optical Properties and Electronic Structure of Amorphous Semiconductors

J. Tauc

*Institute of Solid State Physics of the Czechoslovak Academy of Sciences
Prague, Czechoslovakia*

I. INTRODUCTION

Optical properties of amorphous solids are of considerable interest since glasses of various kinds are very important optical materials. We shall treat in this chapter an example of the absorption spectrum of amorphous material which is particularly simple. We chose amorphous germanium for this study and compared its optical properties with the well-known properties of its crystalline form. The atomic arrangement in amorphous Ge was investigated by X-ray diffraction [1]. The results of these studies can be interpreted in two ways; it is at present impossible to decide between them. In one model amorphous Ge is pictured as composed of small crystallites with linear dimensions of the order 10 Å. In the other model amorphous Ge is considered as a homogeneous medium. The tetrahedra are still the basic units of the structure but the neighboring tetrahedra are irregularly rotated so that the long-range order is lost after several atomic distances. For the interpretation of the optical properties we shall use the latter model here, but the former model would give similar results. We shall show what information concerning the changes of the electronic structure due to the loss of the long-range order can be deduced from the changes observed in the optical properties. This chapter is based on work performed jointly in the Institute of Solid State Physics in Prague and the Institute of Physics in Bucharest and reported elsewhere [1-3].

Let us note that both models of atomic arrangement lead necessarily to the presence of vacancies; these are either concentrated on the boundaries of the crystallites or homogeneously distributed in the solid. The studies of transport properties have shown [2] that these vacancies act as acceptors and are responsible for the *p*-type conductivity of amorphous Ge having a large concentration of holes (10^{18}–10^{19}cm^{-3}). These vacancies may be filled with foreign atoms.

II. EXPERIMENTAL METHOD

Ge can be prepared in an amorphous state only in thin layers. The procedure, carried out in a vacuum (10^{-5}–10^{-6} atm), involved evaporation of very pure Ge on quartz or KCl substrates which were kept at room temperature. Some layers were annealed at 250 °C; they are still amorphous but have somewhat different properties. Annealing at 450 °C converts the layers into a polycrystalline state whose optical properties are essentially those of the crystal. We shall denote nonannealed layers by "a-Ge," the annealed ones by "a_t-Ge." The thickness of the layers was between 800 and 4000 Å. In this range the optical constants did not depend on the thickness of the layers and are therefore believed to be properties of the "bulk" material. The structure of the layers was tested by X-ray diffraction and the thickness was measured by the method of interference fringes.

The optical constants were determined in the range 0.08 to 1.7 eV by the measurement of transmission and reflection; it was necessary, of course, to take proper account of the interference effects in the layers. In regions where the layers were opaque ($\hbar\omega > 1.7$ eV), only the reflection spectrum to be used in the determination of the optical constants from the dispersion relations was measured. However, in our case we met with particular difficulties which made the determination of the optical constants in the region of high absorption difficult. We shall, therefore, return to this problem later and discuss first the three low-energy regions shown in Fig. 1.

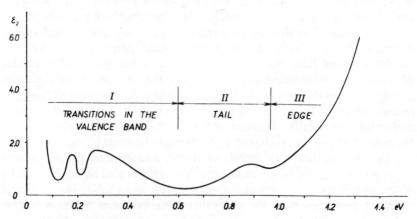

Fig. 1. Absorption in amorphous Ge in the infrared region (after [5]).

III. TRANSITIONS IN THE VALENCE BAND

The spectrum between 0.08 and 0.6 eV shown in Fig. 2 is qualitatively similar to the spectra in p-type crystalline Ge; the peaks in amorphous Ge are shifted toward lower energies. These peaks in crystalline p-type Ge were interpreted as direct (that is, **k**-vector conserving) transitions of holes from the two upper valence bands 1 and 2 into the deep-lying valence band 3 (cf. Fig. 3) and can be approximately described by the formulas

$$\omega^2 \, \varepsilon_2^{13} \sim |\hbar\omega - \Delta E|^{\frac{3}{2}} \exp\left(\frac{m_3}{m_3 - m_1} \frac{\hbar\omega - \Delta E}{kT}\right)$$

$$\omega^2 \, \varepsilon_2^{23} \sim |\hbar\omega - \Delta E|^{\frac{3}{2}} \exp\left(\frac{m_3}{m_3 - m_2} \frac{\hbar\omega - \Delta E}{kT}\right)$$

(5.1)

where ΔE is the spin-orbit splitting and m_1, m_2, and m_3 are the (averaged) effective masses.

As seen in Fig. 4, equation (5.1) describes approximately also the absorption in amorphous Ge (the agreement is better the farther from ΔE). The appropriate values of the parameters ΔE, m_1, m_2, and m_3 for various cases are listed in Table I.

We may now assume that the observed changes of effective masses are due to the transition from the crystalline to the amorphous state; the changes are qualitatively in accord with the relation

$$\frac{1}{m_3} = \frac{1}{2}\left(\frac{1}{m_1} + \frac{1}{m_2}\right)$$

(5.2)

as they should be. This assumption involves the exact conservation of the **k** vector ($\Delta \mathbf{k} = 0$) during the transitions.

Table I.   The Effective Masses in Crystalline and Amorphous Germanium*

	ΔE(eV)	m_1/m	m_2/m
Crystalline Ge	0.29	0.30	0.04
a-Ge	0.195	0.49	0.029
a_t-Ge	0.21	0.34	0.033

* The term m_3 was assumed to be the same as in the crystal, i.e., $m_3 = 0.08 \, m$.

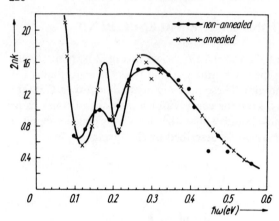

Fig. 2.   Absorption in amorphous Ge in the region 0.08 to 0.6 eV (after [4]).

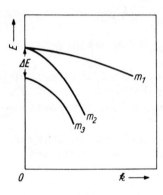

Fig. 3.   Valence band of Ge.

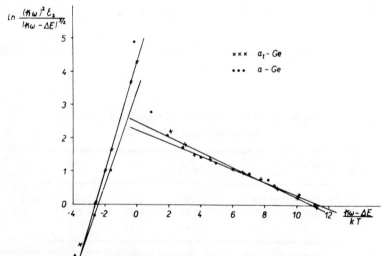

Fig. 4.   Determination of the effective masses (after [5]).

Alternatively, we may assume that the observed changes of effective masses are at least partly fictitious and due to the nonconservation of the **k** vector during the transitions. If we ascribe the whole observed changes of effective masses to this effect, we obtain the upper limit for the uncertainty Δk in the conservation of $|\mathbf{k}|$.

We shall now deduce equation (5.1) under the assumption that **k** is not exactly conserved during the transitions. Let us consider the transitions of holes between the bands 1 and 3 which are assumed parabolic. We have

$$\varepsilon_2^{13} \sim (f_1 - f_3) \text{ (joint density of states)}$$

where f_1 and f_3 are the probabilities that the states in bands 1 and 3 are occupied by holes. Band 3 is practically free of holes; therefore, $f_3 = 0$. For f_1 we have

$$f_1 = \text{const. } \exp\left(-\varepsilon_{p1}/kT\right)$$

where ε_{p1} is the energy of holes in band 1. Energy is conserved during the transitions:

$$\varepsilon_{p3} - \varepsilon_{p1} = \hbar\omega - \Delta E \tag{5.3}$$

where

$$\varepsilon_{p1} = \frac{\hbar^2}{2m_1} k_1^2 \qquad \varepsilon_{p3} = \frac{\hbar^2}{2m_3} k_3^2 \tag{5.4}$$

From equations (5.3) and (5.4) it follows that

$$\varepsilon_{p1} = \left(\frac{m_1}{m_3} \frac{k_3^2}{k_1^2} - 1\right)(\hbar\omega - \Delta E) \tag{5.5}$$

If $k_1 = k_3$, then the first of equations (5.1) is obtained. Let us now assume that $k_3 = k_1(1 + \Delta)$ where Δ is in the interval $(-\Delta_0, \Delta_0)$, $\Delta_0 \ll 1$.

We shall calculate the mean value of f_1 assuming for simplicity that all Δ's in the above-mentioned interval have the same probability:

$$\overline{f_1} = \frac{1}{2\Delta_0} \int_{-\Delta_0}^{\Delta_0} f_1(\Delta)\, d\Delta \tag{5.6}$$

We find that, if $|\hbar\omega - \Delta E|/kT \gg 1$, $\overline{f_1}$ has the same form as in the first of equations (5.1); however, instead of m_1 we have

$$m_{1a} = m_1(1 + 2\Delta_0) \tag{5.7}$$

We have ascribed the factor to m_1 and left m_3 unchanged; this (nonessential) assumption seems plausible in view of the fact that band 3 lies deeper than band 1. A similar reasoning regarding the transition $2 \rightarrow 3$ gives

$$m_{2a} = m_2(1 - 2\Delta_0) \tag{5.8}$$

Table I shows that the difference between the effective masses m_{1a}, m_{2a} and m_1, m_2 observed in amorphous and crystalline Ge, respectively, gives 10^{-1} as the order of magnitude for $2\Delta_0$. In the spectral region considered $k < 10^7$ cm^{-1} and $\Delta k < 10^6$ cm^{-1}. The volume corresponding to Δk in the reciprocal space $B_v = 4\pi/3 \cdot (\Delta k)^3$ is 4×10^{18} cm^{-3}.

How can this result be understood? The theories proposed for the electronic structure of an amorphous solid by Gubanov and others [7,8] consider the difference between the amorphous structure and the crystalline structure with the same short-range order as a perturbation. Assuming that the perturbation of the effective potential in which the electron moves is sufficiently small, they construct the wave functions of an electron (or hole) in an amorphous solid as linear combinations of the Bloch functions in the same band of the corresponding crystal. As during the transitions between the Bloch functions the k-vector is conserved; B_v determined above gives an estimate of the spread of the k vectors of the Bloch functions entering into the linear combinations. If we assume that the wave functions in band 3 are exact Bloch functions, B_v characterized the wave functions in bands 1 or 2 of amorphous Ge.

B_v is a very small part of the reduced Brillouin zone

$$B = (2\pi)^3 \cdot 4/a^3 = 5.6 \times 10^{24} \text{ cm}^{-3}$$

of the crystal: $B_v \sim 10^{-6} B$. This result can be stated in an alternative way using the uncertainty relation $B_v V_v \sim (2\pi)^3$ where V_v is the volume in which the wave functions are localized. In our case we obtain $V_v \sim 10^{-17}$ cm^3. That means that the valence-band wave functions are localized over many millions of atoms; they are therefore practically delocalized and reasonably well described by Bloch functions.

IV. TAIL OF THE ABSORPTION EDGE

The tail near the absorption edge can be separated if an extrapolation is applied to the absorption edge as shown in Fig. 6. In Fig. 5 we see that the tail has a Gaussian shape and the integrated absorption $\int \omega \varepsilon_2 \, d\omega$ increases with temperature; at 300°K it is 1.45 times larger than at 150°K. The tail is not a transition from or into the valence or conduction bands because we would expect in such a case an edge rather than a band and therefore it does not give us information about the distortion of the band edges. It appears likely that the tail is related to the presence of localized defects.

We have suggested that this tail is caused by excitons bound to neutral acceptors which are presumably the vacancies mentioned above. From the temperature dependence of the absorption we have found that their activation energy must be larger than 0.01 eV [5].

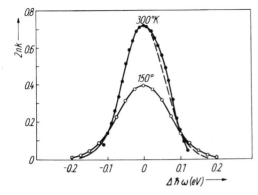

Fig. 5. Temperature dependence of the absorption in the tail. $\Delta\hbar\omega = \hbar\omega - \hbar\omega_m$, where $\hbar\omega_m$ is the maximum of the band ($\hbar\omega_m = E_g - 0.02$ eV) (after [4]).

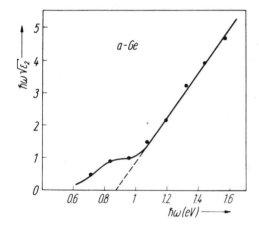

Fig. 6. Absorption edge (after [4]).

Imagine that the following process takes place: An acceptor captures an electron by thermal excitation and the photon excites this electron into a state in which the acceptor binds an electron and a hole. The extent to which such a process occurs should be proportional to the product of the concentration of ionized acceptors n_A and the concentration of holes p.

Let us consider the simplest possibility. In the material there are N_A acceptors/cm^3 with an activation energy E_A and N_D donors/cm^3 with energy levels many kT above the energy level of acceptors; one has then $n_A = N_D + p$. In such a case the de Boer–van Geel formula gives

$$pn_A = (N_A - n_A)\frac{N_v}{\beta}\exp\left(-\frac{E_A}{kT}\right) \qquad (5.9)$$

where N_v is the effective density of states in the valence band, and β is the spin degeneracy of acceptors. In the case that $n_A/N_A \ll 1$ (or $p \ll N_A - N_D$) we obtain from the above, given the ratio 1.45 of integrated absorptions at 300 and 150 °K, $E_A = 0.01$ eV. Of course, we have no evidence that only a small number of acceptors is ionized. It may well be that the compensation is high and n_A/N_A may not be $\ll 1$. In such a case, E_A would be greater than 0.01 eV. Detailed measurements of temperature dependence of the tail are necessary to obtain definite information about the parameters of the statistics.

The suggested explanation of this tail as caused by excitons bound to defects is in principle similar to the explanation of similar tails observed near the fundamental absorption edges of ionic crystals (α and β bands [9]).

V. ABSORPTION EDGE

In Fig. 6, $\hbar\omega \sqrt{\varepsilon_2}$ is plotted against $\hbar\omega$ near the absorption edge for one of the a-Ge samples. It is apparent that the absorption in the tail is followed by an absorption which is represented by a straight line showing that in this region $\omega^2 \varepsilon_2 \sim (\hbar\omega - E_g)^2$. If we extrapolate this line, we obtain at 300 °K $E_g = 0.88$ eV and $E_g = 0.92$ eV for a-Ge and a_t-Ge, respectively.

The dependence of ε_2 on $\hbar\omega$ near the absorption edge suggests that one has here transitions for which only energy but not the $\mathbf{k}$ vector is conserved (indirect transitions). We shall show that such behavior is expected if the wave functions are linear combinations of the Bloch functions as discussed in section III. We shall develop a theory of the optical absorption valid for such kinds of wave functions and compare it with the experimental results. We shall first state more explicitly our basic assumptions.

The wave functions in the valence and conduction bands of the crystal are Bloch functions:

$$|v\mathbf{k}\rangle = V^{-\frac{1}{2}} u_{v\mathbf{k}}(\mathbf{r}) e^{i\mathbf{k}\cdot\mathbf{r}}, \ |c\mathbf{k}'\rangle = V^{-\frac{1}{2}} u_{c\mathbf{k}'}(\mathbf{r}) e^{i\mathbf{k}'\cdot\mathbf{r}} \tag{5.10}$$

normalized in the basic volume V; $\mathbf{k}$, $\mathbf{k}'$ are in the first Brillouin zone.

We assume that the basic volume V contains the same number of atoms in the amorphous as in the crystalline state (for simplicity we neglect the difference of densities of both states).

It is known from both experiment and theory that the valence and conduction bands retain their meaning in the amorphous state. We shall assume that the perturbation describing the change of the crystalline into the amorphous state is such that the wave functions in the valence band of the amorphous state are in zero-approximation linear combinations of the wave functions in the valence band of the crystal; an analogous statement holds for the conduction band. Therefore, the ground state described by a Slater determinant is the same as in the crystal. If we now produce an electron–hole pair in the crystal with wave functions $|v\varkappa >, |c\varkappa' >$ and then apply the perturbation we obtain the wave functions in the amorphous solid. We can label these wave functions with indices $\varkappa, \varkappa'$ and denote them by $|v\varkappa >_a, |c\varkappa' >_a$. As the perturbation is performed in a continuous spectrum we can assume that the energies of the states $|v\varkappa >_a$ and $|v\varkappa >$ measured from the top of the valence band are the same (similarly for the conduction band). The consequence of these assumptions is that the corresponding densities of states are the same in both cases.

The calculation of the absorption in an amorphous solid is based on the general one-electron formula

$$\varepsilon_2(\omega) = \left(\frac{2\pi e}{m\omega}\right)^2 \frac{1}{V} \sum_{i,f} |\mathbf{P}_{if}|^2 \, \delta(E_f - E_i - \hbar\omega) \tag{5.11}$$

where the summation is performed over all initial and final states i, f in the basic volume V. $\mathbf{P}_{if}$ is the momentum matrix element between the wave functions of the final and initial states $|c\varkappa' >_a$ and $|v\varkappa >_a$. Expressing them as linear combinations of Bloch functions in the respective bands, we obtain

$$\mathbf{P}_{if} = {}_a\langle c\varkappa' |\mathbf{p}| v\varkappa \rangle_a = \sum_{\mathbf{k}} \sum_{\mathbf{k}'} {}_a\langle c\varkappa'|c\mathbf{k}' \rangle \langle v\mathbf{k}|v\varkappa \rangle_a \langle c\mathbf{k}' |\mathbf{p}| v\mathbf{k} \rangle$$

$$= \sum_{\mathbf{k}} {}_a\langle c\varkappa'|c\mathbf{k} \rangle \langle v\mathbf{k}|v\varkappa \rangle_a \, \mathbf{p}_{vc}(\mathbf{k}) \tag{5.12}$$

where

$$\mathbf{p}_{vc}(\mathbf{k}) = -i\hbar \frac{1}{\Omega} \int_{\text{cell}} d^3\mathbf{r} \, u_{v\mathbf{k}}^*(\mathbf{r}) \, \text{grad} \, u_{c\mathbf{k}}(\mathbf{r}) \tag{5.13}$$

is the matrix element in the crystal. The summations are performed over all $\mathbf{k}$ vectors in the first Brillouin zone. Ω is the volume of the elementary cell over which the integration is performed.

From equation (5.4) we obtain

$$\varepsilon_2(\omega) = \left(\frac{2\pi e}{m\omega}\right)^2 \frac{2}{V} \sum_{\varkappa} \sum_{\varkappa'} |\sum_{\mathbf{k}} \mathbf{p}_{vc}(\mathbf{k}) \, {}_a\langle c\varkappa'|c\mathbf{k} \rangle \langle v\mathbf{k}|v\varkappa \rangle_a|^2$$

$$\cdot \, \delta \, [E_c(\varkappa') - E_v(\varkappa) - \hbar\omega] \tag{5.14}$$

The summations are taken over all $\varkappa, \varkappa'$ in the first Brillouin zone. The factor 2 respects the doubling of the number of states by spin which is conserved during the transition.

In the crystal we have

$$\langle c\varkappa'|ck\rangle = \delta_{k\varkappa'} \qquad \langle v\mathbf{k}|v\varkappa\rangle = \delta_{k\varkappa}$$

Summing over $\mathbf{k}$ and $\varkappa'$ and replacing then $\sum\limits_{\varkappa}$ by $V/(2\pi)^3 \cdot \int\limits_B d^3\varkappa$ we obtain the usual formula for direct transitions:

$$\varepsilon_2(\omega) = \left(\frac{2\pi e}{m\omega}\right)^2 \frac{2}{(2\pi)^3} \int_B d^3\varkappa \, |\mathbf{p}_{vc}(\varkappa)|^2 \, \delta\left[E_c(\varkappa) - E_v(\varkappa) - \hbar\omega\right] \quad (5.15)$$

Let us now simplify equation (5.14) by an assumption, usually made in the theory of the optical properties of crystals, that $|\mathbf{p}_{vc}(\mathbf{k})|$ is a constant. We put

$$f(\varkappa, \varkappa') = N \left|\sum_{\mathbf{k}} {}_a\langle c\varkappa'|ck\rangle \langle v\mathbf{k}|v\varkappa\rangle_a\right|^2 \tag{5.16}$$

where $N = V/\Omega$ is the number of unit cells in the basic volume V. In a crystal $f(\varkappa, \varkappa') = N\delta_{\varkappa\varkappa'}$. As it is seen from equation (5.16), in an amorphous solid, $f(\varkappa, \varkappa')$ will have a finite maximum for $\varkappa = \varkappa'$ and a certain nonzero width. The double sum $\sum\limits_{\varkappa}\sum\limits_{\varkappa'} f(\varkappa, \varkappa')$ is independent of the actual form of $|v\varkappa>_a, |c\varkappa'>_a$ if these wave functions are constructed as described above and $|\mathbf{p}_{vc}|$ is a constant. In the crystal it gives

$$\sum_{\varkappa}\sum_{\varkappa'} f(\varkappa, \varkappa') = N \sum_{\varkappa}\sum_{\varkappa'} \delta_{\varkappa\varkappa'} = N \sum_{\varkappa} 1 = N^2 \tag{5.17}$$

Therefore, the following sum rule holds

$$\int_B d^3\varkappa \, d^3\varkappa' \, f(\varkappa, \varkappa') = B^2 \tag{5.18}$$

where B is the volume in the reciprocal space of the Brillouin zone. The existence of a sum rule for $f(\varkappa, \varkappa')$ is expected from intuitive reasoning. This sum rule replaces the δ function by a function with a finite width which must obey the same normalization equation.

With the function $f(\varkappa, \varkappa')$ equation (5.14) can be written as

$$\varepsilon_2(\omega) = \left(\frac{2\pi e}{m\omega}\right)^2 \frac{2}{(2\pi)^3} \frac{1}{B} |\mathbf{p}_{vc}|^2 \int_B d^3\varkappa \, d^3\varkappa' \, f(\varkappa, \varkappa') \, \delta\left[E_c(\varkappa') - E_v(\varkappa) - \hbar\omega\right]$$

$$(5.19)$$

We shall now discuss the particularly simple case that near $\varkappa = \varkappa' = 0$ the factor $f(\varkappa, \varkappa')$ is constant, equal to f_0 in a certain part of the Brillouin zone B_c, and zero in the rest. From the sum rule equation (5.18) we determine

f_0 to be B/B_c. In this case in the range considered equation (5.19) gives

$$\varepsilon_2(\omega) = \left(\frac{2\pi e}{m\omega}\right)^2 \frac{2}{(2\pi)^3} \frac{1}{B_c} |p_{vc}|^2 \int d^3\varkappa \, d^3\varkappa' \, \delta \, [E_c(\varkappa') - E_v(\varkappa) - \hbar\omega]$$

$$= \left(\frac{2\pi e}{m\omega}\right)^2 \frac{2}{(2\pi)^3} \frac{1}{B_c} |p_{vc}|^2 \int dE \int d^3\varkappa' \, \delta \, [E_c(\varkappa') - E] \int d^3\varkappa \, \delta \, [E - E_v(\varkappa) - \hbar\omega]$$

$$= \left(\frac{2\pi e}{m\omega}\right)^2 \frac{2}{(2\pi)^3} \frac{1}{B_c} |p_{vc}|^2 \int dE g_c(E) \, g_v(\hbar\omega - E) \qquad (5.20)$$

It is apparent that $\varepsilon_2(\omega)$ is determined by a convolution of the densities of states in the valence and conduction bands $g_v(E)$ and $g_c(E)$ for which energy is conserved. If the energies are measured from the band extrema we have

$$g_v(E_p) \sim E_p^{\frac{1}{2}} \quad g_c(E_n) \sim E_n^{\frac{1}{2}} \quad E_n + E_p = \hbar\omega - E_g$$

We obtain then from equation (5.20)

$$\omega^2 \varepsilon_2(\omega) \sim (\hbar\omega - E_g)^2 \qquad (5.21)$$

in accord with observation.

The suggested approach allows a rough estimate of B_c because $|p_{vc}|$ and g_v, g_c are crystalline quantities and are known in Ge. We assume that the edge observed in amorphous Ge corresponds to $\Gamma_2' \to \Gamma_{25'}$ direct transition in the crystal. We calculate the densities of states with the effective masses $m_p = 0.3$ m and $m_n = 0.04$ m, and we take $p_{vc} = 7 \times 10^{-20}$ g-cm/sec [10]. The slope of the straight line in Fig. 6 is $d\hbar\omega \sqrt{\varepsilon_2}/d\hbar\omega = 6.5$. With these values we obtain from equation (5.20) $B_c \sim 10^{21}$ cm$^{-3} \sim 2 \times 10^{-4} B$.

As $B_c \gg B_v$, B_c gives the reciprocal volume of $\mathbf{k}$ vectors from which linear combinations of the conduction band wave functions in amorphous Ge are formed. Using the uncertainty relation, we find that these combinations describe wave functions localized in real space in a volume

$$V_c = (2\pi)^3/B_c \sim 2 \times 10^{-19} \text{ cm}^3$$

which contains several thousand atoms.

It is seen that the conduction band wave functions are much more localized than the valence band functions; in linear dimensions, the latter extend over at least hundreds of atoms, the former only over a few tens of atoms.

VI. MAIN ABSORPTION BAND

At photon energies larger than 1.8 eV we could measure only the reflectivity. In Fig. 7 the reflectivity of amorphous Ge is compared with that of single-crystal Ge and of polycrystalline layers obtained from amorphous layers by annealing at 450 °C. We see that the sharp structure observed in crystalline material is absent in amorphous Ge. This is in accord with the explanation of sharp structure as due to the long-range order. The valence band plasma region (above 8 eV) seems to be the same as in the crystal.

We attempted to calculate the optical constants from this curve by using the dispersion relations [11], and obtained $\varepsilon_1(\omega)$, $\varepsilon_2(\omega)$ curves published elsewhere [3]. However, it was realized later that the integral

$$\int \omega \varepsilon_2(\omega)\, d\omega$$

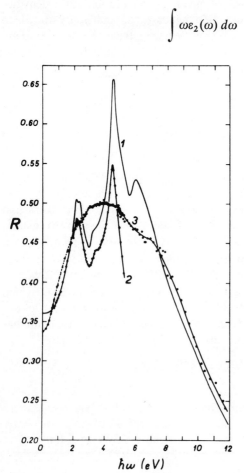

Fig. 7. Reflectivity of amorphous Ge (curve 3) compared with the reflectivity of single crystals (curve 1) and polycrystalline layers (curve 2) (after [3]).

over the whole spectrum is roughly one-half that for the crystal although one expects it should be the same. This is due to the absence in amorphous Ge of the peak near 4.5 eV which is dominant in crystalline Ge. Before entering into an analysis of the spectrum in the fundamental absorption region it should be made clear that this strange feature of the spectrum of amorphous Ge is not due to some experimental error. It is interesting to note that, in a recent paper of Zeppelfeld and Raether [14] on characteristic energy losses of electrons, the peak near 4.5 eV is observed in crystalline Ge and Si, but not in the amorphous state in accordance with preliminary optical results.

VII. CONCLUDING REMARKS

By studying the optical properties of a particularly simple amorphous material we have obtained information about the wave functions near the band extrema. It is hoped that the basic features of this simple analysis may be applicable also to other amorphous materials; however, results reported by Stuke on amorphous Se and Te [12] and preliminary results of Grigorovici on amorphous Si [13] indicate that the situation is probably more complicated. An essential point in our analysis is the assumption that the perturbation caused by the transition to the amorphous state is small; this may not always be true.

There are also many problems connected with the electronic structure which we have not mentioned here. One is the presence of localized states inside the gap. We can only state that our spectra appear not to reveal any presence of such states. Absorption at these states (perhaps because of their small concentration) may be very small and overlapped by other absorptions. Another interesting problem is the change of the gap during the transition from the crystalline to the amorphous state; Keller and Stuke [12] discussed some features of this problem but work on a larger variety of amorphous semiconductors is needed before the situation can be properly understood.

VIII. REFERENCES

1. H. Richter and G. Breitling, *Z. Naturforsch.* **13a**:988 (1958).
2. R. Grigorovici, N. Croitoru, A. Dévényi, and E. Teleman, *Proc. Seventh Conf. Phys. Semicond., Paris, 1964,* Dunod (Paris), 1964, p. 423.
3. J. Tauc, A. Abraham, L. Pajasova, R. Grigorovici, and A. Vancu, *Proc. Intern. Conf. Phys. of Noncrystalline Solids, Delft, 1964,* North-Holland Publishing Company (Amsterdam), 1965, p. 606.
4. J. Tauc, R. Grigorovici, and A. Vancu, *Phys. Status Solidi* **15**:627 (1966).
5. J. Tauc, R. Grigorovici, and A. Vancu, *J. Phys. Soc. Japan (Suppl.)* **21**: 123 (1966).
6. F. R. Kessler, *Phys. Status Solidi* **5**:3 (1964); **6**:3 (1964).

7. A. I. Gubanov, *Quantum Electron Theory of Amorphous Conductors*, Consultants Bureau (New York), 1965.
8. K. Moorjami and C. Feldman, *Rev. Mod. Phys.* **36**:1042 (1964).
9. F. Seitz, *Rev. Mod. Phys.* **26**:7 (1954).
10. M. Cardona, *J. Phys. Chem. Solids* **24**:1543 (1963).
11. J. Tauc, *Progr. Semicond.* **9**:87 (1965).
12. H. Keller and J. Stuke, *Phys. Status Solidi* **8**:831 (1965).
13. R. Grigorovici and A. Vancu (to be published).
14. K. Zeppelfeld and H. Raether, *Z. Physik* **193**:471 (1966).

CHAPTER 6

Optical Constants of Insulators: Dispersion Relations

Manuel Cardona*

Physics Department
Brown University
Providence, Rhode Island

I. INTRODUCTION

The optical behavior of an optically isotropic solid (e.g., a cubic crystal) is determined by the spectral dependence of two parameters: the real and the imaginary part of the refractive index $n = n_r + in_i$ (n_i is usually referred to as k in the literature and called the extinction index), or the real and the imaginary part of the dielectric constant $\varepsilon = \varepsilon_r + i\varepsilon_i$. The two spectral functions which determine the optical behavior are most readily determined by measuring the transmission and the reflection of a plane-parallel slab as a function of frequency [1]. In the region of interband transitions, however, the absorption coefficient reaches very large values (10^5–10^6 cm^{-1}) and the preparation of single-crystal samples thin enough for transmission measurements becomes extremely difficult [2]. Because of the large number of imperfections associated with vacuum-deposited samples, thin films prepared by this method are not very trustworthy for optical measurements, although progress has been made recently by using epitaxial deposition methods [3]. Absorption in the substrate or film backing can also become a problem, especially in the far ultraviolet region, and therefore techniques based exclusively on reflection measurements have been most widely used for the determination of optical constants in the region of electronic interband transitions of metals, semiconductors, and insulators. The same considerations apply to the intraband (or free-carrier) absorption in metals [4], since the corresponding absorption coefficients are also very high.

The reflection techniques fall into two categories. In the first, two independent measurements are performed at a given wavelength; these enable us to determine the two optical constants at that wavelength. An almost

---

* Alfred P. Sloan Fellow.

unlimited number of such measurements can be devised, some of them discussed by Potter in Chapter 16. One can, for instance, measure the reflectivity under oblique incidence for the two normal modes of linear polarization [5]. One can also measure under oblique incidence the ellipticity and orientation of the axes of the polarization ellipse for light linearly polarized at 45° with the plane of incidence [6]. We shall not deal with these techniques in detail in this chapter.

The second category of reflection techniques involves the performance of one optical measurement over the whole spectral region, from 0 to ∞ frequency. The two optical constants can then be determined because of the existence of integral relationships between them, the so-called Kramers–Kronig or dispersion relations [7, 8]. In practice, because of instrumental limitations, measurements from 0 to ∞ frequency are not possible. Measurements are therefore performed over as broad a frequency range as possible and the data are then judiciously extrapolated to zero and to infinite wavelengths.

II. DISPERSION RELATIONS FOR THE DIELECTRIC CONSTANT

The existence of a general relationship between the ε_i and ε_r is a direct consequence of causality: $\varepsilon - 1$ represents the proportionality factor between the *cause* (electric field) and the *effect* (polarization). Under no circumstances can a polarization $\mathbf{P}$ occur *before* the electric field $\mathbf{E}$ is applied, or if

$$\mathbf{E}(t) = 0 \qquad \text{for } t < t_0$$

then also

$$\mathbf{P}(t) = 0 \qquad \text{for } t < t_0$$

We assume that ε is not an explicit function of t, and hence we can always take $t_0 = 0$.

If causality is to hold, $\varepsilon_i(\omega)$ and $\varepsilon_r(\omega)$ (ω is the angular frequency) cannot be chosen independently of each other. This is best illustrated by the following argument by Toll [9]:

Let us consider a field $\mathbf{E}(t)$ as shown in Fig. 1(a): for $t < 0$, $\mathbf{E}(t) = 0$. Figure 1(b) shows the Fourier component $E(\omega_1)$ of $\mathbf{E}(t)$. Let us assume a dielectric constant ε such that $\mathbf{P}(t) = 0$ for $t < 0$. Let us now change the imaginary part of the dielectric constant by removing *all* the oscillators of frequency ω_1 without changing the dielectric constant at other frequencies, i.e., we make $\varepsilon_i(\omega_1) = 0$. Since for $t < 0$ all of the Fourier components of $\mathbf{P}$ cancelled, the elimination of $\alpha E(\omega_1) \sin \omega_1 t$ from $\mathbf{P}$ will introduce a polarization $-\alpha E(\omega_1) \sin \omega_1 t$ for $t < 0$. Hence, if causality is to be preserved, the phases of the other Fourier components will have to change so as to

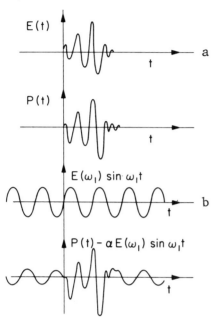

Fig. 1. Illustration of the necessity of dispersion relations between the real and imaginary parts of a linear causal response function (see reference 9).

cancel $-\alpha E(\omega_1) \sin \omega_1 t$ and keep $P(t) = 0$ for $t < 0$. Herein lies the origin of the dispersion relations.

The causal nature of ε can be put on a slightly more symmetric basis than that above. The quantity ε (not $\varepsilon - 1$) can be viewed as the relationship between the normal components of the displacements on both sides of a vacuum–insulator (of dielectric constant ε) boundary. The normal component $D_\perp$ of the displacement $\mathbf{D}$ in the insulator is

$$D_\perp = \varepsilon E_\perp$$

where $E_\perp$ is the normal component of the field (or displacement) in vacuum. For a wave propagating from vacuum to the insulator no displacement $D_\perp$ can occur before $E_\perp$ is applied. Similarly, $1/\varepsilon$ represents the causal relationship between the $D_\perp$ inside and the displacement $E_\perp$ outside: for a wave propagating from the insulator to vacuum no displacement can occur in vacuum before a displacement $D_\perp$ has appeared in the insulator. Therefore, we would expect that any relationships derived from *causality* for ε should also hold for $1/\varepsilon$. Henceforth no explicit vector notation will be used; it is clear that the relations obtained apply to components of $\mathbf{E}$ and $\mathbf{D}$.

Let us consider an arbitrary field which vanishes for $t < 0$, $E(t)$. Let us call $T(t - t')$ the displacement produced at a time t by a δ-function pulse of field applied at time t' (we obviously assume the homogeneity of time, that is, no *explicit* time dependence in ε). The displacement produced by $E(t)$ is then

$$D(t) = \frac{1}{\sqrt{2\pi}} \int_{-\infty}^{+\infty} T(t-t')\, E(t')\, dt' \qquad (6.1)$$

The causality condition, namely, that regardless of the shape of $E(t)$ with $E(t) = 0$ for $t < 0$, $D(t) = 0$ for $t < 0$, is equivalent to $T(t-t') = 0$ for $t < t'$, since there can be no causal response to a δ-function pulse until the pulse is applied.

Let us now assume that $T(t)$ *has* a Fourier transform:

$$\varepsilon(\omega_r) = \frac{1}{\sqrt{2\pi}} \int_{-\infty}^{+\infty} e^{i\omega_r t}\, T(t)\, dt \qquad (6.2)$$

For a system composed of a collection of damped harmonic oscillators (and any material in thermal equilibrium can be decomposed into a summation of damped harmonic oscillators for the purpose of computing ε), $\varepsilon(\omega) - 1$ is the sum of a number of terms of the form

$$\frac{4\pi e^2 N}{m} \frac{1}{\omega_o^2 - \omega^2 - i\omega_c\,\omega} = \frac{4\pi e^2 N}{m} \frac{\omega_o^2 - \omega^2 + i\omega_c\,\omega}{(\omega_o^2 - \omega^2)^2 + \omega^2\,\omega_c^2} \qquad (6.3)$$

where ω_o is the oscillator frequency, $\omega_c > 0$ is the collision or damping frequency, N the density of oscillators, and m their mass. This $\varepsilon(\omega)$ is a well-behaved function for ω real and it can be analytically continued to complex values of ω: the function defined for complex ω as a sum of terms like equation (6.3) is analytic except at the poles:

$$\omega_o^2 - \omega^2 = i\omega_c\,\omega \qquad (6.4)$$

For $\omega = \omega_r + i\omega_i$, equation (6.4) becomes

$$\omega_o^2 - \omega_r^2 + \omega_i^2 = -\omega_c\,\omega_i$$

$$-2\omega_i\,\omega_r = \omega_c\,\omega_r$$

($\omega_r = 0$ is a solution only if $\omega_o = 0$; this case will be treated later). Hence at the poles $\omega_i = -\omega_c/2 < 0$ and the dielectric constant has poles only *below* the real axis. The quantity ε is analytic *above* and on the real axis (provided ω_c and ω_o are not zero; these cases will be treated later). Let us now show that any $\varepsilon(\omega)$ analytic on the real axis and in the upper half plane automatically satisfies the causality requirement. The integral obtained by inverting equation (6.2)

$$T(t) = \frac{1}{\sqrt{2\pi}} \int_{-\infty}^{+\infty} e^{-i\omega_r t}\varepsilon(\omega_r)\, d\omega_r \qquad (6.5)$$

can be calculated for $t < 0$ by drawing the contour in the upper half plane (see Fig. 2). The integral along the half circle tends to zero as the radius tends to infinity and, since there are no poles inside the contour, $T(t) = 0$ for $t < 0$. We have shown that the analyticity of ε in the upper half plane is a sufficient condition for causality to hold. It can be shown [10] that for causality to hold under certain quite general assumptions it is necessary that the continuation of ε be analytic in the *upper half plane*. However, ε can have singular points on the real axis. Since the dielectric constants given by sums of terms like equation (6.3) are always analytic in the upper half plane, we shall not give here the general proof of this condition. No singularities occur on the real axis for a system of *lossy* harmonic oscillators with $\omega_0 \neq 0$. The case of a metal, e.g., of free electrons with $\omega_0 = 0$, will be treated later.

The analyticity in the upper half plane enables us to express ε as the contour integral

$$\varepsilon(\omega_r) - 1 = \frac{1}{2\pi i} \int_c \frac{\varepsilon(\omega') - 1}{\omega' - \omega_r} \, d\omega' \qquad (6.6)$$

around the unbroken contour shown in Fig. 3. The number one has been subtracted from ε so as to ensure that the integral along the half circle vanishes when its radius tends to infinity (for a harmonic-oscillator-type of dielectric constant with *all* ω_0 finite, $\varepsilon - 1 \sim 1/\omega^2$ for $|\omega| \to \infty$ and the integral along the half circle vanishes).

By adding to equation (6.6) the equation

$$0 = \frac{1}{2\pi i} \int_{c'} \frac{\varepsilon(\omega') - 1}{\omega' - \omega_r} \, d\omega'$$

where c' is the broken contour of Fig. 3, we obtain

$$\varepsilon(\omega_r) - 1 = \frac{1}{\pi i} P \int_{-\infty}^{+\infty} \frac{\varepsilon(\omega') - 1}{\omega' - \omega_r} \, d\omega' \qquad (6.7)$$

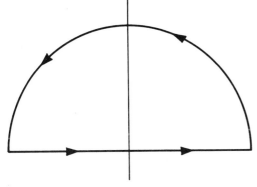

Fig. 2. Integration contour used to prove any response function analytic in the upper half plane is causal.

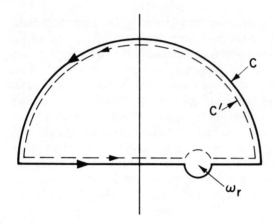

Fig. 3. Integration contours c and c' used to prove equations (6.6) and (6.7).

where P represents the Cauchy principal part. The -1 in the integrand of equation (6.7) can be dropped since

$$P \int_{-\infty}^{+\infty} \frac{1}{\omega' - \omega_r} = 0$$

From equation (6.7) we obtain the dispersion relations for ε_r and ε_i:

$$\varepsilon_r(\omega_r) - 1 = \frac{1}{\pi} P \int_{-\infty}^{+\infty} \frac{\varepsilon_i(\omega')}{\omega' - \omega_r} \, d\omega'$$

$$\varepsilon_i(\omega_r) = \frac{1}{\pi} P \int_{-\infty}^{+\infty} \frac{\varepsilon_r(\omega')}{\omega_r - \omega'} \, d\omega' \tag{6.8}$$

Time-reversal invariance requires $\varepsilon(\omega) = \varepsilon^*(-\omega)$ and hence the integrals in equation (6.8) can be folded into integrals between 0 and $+\infty$. We obtain the following relations by using $\varepsilon_r(\omega) = \varepsilon_r(-\omega)$ and $\varepsilon_i(\omega) = -\varepsilon_i(-\omega)$:

$$\varepsilon_r(\omega_r) - 1 = \frac{2}{\pi} P \int_0^\infty \frac{\omega' \varepsilon_i(\omega')}{\omega^2 - \omega_r^2} \, d\omega'$$

$$\varepsilon_i(\omega_r) = \frac{2\omega_r}{\pi} P \int_0^\infty \frac{\varepsilon_r(\omega')}{\omega_r^2 - \omega'^2} \, d\omega' \tag{6.9}$$

These expressions can be easily generalized to the case $\varepsilon_r(\omega_r) \underset{|\omega| \to \infty}{\to} \varepsilon_\infty \neq 1$ by replacing 1 by ε_∞ in equations (6.9). This case is of interest in the treatment of the lattice vibrations spectrum where one uses for ε_∞ the infrared electronic dielectric constant.

When a magnetic field is present, equation (6.8) can be applied to the dielectric constants for right- and left-circularly polarized light for propagation along the magnetic field B (ε^+ and ε^-). The time-reversal condition, however, becomes $\varepsilon^+(\omega, B) = \varepsilon^{+*}(-\omega, -B)$ and care has to be taken in folding the integrals from $-\infty$ to $+\infty$ to ensure a change in the sign of B. As a result, equations (6.9) *are not correct* [11] for ε^+ and ε^-.

Let us now consider the case in which one of the systems of oscillators has $\omega_o = 0$, that is, the case of a metal. The dispersion relations can then be decomposed into two parts, one containing the "nonpathological" oscillators, analogous to equation (6.8), and another containing $-4\pi Ne^2/m(\omega^2 + i\omega\,\omega_c)$. Since $\omega \cdot \varepsilon(\omega)$ is regular on the upper half plane [9] and on the real axis, and tends to zero like $1/\omega$ for $|\omega| \to \infty$, we can apply to it equations (6.8):

$$\varepsilon_r(\omega_r) - 1 = \frac{1}{\pi\omega_r} P \int_{-\infty}^{+\infty} \frac{\omega'\,\varepsilon_i(\omega')}{\omega' - \omega_r}\, d\omega'$$

$$\varepsilon_i(\omega_r) = \frac{1}{\pi\omega_r} P \int_{-\infty}^{+\infty} \frac{\omega'\,\varepsilon_r(\omega')}{\omega_r - \omega'}\, d\omega \qquad (6.10)$$

or

$$\varepsilon_r(\omega_r) - 1 = \frac{2}{\pi} P \int_{0}^{\infty} \frac{\omega'\,\varepsilon_i(\omega')}{\omega'^2 - \omega_r^2}\, d\omega'$$

$$\varepsilon_i(\omega_r) = \frac{2}{\pi\omega_r} P \int_{0}^{\infty} \frac{\omega'^2\,\varepsilon_r(\omega')}{\omega_r^2 - \omega'^2}\, d\omega' \qquad (6.11)$$

$$= \frac{2\omega_r}{\pi} P \int_{0}^{\infty} \frac{\varepsilon_r(\omega')}{\omega_r^2 - \omega'^2}\, d\omega' + \frac{4\pi\sigma_0}{\omega_r}$$

This last result is readily obtained [8] by subtracting from ε the function $4\pi\sigma_0/\omega = 4\pi Ne^2/m\omega\,\omega_c$ which removes the singularity at $\omega = 0$.

III. DISPERSION RELATIONS FOR THE REFRACTIVE INDEX

It is also possible to derive dispersion relations for $n = n_r + in_i$ although n does not represent a linear causal relationship between two physical quantities. Since $n = \sqrt{\varepsilon}$, n is also analytic in the upper half plane provided ε does not become zero anywhere (branch point of $n = \sqrt{\varepsilon}$). But we stated previously that $1/\varepsilon$ does not have any poles in the upper half plane; hence, n does not have any branch points (zeros) in the upper half plane and we still can apply to it equations (6.8) and (6.9), replacing ε by n, since

$|n-1| \to 1/\omega$ for $|\omega| \to \infty$:

$$n_r(\omega_r) - 1 = \frac{2}{\pi} P \int_0^\infty \frac{\omega' n_i(\omega')}{\omega'^2 - \omega_r^2} \, d\omega'$$

$$n_i(\omega_r) = \frac{2\omega_r}{\pi} P \int_0^\infty \frac{n_r(\omega')}{\omega_r^2 - \omega'^2} \, d\omega' \tag{6.12}$$

These equations are valid provided no *free carriers* are present.

If is easy to see that if n has no poles in the upper half plane, no signal can propagate in the medium at a speed higher than c (the speed of light in vacuum). Hence the dispersion relations of equations (6.12) are an expression of causality in the relativistic sense. Let us consider the integral

$$\dot{E}(t, x) = \int_{-\infty}^{+\infty} E(\omega) \, e^{-i\omega(t - xn/c)} \, d\omega \tag{6.13}$$

which gives the electric field of a plane wave as a function of t and x (the coordinate along the direction of propagation). Let us assume $E(t, 0) = 0$ for $t < 0$, that is, $E(\omega)$ does not have any poles in the upper half plane. Therefore equation (6.13) can be evaluated for $t < x/c$ by using the contour of Fig. 2. Since $n \to 1$ for $|\omega| \to \infty$, the real part of the exponent in equation (6.13) is negative for $|\omega| \to \infty$. Hence, the integral along the half circle vanishes and, since there are no poles inside the contour, equation (6.13) vanishes for $t < x/c$.

IV. DISPERSION RELATIONS FOR THE REFLECTIVITY

Another dispersion relation of interest can be found for the complex amplitude reflectivity. Let us consider the reflectivity at normal incidence (for the sake of simplicity we use the sign which corresponds to the reflectivity for the magnetic field amplitudes)

$$r = r_r + i r_i = \frac{n-1}{n+1} = \rho e^{i\theta} \tag{6.14}$$

where r is an analytic function of ω except, maybe, for $n = n_r + i n_i = -1$ and, therefore, $\varepsilon = 1$. However, n_r is (by definition) positive along the real axis. Since n has no zeros in the upper half plane (branch points), it is not possible to go from $\varepsilon = 1$, $n = 1$ to $\varepsilon = 1$, $n = -1$ without leaving the upper half plane and hence r has no poles in the upper half plane. Since $r \to 0$ for $|\omega| \to \infty$, we can also apply to r_r and r_i the dispersion relations of equation (6.9) replacing $\varepsilon - 1$ by r_r and ε_i by r_i. The fact that no reflected signal can appear before an incident signal is applied (causality) would have led us to the same conclusion.

It is, however, more interesting to obtain dispersion relations between ρ, the magnitude of r obtained in conventional reflection spectra, and the phase shift θ. Since $\log r = \log \rho + i\theta$, we must examine the analyticity of $\log r$ in the upper half plane [12]. We have already shown that r, and thus $\log r$, has no poles in this region. We must also show that $r \gg 0$ (no branch points of $\log r$); $r = 0$ would imply $\varepsilon = 1$ which cannot occur for $\omega_i > 0$ according to equation (6.3). The imaginary parts of ε for $\omega > 0$ are either all positive for $\omega_r > 0$ or all negative for $\omega_r < 0$ and hence they can only cancel for $\omega_r = 0$. For $\omega_r = 0$ the real parts of equation (6.3) are all positive and hence $\varepsilon_r > 1$ for any finite frequency. This conclusion only holds for a *passive* system, that is, a system in its ground state or in thermodynamic equilibrium. If $\omega_c < 0$, that is, if the system is *active*, ε can become one in the upper half plane and the dispersion relations we are about to derive do not hold. For a generalization see reference 9. Equation (6.8) cannot be applied to $\log r$ since $\log \rho \to -\infty$ for $|\omega| \to \infty$ and hence the integral along the upper half circle of our contour does not vanish. We can, however, obtain a suitable dispersion relation by integrating

$$[(1+\omega_r\omega')/(1+\omega'^2)]\,[\log r(\omega')/(\omega'-\omega_r)]$$

along the contours of Fig. 3. The integral along the upper half circle is now zero for $|\omega| \to \infty$ and we obtain (there are poles for $\omega' = i$ and $\omega' = \omega_r$)

$$\pi i \frac{1+\omega_r^2}{1+\omega_r^2} \log r(\omega_r) + 2\pi i \frac{1+i\omega_r}{i+i} \frac{\log r(i)}{(1+i\omega_r)i}$$

$$= P \int_{-\infty}^{+\infty} \frac{1+\omega_r\omega'}{1+\omega'^2} \log r(\omega') \frac{d\omega'}{\omega'-\omega_r}$$

It is easy to see from equation (6.3) that $\log r(i)$ is a real number. From this expression we derive

$$\theta(\omega_r) = \frac{1}{\pi} P \int_{-\infty}^{+\infty} \frac{1+\omega_r\omega'}{1+\omega'^2} \log \rho(\omega') \frac{d\omega'}{\omega'-\omega_r}$$

$$= \frac{2\omega_r}{\pi} P \int_0^\infty \log \rho(\omega') \frac{d\omega'}{\omega'^2-\omega_r^2} \tag{6.15}$$

and

$$\log \rho(\omega_r) = \log \rho(i) + \frac{2}{\pi} P \int_0^\infty \omega' \theta(\omega') \frac{d\omega'}{\omega_r^2-\omega'^2} \tag{6.16}$$

The second of equation (6.15) is used for the determination of optical constants from normal incidence reflection data [13]. In order to remove the $\omega' = \omega_r$ singularity and to calculate this integral with a computer, the integral

$$P \int_0^\infty \log \rho(\omega_r) \frac{d\omega}{\omega'^2 - \omega_r^2}$$

which is identically zero, is subtracted from equation (6.15):

$$\theta(\omega_r) = \frac{2\omega_r}{\pi} P \int_0^\infty \frac{\log \rho(\omega') - \log \rho(\omega_r)}{\omega'^2 - \omega_r^2} \qquad (6.17)$$

This integral is evaluated numerically with a standard polynomial integration formula (e.g., Simpson's rule). The low-frequency limit of experimental data coincides normally with the region in which the optical constants (and hence ρ) become frequency-independent. It is assumed that no frequency variation of ρ occurs below this limit. At high frequencies, ρ becomes small and can be approximated by the result obtained from the asymptotic behavior of ε [equation (6.3)] for $\omega \to \infty$:

$$\varepsilon(\omega) \sim 1 - \frac{2C}{\omega^2} \qquad C \text{ is constant}$$

and

$$\rho(\omega) \sim \frac{C}{\omega^2} \qquad (6.18)$$

C is that quantity which reconciles $\rho(\omega)$ of equation (6.18) with the experimental value of ρ at the high-frequency cutoff ω_M. Equation (6.18) is based on the assumption that the high-frequency cutoff is higher than all relevant oscillator frequencies.

This is not the case in large-band-gap materials (typical cutoff frequencies ω_M correspond to 25-eV photon energies). In such cases equation (6.18) can be modified and the following "*ad hoc*" extrapolation of ρ used:

$$\rho(\omega) \sim \frac{C}{\omega^m} \qquad (6.19)$$

The exponent m can be adjusted to fit directly measured values of ε_i (or the absorption coefficient) near the fundamental absorption edge. In insulators and semiconductors m can be chosen so that the absorption coefficient (and hence ε_i) is zero at frequencies below the fundamental absorption edge.

The integral obtained by substituting equation (6.19) into equation (6.17) (limits ω_M and ∞) cannot be expressed in closed form. It can be evaluated numerically with the aid of the series expansion :

$$P \int_{\omega_M}^{\infty} \frac{\log C\omega'^{-m}-\log \rho(\omega_r)}{\omega'^2-\omega_r^2}\, d\omega' = \frac{1}{2\omega_r} [\log C - \log r(\omega_r)] \log \frac{\omega_M+\omega_r}{\omega_M-\omega_r}$$

$$-\frac{m}{\omega_r}\sum_{l=0}^{\infty} \left(\frac{\omega_r}{\omega_M}\right)^{2l+1} \frac{1+(2l+1)\log \omega_M}{(2l+1)^2} \qquad (6.20)$$

Once the phase shift is evaluated, n_r and n_i are obtained from equation (6.14):

$$n_r = \frac{1-\rho^2}{\rho^2-2\rho\cos\theta+1}$$

$$(6.21)$$

$$n_i = -\frac{[\rho^2(n_r+1)+(n_r-1)]\tan\theta}{1-\rho^2}$$

Figure 4 shows the normal-incidence reflection spectrum of InP. Figure 5 shows the optical constants n_r and n_i obtained from Fig. 4 by the method described above [14].

The analysis of data obtained in recent electroreflectance experiments [15, 16] requires the Kramers–Kronig treatment of normal-incidence reflectivities at the interface between the solid under study and another transparent

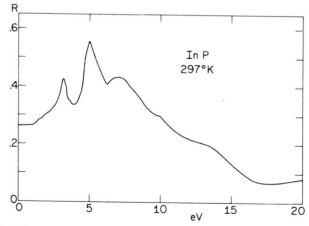

Fig. 4.   Normal-incidence reflection spectrum of InP at room temperature.

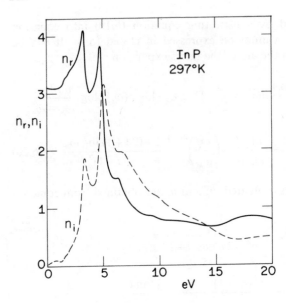

Fig. 5. Optical constants n_r and n_i of InP obtained from the Kramers–Kronig analysis of the data of Fig. 4.

solid or liquid. Equation (6.14) must be replaced by

$$r = \frac{n - n_L}{n + n_L} = \rho e^{i\theta} \tag{6.22}$$

where n_L is the refractive index of the auxiliary solid or liquid. The argument used above to show that r does not have any zeros in the upper half plane does not hold any longer since the equation $n - n_L = 0$ may, in general, have roots for $\omega = \omega_j$ ($j = 1, 2,...$) in the upper half plane. These zeros, and the corresponding singularities in log r, can be removed by multiplying r by the product

$$B(\omega) = \prod_j (\omega - \omega_j^*)/(\omega - \omega_j) \tag{6.23}$$

Functions $B(\omega)$ of the form of equation (6.23) are usually called Blaschke products [8,9]. The function log $[r(\omega) B(\omega)]$ is analytic in the upper half plane (except for $|\omega| \to \infty$) and hence we can apply to it the treatment used in the derivation of equation (6.15). Since the magnitude of the Blaschke product is one, the integrand of equation (6.15) is not modified. We thus obtain instead of equation (6.15):

$$\theta(\omega_r) - i \log B(\omega_r) = \frac{2\omega_r}{\pi} P \int_0^\infty \log \rho(\omega') \frac{d\omega'}{\omega'^2 - \omega_r^2} \tag{6.24}$$

Therefore the determination of $\theta(\omega_r)$ requires the knowledge of the frequencies ω_j at which the zeros of $n - n_L$ occur in the upper half plane.

While the possible appearance of Blaschke products in equation (6.24) cannot be ruled out in general, when the reflection takes place at the interface of two dispersive media, it is possible, however, to show that they are very unlikely to occur whenever the absorption edge of the auxiliary medium occurs much higher in energy than that of the material being studied. Let us assume that the refractive indices n and n_L are due to only one type of oscillators of frequencies ω_o and ω_{oL}, respectively. Under the assumption, quite reasonable in practice, that N and m are roughly the same for both media, the equality of n and n_L implies [equation (6.3)]:

$$\omega_o - \omega^2 - i\omega_c\,\omega = \omega_L^2 - \omega^2 - i\omega_{cL}\,\omega \qquad (6.25)$$

For $\omega = \omega_r + i\omega_i$, separating real and imaginary parts, we obtain

$$\omega_c = \omega_{cL}$$
$$\omega_o = \omega_{oL} \qquad (6.26)$$

Hence if $\omega_{oL} \gg \omega_o$, as in the case of a semiconductor immersed in water, the Blaschke products are not likely to appear in equation (6.24).

Under these conditions equation (6.17) can be used to obtain the variation in the ε_r and ε_i induced by an external modulating agent (an electric field in electroreflectance) from the measured variation in ρ. Equation (6.17) yields, for a small modulation $\Delta\rho$ of ρ,

$$\Delta\theta(\omega_r) = \frac{2\,\omega_r}{\pi}\,P\int_0^\infty \left[\frac{\Delta\rho(\omega')}{\rho(\omega')} - \frac{\Delta\rho(\omega_r)}{\rho(\omega_r)}\right](\omega'^2 - \omega_r^2)^{-1}d\omega'$$

$$\Delta\varepsilon_r = \frac{n_r}{n_L}\,(n_r^2 - n_L^2 - 3\,n_i^2)\,\frac{\Delta\rho}{\rho} + \frac{n_i}{n_L}\,(3\,n_r^2 - n_L^2 - n_i^2)\,\Delta\theta$$

$$\qquad (6.27)$$

$$\Delta\varepsilon_i = \frac{n_i}{n_L}\,(3\,n_r^2 - n_L^2 - n_i^2)\,\frac{\Delta\rho}{\rho} + \frac{n_r}{n_L}\,(3\,n_i + n_L^2 - n_i^2)\,\Delta\theta$$

where n_L is the *real* refractive index of the medium outside the material whose spectrum is being studied. This medium is assumed transparent in the region where $\Delta\rho$ is significant. The quantity n_L equals 1 for air but it is different for the electrolytic techniques recently developed.

Figure 6 shows the electroreflectance spectrum of n-type CdTe obtained by the electrolyte technique. The reflection takes place at the interface between water and the CdTe sample. Figure 7 shows the modulation in ε_r

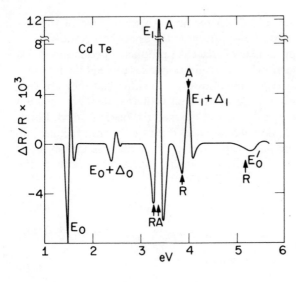

Fig. 6.   Electroreflectance
spectrum of CdTe at room
temperature.

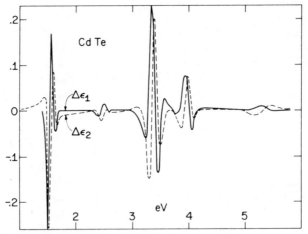

Fig. 7.   Modulations $\Delta\varepsilon_r$ and $\Delta\varepsilon_i$ in the real and imaginary
parts of the dielectric constant induced by an electric field, as
obtained for CdTe from the Kramers–Kronig analysis of the
data of Fig. 6.

and ε_i induced by the electric field, as obtained from the data of Fig. 6 by
using equations (6.27) [17]. It is assumed that $\Delta\rho(\omega) = 0$ for $\omega > 6$ eV. This
assumption is justified by the large energy denominators and the s-like shape
of the peaks in $\Delta R/R$.

V. REFERENCES

1. T. S. Moss, *Optical Properties of Semiconductors*, Butterworth's Scientific Publications (London), 1959.
2. G. Harbeke, *Z. Naturforsch.* **19**:548 (1964).
3. R. B. Schoolar and J. R. Dixon, *Phys. Rev.* **137**:A667 (1965). M. Cardona and D. L. Greenaway, *Phys. Rev.* **133**:A1685 (1964).
4. B. R. Cooper, H. Ehrenreich, and H. R. Phillipp, *Phys. Rev.* **138**:A494 (1965).
5. D. G. Avery, *Proc. Phys. Soc.* **B65**:425 (1952).
6. R. J. Archer, *Phys. Rev.* **110**:354 (1958).
7. H. A. Kramers, *Atti del Congresso Internazionale dei Fisici, Sept. 1927, Como-Pavia-Roma (Nicola Zanichelli, Bologna)* **2**:545 (1928); R. de L. Kronig, *J. Opt. Soc. Am.* **12**:547 (1926).
8. F. Stern, *Solid State Physics, Vol. 15*, in: F. Seitz and D. Turnbull (eds.), Academic Press (New York), 1963, p. 300.
9. J. S. Toll, *Phys. Rev.* **104**:1760 (1956).
10. E. C. Titchmarsh, *Theory of Fourier Integrals*, Clarendon Press (Oxford), 1948 (second edition), pp. 119-128.
11. Y. Nishina, J. Kotodziejczak, and B. Lax, *Phys. Rev. Letters* **9**:55 (1962). J. Kolodziejczak, B. Lax, and Y. Nishina, *Phys. Rev.* **128**:2655 (1962).
12. B. Velicky, *Czech. J. Phys.* **B11**:541 (1961).
13. H. R. Phillipp and E. A. Taft, *Phys. Rev.* **113**:1002 (1959); M. P. Rimmer and D. L. Dexter, *J. Appl. Phys.* **31**:775 (1960); T. S. Robinson, *Proc. Phys. Soc. (London)* **B65**:910 (1952). M. Cardona and D.L. Greenaway, *Phys. Rev.* **133**:A1685 (1964).
14. M. Cardona, *J. Appl. Phys.* **36**:2181 (1965).
15. B. O. Seraphin, "Electroreflectance," in: *Optical Properties of Solids*, S. Nudelman and S. S. Mitra (eds.), Plenum Press (New York), 1968.
16. K. L. Shaklee, F. H. Pollak, and M. Cardona, *Phys. Rev. Letters* **15**:883 (1965).
17. M. Cardona, F.H. Pollak, and K. L. Shaklee, *Phys. Rev.* (to be published).

CHAPTER 7

Electroreflectance

B. O. Seraphin

Michelson Laboratory
China Lake, California

I. INTRODUCTION

In the last two years it has been shown that for a number of insulators, semiconductors, and metals, a strong electric field at or inside the reflecting surface changes the reflectance of a solid in certain narrow ranges of photon energy. One can discuss many aspects of this electroreflectance effect relating it to electro-optics, to surface physics, or to band-structure analysis, to name just a few. This chapter focuses on the contribution of electroreflectance to the determination of the band structure of a solid.

From this viewpoint, one feature of the effect is of particular interest: The sequence of narrow spectral regions in which the electric field causes the reflectance to change most noticeably approximately coincides with the photon energies of interband transition thresholds or, to be more precise, with the photon energies associated with critical points on the interband energy surface in the Brillouin zone. Therefore, if an AC electric field generates the reflectance change, synchronous phase-sensitive detection suppresses the constant background of reflection not affected by the field and amplifies the modulation in the reflected beam resulting from the AC field. This lifts the contribution of a particular critical point out of the large background absorption caused by noncritical areas of the Brillouin zone. Compared to ordinary "static" reflection techniques, this suppression of the background in the "modulated" reflection technique considerably enhances the structure in the optical properties associated with the critical-point spectrum. The correlation of such structure to the spectrum of critical points is one of the major steps in the analysis and the calculation of the band structure of a solid. The replacement of rather poorly resolved structure in the static reflectance by a sequence of pronounced peaks considerably increased the precision and volume of information on which such correlation can be based. These peaks display a characteristic pattern of field shifts and directional and polarization effects. If these features were fully understood, they would probably yield information on type and location of the critical points in the Brillouin zone, in addition to the precise determination of their transition energies established so far.

The fact that the field-induced reflectance change is observed so selectively at only certain photon energies results from the analytic properties of the so-called joint density-of-states function. Representing essentially a measure of the density of the initial and the final states between which the optical interband transition connects, this function possesses an analytic singularity near a critical point. We can visualize this singularity most easily for one type of critical point, the threshold. There will be no optical transition before the photon energy reaches the value of the gap between bands; the joint density-of-states function is zero. Beyond this energy, the absorption probability described by the joint density-of-states function will rise abruptly. If an external parameter is able to modulate this function, it will do so most effectively at the critical point by producing a derivative which is infinite at the analytical singularity that the threshold represents. Although not that easily visualized, a similar singularity is present at critical points of other than the threshold type and can be utilized accordingly. By responding to the modulation in a derivative manner, the contribution of the critical point is lifted from the background of other absorption processes of the same transition energy, which take place in the less interesting areas of the Brillouin zone off the critical points.

There is nothing new, of course, in the superiority of derivative optical techniques in general. For almost a decade, they were used to study transmission at the bottom of the fundamental absorption edge [1], the largest photon energy at which a sufficient amount of light can penetrate through a bulk sample. It was just recently, however, that the importance of the spectral selectivity of a modulation of the joint density-of-states function to reflectance studies was recognized. Operating in reflection rather than in transmission allowed an optical investigation of the region of greatest importance to an analysis of the band structure for photon energies greater than the fundamental edge and made possible much higher resolution than previously obtainable.

The earliest modulated reflectance technique used the electric field as the modulation parameter [2]. It was soon realized, however, that the electric field is just one of a variety of external parameters which affect the density-of-states function and which, if changed periodically, modulate the reflectance in this selective manner. Electroreflectance is thus the oldest member of a growing family of similar modulated reflectance techniques. Its second member is piezoreflectance [3]—modulation by means of a periodic change of stress in the reflecting surface. Just recently, magnetoreflectance [4] was shown to be feasible in conjunction with piezoreflectance. The high resolution of a derivative technique will be common to all this work. The relation, tensorial or nontensorial, between the modulation parameter and the density-of-states function will vary, however, for the different effects, and, in general, similarity of the line shapes cannot be expected. Although tentative analyses are available for both the electroreflectance [5] and piezoreflectance effect [6], no attempt has yet been made to delineate their similarities and differences.

The high spectral resolution of the derivative techniques is only one of their useful properties. Field shifts and anisotropy and polarization effects will probably contribute straightforward information on the symmetry character of a critical point and its location in the Brillouin zone. This type of information can be obtained from static optical measurements only on the basis of circumstantial evidence, while direct experimental criteria are available from the derivative method. In particular, these aspects of the field are presently in the process of rapid progress. As usual in the development of a new technique, the amount of experimental information exceeds the power of the analytical tools available for its analysis. Although the pronounced structure observed by modulated reflectance techniques shows characteristic line shapes, anisotropy and polarization effects, field- and stress-induced shifts, and splittings, we are at present far from a full understanding of their relation to the critical-point spectrum. Early attempts at analysis are encouraging as zero-order approximations but must be considered with caution.

II. EXPERIMENTAL METHOD

Figure 1 shows the experimental setup used for the investigations on semiconductors reported in this chapter [7]. In this method, the electric field at the reflecting surface is produced by utilizing the space-charge layer generally present in the surface of a semiconductor. A DC bias superimposed on the modulating AC field permits the direction and strength of the surface to be adjusted within wide limits. This flexibility is required to exploit fully the information provided by field shifts of the electroreflectance structure. The complete structure is sometimes observed only in a certain range of the DC bias, and satellite structure can be identified. An electrolytic method [8], adapted to electroreflectance studies by Cardona and co-workers [9], produces the electric field at the reflecting surface by utilizing the double layer at the sample–electrolyte interface. Although the penetration depth of the electric field into the sample is typically two orders of magnitude smaller than the penetration depth of the light, it was recently demonstrated that this method is also applicable to metals [10]. The method is rather easy to set up and offers itself to a simultaneous application of stress [11]. It

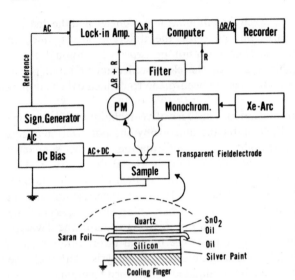

Fig. 1. Block diagram of the experimental setup for the measurement of the electroreflectance effect in silicon. The detail shows a cross section (not to scale) of the sample sandwich.

seems to be restricted to the narrow range of the DC bias between electrolytic deposition and corrosion of the sample, however, and can be used in the temperature range of liquid electrolytes only. Both methods have in common the disadvantage that the space-charge layer as well as the electrolytic double layer generate the modulating electric field normal to the reflecting surface. This implies that polarization experiments are, in general, less conclusive, since the **E** vector of the incident light can only rotate in a plane perpendicular to the direction of the **E** vector of the modulating field.

In the method which uses the electric field in the space-charge layer of a semiconductor, monochromatic light is reflected from a sample sandwich consisting of the reflecting sample surface, a dielectric, and a transparent field electrode. This configuration is similar to the well-known field effect of the surface conductance. The surface potential barrier connected with the space-charge layer extends approximately as far into the sample as the reflected light normally penetrates, therefore exposing it all the way to the electric field. As in the field effect of the surface conductance, direction and average strength of the surface field are adjusted by biasing the field electrode with respect to the sample. Superimposed on this DC bias is the 700-cps modulation voltage which periodically swings the energy bands in the surface region around the DC bias-set point of operation: Photomultiplier and lock-in amplifier detect any synchronous modulation ΔR impressed onto the reflected beam by the modulation of the potential barrier. The normalized modulation depth $\Delta R/R$ is written on a strip-chart recorder as a function of photon energy. It is this normalization of the observed signal which can be accomplished by a variety of experimental tricks that makes electroreflectance work less difficult than a good measurement of the absolute reflectance. The small size of the signal—typically of the order of 10^{-5} for modulation swings of 10^4 V/cm in the surface—seems to make the use of a lock-in amplifier mandatory, however.

III. EXPERIMENTAL RESULTS

Figure 2 demonstrates the spectral selectivity of the electroreflectance effect in the case of germanium [12]. The modulation depth $\Delta R/R$ forms a sequence of sharp spikes on the photon-energy scale at values for which previous band-structure analysis placed critical points. The first spike relates to the fundamental transition, and the second, at 1.09 eV, to this transition's spin-orbit split component [13], which was never seen in reflection before. The big doublet centered at 2.2 eV and separated by only 2/3 of the spin-orbit splitting of the fundamental transition is presumably related to the (A_3-A_1) transition [14]. The structure between 2.8 and 4.0 eV,

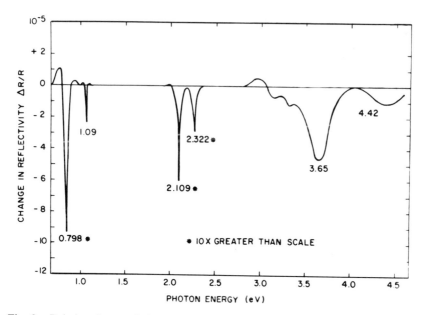

Fig. 2. Relative change of the reflectance, $\Delta R/R$, as a function of photon energy for germanium.

resolved by this technique for the first time, suggests a quadruplet. Pseudo-potential calculations [15] of the band structure place the transition $(\Gamma_{25'}-\Gamma_{15})$ at 3.6 eV. It starts from the same initial level as the fundamental transition in the center of the Brillouin zone, but it has a higher conduction band as the final state. Since both initial and final states are spin-orbit split, this transition would identify itself as a quadruplet in the modulated response. The rather weak response at 4.42 eV is near an extremely strong peak in the reflectance [16]. This indicates that the X and Σ transitions, which were previously being held responsible for this strong reflectance peak, presumably contribute only a fraction, as indicated by the weak derivative response. Theory suggests that the major contribution to the static peak may be associated with a noncritical accumulation of density of states in the Brillouin zone, to which the derivative technique does not respond [17].

Figure 3 shows how the reflectance of silicon, p-type as well as n-type, responds to the electric field in a pronounced structure [18], which was not resolved by previous optical or photoemission studies. At lower temperatures three peaks are observed, which fall in two distinctly different classes, judging from the temperature coefficient of their spectral position. The first peak, moving only slowly with temperature toward larger photon energies, relates apparently to a transition which is "insensitive" to the small changes of the crystal potential resulting from the contraction of the lattice upon cooling. The other two peaks shift with temperature coefficients which are representative for transitions "sensitive" to this change. A similar division of all possible transitions into these two groups, sensitive and insensitive, is found in the calculations of the band structure [19], so that the identification of the electroreflectance peaks is assisted by the precise measurement of the temperature gradients. The disappearance of the middle peak at room temperature suggests its possible association with an exciton [20].

We can only speculate on the reasons for part of the structure inverting its sign upon going from n-type silicon to p-type silicon. The intrinsic

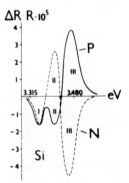

Fig. 3. The field-induced change in the reflectance of n-type and p-type silicon at 215 °K.

optical properties of a semiconductor have, in previous optical experiments, proved insensitive to the doping of the material. A chemically etched silicon surface is n-type, regardless of the conduction type of the material underneath. This indicates the presence of a strong electric field in the inversion layer of a p-type sample and a rather weak field in the accumulation layer of an n-type sample. The complex band structure of silicon in the 3.4-eV region seems to respond differently in the two different cases. Since the edge of the conduction band and the edge of the valence band run very close to parallel over large distances of the Brillouin zone [21], the condition for a critical point—exact parallelism of the two edges—is probably influenced by the strength of the electric field. This would restrict the previous assumption that the existence and the location of a critical point is not affected by small changes in the external parameters.

The near parallelism of the bands in the 3.4-eV region and the resulting cluster of critical points within a narrow range of transition energies is clearly beyond the resolving power of static techniques. This interesting region is presently being investigated in detail by derivative techniques, and may reveal the finer features of this complex band structure [22].

Figure 4 shows the reflectance response as a function of photon energy as induced by an electric field in an n-type sample of gallium arsenide [23]. Two clusters of peaks are seen in the reflectance response, which relate to similar structure in the static reflectance curve, where they are called E_0 and E_1. Each group itself is divided into two subgroups, whose separations follow very precisely the $\frac{2}{3}$ ratio predicted for the spin-orbit splitting in this case. The E_0 group is associated with the fundamental transition at 1.42 eV. It is preceded by a peak which reacts to changes in the temperature or in the modulation field in a manner quite different from the rest of the structure [24].

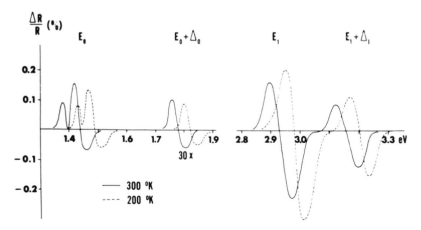

Fig. 4. Electroreflectance structure of GaAs for two different temperatures.

This, its spectral location 40 meV in front of the edge, and the fact that it grows stronger with increasing doping level suggest a connection with an impurity level. Since this peak is observed at the photon energy of injection luminescence and laser emission from gallium arsenide diodes, this particular structure may result from the absorptive counterpart of the emission.

The structure E_1 presumably relates to the $(A_3 - A_1)$ transition [25]. Note that the size of this structure grows upon cooling, in contrast to the fundamental structure, whose size decreases with decreasing temperatures. This is probably the result of two different processes which counteract each other to a different degree for the various parts of the spectrum. First, the modulation swing of the field strength in the surface potential barrier decreases with temperature for a given AC voltage swing. This decreases the signal size. At the same time, however, the lifetime broadening of the electronic states involved in the transition is reduced upon cooling. This reduction is relatively larger at higher photon energies, where the broadening was originally greater because of the more effective scattering. This recovery from lifetime broadening apparently overcompensates the reduction in modulation swing for 3.0-eV photon energy, but not in the fundamental region. Enough parameters are available from such a measurement for an estimate of the Lorentzian broadening parameter. The estimated value is 0.035 eV, about one-third of that previously assumed by theory, lacking any direct experimental information on this parameter. This estimated value is independently confirmed by matching the size of the structure with theoretical calculations.

So far we have focused our attention only on the superior spectral resolution of the electroreflectance technique. In some cases, structure was resolved when static measurements lacked this resolution. Temperature effects, spin-orbit splittings, and lifetime broadening effects can be studied with higher precision. The results, in general, confirmed the knowledge that band-structure analysis had previously derived from static optical measurements.

It was suspected rather early, however, that information on the critical-point spectrum beyond the mere location of the critical point on the scale of photon energies [2] can be derived from the electroreflectance effect. For the full analysis of the band structure, it is necessary to know its location in the Brillouin zone and the topological features of the interband energy surface at the critical point. Essentially two different types of critical points can be expected [26]. For the first type (M_0, M_3), the separation of the bands is at an extremum at the critical point, increasing (M_0) or decreasing (M_3) in whichever direction in k-space one proceeds from the critical point. In line with geometrical classification, such a critical point is called parabolic. For the second type (M_1, M_2), this separation increases in some directions but decreases in others. The interband-energy surface has the symmetry character of a hyperboloid, and the critical point is called the saddle-point type. Since the symmetry properties of the band structure specify certain types of critical points at specific locations of the Brillouin zone, a knowledge of the type derived from experiment complements information on the location.

Three features of the electroreflectance effect in particular are useful for a complete identification of critical points: (1) the spectral shift of the structure occurring when the average electric field in the reflecting surface is changed; (2) orientation effects resulting from the choice of the crystalline orientation of the reflecting surface; and (3) polarization effects observed when the **E** vector of the incident light forms varying angles with some preferred direction (the **E** vector of the modulating electric field or uniaxial stress, to give just two examples). All three features are well documented experimentally. The following sections of this chapter will deal with this aspect of the field.

IV. FIELD SHIFT OF THE STRUCTURE

It was observed early in electroreflectance work that the spectral position of a peak shifts in a systematic manner when the average electric field in the

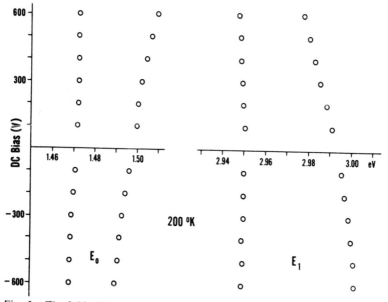

Fig. 5. The field shift of the electroreflectance structure of GaAs, as adjusted in a *p*-type sample at 200 °K by different values of DC bias.

surface potential barrier, adjusted by the DC bias across the sample sand-wich, was changed [7]. This field shift, however, was not in the same direction for all parts of the structure. Some peaks shifted to the blue, others to the red. A typical example of this shift in opposite directions is shown in Fig. 5, which plots the spectral position of the peaks observed in gallium arsenide as a function of the DC bias across the sample sandwich [23, 24]. The structure labelled E_0 and related to the fundamental edge shifts in one direction while the structure in the E_1 group shifts in the opposite direction. The shift of a structure in one direction can be derived from the theory of Franz and Keldysh [27] describing the effect of an electric field on the funda-mental absorption edge, which, as an extremum of the band separation, necessarily represents a critical point of the parabolic type. The absorptive part ε_2 of the complex dielectric constant is, according to this theory, changed by the electric field in a manner shown in Fig. 6 [5]. The period of this oscillation is a function of the electric field, increasing with an increase in the field. Lifetime broadening usually dampens the oscillations, so that only 1–3 peaks are observed experimentally. A change in the electric fields

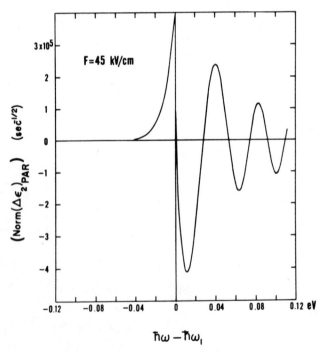

Fig. 6. The normalized change of the imaginary part ε_2 of the dielectric constant, as induced at a parabolic critical point by an electric field of 45 kV/cm.

moves the position of the extrema, and it can be shown that the theory is in reasonably good agreement with the experimentally observed field shift of the fundamental structure [28].

The reason for shifts in both directions became obvious after Phillips expanded the Franz–Keldysh theory in order to include critical points of both the parabolic and the saddle-point type [29]. Phillips demonstrated that under certain assumptions—the electric field being aligned in the principal direction of the interband-energy saddle—the effect of an electric field on such a saddle-point edge could be derived from the similar effect at a parabolic edge in the manner shown in Fig. 7. The oscillatory structure of the field-induced change in ε_2 is now in front of the edge, and a change in the electric field shifts the position of the extrema in the direction opposite to the shift at a parabolic edge. Turning the argument around, the foundation was laid for an experimental criterion which will enable one to derive the type of a critical point from the shift of the associated structure with electric field.

The duality theory illustrated in Fig. 7 forms the basis for the use of the field shift as such an experimental criterion for the identification of critical points. The final formulation of this criterion will be possible only after the theory of the effect is fully developed. More recent calculations

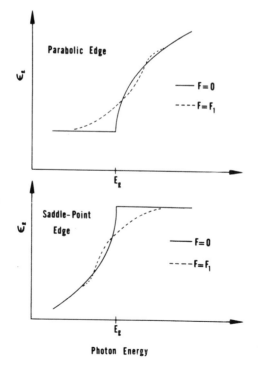

Fig. 7. The field-induced change in ε_2 at a parabolic and a saddle-point edge, according to Phillips' duality theorem.

have shown that the duality theorem is a complete description of the situation only if the electric field is in a direction for which the interband-energy surface has the same curvature as it has in the principal direction [30]. Figure 8 shows a two-dimensional model of the interband-energy surface at a saddle point. If the electric field swings through the direction in which the curvature of this surface is zero, the reduced effective mass inverts sign, the oscillatory structure shown in Fig. 6 inverts its position with respect to the edge, and the field shift goes into the opposite direction. Figure 9 shows a qualitative illustration of the field-induced change in ε_2, if the electric field falls into the angular sections designated $E_{\parallel}$ or $E_{\perp}$ in Fig. 8. An increase in an electric field in the "longitudinal" section dilates the oscillatory structure toward the left and, for an electric field in the "transverse" section, in the opposite direction toward the right. For most critical points outside the center of the Brillouin zone, the interband-energy surface centers around several equivalent locations on different crystal axes. Such equivalent branches will contribute to the field shift in a different manner, depending upon the angle that their principal direction includes with the electric field. For some, the electric field will be in the "longitudinal" cone with the field shift going one way; for others, the electric field will be in the "transverse" cone, shifting in the opposite direction with an increase in the field. The experimentally observed field shift of the structure will result as a super-

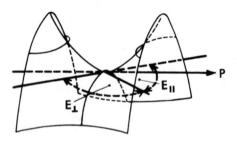

Fig. 8. A two-dimensional model of the interband-energy surface at saddle-point edge. P indicates the principal direction; $E_{\parallel}$ and $E_{\perp}$ designate sections of curvature of the same sign.

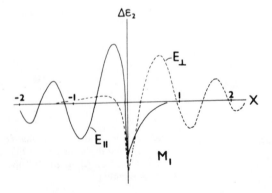

Fig. 9. Schematic diagram of $\Delta\varepsilon_2$ if the electric field falls into the section parallel ($E_{\parallel}$) or transverse ($E_{\perp}$) to the principal direction of a saddle-point edge (after Aspnes [30]).

position of the contributions from the different equivalent branches. No general rule can be expected. With the theoretical treatment completely worked out, however, the usefulness of this field shift is obvious.

V. DIRECTIONAL EFFECTS

The dependence of the size of the reflectance response upon the crystalline orientation of the reflecting surface represents the second helpful aspect in the identification of critical points. It was observed in the case of n-type silicon, for instance, that the first peak in the structure shown in Fig. 3 did not depend upon the direction of the electric field relative to the crystal axes [18]. The second and third peaks, however, changed their magnitude if the direction of the electric field was varied with respect to the crystalline axes by making the reflecting surface coplanar with various crystal planes. This directional dependence, which is shown for the third peak in Fig. 10 [20], supports the assignment of the third peak to a saddle-point edge, for which directional effects can result since a principal axis is oriented in k-space. The lack of directional effects, on the other hand, relates the first peak to a critical point of the parabolic type, for which the coefficients of the interband-energy surface have the same sign in all directions. The simple ratios of the peak size for the different orientations of the reflecting surface probably reflect the topology of the interband-energy surface with respect to the direction of the incident light. These ratios can be explained correctly only if the associated critical point is placed on the proper axis, thereby identifying its location in the Brillouin zone. These directional effects were observed with unpolarized light. Other interesting effects should be observed if polarized light were used. For instance, in germanium and silicon it is believed that peaks in the static reflectance curve are the result of a near degeneracy of two critical points of the saddle-point type which have their

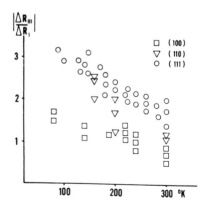

Fig. 10. The absolute value of the ratio of the height of peak III and peak I in silicon as a function of temperature for three different orientations of the reflecting surface.

principal axes aligned in different directions (M_1 and M_2) [31]. Rotation of the plane of polarization will, under these circumstances, vary the contribution of one of these two critical points with respect to the other, possibly identifying such a nearly degenerate complex which is so convenient for the explanation of peaks in the static reflectance.

The full range of information to be gained from the use of polarized light will be available only from sample geometries which do not necessarily align the electric field with the direction of incidence of the reflected light as both variations of the technique presently do by using the space-charge layer inside or at the reflecting surface. Piezoelectroreflectance [11] demonstrates the power of polarization effects in samples under uniaxial stress. In this experiment, not only the peak size varies with the angle between polarization and uniaxial stress, but some peaks are observed to split, which is of value in the analysis of the structure.

VI. ELECTROREFLECTANCE
AND BAND-STRUCTURE ANALYSIS

Although the technique is still in a rather rudimentary state, the existing data on electroreflectance suggest that a large amount of information can be obtained on the electronic structure of solids. Because of the derivative nature of the effect, the results are of high contrast and show a pronounced structure which can be read out with sufficient precision. The information on number and transition energy of critical points already exceeds for some materials the amount of information derived from all other previous optical measurements.

It was mentioned in the preceding sections that the high spectral resolution permits one to observe a characteristic pattern in which the structure responds to changes in size and direction of the electric field, to temperature, crystal orientation, and uniaxial stress. The extent to which this characteristic pattern can be evaluated in terms of type and location of critical points will depend upon a theoretical understanding and a subsequent analysis of the effect.

The basis of such an analysis is provided by the relation between a periodic variation of the electric field and the change of the density-of-states function induced in the neighborhood of critical points by this variation. Since the density-of-states function is proportional to the imaginary part ε_2 of the complex dielectric constant, the reflectance response by which the periodic change of ε_2 is read out will show a definite and significant relation of its phase to the electric field. For instance, it is essential to determine whether an increase in the electric field increases ε_2 or reduces it. Accordingly, the reflectance response will be positive or negative, expressing this fundamental phase relation by the sign of the experimentally observed structure. The significance of the phase relation in terms of absorption leads to a recognition of exciton absorption with respect to interband absorption [7]. Since it is known how exciton absorption responds to the electric field [32], we can, supported by other arguments, rule out the possibility that all structure is caused by exciton effects [33].

The way in which an electric field changes the density of states around a critical point of the parabolic type was extensively studied in the Franz–Keldysh theory. Since the transmission effect could be observed in front of the fundamental absorption edge only, the change of the density of states in the presence of an electric field was never investigated at a saddle-point edge. Stimulated by the observation of the higher interband transitions in the electroreflectance effect, Phillips proved that the effect of the electric field at a saddle-point edge can be derived from the similar effect at a parabolic edge by the simple rotation illustrated in Fig. 6, provided the electric field is parallel to the principal direction of the saddle point.

Only contributions of the change in the density of states to the imaginary part ε_2 of the dielectric constant were considered. However, since real and imaginary parts are connected by the dispersion relation [34] any change $\varDelta \varepsilon_2$ of the imaginary part will induce a related change $\varDelta \varepsilon_1$ in the real part [35, 28]. This contribution will be the decisive one in certain spectral regions. A difficulty of the early analysis was that the well-established Franz–Keldysh change in the absorption coefficient predicts an immeasurably small change in the reflectance only if used in the differential of Fresnel's formula, equation (7.1). This presumably discouraged people from looking for the electroreflectance effect before. If the field-induced change in ε_2 is related by the Kramers–Kronig integral to the real part ε_1, however, a field-induced change $\varDelta \varepsilon_1$ results. This is in satisfactory agreement with the experimental reflectance change, if used in

$$\varDelta R/R = \alpha(\varepsilon_1, \varepsilon_2) \cdot \varDelta \varepsilon_1 + \beta(\varepsilon_1, \varepsilon_2) \cdot \varDelta \varepsilon_2 \qquad (7.1)$$

This equation determines to what extent the two field-induced changes $\varDelta \varepsilon_1$ and $\varDelta \varepsilon_2$ contribute to the change in the reflectance. The coefficients α and β are functions of the photon energy. Their sign and relative magnitude determine the result of the analysis in the different parts of the spectrum. Figure 11 shows α and β as functions of photon energy for the case of silicon in which experimental values of the optical constants are used [36]. There are regions in which $\varDelta \varepsilon_1$ dominates the reflectance response (0 to 3 eV), others in which $\varDelta \varepsilon_2$ dominates (3.5 to 4.0 eV), and regions in which both contribute comparable amounts. The analysis must discuss diagrams of the type of Fig. 11 for every material before assigning experimental structure to critical points.

Now, with the use of the duality theorem, the effect of the electric field on both optical constants ε_1 and ε_2 and at both types of critical points can

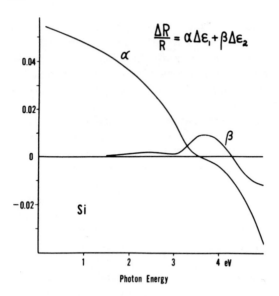

$$\frac{\Delta R}{R} = \alpha \Delta \epsilon_1 + \beta \Delta \epsilon_2$$

Fig. 11. The coefficients α and β in equation (7.1), as calculated from the data in reference 36.

be calculated. Four basic components were obtained, $\Delta\varepsilon_1$ and $\Delta\varepsilon_2$ for each parabolic and saddle-point type of critical point. These four basic components, of which one is shown in Fig. 12 [5], can be normalized so that the spectral position of the critical point enters only as a multiplicative factor. This normalization provides that a mixture of the basic components, appropriate to equation (7.1), can be placed on the scale of photon energy so that the best match with the experimentally observed structure is obtained. Success or failure of this matching procedure will render judgment on the assumption that electroreflectance is or is not a Franz–Keldysh effect observed in reflection at higher interband edges.

Figure 13 shows the result of such a synthesis for the 3.4-eV region in n-type silicon [5]. The dotted line is the experimentally observed structure. The solid line represents the result of an optimal-fit computer calculation. This was obtained by moving the possible combinations of the basic components $\Delta\varepsilon_1$ and $\Delta\varepsilon_2$ along the scale of photon energies, always multiplying with the correct values of α and β. Even on the simple basis of the duality theorem, the synthesis is successful, and we can read out transition energy and type of the critical points which produce this optimal fit. Of course, it is not only the line shape that dictates the selection of critical points. A more powerful guidepost, restricting drastically the number of possible combinations, is the field shift of the structure. Since the experiment shows a shift of the first peak to the right, and the following two peaks to the left with increasing field, only components which reproduce this field shift correctly

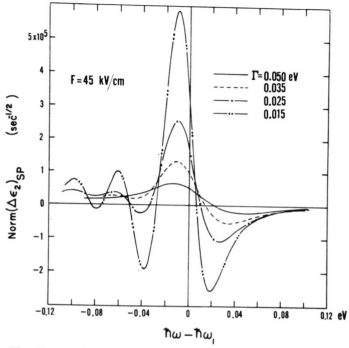

Fig. 12. The normalized change in ε_2 at a saddle-point edge which is exposed to lifetime broadening with different Lorentzian parameters Γ.

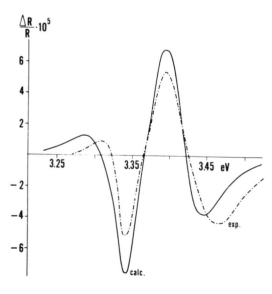

Fig. 13. Comparison between the reflectance response, as experimentally observed in n-type silicon, and the response calculated for a parabolic edge at 3.33 eV and a saddle-point edge at 3.41 eV.

can be used in the synthesis. Figure 14 repeats the solid line of Fig.13, but with the AC field of 15 kV/cm modulating around different points of operation. It is seen that the combination of critical points which leads to a synthesis of the line shape in Fig. 13 also reproduces the field shift properly.

Drawing final conclusions from the particular set of critical points producing the synthesis in Fig. 13 would mean overstressing a procedure which at the present time cannot be more than a zero-order approximation. The duality theorem is a complete description of the situation only for certain directions of the electric field. For critical points with branches at several equivalent sites in the Brillouin zone, the calculation must sum over their individual contributions which differ according to the angle that their principal direction includes with the electric field. In its present state, the theory still includes phase-space integrals left unevaluated, because it is difficult to decide on a cutoff for the integration without being arbitrary. The problem is particularly difficult when several critical points are clustered in a narrow range of transition energies, as seems the case for the 3.4-eV region of silicon.

Although the picture is rather incomplete at the present time, progress is rapid in this field. It can be expected that most of the present shortcomings

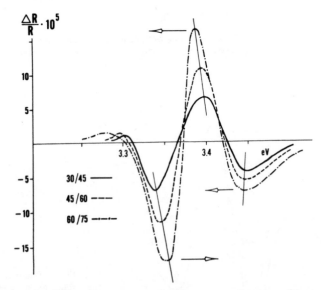

Fig. 14.   Reflectance response of the same arrangement of two critical points as in Fig. 13, calculated for an AC electric field of 15-kV/cm amplitude, modulating around different values of the DC bias, as indicated. Notice that the first peak shifts opposite to the other two, in agreement with experiment.

of the experiment as well as the theoretical interpretation will have improved within the next year or two. Then, modulated reflectance techniques may develop into as powerful a tool in solid-state spectroscopy as magneto-optics and cyclotron resonance.

VII. NOTE ADDED IN PROOF

Since this chapter was prepared, electroreflectance has progressed rapidly. A large number of materials have been investigated including most III–V compounds [37] and some of their alloys [38, 39], II–VI compounds [37, 40–42], rutile [43], and compounds of the type Mg_2Si [44]. Polarization and orientation effects were studied in the electroreflectance spectrum of Ge and Si [45, 46]. Piezoelectroreflectance was further established [47] on Ge, Si, and GaAs. Transverse electroreflectance [48–51] permits configurations $E \parallel F$ and $E \perp F$. Electroreflectance due to free carriers was investigated [52]. The value of electroreflectance for surface studies was established [53]. Magnetoelectroreflectance was further developed [54].

Theory of electroreflectance expanded rapidly. Previous approaches were generalized [55] and new ones added [56, 57]. Electroreflectance at offcenter critical points was investigated [58, 59] and the analytical potential of nonnormal incidence was explored [60]. Line shape effects were discussed [61] and the connection between electroreflectance and electroabsorption established [62, 63]. Magnetoelectroreflectance was studied at saddle points [64].

VIII. REFERENCES

1. T. S. Moss, *J. Appl. Phys.* **32**:2136 (1961); R. Williams, *Phys. Rev.* **126**:442 (1962); J. Lenz and E. Mollwo, *Z. Physik* **176**:536 (1963); M. Chester and P. H. Wendland, *Phys. Rev. Letters* **13**:193 (1964). A. Frova and P. Handler, *Phys. Rev. Letters* **14**:178 (1965).
2. B. O. Seraphin, *Proc. Inter. Conf. Phys. Semicond., Paris, 1964*, Dunod Cie. (Paris), 1964, p. 165; B. O. Seraphin and R. B. Hess, *Phys. Rev. Letters* **14**:138 (1965).
3. W. E. Engeler, H. Fritzsche, M. Garfinkel, and J. J. Tiemann, *Phys. Rev. Letters* **14**:1069 (1965); G. W. Gobeli and E. O. Kane, *Phys. Rev. Letters* **15**:142 (1965).
4. R. L. Aggarwal, L. Rubin, and B. Lax, *Phys. Rev. Letters* **17**:8 (1966).
5. B. O. Seraphin and N. Bottka, *Phys. Rev.* **145**:628 (1966).
6. M. Garfinkel, J. J. Tiemann, and W. E. Engeler, *Phys. Rev.* **148**:695 (1966).
7. B. O. Seraphin, *Phys. Rev.* **140**:A1716 (1965).
8. R. Williams, *Phys. Rev.* **117**:1487 (1960).
9. K. L. Shaklee, F. H. Pollack, and M. Cardona, *Phys. Rev. Letters* **15**:883 (1965); M. Cardona, F. H. Pollack, and K. L. Shaklee, *Phys. Rev. Letters* **16**:644 (1966); K. L. Shaklee, M. Cardona, and F. H. Pollack, *Phys. Rev. Letters* **16**:48 (1966).

10. J. Feinleib, *Phys. Rev. Letters* **16**:1200 (1966).
11. F. H. Pollack, M. Cardona, and K. L. Shaklee, *Phys. Rev. Letters* **16**:942 (1966).
12. B. O. Seraphin, R. B. Hess, and N. Bottka, *J. Appl. Phys.* **26**:2242 (1965).
13. M. V. Hobden, *J. Phys. Chem. Solids* **23**:821 (1962).
14. G. Harbeke, *Z. Naturforsch.* **19a**:548 (1964).
15. D. Brust, J. C. Phillips, and F. Bassani, *Phys. Rev. Letters* **9**:94 (1962).
16. T. M. Donovan, E. J. Ashley, and H. E. Bennett, *J. Opt. Soc. Am.* **53**:1403 (1963).
17. E. O. Kane, *Phys. Rev.* **146**:558 (1966).
18. B. O. Seraphin and N. Bottka, *Phys. Rev. Letters* **15**:104 (1965).
19. F. Hermann and S. Skillman, *Proc. Intern. Conf. Semicond. Phys., Prague, 1960*, Academic Press (New York), 1961, p. 20.
20. J. C. Phillips and B. O. Seraphin, *Phys. Rev. Letters* **15**:107 (1965).
21. M. L. Cohen and T. K. Bergstresser, *Phys. Rev.* **141**:789 (1966); F. Herman, R. L. Kortum, C. D. Kuglin, and R. A. Short, *Quantum Theory of Atoms, Molecules and the Solid State: A Tribute to J. C. Slater*, Academic Press (New York), 1966.
22. B. O. Seraphin, *Phys. Rev.* (in press).
23. B. O. Seraphin, *Proc. Phys. Soc.* **87**:239 (1966).
24. B. O. Seraphin, *J. Appl. Phys.* **37**:721 (1966).
25. M. Cardona and G. Harbeke, *J. Appl. Phys.* **34**:813 (1963); M. Sturge, *Phys. Rev.* **127**:768 (1962); M. Cardona, *J. Appl. Phys.* **32**:2151 (1961); **32**:958 (1961); *Phys. Rev.* **121**:752 (1961); D. L. Greenaway and M. Cardona, *Proc. Intern. Conf. Phys. Semicond., Exeter, 1962*, The Inst. of Phys. and the Phys. Soc. (London), 1962, p. 227; F. Lukes and E. Schmidt, *ibid.*, p. 389.
26. J. C. Phillips, *J. Phys. Chem. Solids* **12**:208 (1960); *Phys. Rev.* **133**:A452 (1964); *Solid State Physics*, F. Seitz and D. Turnbull (eds.), Academic Press Inc. (New York), 1966; H. Ehrenreich, J. C. Phillips, and H. R. Phillip, *Phys. Rev. Letters* **8**:59 (1962); M. Cardona and G. Harbeke, *J. Appl. Phys.* **34**:813 (1963); M. Cardona and D. L. Greenaway, *Phys. Rev.* **131**:98 (1963).
27. W. Franz, *Z. Naturforsch.* **13a**:484 (1958); W. L. Keldysh, *Zh. Eksperim. i Teor. Fiz.* **34**:1138 (1958) [English transl.: *Soviet Phys.—JETP* **7**:788 (1958)]. K. Tharmalingam, *Phys. Rev.* **130**:2204 (1963); J. Callaway, *Phys. Rev.* **130**:549 (1963); **134**:A998 (1964); D. Redfield, *Phys. Rev.* **130**:916 (1963).
28. B. O. Seraphin and N. Bottka, *Phys. Rev.* **139**:A560 (1965).
29. J. C. Phillips, *Proceedings of the International School of Physics "Enrico Fermi,"* Academic Press (New York), 1966, p. 316.
30. D. E. Aspnes, *Phys. Rev.* **147**:554 (1966); **153**:972 (1967).*
31. J. C. Phillips, *Phys. Rev.* **146**:584 (1966).
32. C. B. Duke and M. E. Alferieff, *Phys. Rev.* **145**:583 (1966).
33. Y. Hamakawa, F. Germano, and P. Handler, *J. Phys. Soc. Japan* **21**:111 (1966).
34. F. Stern, *Phys. Rev.* **133**:A1653 (1964).
35. B. O. Seraphin and N. Bottka, *Appl. Phys. Letters* **6**:134 (1965).
36. H. R. Phillip and E. A. Taft, *Phys. Rev.* **120**:37 (1960).
37. M. Cardona, K. L. Shaklee, and F. H. Pollack, *Phys. Rev.* **154**:696 (1967).
38. A. G. Thompson, M. Cardona, and K. L. Shaklee, *Phys. Rev.* **146**:601 (1966).
39. E. W. Williams and V. Rehn, *Phys. Rev.* (to be published).
40. R. Ludeke and W. Paul, *Proc. International Conf. II–VI Compounds, Providence, 1967*, Benjamin (New York), 1967, p. 123.
41. E. Gutsche and H. Lange, *Phys. Status Solidi* **22**:229 (1967).

---

* I am indebted to Dr. Aspnes for his permission to use his calculations prior to publication.

42. R. A. Forman and M. Cardona, *Proc. International Conf. II–VI Compounds, Providence, 1967*, Benjamin (New York), 1967 p. 100.
43. A. Frova, P. J. Boddy, and Y. S. Chen, *Phys. Rev.* **157**:700 (1967).
44. F. Vasquez, M. Cardona, and R. A. Forman, *Bull. Am. Phys. Soc.* **12**:1048 (1967).
45. A. K. Ghosh, *Solid State Commun.* **4**:565 (1966).
46. A. K. Ghosh, *Phys. Letters* **23**:36 (1966).
47. F. H. Pollack and M. Cardona, *Phys. Rev.* (to be published).
48. C. Gaehwiler, *Helv. Phys. Acta* **39**:595 (1966).
49. C. Gaehwiler, *Solid State Commun.* **5**:65 (1967).
50. V. Rehn and D. Kyser, *Phys. Rev. Letters* **18**:848 (1967).
51. R. A. Foreman, D. E. Aspnes, and M. Cardona, *Bull. Am. Phys. Soc.* **12**:658 (1967).
52. J. D. Axe and R. Hammer, *Phys. Rev.* **162**:700 (1967).
53. B. O. Seraphin, *Surface Sci.* **8**:399 (1967).
54. C. R. Pidgeon, S. H. Groves, and J. Feinleib, *Solid State Commun.* **5**:677 (1967).
55. D. E. Aspnes, P. Handler, and D. F. Blossey, *Phys. Rev.* **166**:921 (1968).
56. F. Aymerich and F. Bassani, *Nuovo Cimento* **48**:358 (1967).
57. R. Enderlin and R. Keiper, *Phys. Status Solidi* **19**:673 (1967).
58. N. Bottka and U. Roessler, *Solid State Commun.* **5**:939 (1967).
59. F. Aymerich and F. Bassani, *Phys. Rev.* (to be published).
60. J. E. Fischer and B. O. Seraphin, *Solid State Commun.* **5**:973 (1967).
61. Y. Hamakawa, P. Handler, and F. Germano, *Phys. Letters* **25A**:617 (1967).
62. Y. Hamakawa, P. Handler, and F. Germano, *Phys. Rev.* **167**:709 (1968).
63. Y. Hamakawa, F. Germano, and P. Handler, *Phys. Rev.* **167**:703 (1968).
64. A. Baldereschi and F. Bassani, *Phys. Rev. Letters* **19**:66 (1967).

CHAPTER 8

Infrared Photoconductivity

E. H. Putley

Royal Radar Establishment
Malvern, Worcestershire, England

I. THE INFRARED SPECTRUM

In the majority of topics discussed in this book it is tacitly assumed that we are studying the interactions between electromagnetic radiation and solids in order to obtain a fuller knowledge of the behavior of various solids of interest. Here, we will consider things from the opposite direction and discuss how photoconductive effects may be used to detect and measure infrared radiation.

Before starting to discuss photoconductivity, we must first consider some characteristics of the infrared spectrum. Although the decision to confine the discussion to this spectral region is a somewhat arbitrary one, nevertheless, the majority of the processes of interest involve this radiation so that it should not prove restrictive.

The infrared spectrum is now usually defined as extending from about 0.7 μm (the limit of the visible) to 1000 μm or 1 mm, where it meets the microwave region. It is convenient to divide this into three bands of wavelength, less than 10 μm, 10 to 100 μm, and 100 μm to 1000 μm. It turns out that these subdivisions correspond roughly with the three types of photoconductive process which will be discussed. These three processes are illustrated schematically in Fig. 1. At short wavelengths photons have sufficient energy to raise an electron from the valence to the conduction band of a semiconductor. Between 10 and 100 μm the most important process is the excitation of shallow impurity states in semiconductors, while from about 100 μm to wavelengths greater than 1000 μm a form of photoconductivity associated with free-carrier absorption and hot-electron effects occurs.

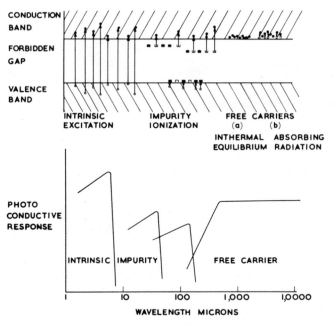

Fig. 1. Photoconductive processes. The upper diagram illustrates the electronic transitions. The lower diagram illustrates the corresponding spectral responses.

II. THERMAL RADIATION

A very significant feature of the infrared spectrum is that the bulk of the thermal radiation from our surroundings falls within it. Thus Fig. 2 shows the flux of photons incident upon a unit area exposed to blackbody radiation at 290°K, which is given by the Planck distribution function,

$$Q_\lambda = (2\pi c/\lambda^4) \left[\exp\left(hc/\lambda kT\right) - 1\right]^{-1} \tag{8.1}$$

When T is 290°K, the maximum of this distribution occurs near 10 μm. This radiation will always be with us (except, perhaps, if our experiment is being carried out in outer space where, even there, the effective blackbody temperature may be 3°K), and its presence must be taken into account when discussing infrared photoconductivity. In some cases the mean level of radiation will affect the properties of the photoconductors, but more important, the fluctuation in the rate of arrival of thermal photons at the surface of the photoconductor will determine the minimum intensity of a weak signal which can be detected against the background. At the radio-frequency end of the spectrum the effect of the background fluctuation can be expressed

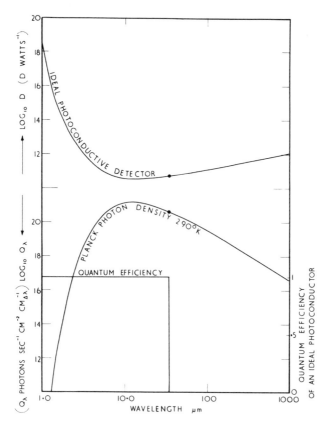

Fig. 2. Thermal radiation. The photon distribution for a black-body at 290 °K is calculated from the Planck distribution [equation (8.1)]. The detectivity for an ideal photodetector at its threshold wavelength is then given by the upper curve. The quantum efficiency is also shown.

in a simple way, but while this is not possible in the infrared it is true that the detectivity or noise equivalent power of a perfect detector is determined by the background radiation. With certain qualifications, the detectivity is independent of the physical mechanism of the detector itself.

The method of calculating the background fluctuation has been discussed in detail by many people, for example, Smith, Jones, and Chasmar [1], Kruse, McGlauchlin, and McQuistan [2], and McLean and Putley [3]. The principal conclusions reached will be summarized here.

If photons could be treated as classical particles, and if at a given point the mean rate of arrival was N per second, the rms fluctuation in this number would be $N^{\frac{1}{2}}$. However, photons in fact obey Bose–Einstein statistics which reduce to classical statistics only if $hc/\lambda > kT$, where λ is the wavelength of the photon. If T is room temperature, this condition will be satisfied for $\lambda \leqslant 10\ \mu m$. The rms fluctuation in this case can then be found by integrating the Planck distribution function to determine the total number of photons arriving at the photoconductor within the band of the spectrum to which it responds.

If the number incident per unit time upon the photoconductor is J_I and of this number a fraction η is absorbed, the rms fluctuation in the number absorbed will be $(\eta J_I)^{\frac{1}{2}}$. The frequency spectrum of this fluctuation will be similar to that for shot noise in a vacuum tube, so that, if the output from the photoconductor is fed into a noiseless amplifier of bandwidth B, the rms noise output from it will be proportional to $(2\eta J_I B)^{\frac{1}{2}}$. The minimum detectable signal can be defined as that producing an output equal to the noise, so that if a flux of signal photons J_s is required, then

$$\eta J_s = (2\eta J_I B)^{\frac{1}{2}} \tag{8.2}$$

The noise equivalent power P_N corresponding to this flux of photons will be

$$P_N = J_s(hc/\lambda_s) \tag{8.3}$$

where λ_s is the wavelength of the signal photons.

Alternatively, a detectivity D can be defined by writing

$$D = 1/P_N \tag{8.4}$$

so that

$$D = \frac{\lambda_s}{hc} \left(\frac{\eta}{2J_I B}\right)^{\frac{1}{2}} \tag{8.5}$$

J_I is calculated for an ideal photoconductor by assuming that it absorbs a fraction η of all photons of wavelengths not greater than the threshold wavelength λ_c (see Fig. 2) of the photoconductor. If we consider an area of $1\ cm^2$ and assume that all radiation emanating over a hemispherical field of view is incident upon the photoconductor, J_I can be evaluated from equation (8.1). Putting $\lambda_s = \lambda_c$ enables the variation of D with λ_c to be calculated. This argument neglects any contribution to the noise from the detector itself. For some types of ideal detector, such as a photovoltaic junction, this is valid, but in a photoconductive device, the mean square fluctuation associated with the recombination of the photoelectrons is equal to that of the background radiation [4]. To take this into account the rms of equation (8.5) must be divided by $\sqrt{2}$, so that for an ideal photoconductor at the threshold wavelength

$$D = \frac{\lambda_c}{2\,hc} \left(\frac{\eta}{J_I B}\right)^{\frac{1}{2}}$$ (8.6)

Assuming $B = 1$ Hz and calculating J_I from the Planck distribution, we obtain the curve for D shown in Fig. 2.

This calculation is sufficiently accurate for wavelengths not greater than 10–$20\,\mu$m but for longer wavelengths a more rigorous calculation employing Bose–Einstein statistics is necessary. The details of this have been discussed by Putley [5, 6] and the curve of D given in Fig. 2 is calculated by the exact method.

The quantity D plotted in Fig. 2 serves as a reference against which the performance of real photoconductors can be compared. Now for the simple case discussed above, $J_I \propto A$ (the area of the photoconductor) so that $D \propto (1/AB)^{\frac{1}{2}}$, and most workers write

$$D = D^*(1/AB)^{\frac{1}{2}}$$ (8.7)

Numerically, the value of D^* will equal the value of D plotted in Fig. 2. In a number of important cases, however, the detectivity of real detectors does not depend on A in this way and therefore the use of D^* (called the specific detectivity) can be misleading. It will not be used here.

Examination of Fig. 2 shows that D is very large at short wavelengths, passes through a minimum near $10\,\mu$m, and then rises steadily as the wavelength is increased further. As the plot of the Planck distribution shows, the density of thermal radiation will be small at the shorter wavelengths, but increases out to about $10\,\mu$m, causing D to fall. For photoconductors responding to the longer wavelengths practically all the room-temperature radiation will be effective so that J_I will be practically constant when λ_c is greater than $10\,\mu$m. D will now start to rise because as λ_c increases the energy associated with a constant number of photons falls.

This result appears to predict that the detectivity should tend to infinity at both very short and very long wavelengths, which is not meaningful. At very short wavelengths, when the number of thermal photons becomes negligible, the discrete nature of the signal photons themselves [3] will determine the minimum detectable power. At very long wavelengths, certain assumptions implicit in the derivation of equation (8.6) break down. For instance, when the dimensions of the photoconductor become of the same order of magnitude as the wavelength, diffraction effects become important. It is no longer valid to assume that Lambert's law applies to the surface. The absorbing surface must be treated in a way similar to that used for a radio antenna [7], and we then find that the detectivity tends to a constant value in the long wave limit which is in agreement with that found for a comparable radio receiver.

III. INTRINSIC PHOTOCONDUCTIVITY

When a semiconductor absorbs photons of energy not less than that separating the bands, an electron will be excited into the conduction band, increasing the number of free electrons, and leaving behind a mobile hole in the valence band. If the semiconductor is exposed to a constant level of exciting radiation, a steady state will be set up in which the concentration of free carriers (both electrons and holes) will exceed its value in thermal equilibrium in the absence of radiation. In the steady state the rate of excitation will be balanced by the rate of recombination. If the radiation is stopped abruptly, the concentration of free carriers will decay down to the thermal equilibrium value at a rate determined by the recombination processes, but which can often be characterized by a recombination time τ.

The optical properties of semiconductors near the intrinsic energy gap have now been studied in great detail. All we need recall at this point is that at photon energies just greater than that of the energy gap the absorption coefficient α becomes very large, rising rapidly to values greater than 1000 cm^{-1}. This means that the bulk of the incident photons will be absorbed in a thin surface layer of the material and therefore that intrinsic photo-conductive effects can be studied most effectively in very thin specimens (typically of thickness $d \sim 1/\alpha$).

So far we have pointed out that the intrinsic excitation leads to an increase in conductivity. It may also produce other electrical effects. Suppose we have a p–n junction biased for negligible current flow. If photons are absorbed in the barrier region, the resulting electron–hole pairs will split up, one travelling to each electrode, thus increasing the current through the junction and so enabling the presence of the radiation to be observed. A second method of splitting the electron–hole pair is by applying a magnetic field parallel to the surface being illuminated. If the thickness of the sample is somewhat large compared with the reciprocal absorption coefficient, the excess free carriers at the surface will set up a diffusion current into the interior. The diffusing carriers will be deflected by the magnetic field and will produce a potential difference perpendicular to the direction of the magnetic field. This effect is an optical analog of the Nernst effect and is called the photoelectromagnetic or PEM effect.

Probably the best way to amplify these remarks is to discuss an actual material. A suitable one for more detailed discussion is InSb. In this III–V compound the energy gap at room temperature is about 0.18 eV making the threshold wavelength about 7 μm. At a wavelength of 6 μm the absorption coefficient is about 3000 cm^{-1} and rises further as the wavelength is reduced. At room temperature the thermally excited intrinsic carrier concentration is 1.6×10^{16} cm^{-3}. On cooling to 77°K, the energy gap increases to about 0.23 eV (equivalent to 5.5 μm) while the thermally excited intrinsic carrier concentration falls to 2×10^{9} cm^{-3}. This latter value is so small that even in the purest material the impurity concentration will be several orders greater. Figure 3 shows the spectral response of the photoconductivity in InSb at three temperatures [8].

If we assume that the recombination processes can be described by the exponential law $n = n_0 \exp(-t/\tau)$, solution of a simple differential equation shows that if photons arrive at the surface of a thin plate of InSb at a rate of J per second, and of these a fraction η are absorbed, then the number of free electrons in the sample will be increased by N where

$$N = \eta \tau J \tag{8.8}$$

In this model there will be an equal increase in the number of free holes. If the volume of the sample is v there will be an increase in conductivity

$$\Delta\sigma = \eta \tau J e (\mu_e + \mu_h)/v \tag{8.9}$$

where μ_e and μ_h are the mobilities of electrons and holes, respectively, and e is the electronic charge. We can express the change in conductivity in terms of the power absorbed by noting that the power flowing into surface is

$$Q = h\nu J \tag{8.10}$$

Also, the conductivity σ in the absence of radiation is

$$\sigma = ne(\mu_e + \mu_h) \tag{8.11}$$

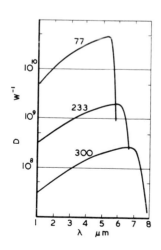

Fig. 3. Spectral response of InSb photoconductor at 77, 233, and 300°K [8].

where n is the intrinsic carrier concentration. Thus the relative change in conductivity is

$$\Delta\sigma/\sigma = (\eta\tau Q)/(h\nu n\upsilon) \qquad (8.12)$$

This shows that η should be as large as possible while n and υ should be small. The relative effect also increases as τ is increased, but τ will probably be fixed by other considerations. This simple result demonstrates that the photoconductive effect will tend to increase on lowering the temperature, since n will fall. Since the area of the sample will be determined by optical factors, the only disposable dimension will be the thickness d. It must be large enough to ensure that the bulk of the radiation is absorbed and will be optimal when $d \sim 1/\alpha$. If the only significant absorption process is the photoconductive one (that is to say that processes such as lattice band absorption can be neglected), then η will be determined by the reflection coefficient of the surface, the absorption coefficient, and the thickness.

Thus

$$\eta = (1-R)\,(1-e^{-\alpha d})/(1-Re^{-\alpha d}) \qquad (8.13)$$

so that for $\alpha d \gg 1$

$$\eta = 1-R$$

while for $\alpha d \ll 1$

$$\eta = \alpha d$$

This shows that R should be as small as possible. For pure semiconductors R will be determined by the dielectric constant K

$$R = \frac{(1-K^{\frac{1}{2}})^2}{(1+K^{\frac{1}{2}})^2} \qquad (8.14)$$

so that, for InSb, $R \sim 0.35$. This means that without special treatment over 50% of the incident radiation will be absorbed, but by blooming the surface this factor can be improved upon.

Let us now discuss τ in more detail. There are several processes by which the photoexcited electron can combine with the hole. The first one is the inverse of the exciting process—radiation recombination in which the excess energy of the electron is liberated as a photon. Similarly, the energy could be dissipated in lattice vibrations. A third process, Auger recombination, occurs if two electrons collide with a hole. One recombines with the hole while the second acquires as kinetic energy the energy given up by the recombining electron. In these three processes the electron and hole recombine directly but there are also important processes involving impurity centers [7]. Suppose an electron leaves the conduction band and occupies a vacant impurity center. The trapped electron may still be able to recombine with a vacant hole if the hole can also interact with the impurity center but frequently the probability of this happening may be very small. In this event, the free hole cannot be filled until the electron becomes thermally excited from the impurity center into the conduction band. Recombination can then occur via one of the direct processes. In this situation, the lifetime for holes would be longer than that for electrons.

To decide which processes occur in a particular case requires a combination of detailed theoretical and experimental studies. Calculations of the radiation recombination process in InSb [9] show that the lifetime associated with this process should be about 1 μsec. However, measurement of the lifetime at room temperature [10, 11] where the material is intrinsic (and therefore effects associated with impurity centers should not be important) gives a much shorter value ($\sim 10^{-8}$ sec). At temperatures greater than 200°K the lifetime values are independent of the sample, confirming that the recombination process is an intrinsic one and the measured values agree well both in magnitude and in the rate of change with temperature with that calculated for the Auger process [12]. The radiation recombination lifetime is both too long and does not have the correct temperature dependence. Recombination involving phonon interaction is an even less likely process since multiple phonon interactions would be required.

As the temperature is lowered from 300°K (see Fig. 4), the lifetime rises approximately as exp (E_G/kT) and in a similar way for all samples, but near 200°K the lifetime passes through a maximum and its value now depends upon the particular sample. It is possible by studying both the photoconductive and PEM effects (for detailed analysis see Blakemore [13]) to calculate independently the lifetimes for electron and holes. In the intrinsic region, these are equal, but below 200°K this is no longer true. At first both the electron and hole lifetimes fall at the same rate, but below about 170°K the hole lifetime starts to rise again. The electron lifetime continues to fall but flattens out below 150°K to a value less than 10^{-9} sec.

This behavior can be accounted for by the effects of impurity states in the band gap. The nature of these states is still not fully understood for InSb but there appear to be shallow acceptor levels located about 0.007 and

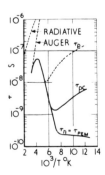

Fig. 4. Recombination processes in InSb (see [9-11]).

0.05 eV above the valence band and at least one level lying nearer the center of the gap. Impurity states can affect the recombination processes in two ways. A level near the center of the band gap having a large capture cross section for both holes and electrons will assist in the recombination process, and this is what is happening in the intermediate temperature region. As the temperature rises these levels will tend to become occupied with thermally excited carriers so the lifetime will increase. When the temperature becomes sufficiently high, the concentration of carriers thermally excited across the gap will become large enough for the intrinsic Auger process to take over; the lifetime starts to fall again because the concentration of excited carriers is increasing and with it the probability of an electron–electron collision leading to recombination. At very low temperatures, however, the shallow acceptor levels become occupied. These levels do not have a large capture cross section for electrons. The electrons will fall into the level near the center of the gap, and this process will be responsible for the very short electron lifetime. The rate at which the holes can recombine with these electrons will now be determined by the rate at which the holes are released from the acceptor levels into the valence band when they can recombine with electrons at the deep levels. Although this model accounts for the type of behavior observed, the results of quantitative calculations are not entirely satisfactory. One of the difficulties standing in the way of an exact analysis is that one cannot control the concentration and type of centers since they are associated with residual defects or impurities present in even presently available material.

Figure 4 illustrates these processes. It shows calculated values for the radiation [9] and the Auger lifetime [12] and measured values (by Zitter, Strauss, and Attard [10]) of the photoconductive and PEM lifetime for a sample of InSb with a net acceptor concentration of 6×10^{15}. The values for the lifetimes of the holes and electrons found from the analysis of the experimental data are also shown.

Having described the recombination processes in one particular case, we now want to consider the random fluctuations which will determine the minimum detectable power. In any real situation there will be several sources of noise, some of which will be associated with gross imperfections in the apparatus. We might imagine these could be eliminated if we were clever enough, but there will be others of a more fundamental nature associated with the effect being studied. In photoconductivity we can consider two (in addition to fluctuation in the incident radiation). The first one is the fluctuation associated with the generation and recombination of free carriers (generation–recombination noise). In the absence of external radiation there will be a certain concentration n of thermally excited intrinsic or impurity carriers. This number will fluctuate [14] and this fluctuation will be one of the factors limiting the detectable change in n produced by the radiation we wish to observe. In semiconductors the electron concentration is usually so small that it can be treated by classical statistics. Thus, if the mean number present in a sample in thermal equilibrium is N_d, the rms fluctuation will be $N_d^{\frac{1}{2}}$ and if a steady current I is passing, the power spectrum of the current fluctuation will be

$$P(f) = \frac{4 I^2 \tau}{N_d (1 + \omega^2 \tau^2)} \tag{8.15}$$

For a near intrinsic conductor, a more exact expression is

$$P(f) = 4 I_0^2 \frac{(b+1) N_d P_d}{(b N_d + P_d)^2 (N_d + P_d)} \frac{\tau}{1 + \omega^2 \tau^2} \tag{8.16}$$

If the resistance of the specimen is R and the amplifier bandwidth B, the noise voltage produced will be

$$v_{gr} = \frac{2 I \tau^{\frac{1}{2}}}{N_d^{\frac{1}{2}}} B^{\frac{1}{2}} R = 2 IR \left(\frac{B}{G}\right)^{\frac{1}{2}} \tag{8.17}$$

where $G = N_d/\tau$ is the rate of generation. (This assumes B is centered at a low frequency so that $\omega^2 \tau^2 \ll 1$.) The second noise process which will be with us whatever we do will be the Johnson noise associated with the resistance R:

$$v_J = (4 k T R B)^{\frac{1}{2}} \tag{8.18}$$

In the near infrared it is easy to compare these sources of noise with that associated with photon fluctuations. Here photons can be treated as classical particles and an expression with a form similar to equation (8.17) can be obtained as follows:

$$v_{ph} = 2 IR \left(\frac{B}{\eta J}\right)^{\frac{1}{2}} \tag{8.19}$$

where J is the rate of arrival of photon as calculated from the blackbody distribution function and ηJ is the rate of generation of optically excited carriers. In fact this can be combined with the thermal generation noise G and written

$$v_{ph+gr} = 2 IR \left(\frac{B}{\eta J + G}\right)^{\frac{1}{2}} \tag{8.20}$$

If the photon fluctuation is to be dominant, we must make $\eta J > G$ and also equation (8.20) > equation (8.18). Consider the ratio of these equations:

$$\frac{v_J}{v_{ph+gr}} = \frac{(4 k T R B)^{\frac{1}{2}}}{2 IR [B/(\eta J + G)]^{\frac{1}{2}}} = [k T n e \mu (\eta J + G) a/l]^{\frac{1}{2}}/I \tag{8.21}$$

To make this quantity small, it is desirable to keep T small, which also reduces G. We want to make μ small, which is why p-type semiconductors rather than n-type ones are used for making intrinsic detectors. The bias I is made as large as possible, but it will be limited by other current-dependent noise processes, such as contact noise, or by heating effects.

With InSb we find that at room temperature equation (8.21) cannot be made small and the limit to the minimum detectable power is set by Johnson noise. At 77 °K, however, G becomes very small while ηJ is independent of temperature; we then approach closely to the condition in which the limiting fluctuation is set by the background radiation for a detector with a fairly wide field of view. If the field of view is restricted, or a cooled optical filter is introduced, we will find that the nonideal form of noise will usually limit the performance.

Having talked at some length on InSb, the discussion of intrinsic photoconductors will be concluded by mentioning some other semiconductors. The lead salts PbS, PbSe, and PbTe were the first group of compounds in which infrared photoconductivity was extensively studied. The first work on PbS was reported in 1902 but the group did not attract a great deal of attention until World War II when there was a considerable interest in military applications of infrared. This interest led, in the years immediately after the war, to a very detailed study of the fundamental properties of these compounds, some of the results of which are described in this volume by Dixon (Chapter 3). However, although the photoconductive effects are still of practical importance and much detailed study has been made of them, it is still not possible to analyze them in the same detailed way that has been achieved with InSb. The reason is that for photoconductive studies one needs good single crystals containing excess carrier concentration not greater than 10^{15} cm^{-3}. In the best lead-salt single crystals, the carrier concentration is greater than 10^{16}, more typically 10^{17}–10^{18}. Photoconductive devices are produced from polycrystalline chemically deposited or evaporated layers [15] in which by suitable treatment the desired low carrier concentration is obtained. Typical spectral-response curves are shown in Fig. 5. In the interpretation of the behavior of these layers it is difficult to disentangle the bulk, surface, and contact effects. It may prove possible to resolve this problem with modern developments in the growth of epitaxial layers, but so far this has not been done.

The final point to discuss is the long wave limit for intrinsic photoconductivity. In both InSb and PbSe the long wave limit is about 7 μm. Considerable efforts have been made to discover materials with even smaller energy gaps to permit operation to 10 μm or beyond. Several materials which might satisfy this requirement are known. These include the unstable grey tin (0.08 eV) where photoconductivity has been reported to beyond 16 μm and alloys of HgTe and CdTe where alloys containing 10–20% CdTe have

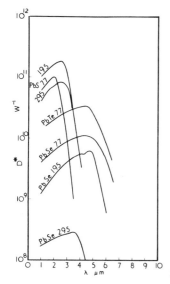

Fig. 5. Spectral response of PbS, PbSe, and PbTe photoconductors (see [15] for further discussion).

energy gaps less than 0.15 eV and where photoconductivity beyond 10 μm has been observed [16]. Another variable-gap mixed compound that has been studied is Mg_2Pb–Mg_2Sn, but this is an unstable material. Recently there have been reports of laser action beyond 10 μm in PbTe–SnTe alloys [17]. Both PbTe and SnTe have energy gaps greater than 0.12 eV but in alloys containing roughly equal amounts of Pb and Sn the energy gap falls to zero.

The difficulty with all these narrow gap compounds is that, apart from problems of chemical stability, no one has really succeeded in producing them with the degree of purity required to enable efficient photoconductive devices to be produced simply. It was the realization of these difficulties which provided much of the impetus behind the study of the extrinsic photoeffects.

IV. EXTRINSIC PHOTOIONIZATION PHOTOCONDUCTIVITY

While it has proved difficult to prepare small-gap semiconductors in a very pure form, Ge is probably the purest solid ever prepared, the best sample containing only about 1 in 10^{10} atoms of electrically active impurities. The extrinsic carriers are associated with foreign atoms rather than deviations from stoichiometric composition, important in most compound semiconductors. The properties of over twenty different impurity atoms have been studied in detail. Germanium is a member of Group IV of the periodic table and atoms from Group V have one more outer shell electron than the Ge atom. Atoms from Group V can be substituted for Ge atoms and then the extra outer electron together with its neutralizing unit of nuclear charge behaves as a pseudo-hydrogen atom. The ionization energy can be estimated from the Bohr atom model modified by the dielectric constant and effective mass appropriate to Ge as follows:

Ionization energy

$$\varepsilon = 13.6 \left(\frac{m^*}{m} \right) \frac{1}{K^2} \quad \text{eV} \tag{8.22}$$

Bohr radius

$$a = 5.29 \times 10^{-9} \left(\frac{m}{m^*} \right) K \quad \text{cm} \tag{8.23}$$

For Ge

$$a = 4.2 \times 10^{-7} \text{ cm}$$

$$\varepsilon = 10^{-2} \text{ eV}$$

From equation (8.22)

$$\left(\frac{m^*}{m} \right) = \frac{\varepsilon K^2}{13.6}$$

$$a = 5.29 \times 10^{-9} \frac{13.6}{\varepsilon K^2} \cdot K$$

$$= 7.2 \times 10^{-8} \left(1/\varepsilon K \right)$$

The value obtained is roughly 0.01 eV, corresponding to a wavelength of about 100 μm. Detailed measurements and calculation show that the exact value depends on the impurity, but confirm that this simple model gives the right order. Group III atoms behave in an analogous way, except that they contribute an electron deficit or hole. The behavior of other types of atoms is more complex; some can be treated as pseudo-helium atoms while others have more complex behavior. The behavior of these is illustrated in Fig. 6.

Figure 6 shows that, apart from Group III and V impurities, others have at least two states of ionization and the ionization energies are mostly much greater than those for the III and V impurities. Thus it is possible to select impurities whose ionization energies are equivalent to wavelengths ranging from less than 10 μm to beyond 100 μm [6].

Clearly impurity photoionization has many features in common with the intrinsic process, and we will now compare them in more detail.

First consider the absorption coefficient. The absorption coefficient for impurities is much smaller than that for lattice absorption. It can be estimated

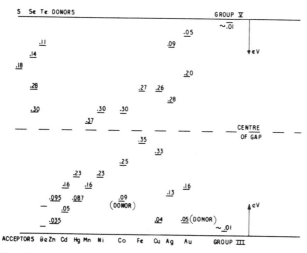

Fig. 6. Impurity levels in Ge (energies measured from nearer band edge).

by using a scaled expression for the absorption cross section of a hydrogen atom. Burstein, Picus, and Sclar [18] showed that for a hydrogenic impurity

$$\sigma_v = \frac{8.28 \times 10^{-17}}{\varepsilon \sqrt{K}} \left(\frac{m}{m^*}\right) \left(\frac{\varepsilon}{hv}\right)^{\frac{8}{3}} \ cm^2$$

with $hv \geqslant \varepsilon$. Putting $hv \sim \varepsilon$ will give a maximum value for σ_v and if (m/m^*) is eliminated using equation (8.22), then

$$\sigma_v = \frac{8.28 \times 10^{-17} \times 13.6}{\varepsilon^2 \, K^{\frac{3}{2}}} \ cm^2$$

For Ge, K is 16 and σ_v equal to $1.1 \times 10^{-18} \, \varepsilon^{-2}$. Although this result might only seem plausible for hydrogenic impurities, it is in order-of-magnitude agreement with a number of deep-lying impurities [6] (see Table I).

Table I

Impurity	As	Zn	Cu	Au		Ni
σ_v (measured, $\times 10^{16}$)	40	16	8.5	0.1	0.9	2
σ_v (calculated, $\times 10^{16}$)	68	9	7	0.4		0.2
ε (eV)	0.0127	0.035	0.04	0.16		0.23

If the concentration of centers is N_I the absorption coefficient will be

$$\alpha = \sigma_v N_I$$

There are limits to the attainable values for N_I. Apart from considerations of solubility, the concentration must be sufficiently dilute that neighboring impurities do not interact. Thus we must have

$$(N_I)^{-\frac{1}{3}} > a$$

since $(N_I)^{-\frac{1}{3}}$ will be of the order of the distance apart.

This means that the concentration must be not more than 10^{15}–10^{16}. If we have a concentration of 10^{15} cm^{-3} of As atoms, α will be

$$\alpha = 40 \times 10^{-16} \times 10^{15} = 4 \text{ cm}^{-1}$$

and similar results will be obtained for other impurities. So we see at once that the absorption coefficient is orders smaller than that for intrinsic photoconductivity.

Because the density of impurity centers will always be a small fraction of the density of Ge atoms, the cooling requirements will be more stringent. While with an intrinsic material it is possible to observe photoconductivity at high temperatures where the intrinsic carrier concentration is substantial, at corresponding temperatures the impurity center will be completely thermally ionized and its photoeffects will be negligible. Another important difference between the two processes is that only one type of free carrier is produced in the extrinsic process. This means that p–n junctions cannot be used nor does the PEM effect occur.

There are some points of similarity between the recombination processes. In addition to optical generation, the impurity center can be ionized by thermal processes and by impact from energetic electrons. Recombination by the inverse of all these processes must be considered. That associated with the thermal processes will be the most important at low electric fields, but, as the electric field is increased, impact ionization and the inverse recombination process (which is an Auger effect) will dominate the behavior and then the photoconductivity will fall. The lifetimes associated with these processes are usually less than 10^{-6} sec.

So far it has been assumed that the shallowest state of a given impurity is being ionized. It is possible to observe photoeffects with the higher states, but to do this a compensating impurity must be added in just sufficient concentration to ionize all the lower states.

Photoionization effects have been observed in Si, Ge, and p-type InSb. It is only with Ge, however, that materials can be prepared with the requisite purity for these effects to be studied in great detail. By choosing suitable impurities, such as Ge, Zn, Cu, and Hg (although others may be chosen),

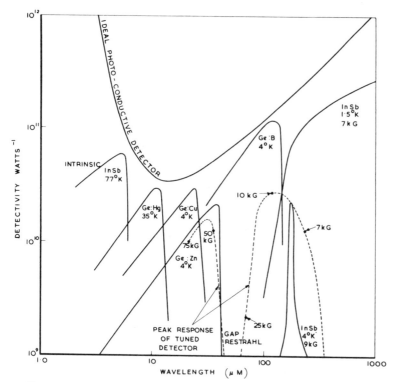

Fig. 7. Spectral responses of extrinsic Ge and of InSb photoconductors.

the photoconductive threshold may be varied between 120 and 13 μm. The most suitable one can then be selected for a particular application. Some examples of the spectral response are shown in Fig. 7.

V. FREE-CARRIER PHOTOCONDUCTIVITY

At room temperature, the electrons in the conduction band of a semiconductor are so tightly coupled to the crystalline lattice that when they absorb energy from infrared radiation it is rapidly shared with the lattice. If any significant change occurs it will be a warming up of the whole crystal giving a bolometric effect. In fact this was proposed some years ago by Novak [19] who studied the effect in InSb at room temperature. More recently it has been found that room-temperature InSb photocells can be used in this way to detect 10-μm CO_2 laser radiation (well beyond the normal photo cutoff, but of course a very slow effect). When high-mobility semiconductors, such as Ge, InSb, or GaAs, are cooled to 4°K, the situation becomes somewhat different. The electron–lattice coupling now becomes much weaker and it is now very easy to set up a steady state in which the average electron energy is well above the thermal equilibrium value. This is usually done by applying a large electric field. Large is a relative term since this effect can be observed in n-type Ge with fields of a few V/cm, while in n-type InSb, effects occur at fields less than 1 V/cm. Since the electron mobility depends on the mean energy, an increase of energy (heating of the carriers) produces a change in mobility, which may either decrease or increase, depending on which type of scattering process is dominant. At very low temperatures, ionized impurity scattering is usually the dominant process, so the mobility will increase. As a result the conductivity increases, although the number of free carriers is not changed.

The process of carrier heating is essentially a DC effect. How does this help us to detect far infrared radiation? Classical theory of the electrical conductivity shows a variation with frequency as

$$\sigma(\omega) = \sigma_0(1+\omega^2\tau_e^2)^{-1} \tag{8.24}$$

where σ_0 is the zero-frequency conductivity and τ_e the dielectric relaxation time. Also

$$\sigma_0 = ne^2\tau_e/m^* \tag{8.25}$$

At very high frequencies where the conduction component of the current is small compared with the displacement current, we obtain for the absorption coefficient

$$\alpha = 4\pi\sigma/c(K)^{\frac{1}{2}} \tag{8.26}$$

This shows that where $\omega\tau < 1$ (provided σ is not too large), α will be a constant, but at high frequencies ($\omega\tau > 1$) α will vary as ω^{-2} or λ^2. Thus, free-carrier absorption will be large at long wavelengths and fall off as the wavelength is reduced. To put some numbers into these equations, consider n-type InSb at $4°$K [20]. In a pure sample we could have $n = 5\times10^{13}$, $\mu = 10^5$ cm^2/V-sec. Since $\mu = e\tau/m^*$ and $m^* \sim 1.4\times10^{-2}\,m$, then $\tau_e = 8.5\times10^{-13}$ sec gives $\omega\tau_e = 1$ for $\lambda = 1.6$ mm. Thus for wavelengths of the order of 1 mm we can still use the DC conductivity to discuss the behavior, but would expect the free-carrier absorption to start falling off as λ is reduced below 1 mm. We find for the values quoted at 1 mm that $\alpha = 22$ cm^{-1}, while at 100 μm it has fallen to 0.30 cm^{-1}.

This process may be used to measure radiation in exactly the same way that a microwave bolometer is used to measure microwave power. If a large DC current is passed through the specimen to bias it to a point where the current–voltage characteristic is nonlinear, the absorption of a small amount of far infrared radiation will change the operating point and thus can be detected. I have called this an "electronic bolometer." It differs from the usual type of bolometer in that since the electrons are not tightly coupled to the lattice the response time is much shorter and is in fact less than 1 μsec. Quantitatively, the nonohmic behavior at low fields can be written as

$$\sigma = \sigma_0(1 + \beta E^2) \tag{8.27}$$

The responsitivity to absorbed power has been shown to be [21]

$$R = \beta V/v\sigma \tag{8.28}$$

where V is applied voltage and v is crystal volume, while the response time will be

$$\tau = \tfrac{3}{2}(k/e)\,\beta\,\frac{dT}{d\mu} \tag{8.29}$$

At the actual operating fields equation (8.27) is not a good approximation but equations (8.28) and (8.29) can still be used by writing

$$\beta = 1/\sigma\; d\sigma/d(E^2) \tag{8.30}$$

The quantities β and $dT/d\mu$ can be measured by DC methods and used to calculate R and τ which can then be compared with directly measured values. A typical value for β is $\sim 25\ \mathrm{cm^2/V^2}$; this gives $R \sim 650\ \mathrm{V/W}$ and $\tau \sim 0.4\ \mu\mathrm{sec}$, in reasonable agreement with measured values.

Although I have called this an electronic bolometer, it is perhaps better to look on it as a photoconductive effect in which the radiation changes the mobility rather than number of free carriers.

We must now discuss the effect of a magnetic field. The present available n-type InSb in the absence of a magnetic field does not show carrier freeze out. Because of the low effective mass the Bohr radius is so large that even with impurity concentrations of $10^{14}\ \mathrm{cm^{-3}}$ (the lowest available) the centers overlap so much that the impurity states merge with the conduction band. Application of an induction of a few kilogauss splits off the impurity states. One might expect this to reduce the free-electron photoconductivity and replace it with a photoionization process, but it does not happen exactly in this manner. At first, the main effect is to reduce σ by two to three orders of magnitude so that R in fact increases; τ becomes slightly shorter. Eventually, the magnetic field, by removing free carriers, reduces the absorption to a point where R falls. When this happens, a new type of photoeffect appears [22]. This is in the form of a resonance line located near the cyclotron frequency

$$\omega_e = eB/m^* \tag{8.31}$$

The actual transition involved is not the free cyclotron transition between Laudau levels but rather between the bound levels. Evidence for

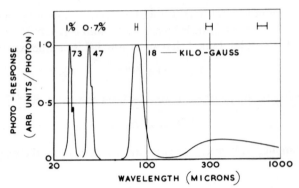

Fig. 8.   Resonant photoconductive effects in InSb [22].

this is given by the structure which can be seen on the line and which can be accounted for by transition to the excited states of the higher impurity levels. Typical resonance lines are shown in Fig. 8, while Fig. 7 shows both the broad band and the resonant photoeffect. The reason for the disappearance of the effect near 50 μm is that the reststrahlen band occurs near here. These photoeffects in InSb are most important at wavelengths longer than 100 μm, beyond the long wave limit of extrinsic Ge photoconductivity. They may be used to wavelengths longer than 1 mm where microwave techniques take over. In the region 100 μm to 1 mm they provide the most convenient detectors where both speed and sensitivity are required.

VI. REFERENCES

1. R. A. Smith, F. E. Jones, and R. P. Chasmar, *The Detection and Measurement of Infra-red Radiation*, Oxford University Press (Oxford), 1957.
2. P. W. Kruse, L. D. McGlauchlin, and R. B. McQuistan, *Elements of Infrared Technology*, John Wiley & Sons (New York), 1962.
3. T. P. McLean and E. H. Putley, *Roy. Radar Establishment J.*, No. 52, p. 5, April 1965.
4. C. T. J. Alkemade, *Physica* 25:1145 (1959).
5. E. H. Putley, *Infrared Phys.* 4:1 (1964).
6. E. H. Putley, *Phys. Status Solidi* 6:571 (1964).
7. A. Rose, *Concepts in Photoconductivity and Allied Problems*, Interscience (New York), 1963.
8. F. D. Morten and R. E. J. King, *Appl. Opt.* 4:659 (1965).
9. D. W. Goodwin and T. P. McLean, *Proc. Phys. Soc. (London)* B69:689 (1956).
10. R. N. Zitter, A. J. Strauss, and A. E. Attard, *Phys. Rev.* 115:266 (1959).
11. R. A. Laff and H. Y. Fan, *Phys. Rev.* 121:53 (1961).
12. A. R. Beattie and P. T. Landsberg, *Proc. Roy. Soc. (London)* A249:16 (1958).
13. J. Blakemore, *Semiconductor Statistics*, Pergamon Press (Oxford), 1962.
14. K. M. Van Vliet, *Proc. I.R.E.* 46:1004 (1958).
15. E. H. Putley, *J. Sci. Instr.* 43:857 (1966).

16. P. W. Kruse, *Appl. Opt.* **4**:687 (1965).
17. J. D. Dimmock, I. Melngailis, and A. J. Strauss, *Phys. Rev. Letters* **16**:1193 (1966).
18. E. Burstein, G. Picus, and N. Sclar, *Atlantic City Photoconductivity Conference*, John Wiley & Sons (New York), 1956, p. 353.
19. R. Novak, *Fifth Conference of the International Commission for Optics*, Stockholm, 1959.
20. E. H. Putley, *Appl. Opt.* **4**:649 (1965).
21. Sh. M. Kogan, *Fiz. Tverd. Tela* **4**:1891 (1962); *Soviet Phys. Solid State* (*English Transl.*) **4**:1386 (1963).
22. M. A. C. S. Brown and M. F. Kimmitt, *Infrared Phys.* **5**:93 (1965).

CHAPTER 9

Excitons

S. Nikitine

University of Strasbourg
Strasbourg, France

The title of this chapter is "Excitons." However, within the limitation of a single chapter, it will be impossible to give a detailed account of this broad subject, even making use of Dr. Reynolds' chapter dealing with the important II–VI compounds (Chapter 10). Therefore, this chapter will present selected topics within the subject area, rather than a general view. For our purpose, it will be necessary to give only an introduction to the theory of excitons. A detailed treatment can be found in the author's review article [1], the excellent monograph of Knox [2], as well as the original papers of Elliott [3], Haken [4], and others quoted in the first two references.

Early experimental work stimulated theoretical activity. This, in turn, stimulated new experimental research to provide better comparisons between observation and theory. Examples of this development are the research on Cu_2O and CuCl. A large amount of work on these substances has been carried out in the author's laboratory. All results reported were obtained a few years ago: a small part of these will be summarized here. Recent work on the calculation of band structure of Cu_2O and CuCl and comparisons with observed spectra will also be discussed to provide a modern aspect of the subject.

It should be noted that, in addition to the material concerning Cu_2O that was presented at the symposium on which this volume is based, this chapter also contains material concerning CuCl which was added to provide a second example. Also, it should be mentioned that only a limited view of exciton spectra and solid-state spectroscopy is reported here. It should be emphasized that solid-state spectroscopy has developed within the last fourteen years into a very large chapter in the book of spectroscopy.

I. THEORETICAL BACKGROUND

A. Introduction

It is well known that analysis of atomic and molecular spectra has made possible an understanding of the structure and properties of atoms and molecules. Expectation of an analogous development in the field of solid state about fourteen years ago led to the undertaking of a substantial effort in several laboratories for the development of a solid-state spectroscopy. It was hoped that this would lead to important new information on solids and, in particular, on their band structures. This program met with considerable experimental difficulty, however, and progress is achieved rather slowly.

We know now that solid-state spectroscopy should be divided into several sections according to the kind of spectra observed. The Strasbourg group has been especially involved in the investigation of the spectra of semiconductors and insulating crystals with rather strong covalent character. Accordingly, our interest will be focused on this section of solid-state spectroscopy. It must be emphasized, however, that this part of spectroscopy (though very important) is not representative of all solids.

B. Excitons

The classification of spectra of our interest is justified both by experiment and theory. The electronic states in our crystals are successfully described by the Bloch model in which electron states form bands which are separated by an energy gap. For insulators or intrinsic semiconductors at low temperature, the highest occupied energy band (valence band) is just filled with electrons and the next highest band (first conduction band) is empty. Let the energy gap between these bands be E_g.

Frenkel has shown that the absorption of light in such a crystal involves two processes. If the light quantum $\hbar\omega$ of the incident light has sufficient energy, electrons can be transferred from the valence band to the conduction band. This band-to-band transition has a lowest energy limit $\hbar\omega_\infty = E_g$. The electron in the conduction band and the hole in the valence band are free and both contribute to the conductivity of the crystal. So the crystal, an insulator in the dark, becomes a conductor under the effect of light of suitable frequencies. This effect is known as photoconductivity. From the

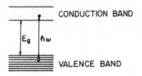

Fig. 1.   Schematic representation of photoconductivity.

atomic point of view, this effect corresponds to the transition of electrons from a bound to a free state (Fig. 1). Frenkel [5] and later Peierls [6] have suggested that electrons and holes could be created in a bound state in which a rather small coulomb interaction exists between them. This bound pair corresponds to an excited state of the crystal and is called an exciton. This excited state of the crystal can be created in any position in the crystal and therefore can travel across the lattice without expending any energy. However, no current is carried by excitons since a pair of charges of both signs is transported.

It is evident that in an exciton the binding energy is quantized as in a hydrogen atom, and therefore the exciton admits a series of energy states. It can be shown that the mechanism for creation of unbound pairs by absorption of light leads to a continuous absorption spectrum having a minimum energy limit $(\hbar\omega_\infty)$. The mechanism for creation of bound pairs corresponds to a series of sharp absorption lines converging to the above-defined energy limit $(E_g = \hbar\omega_\infty)$. In this model, the ground state of the system corresponds to the total energy of the unperturbed crystal in which no free charge carriers (electrons or holes) or bound electron–hole pairs are supposed to exist. The excited state of the crystal lattice is formed of an exciton in an otherwise unperturbed crystal. The photoconducting state is observed when a free electron and a free hole are present in an unperturbed crystal. This state can be obtained either by absorption of a photon $\hbar\omega > \hbar\omega_\infty$ or by creation of an exciton and a subsequent dissociation of this exciton which requires energy taken from the crystal. It is also evident that the series of excited quantum states of the exciton converge to the above limit.

We are going to focus our interest on spectra corresponding to the creation of excitons. It may be helpful to give here a short explanation of the classification of spectra before entering into detailed discussion of this process. Excitons are divided into two groups, "localized" and "delocalized" excitons. In the first group, the excitation is strongly localized, confined practically to one atom. In this case the absorption lines are sometimes strongly broadened into bands. These spectra are of strongly ionic compounds such as alkali halides. In molecular crystals, the excitation is also localized to a molecule and is characteristic of the molecule rather than the crystal. The molecular spectrum is perturbed only by the crystal field. In this case, the band description of the energy states of electrons may not be an adequate one and it is not possible to consider the exciton as an electron–hole pair. A similar situation is met in spectra of rare earths dissolved in a crystal. In delocalized excitons, the hole and the electron are loosely bound and the orbits are large. The region of excitation concerns a rather large volume of the crystal. In this case, the exciton lines are rather sharp. These spectra are observed in crystals with a rather strong covalent character.

Two essential properties of excitons are to be considered: (1) spectroscopic properties of crystals associated with the optical formation of excitons, and (2) energy transport associated with the migration of excitons, without any transport of mass or charge.

A very large amount of data has been obtained on exciton spectra, but no direct experimental confirmation of the migration of excitons has been obtained. Therefore, discussions here will concentrate principally on the spectroscopic properties of excitons. However, the general shape of the spectrum can also be of great interest for assignments of transitions to be compared with band structure suggested by calculations. Thus, interest will not be focused exclusively on exciton spectra.

C. Elementary Theory

The theory of excitons involves rather long calculations which will be omitted in this section; they are given in a rather descriptive way in the following sections. A very simple model, however, shows that the interaction between the electron and the hole is an electrostatic coulomb one. The problem is then similar to that of a hydrogen-like atom, and the exciton states are quantized. However, it is necessary to take into account the dielectric constant of the crystal. This takes care of the interaction of all atoms of the crystal with the exciton. Thus, the binding energy of the exciton is then given by the hydrogen-like relation derived by Mott and Gurney [7]

$$E_b = E - E_c = - \frac{e^4 \mu_{ex}}{2 \hbar^2 n^2} \frac{1}{\varepsilon^2}$$

where E_b is the binding energy, E is the energy of the exciton,† E_c is the energy of the lowest state of the conduction band, ε is the dielectric constant, and

$$\mu_{ex} = \frac{m_h^* m_e^*}{m_h^* + m_e^*}$$

It will be shown that, because of drastic selection rules, the spectrum of formation of excitons corresponds to a series of sharp lines given by the formula

$$\nu = \nu_\infty - \frac{R_{ex}}{n^2} \ \mathrm{cm}^{-1}$$

---

† The energy state of the exciton should be in some cases corrected for the kinetic energy of the exciton which is $\hbar^2 k^2 / 2M_{ex}$. In the optical formation of an exciton, this term is usually neglected (k is the wave vector and M_{ex} is the total mass of the exciton).

where

$$R_{ex} = \frac{R\mu_{ex}}{\varepsilon^2 m_0}$$

$$v = \left(\frac{E-E_v}{hc}\right) \text{cm}^{-1}$$

$$v_\infty = \frac{E_g}{hc} \text{cm}^{-1}$$

where R is the atomic Rydberg constant and R_{ex} is the exciton constant defined above. E_v is the energy of the highest state of the valence band. It can also be shown that the radius of an excitonic orbit is

$$r_{ex}(n) = a_0 \frac{m_0}{\mu_{ex}} \varepsilon n^2$$

where a_0 is the radius of the first orbit of the hydrogen atom. Thus, if $\varepsilon \sim 10$ and $m_0/\mu \sim 2$, then the radius of an exciton is roughly about 20 times that of the hydrogen atom, or 10 Å. R_{ex} is also much smaller than R, since $R_{ex} \sim R/200$ or of the order of 550 cm^{-1}. In excitonic line spectra, the lines in the series are very close to one another. For $v > v_\infty$ the spectrum is a continuum and this absorption corresponds to the band-to-band transition and to the onset of photoconductivity.

Considerable progress has been made recently in the calculation of E_g, which was possible a few years ago for only a few exceptional crystals. Probably it will be possible to calculate v_∞ approximately *a priori* in the near future. Actually, however, it happens that E_g usually cannot be calculated very accurately, so the experimental value of E_g is used and the parameters entering in the theory are adjusted to obtain a good fit. Therefore, the measurement of E_g has great importance. Unfortunately, this measurement is never very accurate either.

D. A More Rigorous Approach to the Theory of Excitons*

In a more rigorous treatment one has to consider that the electron and the hole are under the influence of each other and of all other atoms or ions of the crystal. The wave function of the exciton is given then by the many-body Schrödinger equation in the tight binding approximation:

$$\left[\sum_1^N -\frac{\hbar^2}{2m_0}\nabla_i^2 + \sum_1^N V(\mathbf{r}_i) + \sum_{i \neq j}\frac{e^2}{e\mathbf{r}_{ij}}\right]\psi(\mathbf{r}_1 \dots \mathbf{r}_N) = E\psi(\mathbf{r}_1 \dots \mathbf{r}_N)$$

---

* This section as well as the following two sections may be omitted by the reader if he is not interested in the theoretical part of the subject.

According to Heitler and London, the solution for the unexcited state is

$$\psi = \begin{bmatrix} a_1(\mathbf{r}_1) & a_2(\mathbf{r}_1) & \ldots & a_N(\mathbf{r}_1) \\ a_1(\mathbf{r}_2) & a_2(\mathbf{r}_2) & \ldots & a_N(\mathbf{r}_2) \\ \multicolumn{4}{c}{\ldots\ldots\ldots\ldots\ldots\ldots\ldots} \end{bmatrix} \frac{1}{\sqrt{N!}}$$

where a_i is the atomic wave function and $\mathbf{r}_i$ is the coordinate of the ith electron. The excited state is

$$\psi_i = \frac{1}{\sqrt{N!}} \begin{vmatrix} \ldots & a_i'(\mathbf{r}_1) & \ldots \\ \ldots & a_i'(\mathbf{r}_2) & \ldots \\ \multicolumn{3}{c}{\ldots\ldots\ldots\ldots} \end{vmatrix}$$

where the a_i''s are wave functions of excited atoms. However, this solution is not stable since the excited atom can be created at any point in the crystal and can move without any loss of energy. Thus, a linear combination of these solutions has to be formed. The coefficients in this linear combination are transition probabilities to a particular excited state and the excited state is now

$$\psi_e = \frac{1}{\sqrt{N}} \sum \exp(i\mathbf{k} \cdot \mathbf{n}) D$$

where D is the determinant formed above, $\mathbf{n}$ describes the position of the nth atom, and $\mathbf{k}$ is a wave vector. This formula describes an excitation wave. Proceeding in this manner causes some difficulties since the wave functions of the different atoms are not orthogonal. These can be reduced by replacing atomic functions by Wannier functions [8]. The Wannier functions can be expressed as an expansion in Bloch functions and *vice versa*.

$$W(\mathbf{r} - \mathbf{n}) = \frac{1}{\sqrt{N}} \sum_k \exp(-i\mathbf{k} \cdot \mathbf{n}) \exp(i\mathbf{k} \cdot \mathbf{r}) U_k(\mathbf{r})$$

$$\exp(i\mathbf{k} \cdot \mathbf{r}) U_k(\mathbf{r}) = \frac{1}{\sqrt{N}} \sum_n \exp(i\mathbf{k} \cdot \mathbf{n}) W(\mathbf{r} - \mathbf{n})$$

By use of these properties it can be shown that the wave functions both of the ground and excited states can be written by forming determinants from Wannier or Bloch functions. If $U_{vkj}(\mathbf{r})$ is a Bloch function of the valence band, the ground state is

$$\psi = \frac{1}{\sqrt{N!}} \begin{vmatrix} \exp(i\mathbf{k}_1 \cdot \mathbf{r}_1) U_{vk_1}(\mathbf{r}_1), & \exp(i\mathbf{k}_1 \cdot \mathbf{r}_1) U_{vk_2}(\mathbf{r}_1), & \ldots \\ \ldots (i\mathbf{k}_1 \cdot \mathbf{n}_2) \ldots (\mathbf{n}_2), & \ldots (i\mathbf{k}_1 \cdot \mathbf{n}_2) \ldots (\mathbf{n}_2), & \ldots \\ \vdots & \vdots & \vdots \end{vmatrix}$$

It can be shown in a generalization by Slater and Shockley [9] that determinants for the excited state can be formed with the hole on the nth atom and the electron on the mth. These determinants $D(\mathbf{mn})$ now can be generalized and written in forms of Bloch determinants

$$D(\mathbf{m\ n}) = \frac{1}{N} \sum_{k_1} \sum_{k_2} \exp{(i\mathbf{k}_1 \cdot \mathbf{m} - i\mathbf{k}_2 \cdot \mathbf{n})}\, B(\mathbf{k}_1\, \mathbf{k}_2)$$

$$B(\mathbf{k}_1\, \mathbf{k}_2) = \frac{1}{N} \sum_{m} \sum_{n} \exp{(-i\mathbf{k}_1 \cdot \mathbf{m} + i\mathbf{k}_2 \cdot \mathbf{n})}\, D(\mathbf{m\ n})$$

Here $B(\mathbf{k}_1 \mathbf{k}_2)$ is a determinant formed from Bloch functions in which an electron with a wave vector $\mathbf{k}_1$ is removed from the valence band and replaced by an electron in the conduction band with a wave vector $\mathbf{k}_2$.

A final solution of the problem requires that a linear combination of $B(\mathbf{k}_1 \mathbf{k}_2)$ be formed such that

$$\psi_{ex} = \sum_{k_1 k_2} f(\mathbf{k}_1\, \mathbf{k}_2)\, B(\mathbf{k}_1\, \mathbf{k}_2)$$

The value of $f(\mathbf{k}_1 \mathbf{k}_2)$ is related to a two-particle exciton wave function by the relation

$$\phi(\mathbf{r}_1 \mathbf{r}_2) = \frac{1}{V_0} \int \exp{(-i\mathbf{k}_1 \cdot \mathbf{r}_1 + i\mathbf{k}_2 \cdot \mathbf{r}_2)}\, f(\mathbf{k}_1\, \mathbf{k}_2)\, d\mathbf{k}_1\, d\mathbf{k}_2$$

where V_0 is the volume of the unit cell. The function ϕ is the solution of the two-particle Schrödinger equation

$$\left(-\frac{\hbar^2}{2m_e^*} \nabla_e^2 - \frac{\hbar^2}{2m_h^*} \nabla_h^2 - \frac{e^2}{\varepsilon r_{eh}} \right) \phi_{ex} = E_{ex}^b\, \phi_{ex}$$

where ε is the dielectric constant introduced by Haken in an elegant way, m_e^* and m_h^* are the masses of the electron and the hole, respectively, and r_{eh} is their distance of separation. This is the same result as obtained in the model of Mott. However, it is possible now to complete the theory in calculating the transition probabilities. Finally, it should be mentioned that these considerations are restricted to the case when both bands are of spherical symmetry in the vicinity of $\mathbf{k} = 0$.

E.   The Oscillator Strength

The oscillator strength is given by the relation

$$f = g_q\, |Q_{0e}|^2$$

$$Q_{0e} = \int \psi_{ex}^*\, \mathbf{e}\, \mathrm{grad}\, \psi_0\, d\tau$$

where $g_q = \hbar/3\pi m_0 \omega_e$ and $\mathbf{e}$ is the unit vector along the electric field of the exciting light. The term ψ_{ex} is given by the relation obtained in the last section and ψ_0 is the wave function of the ground state given in its determinant form. If we express both functions in terms of Bloch functions, a number of integrals are going to vanish because of the orthogonality of the Bloch functions. The integrals that remain are

$$Q_{0e} = \sum_{\mathbf{k}_1 \mathbf{k}_2} f(\mathbf{k}_1 \mathbf{k}_2) \int \exp(-i\mathbf{k}_1 \cdot \mathbf{r}) \, U_{c\mathbf{k}_2} \, \mathbf{e} \, \mathrm{grad} \, \exp(i\mathbf{k}_2 \cdot \mathbf{r}) \, U_{v\mathbf{k}_1} \, d\tau$$

The integral is different from zero only if $\mathbf{k}_1 = \mathbf{k}_2$, which is the $\varDelta\mathbf{k} = 0$ selection rule. When the gradient is calculated, one of the integrals is zero and the second is

$$\sum_{\mathbf{k}} f(\mathbf{k}) \int U_{c\mathbf{k}}^* \, \mathbf{e} \, \mathrm{grad} \, U_{v\mathbf{k}} \, d\tau$$

Now, one can write $-\mathbf{k}_2 + \mathbf{k}_1 = \mathbf{k}$. Since $\mathbf{k}_1 = \mathbf{k}_2$, the vector $\mathbf{k}$ is 0, which is a second condition from the selection rules $\varDelta\mathbf{k} = 0$ and $\mathbf{k} = 0$. Furthermore

$$\phi(\mathbf{r}_1 \mathbf{r}_2) = \frac{\exp(i\mathbf{k} \cdot \mathbf{R})}{\sqrt{V_0}} \, \psi(\mathbf{r}_{eh})$$

if the wave function is referred to at the center of gravity. Again ψ can be written as

$$\frac{\psi(\mathbf{r}_{eh})}{\sqrt{V_0}} = \sum \frac{\exp i\mathbf{k}_{12} \cdot \mathbf{r}_{eh}}{V_0} \, f(\mathbf{k}_{12})$$

where we take into account that $\mathbf{k} = 0$ and write $\mathbf{k}_{12}$ for $\frac{1}{2}(\mathbf{k}_1 + \mathbf{k}_2)$. This relation will make it possibile for us to calculate the different transition probabilities.

F.   Selection Rules

In the above treatment we have neglected the wave vector of light. Accordingly, $\mathbf{k}$ is not exactly zero. In fact one has to consider Q_{0e} for values of $\mathbf{k} \sim 0$, but not necessarily zero. Therefore, an expansion of Q for small values of $\mathbf{k}$ gives

$$Q_{0e} = \sum_{\mathbf{k}} f(\mathbf{k}) \int U_{v0}^* \, \mathbf{e} \, \mathrm{grad} \, U_{c0} \, d\tau$$

$$+ \sum_{\mathbf{k}} \mathbf{k} \, f(\mathbf{k}) \, \mathrm{grad} \int U_{v0}^* \, \mathbf{e} \, \mathrm{grad} \, U_{c0} \, d\tau$$

$$+ \sum_{ij} \mathbf{k}_i \mathbf{k}_j \, f(\mathbf{k}) \, \frac{\partial^2}{\partial \mathbf{k}_i \partial \mathbf{k}_j} \int U_{v0}^* \, \mathbf{e} \, \mathrm{grad} \, U_{c0} \, d\tau$$

where U_{v0} represents the Bloch wave function for the valence band, U_{c0} is the same for the conduction band, and $\mathbf{k} = 0$. The different terms of the expansion are usually not simultaneously zero. We are going to call the transition for which the first term is different from zero a first-class transition. When the second term is different from zero, the first is usually zero (for crystals with a center of symmetry). These are named second-class transitions. They are weakly forbidden transitions but still correspond to the dipole approximations. Very little is known about the third term, except that it is very small in comparison to the others. It can be seen from the above formula that the selection rules are essentially governed by the symmetry of the U function. Furthermore, if the excitonic spectrum is dichroic, this depends on the band structure.

G. The First-Class Spectra

A first-class spectrum corresponds to the case when the first term of the expansion is different from zero. It can be shown that in this case the second term is usually zero. The transition corresponds to $\Delta\mathbf{k} = 0$ and $\mathbf{k} = 0$. In this case, assuming that the conduction band has a spherical minimum at $\mathbf{k} = 0$ and that the valence band has a maximum at this point, we can show that the matrix element is

$$Q_{0e} = \sqrt{V_0}\, \psi_{nlm}^{(0)} \int U_{v0}^* \, \mathbf{e}\, \mathrm{grad}\, U_{c0}\, d\tau$$

where V_0 is the unit cell, $\psi_{nlm}(0)$ is a hydrogen-like wave function of the exciton taken for $\mathbf{r} = 0$, U_{v0} and U_{c0} are, respectively, the Bloch functions of the highest state of the valence band and the lowest state of the conduction band for $\mathbf{k} = 0$. This matrix element is different from zero only for s states. Thus, in this case, excitons are formed in spectroscopic S states. Now, if a_{ex} is the radius of the first exciton orbit

$$\psi_{nlm}^{(0)2} = \frac{1}{\pi a_{ex}^3 n^3}$$

The oscillator strength per unit cell is then

$$f_V = \frac{V_0}{\pi a_{ex}^3 n^3}\, g_q \left(\int U_{v0}^* \, \mathbf{e}\, \mathrm{grad}\, U_{c0}\, d\tau \right)^2$$

or per atom

$$f_a = \frac{V_p}{\pi a_{ex}^3 n^3} f_B$$

where $V_p = V_0/p$, p being the number of atoms per unit cell, and where

$$f_B = g_q \left(\int U_{v0}^* \, \mathbf{e} \, \text{grad} \, U_{c0} \, d\tau \right)^2$$

for

$$g_q = \frac{h}{3 \pi m_0 \omega_e}$$

Thus, it is expected in this case that a series of lines must be observed with the wave numbers corresponding to the formula

$$v_n = v_\infty - \frac{R_{ex}}{n^2}$$

where $n = 1, 2, 3, 4, \ldots$.

The intensity of the lines decreases as n^{-3}, and the absolute value of f_a can be expected to be

$$f_a \simeq 2 \times 10^{-3} \frac{1}{n^3} f_B$$

This is obtained assuming that $\varepsilon = 10$, $m_e^* = m_h^*$, $\mu = 0.25 \, m_0$, $a_{ex} \sim 20 \text{Å}$, and $V_p \sim 50 \text{ Å}^3$. Since f_B in the tight binding approximation is approximately 1, it is expected that f_a should be of the order of magnitude of 2×10^{-3}. This absorption may appear to be weak. However, with a line width of 10 or 5 Å, the maximum of the absorption coefficient K_{max} in the line (for the visible) is expected to be about 4 or $2 \times 10^5 \text{ cm}^{-1}$, respectively, which is a very high value. In order to measure the profile of the absorption line, one should use thin platelets of about $x \sim 10^{-5}$ cm or 0.1 μ. This is indeed a very small thickness for single crystals.

Notice also that the second line ($n = 2$) is 8 and the third line ($n = 3$) is 27 times weaker than the first line. For this reason and because of other drastic conditions, it is not likely that lines of much higher order could be observed. Furthermore, as lines of higher quantum order are very close to one another, they will probably overlap and form a continuum.

H. The Continuum

On the high-energy side, the series converge to a continuum. Elliott [3] has shown that the continuum is of two kinds: (1) the "overlap continuum" and (2) the real continuum. The overlap continuum is composed of the super-position of lines of higher quantum number. Elliott has calculated this continuum with the assumption that the overlapping lines are narrow.

In fact, the width of the lines is comparable to the width of the overlap continuum. This assumption is probably too rough. The absorption coefficient under this assumption is

$$n_0 K = 8 \pi \omega_0 (Z_B)^2 \varepsilon / a_{ex}^2 c$$

which is practically a constant. Here

$$Z_B = \int U_{0v}^* z U_{0c} \, d\tau$$

and n_0 is the refractive index. The absorption coefficient is of the order of 10^5 for the same conditions as above. For this continuum, $v < v_\infty$.

The true continuum where $v > v_\infty$ has been calculated by Elliott and is given by

$$n_0 K = \frac{2 \pi \omega_0 e^2 |Z_B|^2 \alpha \exp(\pi \alpha) (2\mu)^{\frac{3}{2}} E^{\frac{1}{2}}}{c \hbar^3 \, sh \pi \alpha}$$

where

$$\alpha = \left(\frac{R_e \, ch}{E} \right)^{\frac{1}{2}}$$

and

$$E = h v - E_g$$

In this continuum the interaction between hole and electron is taken into account but the binding energy is smaller than the kinetic energy. Therefore, this is not identical with the usual band-to-band transition, though it involves a similar mechanism. Furthermore, it can be seen that both continua tend to the same value when $v \to v_\infty$. For $E \gg R_e ch$, K varies as $E^{1/2}$ as in the case of band-to-band transitions.

I. Second-Class Transitions [3]

In some transitions, the first term of the expansion of the matrix element for the transition moment may be zero. In this case the second term of the expansion may be different from zero. The oscillator strength is then given by the formula

$$f_a = g_q V_p \left[\frac{\partial \psi(0)}{\partial r} \right]^2 \left(\frac{\partial}{\partial k} \int U_{0v}^* \, e \, \text{grad} \, U_{0c} \, d\tau \right)^2$$

This corresponds to another type of transition, which is weakly forbidden. It should be emphasized that these transitions are still electrical dipole transitions. Therefore we suggest giving them the name of second-class rather than "forbidden" transitions. Only for p states is $[\partial\psi(0)/\partial r] \neq 0$. Thus, in this case, in a spectroscopic P state, excitons are formed. Accordingly, a hydrogen-like series of lines is expected. However, the transition to the $n = 1$ exciton state is not possible since there is no P state corresponding to $n = 1$. Now

$$\left[\frac{\partial\psi(0)}{\partial r}\right]^2 = \frac{n^2-1}{n^5}\frac{1}{3\pi a_{ex}^5}$$

The oscillator strength can thus be written

$$f_a = \frac{n^2-1}{n^5\varepsilon^5}\cdot C_2$$

and it can be seen that with $\mu/m_0 = \frac{1}{2}$, the value of C_2 is roughly of the order of unity. It follows for $n = 2$ (first line of the series) and $\varepsilon \sim 10$ that $f_a \simeq 10^{-6}$; and for $\Delta\lambda_{\frac{1}{2}} \simeq 10$ Å and $\lambda \simeq 5000$ Å that $n_0 K_{max} \simeq 10^2\,\mathrm{cm}^{-1}$. The absorption is thus expected to be roughly 10^3 times weaker for second-compared with first-class transitions. This happens to be a fortunate circumstance since in this case crystals are required to be only 10^{-2} to 10^{-3} cm thick. These are much easier to obtain than the 10^{-5} cm thick crystals needed in the case of the first-class transitions.

J. The Second-Class Continuum

It can be shown that in this case two kinds of continua again must be considered. The overlap continuum is given by the formula

$$n_0 K = \frac{8\pi\omega_e}{3}(Z_B)^4\,\varepsilon\,[1+(h\nu-E_g)/R_e\,c\hbar]/ca_{ex}^4$$

This gives a value of about $n_0 K = 5\times10^2$. The true continuum can be shown to be

$$n_0 K = \frac{2\pi\omega_0\,e^2\,|Z_B|^4\,(1+\alpha^2)\,\alpha\,\exp(\pi\alpha)\,(2\mu)^{\frac{3}{2}}\,E^{\frac{3}{2}}}{3\,c\hbar^5\,sh(\pi\alpha)}$$

where E is $h\omega-E_g$ and is > 0. The transition between both continua is again smooth.

It has to be noted that as $h\omega-E_g$ is not very small with respect to $R_e\,c\hbar$ in the first continuum, $n_0 K$ increases slightly with $h\nu$; whereas in the second continuum for E large, $n_0 K$ increases with $E^{3/2}$. Here again the coulomb interaction has been taken into account.

K. Indirect Transitions [3]

The theory for indirect transitions is also applicable to transitions which involve phonon cooperation but are not really indirect. Suppose the minimum of the conduction band is off the $k = 0$ position and corresponds to $k = k_0$ while the valence band has a maximum at $k = 0$. Furthermore, assume that ω_{k_0} is the circular frequency of the phonon of wave vector k_0. Elliott [3] has shown that indirect transitions can take place with cooperation of phonons of frequency ω_{k_0} (Fig. 2). If E_g is the energy gap in this case, two absorption edges will arise from the first exciton level

$$E_g - R_{ex} \, ch \pm \hbar\omega_{k_0}$$

If it is assumed that $E = \hbar\omega - E_g \pm \hbar\omega_{k_0}$, the absorption coefficients corre-

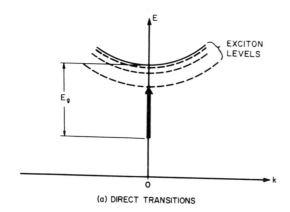

(a) DIRECT TRANSITIONS

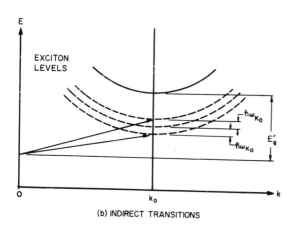

(b) INDIRECT TRANSITIONS

Fig. 2. Direct (a) and indirect (b) transitions.

sponding to the transitions to the first exciton state are of the form

$$K_1 = A(R_{ex}\,ch + E)^{\frac{1}{2}}\,N$$
$$K_2 = A(R_{ex}\,ch + E)^{\frac{1}{2}}\,(N+1)$$

with phonon cooperation in each edge independent of the other. Clearly, these edges form continua, as phonons of all suitable energies are available or can be given to the lattice. These expressions can be written as

$$K_1 = a(\omega - \omega_1)^{\frac{1}{2}}\,N$$
$$K_2 = a(\omega - \omega_2)^{\frac{1}{2}}\,(N+1)$$

with $\omega_2 - \omega_1 = 2\omega_{k_0}$ and the number of available phonons N is

$$N = 1/[\exp(\hbar\omega_{k_0}/kT) - 1]$$

It appears from these formulas that as N tends to zero for $T \to 0$, $K_1 \to 0$ but $K_2 \to a(\omega - \omega_2)^{1/2} \neq 0$.

The quantities

$$p_1 = \frac{K_1}{(\omega - \omega_1)^{\frac{1}{2}}} \begin{cases} \dfrac{akT}{\hbar\omega_{k_0}} & T \to \infty \\[3mm] \exp\left(-\dfrac{\hbar\omega_{k_0}}{kT}\right) & T \text{ small} \end{cases}$$

$$p_2 = \frac{K_2}{(\omega - \omega_2)^{\frac{1}{2}}} \begin{cases} a\left(\dfrac{kT}{\omega_{k_0}\hbar} + 1\right) & T \text{ big} \\[3mm] a\left[\exp\left(-\dfrac{\hbar\omega_{k_0}}{kT}\right) + 1\right] & T \text{ small} \end{cases}$$

Figure 3 illustrates schematically first-class, second-class, and indirect transition spectra as predicted by the theory.

L. Spin Orbit Splitting

We have seen that excitons in S or P states can be formed. These are spectroscopic states representative of the relative motion of a hole and an electron.

The charge carriers in an exciton are nearly a free hole in the valence band and nearly a free electron in the conduction band. Now the hole and electron in both respective bands have a spin and may also have an orbital momentum. These factors introduce the following considerations.

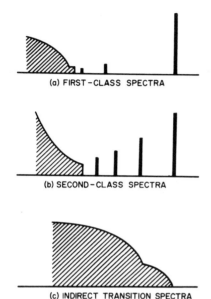

(a) FIRST-CLASS SPECTRA

(b) SECOND-CLASS SPECTRA

(c) INDIRECT TRANSITION SPECTRA

Fig. 3. The aspect of (a) first-class, (b) second-class, and (c) indirect transition spectra as predicted by the theory.

Consider for example CuCl. A band structure calculation reveals that the highest valence band has the character of $3d$ Cu^+ electrons, although with an appreciable superposition of $3p$ Cl^- character. It will be seen that the symmetry of this band is Γ_{15}, which in the tetragonal field of zinc-blende structure has a p character. The conduction band is a Γ_1 band made up of $4s$ Cl^- and $4s$ Cu^+ orbitals. From the atomic point of view, the $d \rightarrow s$ transition on Cu atoms is forbidden; thus, the transition is allowed because of $p \rightarrow s$ transitions on Cl^-. In any event, this transition has a p–s character from both the atomic and group-theoretical points of view. In fact, in the tight binding approximation, the hole in the valence band of Cl is just a neutral Cl atom. This has the configuration $3p^5$ and has consequently an overall momentum $j = 1/2$ or $3/2$. The lowest spectroscopic state in CuCl is a $p_{3/2}$ state. Both atomic states have different energies, so that the valence band splits into two bands.

For the case considered, the electron in the conduction band is certainly in an s state. The overall exciton state is an S state. It is probable that the jj coupling is more adequate for excitons. The following exciton states in order of decreasing energies are

$$(3/2, 1/2)1 \quad (3/2, 1/2)2 \quad (1/2, 1/2)1 \quad (1/2, 1/2)0$$

The ground state of the unexcited crystal is certainly an $^1S_0$ state. The transitions then are possible only to the first and third states with $\Delta J = 1$. The transition to the second state is forbidden because $\Delta J \sim 2$ and the transition to the last state is also forbidden because of the $j = 0 \rightarrow j = 0$ interdiction.

The magnitude of the spin orbit splitting can be calculated. In the case of CuCl this calculation must take into account the mixed character of the valence band. It is shown that this mixed character can produce considerable perturbations

$$\Delta E_{CuX} = \lambda \left[\alpha \Delta E_{X-} - (1-\alpha) \Delta E_{Cu+} \right]$$

Here λ is a coefficient, α is the percentage of halogen in the mixing, while ΔE_{X-} and ΔE_{Cu+} are, respectively, the spin orbit splitting of the two components. The above formula was suggested empirically be Cardona [11] and obtained theoretically since by Song [12].

M. The Interaction with the Lattice

The interaction with the lattice has been studied by several authors and has an influence on two important factors.

1. The Shape of the Lines and the Line Width

This has been studied by Toyozawa [13]. The form of the exciton lines has been studied as a function of temperature and of the coupling with the lattice vibrations. The theory is rather complicated and the approximations were worked out in a satisfactory way only quite recently, so we are not going to analyze it here. The general result is that, if the coupling is weak and the temperature low, the shape of the lines is Lorentzian though dissymmetrical. The dissymmetrical form depends on the existence of lower or higher states. Thus, in a case when a lower state exists, the lines will have stronger wings on the low-energy side. The interaction of two lines which are close to one another is also important and can even lead to so-called negative absorption. However, a quantitative comparison with experiment has not yet been made. It is not quite certain that the theory is in a sufficiently advanced state to be compared with the experimental data.

The interpretation of the shape of the line is certainly very important. Very useful data will be available as soon as this interpretation is possible. Some important results have been obtained recently in the present author's group with Merle [14].

2. The Dielectric Constant Involved in Exciton Theories

Notice that a dielectric constant is involved in the theory of the exciton states. This dielectric constant was first empirically introduced by Mott and Gurney [7] in order to take care of the interaction with the crystal. It has been subsequently shown by Haken and Schottky [15] that this assumption can be established in a rigorous way.

Haken [4] and Meyer [16] have shown that the term of potential energy in the Schrödinger equation should be written as

$$-\frac{e^2}{\varepsilon_0\,r}+\frac{e}{r}\left(\frac{1}{\varepsilon_0}-\frac{1}{\varepsilon_s}\right)\left[1-\frac{\exp(-w_1\,r)+\exp(-w_2\,r)}{2}\right]$$

In this formula ε_0 and ε_s are the optical and static dielectric constants, respectively, and

$$w_i = (2\,m_i\,\omega_0/\hbar)^{\frac{1}{2}}$$

where m_i is the mass of the electron or the hole and ω_0 is the circular frequency of the longitudinal optical vibration of the lattice.

A rough rule can be deduced from this formula. If $r \gg w^{-1}$, the exponential can be neglected and the expression becomes $-e^2/\varepsilon_s r$. If, however, $r \ll w^{-1}$, the formula becomes $-e^2/\varepsilon_0 r$. This means roughly that for the first orbit $n = 1$ and ε_0 must be used in first approximation. For orbits of higher order $n = 2, 3$, etc., ε_s must be used. This is, however, not always a good approximation. In many cases the general formula has to be used. This is not quite simple as the radius of the orbit depends on the dielectric constant and *vice versa*.

II. EXPERIMENTAL INVESTIGATION OF EXCITON SPECTRA

A. Introduction

Different substances showing spectra of nonlocalized excitons have been studied by the Strasbourg group. Experimental results were compared with Elliott's theory and found to be in quite good agreement. The theory describes the experimental observations even in some detail. Conversely, this good agreement proves that the observed spectra are exciton spectra and provides a confirmation of the existence of excitons.

Discussions here will deal with two examples of experimental results: Cu_2O and $CuCl$. Since the band structure of these compounds was recently calculated, we would like to describe the spectra of these substances from the point of view of band structure calculations.

B. The Spectrum of Cu_2O

1. *Experimental Conditions*

It will be impossible in this chapter to give details of the experimental techniques used in the preparation of samples and spectroscopic or spectrometric measurements. The spectroscopic techniques contain no great difficulties. However, the major problem in solid-state spectroscopy remains in the preparation of the samples of known purity and optically suitable quality and thickness. When this problem is solved for a given substance the rest is rather simple.*

2. Early Work on Cu_2O

The spectrum of Cu_2O is remarkable since all the classes of transitions predicted by the theory for exciton spectra are observed in different parts of the visible. The first observations of exciton spectra were made by Hayashi Mazakuzu [17] and published in 1950 and 1952. Gross and Karryeff [18] made similar observations about six months later and reported them in 1952. The Strasbourg group was involved in similar work on HgI_2 as early as 1951. However, this case happened to be very complicated and the first publication appeared in 1954 [19]. However, a parallel work was started on Cu_2O in 1952 and reported in November 1953 [20]. The spectra of Cu_2O have been investigated since by the Leningrad and Strasbourg groups in parallel work.

3. General Properties of the Spectrum of Cuprous Oxide

The spectrum of cuprous oxide begins in the red with two absorption edges 208 cm^{-1} apart (Fig. 4). On the high-energy side of the edges, the absorption is continuous and increases with the wave number. At low temperatures a line spectrum appears on top of the red continuum. The lines belong to two series named the "yellow" (Fig. 5) and the "green" (Fig. 6) series. These two series are separated by a yellow continuum strongly increasing from the yellow to the green series.

Both series are hydrogen-like. In the yellow series nine lines have been observed in Strasbourg, while only three lines are observed in the green series. At the temperature of 4.2 °K, the series is well represented by the formulas

$$\nu_n^y = 17{,}525 - \frac{786}{n^2} \text{ cm}^{-1} \qquad n = 2, 3, 4, \ldots$$

$$\nu_n^g = 18{,}588 - \frac{1242}{n^2} \text{ cm}^{-1} \qquad n = 2, 3, 4, \ldots$$

A very weak line has been identified as the $n = 1$ line of the yellow series at $\lambda_1^y = 6097.7$ Å (16,399.5 cm^{-1}) [21]. This line is 339.5 off the position given

---

* The quantitative interpretation of measurements on exciton lines may lead to considerable difficulties, however, when the absorption is very strong and when sufficiently thin samples are too difficult to prepare.

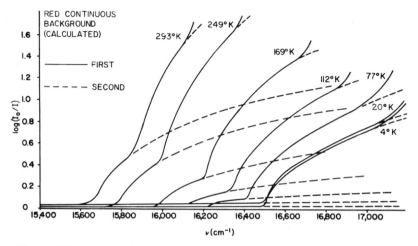

Fig. 4. The two absorption edges of the low-energy part of the absorption spectrum of Cu_2O at different temperatures. Note that the low-energy edge disappears at very low temperatures. The dotted curves are calculated on the basis of Elliott's theory.

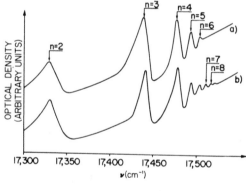

Fig. 5. The yellow series of absorption lines of Cu_2O at $4.2°K$. (a) A sample of good quality. (b) An exceptional sample. Note that the $n = 9$ line is observed. This is a densitometric record of a photographic plate.

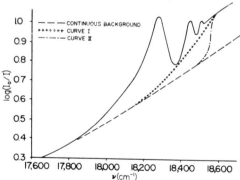

Fig. 6. The green series of absorption lines of Cu_2O at $4.2°K$; the different interrupted lines are extrapolations for construction of the continuum on top of which the lines are observed.

by the formula on the low-energy side. This agrees with Haken's [4] calculations (Fig. 7).

Another line at λ_1^y = 5829 Å at 4.2 °K has been recently observed and could be the $n = 1$ line of the green series [22]. Both series are 1063 cm^{-1} apart. In the region described above, the absorption is rather weak. However, it becomes very strong on the high-energy side of the green series. Several lines can be seen in the blue and indigo [23].* These lines are very broad and strong. They form two distinctly different spectra. It was not possible to find more than one or two lines in each spectrum. These lines are separated again by about 1000 cm^{-1}. When the sample is prepared on a glass or quartz substrate, the lines exhibit a splitting (probably on account of tensions) (Fig. 8). They are rather difficult to observe, requiring very thin films without any substrate to prevent tensions. This is very difficult to obtain. Furthermore, it is necessary that the crystallites in the film be rather large, say, about 10^4 Å or more. In spite of these difficulties, some reliable data are available at present. In this part of the spectrum, thin as well as thick samples exhibit strong reflection anomalies similar to those observed by Reiss and Nikitine on other substances (Fig. 9). A rather complicated spectrum is observed in the ultraviolet [24].

* All the exciton lines are rather broad. The line width is a few angstroms for some lines of some substances, but can be 20−50 Å for other lines and substances. Though these lines are much broader than atomic lines, it has become common to call them "extension lines."

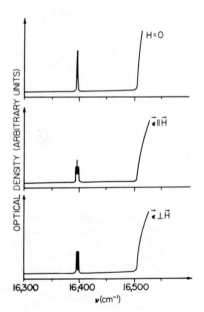

Fig. 7.   The forbidden quadrupole $n = 1$ line of the yellow series at 4.2 °K. This line is very weak and narrow; it does not appear in Fig. 4. The Zeeman effect of this line shown in this figure is not discussed in the text but has been studied by the Leningrad and Strasbourg groups.

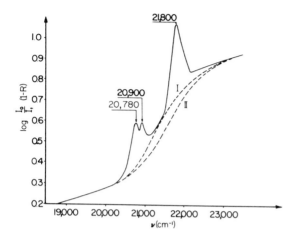

Fig. 8. Absorption spectrum of very thin films of Cu_2O at 4.2 °K in the blue and indigo parts of the spectrum.

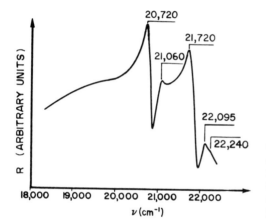

Fig. 9. Reflection anomalies in the same part of the spectrum as the lines represented in Fig. 8, but obtained from the surface of thick crystals. On thin films the anomalies are similar but less pronounced.

It is obvious that absorption data on cuprous oxide suggest that: (1) the two red edges are transitions with phonon cooperation; (2) the yellow and green series are second-class transitions separated by a spin orbit splitting; (3) the absorption lines in the blue and indigo are exciton first-class transitions from the same valence band to another conduction band. In order to establish these points a quantitative investigation of these spectra was necessary.

4. Quantitative Investigation of the Red Edges [26]

The red absorption is composed of two edges beginning at v_1^r and v_2^r. For $v_1^r < v < v_2^r$, $K = K_1$ and for $v > v_{2b}$, $K = K_1 + K_2$ or $K_2 = K - K_1$. A first consideration would be to test experimentally the relations given by the theory.

$$K_1 = a(v - v_1^r)^{\frac{1}{2}} N$$

$$K_2 = a(v - v_2^r)^{\frac{1}{2}} (N+1)$$

However, Elliott has shown that on account of the blue absorption which is temperature-independent, while the red edges shift to lower energies when the temperature increases, a is variable with temperature. Therefore,

$$K_1 = \frac{bN(v - v_1^r)^{\frac{1}{2}}}{[E_{b1} - (E_1 - \hbar\omega_{k_0})]} = p_1(v - v_1^r)^{\frac{1}{2}}$$

$$K_2 = \frac{b(N+1)(v - v_2{}^r)^{\frac{1}{2}}}{E_{b1} - (E_1 + \hbar\omega_{k_0})} = p_2(v - v_2{}^r)^{\frac{1}{2}}$$

where $E_{b1} = cte$, $E_1 = f(T)$, $\omega_{k_0} = (\omega_2{}^r - \omega_1{}^r)/2$, b is a constant, and the denominator is variable with T.

The variation of $K_i{}^2$ with v for different temperatures is given in Figs. 10 and 11. It can be seen that the slope $p_1{}^2$ tends to zero for 4.2°K and that $p_2{}^2$ tends to a finite value. It should in particular be noted that K_1 vanishes with T. No phonons are available as $T \to 0$. There is no doubt that the red

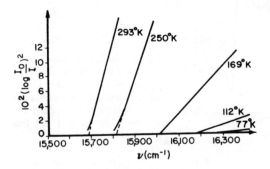

Fig. 10. Variation of $K_1{}^2$ versus v for different temperatures.

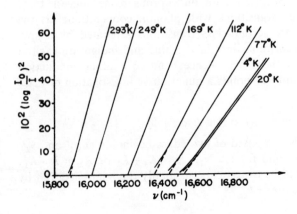

Fig. 11. Variation of $K_2{}^2$ versus v for different temperatures.

edges are due to indirect transitions to the $n = 1$ exciton state ($\pm \hbar \omega_{k_0}$), the direct transition to $n = 1$ being forbidden in the dipole approximation. The agreement with Elliott's theory is quite good.

5. *Oscillator Strength of the Lines in the Yellow and Green Series (Second-Class Transitions)*

   a. *Yellow Series* [26]. The oscillator strengths per unit cell of the lines $n = 2$, $n = 3$, and $n = 4$ have been determined by the method of Krawetz. The last is however not an accurate determination because the indeterminancy in the background must be substracted. It has been found that at $4.2°K$, $f_2^y = 3 \times 10^{-6}$; $f_3^y = 1 \times 10^{-6}$; $f_4^y = 0.33 \times 10^{-6}$; $f_3/f_2 = 0.36$; and $f_4/f_2 = 0.11$.

   The theoretical values for second-class transitions are: $f_3/f_2 = 0.34$ and $f_4/f_2 = 0.14$. Furthermore, it will be remembered that the order of magnitude predicted by the theory for second-class transitions agrees well with the measured data. Thus, the agreement is very good. Furthermore, it is also possible to calculate $r_2^y = 36$ Å; $\mu^y = 0.52 \, m_0$; $E_g^y = 17,525 \, cm^{-1}$.

   b. *Green Series* [27]. The following values have been obtained for the first two lines of the green series $f_2 = 3.5 \times 10^{-5}$; $f_3 = 0.5 \times 10^{-5}$; $f_3/f_2 = 0.14$; and $f_2^g/f_2^y = 11$. It must be mentioned that the determination of f_3 for the green series is not very accurate and therefore the ratio f_3/f_2 is very likely to be off the theoretical value. As above, calculations show that $r_2^g = 24$ Å; $\mu^g = 0.8 \, m_0$; $E_g^g = 18,588 \, cm^{-1}$; and $f_2^g/f_2^y = 11$. The ratio is approximately the same as $(R_e^g/R_e^y)^5 = 9.1$. The yellow and green continua are also of the theoretical type. There is no doubt that the yellow and green series are second-class exciton transitions. The absolute value of the oscillator strengths, the relative variation of these with n, the fact that the series begin with $n = 2$, and finally the yellow and green continua are all in quite good agreement with the prediction of the theory.

6. *The* n $= 1$ *Line of the Yellow Series*

   The $n = 1$ line of the yellow series is of great interest and much work has been carried out on this line [21]. It has been shown that this forbidden line has a quadrupole character and is dichroic. Considerable work is now being done by the Strasbourg group on this line, since it appears that the controversy between Elliott and Dahl on the spin orbit splitting can be resolved from experimental evidence on this line. The line is off its theoretical position in the direction predicted by Haken. All the described spectra strongly shift to low energies when the temperature rises. This important property will not be discussed here. However, the temperature shift of the two edges and the $n = 1$ line is shown in Fig. 12. Any further discussion is beyond the scope of this chapter.

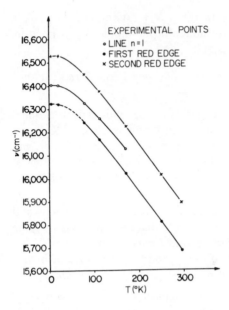

Fig. 12. Temperature shift of the two absorption edges and of the $n = 1$ line.

7. The First-Class Exciton Transition in Cu_2O [23, 28]

In the blue and blue-violet (indigo) part of the spectrum, the absorption becomes very strong. This requires that very thin samples be prepared, which is not a simple matter. However, the samples obtained on a SiO_2 substrate have been investigated for absorption. In addition, both thick samples without support and thin films with substrate have been studied for reflection. The chemical composition of the films has been studied both by gravimetric and electrical procedures. Cu_2O is usually obtained with some excess of oxygen. However, our samples do not differ substantially from the theoretical proportion.

In this part of the spectrum two strong absorption bands have been observed 1000 cm^{-1} apart [23]. Strong reflection anomalies are superimposed on these absorption peaks. The reflection spectra of the two kinds of samples are similar. The absorption peaks observed only with thin films are at $\lambda_{BI} = 4820$ Å and $\lambda_{Ind} = 4604$ Å. At 4.2°K the peaks are rather broad and the separation is not very accurately measured, but is almost the same as for the second-class transition in the yellow and green. From these thin films, the reflection and absorption spectra show a second line which is almost certainly due to splitting by tension caused by the substrate. In the best films, a very weak structure indicated a second peak on the high-energy side of the strong lines. The oscillator strengths of both peaks are, respectively, $f_{BI} = 6 \times 10^{-3}$ and $f_{Ind} = 10^{-2}$, as reported recently by Daunois et al. [28]. The samples used for absorption measurements were about 1000 Å thick, and more. It is most likely that these peaks be identified as first-class exciton peaks.

The transitions almost certainly involve the same valence band (split by spin orbit interaction) as above, but also a second higher conduction band (Elliott and Nikitine) [23]. They are not shifted with temperature. This behavior is opposite to that of the red, yellow, and green absorption. These transitions can also be studied by reflection from good surfaces of thick crystals, where very strong anomalies are observed. They are in agreement with first-class transitions, observed with other substances. With thin films, the anomalies are similar but less pronounced and correspond well to the absorption lines.

8. The High-Energy Absorption of Cu_2O

In the ultraviolet the spectrum of Cu_2O becomes very complicated and the absorption still stronger. These measurements have been carried out by Mme. Brahms [24]. Samples of thickness down to 400 Å had to be used. The reflection anomalies of thick samples still correspond very well to the absorption singularities of the thin samples. Thus their composition should still be comparable to somewhat thicker samples.

Absorption and reflection spectra are shown in Figs. 13 and 14, respectively. A broad maximum labeled A is observed at 3.62 eV. This maximum is rather flat at room temperature but becomes much higher and sharper at low temperatures.

Considerable depressions are observed in both the absorption and the reflection curves between 3.65 and 4.2 eV. The absorption increases strongly with decreasing wavelength and shows a group of maxima between 4.3 and

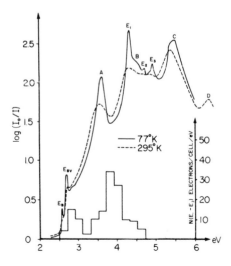

Fig. 13.   High-energy absorption spectrum from a thin film of Cu_2O. The absorption in the yellow and green parts of the spectrum is too weak and does not appear in this diagram.

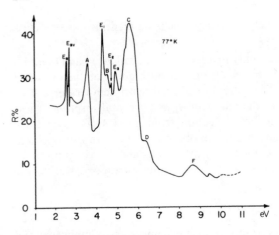

Fig. 14. Reflection from a thick single crystal of Cu_2O. The reflection spectra from thin films are similar but the anomalies are less pronounced.

5.5 eV. When the temperature is lowered to 77°K, there is observed also a sharp peak E_1 at 4.33 eV, a shoulder at 4.48 eV, followed by several weaker but sharp peaks E_2 and E_3 at 4.74 and 4.86 eV, respectively. The half-width of E_1 is of the order of 0.04 eV. Half-widths for E_2 and E_3 are observed to be about 0.02 eV or less. Two strong maxima are found in the far ultraviolet part of the spectrum: C at 5.53 eV with signs of structure at 5.39 and 5.84 eV; finally, D at about 6.5 eV of much lower intensity.

Interpretation of these peaks is not simple. The band structure of Cu_2O having been calculated by Dahl and Switendick [29], however, it now seems reasonable to hope for some progress. Nevertheless, the sharp peaks E_1, E_2, and E_3, strongly temperature-dependent, are likely to be related to exciton formation. The rather large value of the line width can be explained by strong inter- and intraband diffusion, different bands being quite close to these exciton levels. Though A depends strongly on the temperature, its width seems too wide for an exciton transition.

9.   *Band Structure of Cu_2O*

The band structure of Cu_2O has been calculated by Dahl and Switendick [29] by the APW method. First, it is shown that bands formed with oxygen orbitals do not contribute to the optical absorption in the visible and the ultraviolet. These bands are much deeper than the bands formed with 3d copper orbitals. Accordingly, the transitions due to oxygen are probably in the far ultraviolet.

Second, it is shown that the 4s electrons of Cu^+ form in the unit cell a Γ_1 and Γ'_{25} band, with Γ_1 being much lower than the valence band and Γ'_{25} being much higher than the first conduction band. This results in one of the bands formed with $3d_{z^2}$ orbitals being empty. Therefore the energy gap is located between a Γ_1 band and a Γ'_{25}, both formed with $3d_{z^2}$ orbitals but

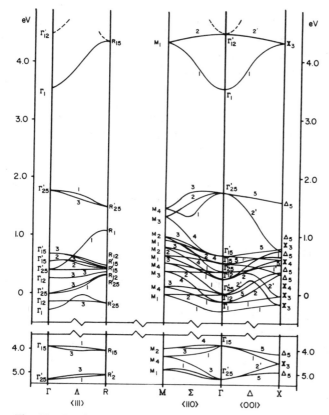

Fig. 15. Band structure of Cu₂O as calculated by Dahl and Switendick.

split by the crystal field and Γ_1 empty. The next higher conduction band is a Γ'_{12} band. The highest Γ'_{12} valence band is split by spin orbit interaction and gives a $\Gamma_8{}^+$ and $\Gamma_7{}^+$ band; the Γ_1 conduction band becomes with spin consideration a $\Gamma_6{}^+$ band. A large number of bands deriving from the $3d$ copper orbitals are located below the Γ'_{25} band. This situation is shown in Fig. 15.

10. Interpretation [³⁰]

At first sight the agreement between theory and experiment is striking. However, a detailed examination leads to some discrepancies.

The yellow and green series correspond to a forbidden transition between the highest occupied band Γ'_{25} and the conduction band Γ_1. The transition is forbidden because of the pronounced d character of the corresponding orbitals. However, band overlapping weakens this interdiction somewhat.

With spin orbit coupling, the valence band is split into a Γ_8^+ and a Γ_7^+ band, the first being the highest; the conduction band becomes a Γ_6^+. If the spin orbit coupling is disregarded, the calculated energy gap is 1.77 eV. Observed transitions are considerably larger, being 2.17 and 2.305 eV, respectively. Opinion is that the calculated Γ_1 level is about 0.5 eV too low.

The first allowed transition is the $\Gamma_{25}' \rightarrow \Gamma_{12}'$ with a calculated energy difference of 2.7 eV. Experimental values for the blue and blue-violet exciton peaks are 2.58 and 2.71 eV, respectively. This fit is surprisingly good. Thus Γ_{12}' is correctly situated above the Γ_{25}' state.

The attribution of observed spectral singularities at higher energies to definite band-to-band transitions is complicated on account of considerable overlapping of a large number of transitions. A few suggestions can be made, however:

1. The general aspect of the histogram resembles that of the ultraviolet spectrum. However, the energy scale does not fit well with observations. The two maxima of the histogram should be shifted to higher energies by as much as 0.6 eV in order to fit well the experimental maxima at about 3.6 and 4.3 eV.

Nevertheless, the similarity of the calculated density of states and the experimental data allows one to suppose that the maxima on the lower-energy side, i.e., A, B, E_1, E_2, and E_3, should correspond to transitions from d copper states to the $M_1 \Gamma_{12}' X_3$, R_{15} conduction band at the corresponding symmetry points of the Brillouin zone. This is also in agreement with the temperature shift of the peaks.

2. The E_1, E_2, E_3 peaks could be due to exciton formation. The intense E_1 peak probably originates from a transition at the X point. The strongly temperature-dependent peak A could be associated with exciton formation (at the M point). However, its width is unusually large, which is rather in contradiction with an exciton assignment.

3. The very intense absorption above 5 eV (C, D maxima) could be related to transitions from the d copper valence band states to a higher excited state, which should lie about 2 eV above the $M_1 \Gamma_{12}' X_3$ conduction band. In this case the transitions should take place at the M and X points and are forbidden at the Γ point having a Γ_{25}' representation.

4. Reflectivity (Fig. 14) as well as absorption decrease towards much higher energies. The reflectivity maxima in the 9-eV region could be due to transitions from oxygen $2p$ states to the $M_1 \Gamma_1 X_3$ conduction band. Dahl's calculations give too large a separation between the $3d$ copper valence band and the $2p$ oxygen band.

A last point may be extremely important. The red edges and the yellow and green series are strongly shifted to lower energies when temperature is increased. The exciton peaks E_B and E_{BV} do not shift very much. The assignment predicts that the first transition is $\Gamma'_{25} \to \Gamma'_{12}$. These two bands arise from $3d_{z^2}$ orbitals of copper and are split by the crystal field. This energy interval is likely to be rather sensitive to temperature variation. Conversely the E_B and E_{BV} exciton peaks are related to a $\Gamma'_{25} \to \Gamma'_{12}$ transition, in which Γ'_{12} is formed mostly of Cu $4p$ orbitals. The energy interval between these two states is likely not to vary strongly with temperature. Shifting of some high-energy peaks is observed and will allow identification of the transitions. This point needs, however, further investigation.

In conclusion, it can be stated that, though the quantitative agreement between theory and experiment is not quite good, the band structure calculation of Dahl is of great help in understanding the absorption of Cu_2O. Great progress in interpretation has been made, and it is likely that this will improve in the near future.

C. The Spectra of Copper Halides

1. General Properties of Spectra of Copper Halide

Copper halides have been studied extensively by the Strasbourg group [31]. Figure 16 shows the surface of a sample of CuCl, in which rather large grains can be seen. Samples of such quality are essential to observe details in the structure of the spectrum.

Figures 17, 18, and 19 present the absorption spectra of CuCl, CuBr, and CuI, respectively, at 4.2°K. All these substances show pronounced first-class exciton spectra. Strong anomalies are observed in the reflection spectrum in the vicinity of the strong absorption peaks [32]. One of them is shown in Fig. 20. The refractive index has been measured on both sides of the exciton peaks (Fig. 21) [33]. It is of interest to note that the absorption, reflection, and refractive indices do not seem to be related by a classical relation as suggested by Biellmann and Ringeissen.

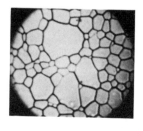

100 μ

Fig. 16. Thin film of CuCl showing single-crystal domains. The quality of the sample is essential for observation of details of the structure of an absorption spectrum.

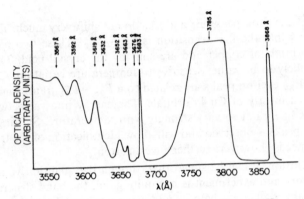

Fig. 17. Absorption spectrum of CuCl at 4.2 °K.

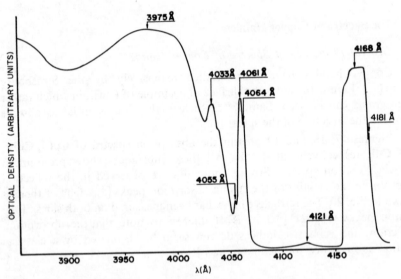

Fig. 18. Absorption spectrum of CuBr at 4.2 °K.

We will now focus our interest on the spectrum of CuCl. K. S. Song recently calculated the band structure of this substance. The CuCl spectrum will be described in some detail and then compared with the calculations of Song.

2. The Absorption Spectrum of CuCl

The spectrum of CuCl was investigated some time ago by Reiss [34] and by Ringeissen [35]. Recently, however, new important results were

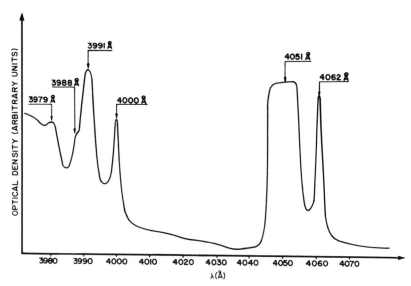

Fig. 19.   Absorption spectrum of CuI at 4.2 °K.

obtained when samples of very good quality became available. A summary of these results will be given here [36].

The absorption spectrum of CuCl consists of two groups of peaks. The peaks of lower energy are sharp, while those of higher energy are diffuse. These peaks are interpreted as two exciton line series separated by spin orbit splitting.

The sharp lines form a series which is not exactly hydrogenic although the higher terms seem to converge to a limit. These lines $\lambda_1(f) = 3868$ Å, $\lambda_2(f) = 3683$ Å, $\lambda_3(f) = 3663$ Å have the distinct character of a first-class exciton series. The absorption is strong and the oscillator strength of the first line is $f = 4.5 \times 10^{-3}$.

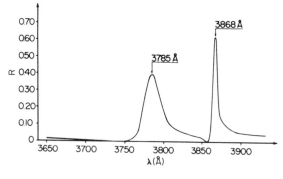

Fig. 20. Reflection anomalies (excitonic residual rays and "missing" rays) of CuCl at 4.2 °K. This is a reflection from the interface surface, fused SiO_2–CuCl (after Ringeissen [35]).

Ringeissen suggests associating with the second and third lines, a satellite line $\lambda_2(s) = 3652$ Å and the shoulder at $\lambda_3(s) = 3632$ Å. These lines are separated from the two original lines by 232 cm^{-1}, the optical longitudinal phonon being 220 cm^{-1}. Furthermore, these two satellite lines shift in the same way as the original lines when the sample is doped with bromine. They also disappear with $\lambda_2(f)$ and $\lambda_3(f)$ when the grain of the sample becomes too small. The diffuse lines $\lambda_1(d) = 3785$ Å, $\lambda_2(d) = 3620$ Å, $\lambda_3(d) = 3592$ Å were initially thought to form a second exciton series of lines. It is almost certain that the first line is an exciton line. There is only a very weak background absorption, both sides of the line and its oscillator strength being high ($f = 11.4 \times 10^{-3}$). However, recently, Ringeissen reported that with very good samples three more peaks appear at $\lambda_4(d) = 3565$ Å, $\lambda_5(d) = 5537$ Å, and $\lambda_6(d) = 3508$ Å. He notes also that λ_2 upward from all these peaks are roughly equidistant, 220 cm^{-1} apart. The spectrum has obviously an oscillatory character beginning at $\lambda_3(d)$. All these peaks are superimposed on the continuum of the sharp series (Fig. 22).

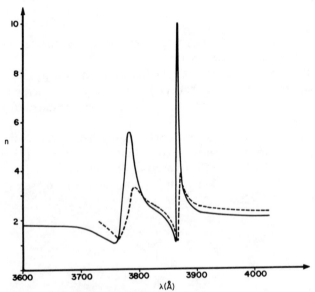

Fig. 21. Refractive index of CuCl as determined by Ringeissen [35] at 4.2 °K. The dotted curve was obtained by Reiss, Peters, *et al.* using the channelled spectra method on evaporated films. The solid curve has been obtained by Ringeissen on better samples and single crystals. The curve is obtained from reflection and absorption measurements. Note the difference which is as yet unexplained.

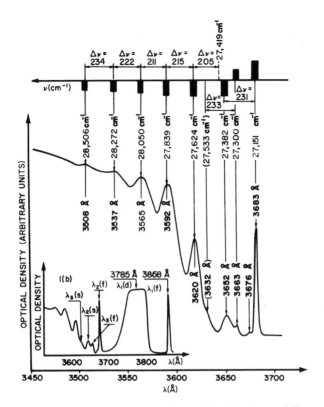

Fig. 22. Oscillatory structure from a thin film of a specially good quality at 4.2 °K. The upper diagram represents exciton absorption lines and the lower suggested vibronic satellites.

3. Interpretation

It is tentatively suggested that a new absorption mechanism appears here (Fig. 23).

The absorption in a continuum is likely to be strongly reinforced when the usual band-to-band transition is in resonance with a second absorption mechanism. This second absorption involves the excited state becoming a sharp exciton state by emission of a given phonon, for example, the LOP. Via this mechanism, a satellite line now appears.

The $\lambda_2(s)$ and $\lambda_3(s)$ lines could correspond to such an absorption mechanism. The $\lambda_1(f)$ and $\lambda_1(d)$ have no satellite lines because the continuum is too far. No low-energy satellites appear for the same reason.

One can invoke the same mechanism suggesting, however, that several identical phonons may be emitted simultaneously or successively. The

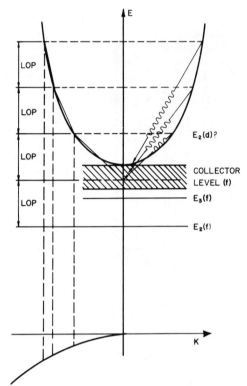

Fig. 23.   The mechanism suggested for absorption satellite formation.

oscillatory structure of the spectrum may be tentatively identified with this process.

One can also attempt to assign the sharp final state considered in this process (the collector level) with the following considerations. (1) The separation between the first peak and the limit of the series of sharp lines is very close to the LOP. It is difficult to estimate if the overlap continuum is sufficiently narrow to be responsible for the above mechanism. This is also not very likely because when the grain of the sample is small in order to prevent the appearance of line $\lambda_3(f)$, the first two lines of the oscillatory structure are observable. (2) The question could be raised as to whether the exciton level corresponds to $\lambda_2(f)$. This again is not very likely since the energy differences between $\lambda_2(f)$, $\lambda_2(s)$ and the first line of the oscillatory structure have higher values than the LOP (231 and 240 cm^{-1}). If, however, in spite of this difficulty, the conclusion should be given further consideration, then a second oscillatory structure should be observed from the line $\lambda_3(f)$. (3) Finally, it can be suggested that the first line of the oscillatory structure could be a second exciton line of a diffuse series. The corresponding level in this case could be considered as a "collector" level. It is difficult at present to decide in favor of one of these possibilities. However, Dr. K. S. Song† has recently calculated the electron and hole masses from band structure. It is borne out in the calculations that $m_e^*/m_h^* \sim 0.03$ or less, with the hole having a large mass of about 15 m_0. If this result is confirmed, the excitons in CuCl should be localized. It is well known that in this situation the coupling with lattice vibration may become large. Therefore, the calculations of Dr. K. S. Song are likely to help us understand why vibronic spectra can appear in CuCl.

It is interesting to note that the apparent dispersion in the energy differences between the peaks of the vibrational structure can be interpreted in terms of valence band curvature as illustrated on the left of Fig. 23. A remark of additional interest is also that the possibility of such an absorption mechanism was considered right at the beginning of our research on exciton spectra. An indication in favor of such a mechanism had been suggested based upon photoelectric measurements on CdS, InSb, and GaSb [37].

4.   *Oscillatory Structure of the Response Curve of the Photoconductivity of CuCl*

An effect which seems to be strictly related to the above-described absorption mechanism has been reported recently by Coret *et al.* [38]. It is observed with samples having a very high reflectivity. These samples have been obtained with the zone fusion technique and melt on the surface of a SiO₂ optically finished plate. When separated from the substrate, the surface of the sample has a very high reflectivity.

These samples have a substantially different photoresponse curve than the evaporated and recrystallized samples used in previous experiments. The photoresponse curve is given in Fig. 24. We will not discuss here details of the behavior of this sample in the part of the spectrum where exciton absorption is observed. It is well known that the photoresponse curve can be very complicated in this part of the spectrum.

A striking property of the photoresponse spectrum (Fig. 24) on the high-energy side is that it contains an oscillatory structure similar to the structure observed in absorption. However, the maxima of absorption correspond to the minima of the photoelectric response curve. The minima are separated by 216 cm⁻¹. The curve becomes somewhat complicated around $\lambda_3(f)$ and will be discussed in more detail elsewhere, but on the whole the two structures correspond well.

Clearly, the minima of the photoelectric curve should correspond to the maxima of the absorption curve. If by the mechanism of absorption the

† Private communication; to be published by Dr. K. S. Song.

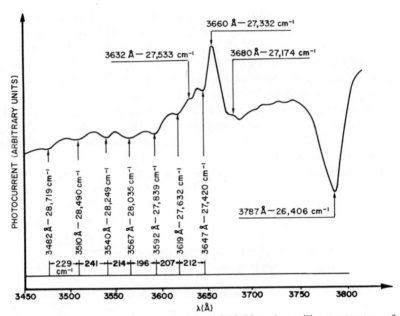

Fig. 24. Photoresponse curve of CuCl at 4.2 °K. Note the oscillatory structure of
the high-energy part in which the minima correspond fairly well with the maxima
of Fig. 23. The low-energy part is complicated on account of the superposition
of exciton lines.

absorbed energy is stored in an exciton level after emission of one or several
phonons, no free electrons and holes can be formed by absorption of light
in this part of the spectrum. This will not be the case for absorption in the
spectrum adjacent to the maxima since no direct conversion with phonon
emission will be possible. This explains why the minima in the photoconductivity curve correspond to the maxima in the absorption curve in the
oscillatory structure. It is also the first example of simultaneous observation
of oscillatory structure in both the absorption and photoresponse curve.

5.   *Summary of Absorption in CuCl Crystals at Low Temperatures*

It has been seen that CuCl certainly exhibits two exciton absorption
spectra. They are separated by 0.069 eV because of spin orbit interaction.
The sharp series contain three lines. The diffuse series contain only one
evident exciton line, but on top of the continuum of the sharp series, an
oscillatory structure discussed above appears very distinctly. Some of these
lines may be related to the higher-order diffuse exciton lines. The oscillatory
structure in the absorption curve corresponds to a similar structure in the
photoresponse curve, the minima of this latter curve corresponding to the
maxima of the absorption curve.

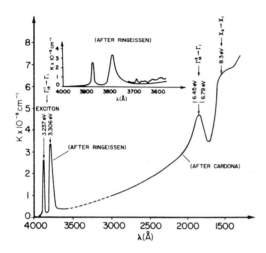

Fig. 25. The general shape of the spectrum of CuCl up to about 10 eV. The high-energy part is taken from measurements of Cardona. The low-energy part was measured by the Strasbourg group.

The high-energy spectrum of CuCl at low temperatures has been studied by Cardona [39] using samples prepared by a different technique. This spectrum is rather simple and is represented in Fig. 25. It might be of interest to note that solid-state spectroscopy is very sensitive to the quality of samples. The spectra of CuCl have been studied by the Strasbourg group for a long time. However, in the last months (by succeeding in the preparation of very good recrystallized samples) it has become possible to make new improved observations and enhance to a great extent our understanding of these spectra.

Finally, we will give an account of the calculations of Dr. Song* on the band structure of *CuCl*.

6. Band Structure of CuCl

The band structure of CuCl has been calculated by K.S. Song, Institut de Physique, Strasbourg [40]. The L.C.A.O. method has been used to calculate the valence bands (VB) and the OPW method to calculate the conduction bands (CB).

For the VB, the Hartree atomic orbitals have been used, $3d$ of Cu^+ and $3s$ and $3p$ of Cl^-. The potential used was made up of Hartree ionic potentials, corrected by the Madelung potential. For the CB, Slater's exchange potential corrected by an effective dielectric constant was used. This constant is taken as an adjustable parameter so as to reproduce the fundamental energy gap at Γ.

---

* The author is grateful to Dr. K. S. Song for permission to give a short account of the calculation he has carried out prior to its publication.

The calculations show that a strong mixing of $3p\,Cl^-$ and $3d\,Cu^+$ orbitals takes place. The first VB has a mixed character of $3d\,Cu^+$ with a considerable participation of $3p\,Cl^-$ orbitals. The second VB has a stronger $3p\,Cl^-$ character with a participation of $3d\,Cu^+$ character. Finally, the third VB has predominantly the character of $3s\,Cl^-$ orbitals. These three VB's are rather flat and are widely separated (about 2 eV between the first and second band). The third is much lower and in fact is not considered for optical spectra.

From the tight binding point of view, the first CB is made up of $4s$ orbitals, the next band of $4p$ orbitals. These two bands should cross. Since this is not allowed on account of their symmetry, a second minimum is formed in the lowest band at the X_3 point and a maximum at a Δ_1 point. The upper band has a minimum at a Δ_1 point and a second maximum at the X_1 point. The band structure of CuCl with the corresponding irreducible representations is shown in Fig. 26.

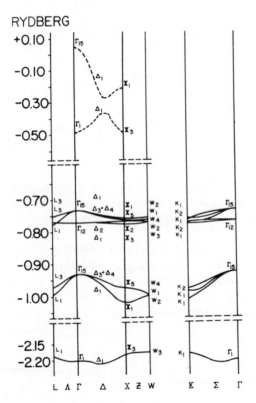

Fig. 26. Band structure of CuCl according to K. S. Song.

7. Comparison with Experiment

The first allowed transition is $\Gamma_{15}^{1V} \to \Gamma_1^{1C}$ (V for VB and C for CB) at 3.429 eV. The effective dielectric constant has been taken as $\varepsilon = 2.571$. This value cannot be compared with an experimental constant. The spin orbit splitting of the valence band Γ_{15}^{1V} in Γ_7 and Γ_8 is found to be 0.066 eV with the experimental value being 0.069 eV. As suggested by Cardona, the Γ_8 is taken as the lowest one. The next allowed transition $X_5^{1V} \to X_1^{3C}$ at 3.836 eV is a factor of 10 weaker than the $\Gamma_{15}^{1V} \to \Gamma_1^{1C}$ transition. Since this transition is quite close to the above transition, it is not clear why no experimental evidence appears in the spectrum. Furthermore, an indirect transition should be observed involving the second minimum of the conduction band. However, this transition might be very weak. The next transition should be from the second Γ_{15} VB to the first Γ_1 CB. The calculated value is 6.131 eV; the values observed by Cardona are 6.1 to 6.79 eV. The next transition should be $X_5^{1V} \to X_1^{1C}$ at 7.549 eV. The observed value is 8.3 eV. This discrepancy of about 0.75 eV is rather large. Finally the transition $\Gamma_{15}^{1V} \to \Gamma_{15}^{1C}$ is calculated to be 10.585 eV. Cardona finds 10 eV.

The assignments of Cardona do not fit the calculations of Song for the third and the last transitions. The absorption curve of CuCl constructed from measurements of the Strasbourg group and of Cardona is compared with the assignments born out of Song's calculations and is represented in Fig. 25.† It appears from Song's calculation that m_e^*/m_h^* is rather small and could be about 0.01 to 0.03. The excitons are then localized. This could be in favor of strong coupling with the lattice and would explain the vibronic structure.

These two examples show that great progress has been made in understanding both the absorption spectra and the band structure of some substances. It is to be noted, however, that the calculations are difficult and involve many approximations. A good fit with experiment can be obtained only by adjusting theoretical parameters. A completely self-consistent calculation is not yet possible. Furthermore, it is necessary to have at hand very accurate absorption measurements. However, the approach used seems a good way of proceeding. With the restrictions noted above, progress is very satisfactory.

III. REFERENCES

1. S. Nikitine, *Progr. Semicond.* **6**:235 (1962).
2. R. S. Knox, "Theory of Excitons," in: *Solid State Physics, Supplement* 5, Academic Press (New York).

---

† Recent calculations of Dr. Song have changed some of the values given above and added some more allowed transitions. A more detailed discussion will be given later by Dr. Song. The author is thankful to Dr. Song for numerous interesting discussions and information.

3. R. J. Elliott, *Phys. Rev.* **108**:1384 (1957); see also R. J. Elliott and R. Loudon, *J. Phys. Chem. Solids* **8**:382 (1959).
4. H. Haken, *Fortschr. Physik* **38**:271 (1958); *Halbleiterprobleme, Vol. II*, W. Schottky (ed.), Vieweg (Berlin), 1955; *J. Phys. Chem. Solids* **8**:166 (1959); *Z. Physik* **15**:223 (1959).
5. J. Frenkel, *Phys. Rev.* **37**:17 (1931).
6. R. Peierls, *Ann. Phys.* **13**:905 (1932).
7. N. F. Mott and R. W. Gurney, *Electronic Processes in Ionic Crystals*, Oxford University Press (London), 1953.
8. G. H. Wannier, *Phys. Rev.* **52**:191 (1937).
9. J. C. Slater and W. Shockley, *Phys. Rev.* **50**:705 (1936).
10. G. Dresselhaus, *J. Phys. Chem. Solids* **1**:14 (1955).
11. M. Cardona, *Phys. Rev.* **129**:69 (1963).
12. K. S. Song, *J. Phys.* **28**:C3-43 (1967).
13. Y. Toyozawa, *Progr. Theoret. Phys.* (*Kyoto*) **20**:53 (1958); *J. Phys. Chem. Solids* **8**:289 (1959).
14. J. C. Merle, M. Certier, and S. Nikitine, *J. Phys.* **28**:C3-92 (1967).
15. H. Haken and W. Schottky, *Z. Physik. Chem.* (*Frankfurt*) **16**:218 (1958).
16. H. J. G. Meyer, *Physica* **22**:109 (1956).
17. M. Hayashi and K. Katzuki, *J. Phys. Soc. Japan* **5**:380 (1950); M. Hayashi, *J. Fac. Sci. Hokkaido Univ.* **4**:107 (1952).
18. E. F. Gross and N. A. Karryeff, *Compt. Rend. Acad. Sci. URSS* **84**:261, 471 (1952). For the numerous and very important publications of the Leningrad group, see the review articles: E. F. Gross, *Nuovo Cimento Suppl.* **3**:672 (1956); *Advan. Phys. Sci.* (*Moscow*) **63**:575 (1957); **76**:422 (1962).
19. S. Nikitine, Mme. L. Couture, M. Sieskind, and G. Perny, *Compt. Rend. Acad. Sci.* (*Paris*) **238**:1987 (1954).
20. S. Nikitine, G. Perny, and M. Sieskind, *Compt. Rend. Acad. Sci.* (*Paris*) **238**:67 (1954); *J. Phys. Radium* **13**:18S (1954), communication at the Soc. Fr. de Phys., presented in November 1953; S. Nikitine, R. Reiss, and G. Perny, *Compt. Rend. Acad. Sci.* (*Paris*) **240**:505 (1955).
21. E. F. Gross and A. A. Kaplyanski, *Dokl. Akad. Nauk SSSR* **132**:98 (1960); J. B. Grun, *Rev. Opt.* **41**:439 (1962); S. Nikitine, J. B. Grun, M. Certier, J. L. Deiss, and M. Grosmann, *Proc. Intern. Conf. Phys. Semicond. Exeter*, 1962, 441; S. Nikitine, J. B. Grun, and M. Certier, *Phys. Matière Condensée* **1**:214 (1963); M. Certier, J. B. Grun, and S. Nikitine, *J. Phys. Radium* **25**:361 (1964).
22. J. L. Deiss, J. B. Grun, and S. Nikitine, *J. Phys. Radium* **24**:206 (1963).
23. J. B. Grun, M. Sieskind, and S. Nikitine, *J. Phys. Chem. Solids* **21**:119 (1961).
24. S. Brahms and S. Nikitine, *Solid State Commun.* **3**:209 (1965).
25. J. B. Grun, M. Sieskind, and S. Nikitine, *J. Phys. Chem. Solids* **19**:189 (1961).
26. S. Nikitine, J. B. Grun, and M. Sieskind, *J. Phys. Chem. Solids* **17**:292 (1961).
27. J. B. Grun, M. Sieskind, and S. Nikitine, *J. Phys. Radium* **22**:176 (1961).
28. A. Daunois, J. L. Deiss, and B. Meyer, *J. Phys.* (*Paris*) **27**:142 (1966).
29. J. P. Dahl and A. C. Switendick, *J. Phys. Chem. Solids* **27**:931 (1966).
30. S. Brahms, S. Nikitine, and J. P. Dahl, *Phys. Letters* **22**:31 (1966).
31. S. Nikitine, Mme. L. Couture, G. Perny, and R. Reiss, *Compt. Rend. Acad. Sci.* (*Paris*) **241**:629 (1955); S. Nikitine, *J. Phys. Radium* **17**:817 (1956); S. Nikitine, R. Reiss, and G. Perny, *Compt. Rend. Acad. Sci.* (*Paris*) **242**:1588, 2540 (1956); S. Nikitine, *Progr. Semicond.* **6**:235 (1962); S. Nikitine, S. Lewonczuk, J. Ringeissen, and K. S. Song, *Compt. Rend. Acad. Sci.* (*Paris*) **262**:1506 (1966).
32. S. Nikitine and R. Reiss, *Compt. Rend. Acad. Sci.* (*Paris*) **242**:238 (1956); *J. Phys. Chem. Solids* **16**:237 (1960).

33. S. Nikitine and R. Reiss, *J. Phys. Radium* **17**:1017 (1956); R. Reiss, *Rev. Universelle Mines* **XV**: 9 (1956); *Cahiers Phys.* **13**:129 (1959).
34. R. Reiss and S. Nikitine, *Compt. Rend. Acad. Sci.* (*Paris*) **250**:2862 (1960); R. Reiss, *Cahiers Phys.* **13**:129 (1959); S. Nikitine, *Progr. Semicond.* **6**:235 (1962).
35. S. Nikitine and J. Ringeissen, *J. Phys. Radium* **26**:171 (1965); S. Nikitine, J. Ringeissen, and M. Certier, *Acta Phys. Polon.* **XXVI**:745 (1964); S. Nikitine, J. Ringeissen, and C. Sennett, *Septième Congrès Intern. de Physique des Semiconducteurs* (*Recombinaison Radiative dans les Semiconducteurs*) Paris, 1964, p. 279.
36. J. Ringeissen, S. Lewonczuk, A. Coret, and S. Nikitine, *Phys. Letters* **22**:571 (1966).
37. D. C. Reynolds, C. W. Litton, and T. C. Collins, *Phys. Status Solidi* **9**:645 (1965).
38. A. Coret, R. Lévy, J. B. Grun, A. Mysyrowicz, and S. Nikitine, *Phys. Letters* **22**:576 (1966).
39. M. Cardona, *Phys. Rev.* **129**:69 (1963).
40. K. S. Song, *J. Phys.* **28**:195 (1967).

CHAPTER 10

Excitons in II–VI Compounds

D. C. Reynolds

Aerospace Research Laboratories
Wright-Patterson Air Force Base, Ohio

I. INTRODUCTION

The optical properties of II–VI compounds have been the subject of considerable study for many years; indeed, in the older published literature, there is a wealth of optical data, some of which are related to the fundamental properties of those materials. It has only been in recent years, however, that even the older data have been associated with the exciton structure. The more recent (mostly since 1959) magneto-optical data (spectral data obtained at high magnetic fields and low temperatures) are very closely related to the fundamental properties of the materials; these data have come to be very well understood in terms of excitons in these compounds. In fact, the isolation and subsequent elucidation of many properties fundamental to these materials has frequently been possible because of a better understanding of the intrinsic-exciton structure.

II. INTRINSIC-EXCITON SPECTRUM AND BAND STRUCTURE

A detailed study of the absorption edge of CdS in the temperature range 90–340°K was made by Dutton [1]. The dichroism reported by Gobrecht and Bartschat [2] and also by Furlong and Ravilious [3] was clearly demonstrated. The reflection spectra observed at 81°K are shown in Fig. 1. Dutton was unable to account for the source of the dichroism but felt that the absorption near the edge was due to exciton transitions. Gross and co-workers [4-6] studied the absorption near the edge in CdS at 4.2°K. They attributed the short-wavelength intrinsic absorption lines to exciton transitions. Birman [7] attacked the problem from a theoretical point of view. Assuming a tight binding approximation in conjunction with group

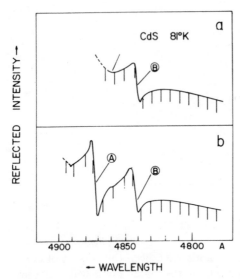

Fig. 1. Reflection spectra. Reproduction of original data records giving reflected light intensity as a function of wavelength for light (a) $E \| c$ and (b) $E \perp c$. The upturn at the left of each curve marked by heavy arrows is due to additional reflection from the second surface, since the crystal becomes transparent at these points. The vertical pips are wavelength markers. Note the similarity of anomaly B in the two polarizations (after Dutton).

theory, he arrived at the irreducible representations, band symmetries, and selection rules for the zinc-blende and wurtzite structures. If we consider the absorption (emission) of electromagnetic radiation by atoms, the probability of the occurrence of a transition between two unperturbed states ψ_i and ψ_f as caused by the interaction of an electromagnetic radiation field and a crystal is dependent on the matrix element

$$\int \psi_f^* \, H_{\text{int}} \, \psi_i \, d\tau \tag{10.1}$$

where H_{int} is the dipole moment operator

$$H_{\text{int}} = \frac{e\hbar}{imc} \, \mathbf{A} \cdot \mathbf{\Delta} \tag{10.2}$$

In order for an electric dipole transition to be allowed, the above matrix elements between the initial and final states must be nonzero.

In the case of transitions between two states of an atom (which is in a crystalline field), the initial and final states of the atom are characterized by irreducible representations of the point group of the crystal field. Also the dipole moment operation must transform like one of the irreducible representations of the group. If we denote the representations which correspond to the initial and final states of the transition and to the multipole radiation of order $S(S = 0$ for electric dipole radiation) Γ_i, Γ_f, and $\Gamma_r^{(S)}$, respectively, then the matrix element in equation (10.1) transforms under rotations like the triple direct product

$$\Gamma_f \times \Gamma_r^{(S)} \times \Gamma_i \tag{10.3}$$

The selection rules are then determined by which of the triple direct products in question do not vanish.

The dipole moment operator for electric dipole radiation transforms like x, y, or z, depending on the polarization. When the electric vector ξ of the incident light is parallel to the crystal axis, the operator corresponds to the Γ_1 representation. When it is perpendicular to the crystal axis, the operator corresponds to the Γ_5 representation. The characters of the representations corresponding to these symmetry operations are given in Table I.

In the tight binding approximation, the wave functions at the center of the Brillouin zone are taken to be linear combinations of atomic wave functions. In CdS it is assumed that the bottom of the conduction band is formed from the $5s$ levels of Cd, and the upper valence bands are formed from the $3p$ levels of sulfur. Since the crystal has a principal axis, the crystal field removes part of the degeneracy of the p levels. Thus, disregarding spin-orbit coupling, the following decomposition at the center of the Brillouin zone is obtained:

conduction band $\qquad S \to \Gamma_1$

valence band $\qquad P_x, P_y \to \Gamma_5$

$\qquad\qquad\qquad\quad P_z \to \Gamma_1$

Table I. Character Table for the Wurtzite Structure

	E	$\bar{E}$	C_2 $\bar{C}_2$	$2C_3$	$2\bar{C}_3$	$2C_6$	$2\bar{C}_6$	$3_{\sigma d}$ $3_{\bar{\sigma} d}$	$3_{\sigma v}$ $3_{\bar{\sigma} v}$
Γ_1	1	1	1	1	1	1	1	1	1
Γ_2	1	1	1	1	1	1	1	-1	-1
Γ_3	1	1	-1	1	1	-1	-1	1	-1
Γ_4	1	1	-1	1	1	-1	-1	-1	1
Γ_5	2	2	-2	-1	-1	1	1	0	0
Γ_6	2	2	2	-1	-1	-1	-1	0	0
Γ_7	2	-2	0	1	-1	$\sqrt{3}$	$-\sqrt{3}$	0	0
Γ_8	2	-2	0	1	-1	$-\sqrt{3}$	$\sqrt{3}$	0	0
Γ_9	2	-2	0	-2	2	0	0	0	0

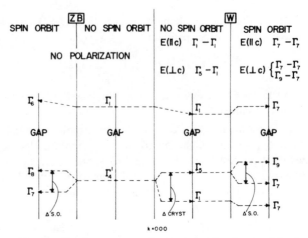

Fig. 2.   Band structure and selection rules for zinc-blende
and wurtzite structures. Crystal splittings and spin-orbit
splittings are indicated schematically. Transitions which are
allowed for various polarization of photon electric vector
with respect to crystal c axis are indicated (after Birman).

Introducing the spin doubles the number of levels. The splitting caused
by the presence of spin is represented by the inner products

$$\Gamma_5 \times D_{\frac{1}{2}} \rightarrow \Gamma_1 + \Gamma_5$$

$$\Gamma_1 \times D_{\frac{1}{2}} \rightarrow \Gamma_1$$

and the band structure at $K = 0$ along with the band symmetries and
selection rules is shown in Fig. 2. This band structure could readily account
for the dichroism observed in CdS and other wurtzite-type structures.

A. Intrinsic-Exciton Spectrum of CdS

Thomas and Hopfield [8] studied the reflection spectra from CdS
crystals at 77 and 4.2 °K. Their data at 77 °K were in good agreement with
those obtained by Dutton. They observed one reflection peak at 2.544 eV
(4873 Å) active only for $E \perp c$ and another at 2.559 eV (4844.5 Å) active
in both modes of polarization. These correspond to transitions involving
the two top valence bands. They also observed a third and broader peak at
2.616 eV that Dutton did not observe. This peak was active for both modes
of polarization and is associated with transitions from the third valence
band.

They repeated the measurements at 4.2°K and found a more complex reflection spectrum. Densitometer traces are shown in Fig. 3. In addition to the parent transitions A, B, and C involving the three valence bands, additional structure A' and B' was observed. The polarization properties of A' and B' indicate that they are associated with the transitions A and B, respectively. The energy separations between $A - A'$ and $B - B'$ are identical

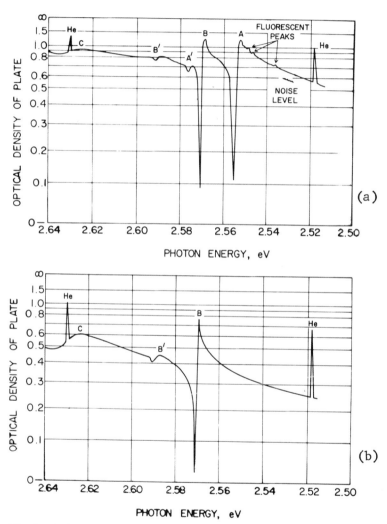

Fig. 3.   Reflection of CdS at 4.2°K for light polarized with (a) $E \perp c$ and (b) $E \| c$. These are traces taken from microphotometer recordings of Kodak 103-0 plates. The noise level is as indicated in Fig. 3 (a). The helium calibration lines indicate the resolution used (after Thomas and Hopfield).

and equal to 0.021 eV. It is reasonable to assume that A and B correspond to the ground-state excitons associated with the first and second valence bands, respectively. The weaker transitions A' and B' result from the $n = 2$ states of the parent transitions. If the excitons are hydrogen-like, an estimate of the binding energy can be made ($4/3 \times 0.021 = 0.028$ eV). Using this value for the binding energy and the expression [9], we obtain

$$G = \frac{e^4 \mu}{2 \hbar^2 \varepsilon^2} \tag{10.4}$$

The reduced exciton mass was calculated to be 0.18. From the analyses of their reflection data, Thomas and Hopfield deduced many of the parameters relating to the band structure (Fig. 4) of CdS as follows:

$$E_g - E_{\text{exciton } A} = 2.554 \text{ eV}$$
$$E_g - (E_B - E_A) - E_{\text{exciton } B} = 2.570 \text{ eV}$$
$$E_g - (E_C - E_A) - E_{\text{exciton } C} = 2.632 \text{ eV} \tag{10.5}$$

where $E_{\text{exciton } A}$ represents the binding energy of exciton A and similar notations are used for B and C. If it is assumed that $E_{\text{exciton } A} = E_{\text{exciton } B}$, then

$$E_A - E_B = 0.016 \text{ eV} \tag{10.6}$$

and, with $E_{\text{exciton } A} = 0.028$ eV, the band gap is $E_g = 2.582$ eV.

In order to determine $(E_C - E_A)$ it was necessary to compute $E_{\text{exciton } C}$ which they did using a quasi-cubic model based on the similarity between the wurtzite and zinc-blende structures. The values obtained are

$$E_{\text{exciton } C} = 0.026 \text{ eV} \qquad E_A - E_C = 0.073 \text{ eV} \tag{10.7}$$

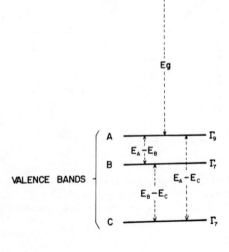

Fig. 4. Energy band structure of CdS at $K = 0$. The levels A, B, and C refer to the hole bands from which the exciton causing reflections A, B, and C arise. The band symmetry is given at the right (after Thomas and Hopfield).

B. Magneto-Optical Effects in the Exciton Spectrum

The effects of external static magnetic fields have been studied by two groups, Hopfield and Thomas[10] and Wheeler and Dimmock[11]. Hopfield and Thomas made a detailed study of the exciton spectra in CdS. They observed two overlapping exciton series. The absorption lines observed for light polarized with $E \perp c$ and $E \parallel c$ are shown in Tables II and III, respectively. The lines identified as $I_\alpha (\alpha = 1, 2, ...)$ are impurity lines.

In the orientation $E \perp c$ the ground state of the A exciton series is the transverse exciton. It is seen that the $n = 2$ state of the A exciton series overlaps the ground state of the B exciton series. For orientation $E \parallel c$, the lowest-energy intrinsic transition is the Γ_6 exciton in which the hole and electron spins are parallel. This exciton splits in a magnetic field in the orientation $(H \parallel c)$ with a g value of 2.93. In the orientation $E \parallel c$, the lowest energy A exciton is the longitudinal exciton. The absorption coefficient for this line was measured as a function of angle of incidence. As the crystal was rotated away from normal incidence about an axis normal to the c axis, the absorption coefficient increased rapidly. The effect was evident for rotations of less than 10°. The deviation from normal incidence permits a component of the E vector in the direction perpendicular to c and thus interacts with the allowed Γ_5 transition. The exciton propagates in the direction of the photon which is almost in the direction of the polarization vector of the exciton which results in a longitudinal exciton [12].

The $n = 2$ state of A has two components as shown in Table III. A densitometer trace of the two components is shown in Fig. 5(c). When viewed in a magnetic field of 31,000 G, $H \parallel c$, $q \perp H$ (q is the wave vector of the exciting light) the line designated as P_z shows a diamagnetic shift

Table II. A Summary of the Data for Some of the Absorption Lines Seen in Light Polarized Perpendicular to the Hexagonal c Axis $E \perp c$ at 1.6 °K

Line	Position (Å)	Position (eV)	Approximate apparent width × 10³ (eV)	Crystal
I_1	4888.6	2.5359	0.3	GEB 3
I_2	4868.7	2.5463	0.5	CEA 1
I_3	4859.0	2.5513	0.3	GEA 1
$A_{n=1} (S_T)$	4854.5	2.5537	3.5	GEA 1
I_4	4837.7	2.5626	1.7	GEB 3
$B_{n=1}$	4826.4	2.5686	3.5	GEA 1
$A_{n=2}$	4812.9	2.5758	3.0	GEA 1

Table III.   A Summary of the Data for Some of the Absorption Lines Seen in Light Polarized Parallel to the Hexagonal c Axis $E \parallel c$ at 1.6 °K

Line	Position		Approximate apparent width $\times 10^3$	Crystal
	(Å)	(eV)	(eV)	
$A(1\ S\ \Gamma_6)$	4857.0	2.5524	0.1	GED 7
$A_n(1\ S_L)$	4852.9	2.55455		GED 7
I_4	4837.7	2.5626	1.7	GEB 3
$B_{n=1}$	4826.1	2.5687	4.3	GEA 1
$A_{n=2}$	4814.21	2.57508	0.1	GED 7
	4812.96	2.57575	0.4	GED 7
$A_{j=3}$	4805.46	2.57977	0.5	GED 7
$A_{n=4}$	4803.28	2.58094	0.3	GED 7
$B_{n=2}$	4784.9	2.59085	3.5	GEA 1

and also splits with a g value of 0.62 ± 0.06. This splitting along with the magnetic behavior of the other component of the $n = 2$ state is shown in Fig. 6. A total of seven lines are derived from the two zero-field components. Absorption measurements were made at the peak magnetic field as the crystal was rotated about an axis normal to the c axis. As the crystal was rotated away from normal incidence, the line marked S_L rapidly increased in intensity. The results in Fig. 6 are shown for the orientation of the crystal c axis 5° away

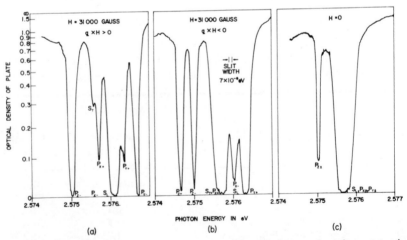

Fig. 5.   Microphotometer trace of Zeeman effects of the $n = 2$ exciton states in CdS crystal GED 8 at 1.6 °K with $c \perp H$, $H \parallel x$, $q \perp H$, and $q \perp c$ where q is here the wave vector of the exciting photon (after Hopfield and Thomas).

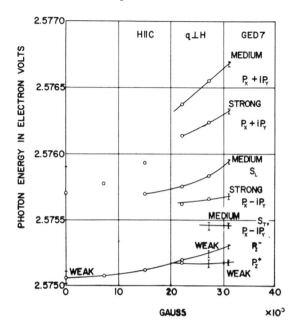

Fig. 6. The Zeeman effect at 1.6 °K for $H \| c$. At $H = 0$ the limits indicate the apparent line widths; otherwise they indicate the approximate uncertainty in the line positions (after Hopfield and Thomas).

from normal incidence. These rotational experiments shown that S_L is a longitudinal exciton derived from the S state active for light polarized with $E \perp c$.

Observation of the line $(S_T, P_x - iP_y)$ in the same rotational experiment described above showed that it also increased in intensity with angle of rotation, but not as much as S_L. Also the intensity did not go to zero at normal incidence as was observed for S_L. This intensity behavior indicates that the transverse exciton is contained in this line. However, since it does not vanish at normal incidence it must contain another state. This state must be a P state. The remaining state must be the lowest-energy P state having zero angular momentum about the c axis and denoted as P_z.

The magnetic field behavior of the two components of the $A_{n=2}$ state for the orientation $H \perp c$, $q \| H$ is shown in Fig. 7. In this field orientation the P_y and P_z states are mixed. The P_z states are moved to lower energy and increase in intensity because of mixing with the more intense P_y states. The splitting of the P_z state is increased over the splitting for the $H \| c$ orientation. In this orientation the spin magnetic moment of the hole does not interact with H; therefore, the spin magnetic moment of the exciton is just that of the electron in this case so the splitting is due solely to the electron. In the $H \| c$ case the spin magnetic moment of the exciton arises from the difference of the electron and hole spin moments. The P_y

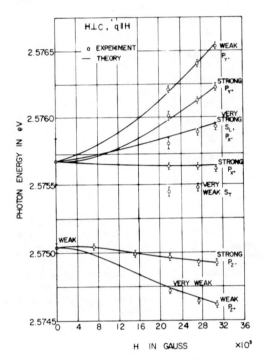

Fig. 7. The Zeeman effect at 1.6 °K for $H \perp c$ and $q \parallel H$, i.e., the light is here travelling parallel to the magnetic field. In this diagram the lines represent theory; the mass values A have been used and the following values are employed: Rydberg constant $= 0.027$ eV, $\gamma = 0.222$, $g_{e\perp} = 1.73$, and $g_{h\perp} = 8.3$. As before, the limits on the experimental points at $H = 0$ indicate apparent line width; otherwise they indicate the approximate uncertainty in the line positions (after Hopfield and Thomas).

states move to higher energy with increasing field, and the P_x states split and show a diamagnetic shift. S_L is degenerate with P_x, whereas the S_T state is resolved in this orientation. Hopfield and Thomas analyzed the data theoretically. From Fig. 7 the splitting of the higher $n = 2$ state into four components establishes it as the $P_x P_y$ state. The lower-energy spin doublet is the P_z state. Using a Rydberg constant of 0.0270 eV and a mass anisotropy γ of 0.222, they calculated the following energy levels which are compared with the experimentally determined values for some of the same levels (in eV).

State	Theory	Experiment	State	Theory	Experiment
$1S$	0.0270	0.0298	$3P_{\pm 1}$	0.0029	
$2P_0$	0.0071	0.0075	$3D_{\pm 1}$	0.0030	
$2P_{\pm 1}$	0.0065⎫		$3D_{\pm 2}$	0.0029	0.0028
$2S$	0.0067⎭	0.0069	$4P_{\pm 1}$	0.0016⎫	
$3P_0$	0.0032		$4D_{\pm 1}$	0.0017⎬	0.0017
			$4D_{\pm 2}$	0.0016⎭	

The agreement between theory and experiment is quite good.

The P_0 state splits into two components in a magnetic field ($H \parallel c$) as shown in Fig. 6. The $\Gamma_5 P_z$ state has antiparallel spins; therefore, the g value should be $|g_{e\parallel} - g_{h\parallel}|$. The $\Gamma_6 P_z$ state has parallel spins and the g value should be $|g_{e\parallel} + g_{h\parallel}|$. Since the Γ_6 transition is forbidden, the observed g value of $g = 0.62 \pm 0.06$ is associated with the difference of the g values. The g value of the $1S\Gamma_6$ state was determined to be 2.93 ± 0.03. The spin-orbit coupling in CdS is small, and the conduction band is treated as S-like; therefore, it is likely that the electron g value will be near -2.0. From the measured g values, the most reasonable electron and hole g values would be

$$g_{e\parallel} = -1.78 \pm 0.05 \qquad g_{h\parallel} = -1.15 \pm 0.05 \qquad (10.8)$$

This electron g value agrees with that obtained by Lambe and Kikuchi [13] from spin-resonance measurements. Hopfield and Thomas were further successful in analyzing the higher excited states from which g values and effective masses of the electrons and holes both parallel and perpendicular to the c axis were determined.

In Fig. 5(a) and (b) [14] for the orientation $H \perp c$, $q \perp H$, and $q \perp c$ it is seen that two different spectra are obtained for two directions of H. In Fig. 5(a), P_z-, P_x-, and P_y- are more intense than the positive counterparts. The minus sign corresponds to the electron-spin orientation parallel to the magnetic field and the plus denotes antiparallel. When the magnetic field is reversed, the intensity ratios reverse [Fig. 5(b)]. In crystals of the wurtzite structure, which do not have inversion symmetry, time reversal reverses the sign of the magnetic field, but leaves the selection rules for infinite-wavelength, plane-polarized light unchanged. In zero magnetic field, the four states $P_{x\pm}$ and $P_{y\pm}$ (eight states including hole spin) are degenerate. Thomas and Hopfield have shown by utilizing group theory that one linear combination of these eight states (derived from Γ_1) has an optical matrix element in the limit $q \to 0$ and that one linear combination (derived from Γ_5) has an optical matrix element proportional to q_y. Utilizing these linear combinations, they computed the optical matrix elements for states $P_{x\pm}$. They found for the respective states having spins in the positive and negative x directions

Matrix elements for spin $+x = A + q_y B$

Matrix elements for spin $-x = A - q_y B$

The absolute magnitudes of these two matrix elements are different as long as both A and B are nonzero. Reversing the direction of q_y reverses the relative magnitudes of the two matrix elements. One state (say, with spin in the positive x direction) has an energy corresponding to P_{x+} or P_{x-}, according to the direction of H. Reversing H therefore interchanges the energy levels of the two spin orientations without altering the optical matrix elements.

C. Exciton Structure and Zeeman Effects in CdSe

The optical absorption and reflection spectra of CdSe were studied by Wheeler and Dimmock [11]. The identification and interpretation of the spectra were aided by the Zeeman structure it displayed. The spectra were analyzed in terms of the theory to obtain the band parameters at $K = 0$.

Three nonoverlapping exciton series resulting from the three valence bands of the wurtzite structure were identified. Here one sees a significant difference between CdSe and CdS. In the case of CdS the splitting between the two top valence bands is less than the binding energy of the exciton. This leads to more valence band mixing, which is evidenced by the fact that unallowed transitions are observed in CdS. The agreement between theory and experiment is in general good for CdSe. The significant band parameters obtained by Wheeler and Dimmock are given in Table IV.

Table IV. Significant Band Parameters

State	Experimental (cm^{-1})	Calculated (cm^{-1})	Mass parameters
Series 1 *			
$1S$	14,727 ±1	14,734 ±6	$m_{ex}^* = 0.13\pm0.01\ m$
$2P_0$	14,818.6±0.3		$m_{ez}^* = 0.13\pm0.03\ m$
$2P_{\pm1}$	14,822.5±0.2		$m_{hx}^* = 0.45\pm0.09\ m$
$3P_{\pm1}$	14,839 ±1	14,838.0±1.5	$m_{hz}^* \geqslant m$
Series 2 †			
$1S$	14,931$\pm$ 3		
$2S-2P_{\pm1}$	15,032$\pm$ 2		
$2P_0$	15,022$\pm$ 3		
Calculated series limit	15,050±15		
Series 3 ‡			
$1S$	18,218±10		
Calculated series limit	18,340±20		
Crystal field splitting	200±15		
Spin-orbit splitting	3,490±20		

* From $H\|c$ diamagnetic shift, $\mu_x = 0.100\pm0.005$. From zero-field positions, $n_1 = 2$ states. Hence $\alpha = 1 - \mu_x\varepsilon_x/(\mu_x\varepsilon_z) = 0.32\pm0.02$; $\mu_x/\mu_z = 0.75\pm0.04$; and $\mu_z = 0.13\pm0.01$. From $H\perp c$ diamagnetic shift, $\mu_x/\mu_z = 0.77\pm0.04$; $\mu_z = 0.13\pm0.01$; $R_y = 106\pm5$ cm^{-1}; $a_0 = 54$ Å; $E_{gap} = 14,850.5\pm2.0$ cm^{-1}.

† $R_y = 120\pm10$ cm^{-1}; $\mu_x = 0.11\pm0.01$; $m_{ex}^* = 0.13\pm0.01\ m$; and $m_{hx}^* = 0.9\pm0.2\ m$.

‡ Electron g values : $|g_{ex}| = 0.51\pm0.05$; and $|g_{ez}| = 0.6\pm0.1$. There is some evidence that the electron g values are negative. The Zeeman splitting between $2P_{\pm1}$, $\Gamma_1-\Gamma_2$, and Γ_5 states indicates a negative g_{ez}.

D. Exciton Spectrum of ZnO

Thomas [15] was first to investigate the fundamental exciton structure of ZnO by absorption and reflection measurements. At 4.2°K he observed three reflection peaks, peaks A and B being active in the polarization mode $E \perp c$ while peak C is active for $E \parallel c$. These peaks are identified with the ground-state exciton for the three valence bands in ZnO. Three additional peaks are observed and identified as A', B', and C'. These peaks are all on the short-wavelength side of the respective A, B, and C peaks by approximately the same energy and are associated with the $n = 2$ state of the respective ground-state excitons. The energy positions of these peaks are given in Table V.

In absorption, Thomas observed a line in the $E \parallel c$ orientation at the energy of the A exciton line. This indicates a mixing of the A and C bands in ZnO. Here the crystal field splitting dominates. In the case of CdS the spin-orbit splitting dominates so that mixing occurs in the B and C bands so that the intensity of these two peaks is essentially equal in both modes of polarization. The interpretation of the experimental results of polarization in ZnO places the Γ_7 valence band above the Γ_9 valence band. These are the $P_{x,y}$ valence bands whose degeneracy is lifted primarily by spin-orbit interaction.

This aspect of the experiment is not in agreement with the theory of Hopfield [16]. Placement of the Γ_7 band above the Γ_9 band places the wrong sign on the spin-orbit energy.

More recently Park et al. [17] have examined the exciton structure of ZnO and their interpretation of the results places the Γ_9 band above the Γ_7 band. The essential difference between the work of Thomas and the latter

Table V. The Energy Positions of Reflection Peaks in Exciton Spectrum of ZnO at 4.2°K

Oscillator	Position (eV) (± 0.0005 eV)	Difference (eV)	Oscillator strength	Full width at half-height (eV)
A	3.3768	0.0457	13×10^{-4}	0.0015
A'	3.4225		3.5×10^{-4}	0.003
B	3.383	0.0445	45×10^{-4}	0.0022
B'	3.4275		2.2×10^{-4}	0.0025
C	3.4215	0.0435	60×10^{-4}	0.0017
C'	3.465		15×10^{-4}	0.009

work is in line assignments. Thomas interprets the lines I_b (3669.66 Å) (labeled A_L by Thomas) as the ground-state A-exciton transition. Park *et al.* interpret this line as a bound-exciton line resulting from an ionized donor complex. The strength and reproducibility of the line in many crystals is such that one is likely to interpret the line as being of an intrinsic nature. However, the intensity of the line I_b is always less than that of the 3653.42 Å (A, $n = 1$) line.

Since a bound-exciton complex is made up of an impurity atom (or lattice defect) to which is bound an intrinsic (free) exciton and since the intrinsic exciton is a property of the crystalline host lattice, a question naturally arises concerning the relative intensities of absorption lines that derive from these two types of excitons. This is especially so when the bound-exciton lines are observed to be nearly as intense as the intrinsic-exciton lines, a frequent observation in both the absorption and reflection spectra of the II–VI compounds. A case in point is shown in the ZnO reflection anomaly of Fig. 8, where, from a comparison of reflectivity minima, one observes an "impurity"-exciton line intensity comparable to that of the intrinsic exciton. In fact, the intensity of the line I_b is observed to be within an order of magnitude of the intrinsic-exciton intensity in a large number of ZnO crystals. Still another example of strong "impurity"-exciton lines has been observed in the reflection spectra of CdSe and CdS platelets, where bound-exciton peaks have been observed with intensities comparable to those of the intrinsic-exciton reflection anomalies. In support of the experimental observations and in spite of our instinctive notions regarding line intensities, recent work has rather strikingly shown on purely theoretical grounds that it is reasonable to expect bound-exciton line intensities comparable to those of the intrinsic excitons in spectra of semiconducting solids.

The theory of "impurity" or defect absorption intensities in semiconductors has been studied by Rashba [18]. By use of the Fredholm method [19], he finds that, if the absorption transition occurs at $k = 0$ and if the discrete level associated with the impurity approaches the conduction band, the intensity of the absorption line increases. The explanation offered for this intensity behavior is that the optical excitation is not localized in the impurity but encompasses a number of neighboring lattice points of the host crystal. Hence, in the absorption process, light is absorbed by the entire region of the crystal consisting of the impurity and its surroundings.

In an attack on the particular problem of excitons which are weakly bound to localized "impurities," Rashba and Gurgenishvili [20] derived the following relation between the oscillator strength of the bound exciton F_d and the oscillator strength of the intrinsic exciton f_{ex}, using the effective-mass approximation

$$F_d = (E_0/|E|)^{3/2} f_{ex} \tag{10.9}$$

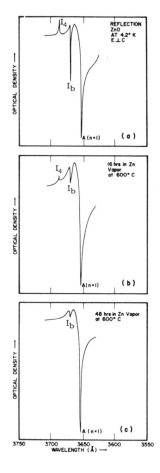

Fig. 8. A densitometer trace of the reflection spectra from (a) a vapor-grown platelet, (b) a flux-grown bulk crystal, and (c) a hydrothermally grown bulk crystal of ZnO at 4.2 °K for $E \perp c$. The variation of the strength of I_b (labeled A_L by Thomas) in different crystals of ZnO is to be noted. In (a) the I_b line is most pronounced, in (b) the strength of I_b is considerably reduced, and in (c) it is completely absent. Note also in (a) the reflection anomalies due to the bound-exciton lines I_g (3690.22 Å) and I_h (3694.93 Å). Similar anomalies due to the bound-exciton lines were also observed in the reflection spectra of CdSe and CdS crystals.

where $E_0 = (2\hbar^2/m)(\pi/\Omega_0)^{2/3}$, E is the binding energy of the exciton to the impurity, m is the effective mass of the intrinsic exciton, and Ω_0 is the volume of the unit cell. If we now substitute the ZnO parameters for the line I_b (3669.66 Å) into equation (10.9), we find that F_d exceeds f_{ex} by more than four orders of magnitude ($F_d/f_{ex} > 10^4$). Now let us assume an impurity concentration of $N_d \sim 10^{17}/\text{cm}^3$ (a plausible assumption for platelet samples) and also make the usually valid assumption of $N_{ex} \approx 10^{22}$ for the intrinsic-exciton concentration. From these assumptions and the further assumption that the line intensities are equal to a simple product of density and oscillator strength ($I_d \approx N_d F_d$), it is easy to see how "impurity"- or bound-exciton intensities which are within an order of magnitude of those of the intrinsic-exciton lines can be realized [compare the I_b line strength with that of the intrinsic exciton (Fig. 8)]. An inspection of equation (10.9) reveals that, as the intrinsic exciton becomes more tightly bound to the associated center, the oscillator strength (and hence the intensity of the exciton-complex line) should decrease as $(1/E)^{3/2}$. This fact is nicely illustrated in a comparison of the I_h and I_g line intensities with that of I_b (see Fig. 8). Note that the lines I_h and I_g, whose line strengths (as measured from reflectivity) are much less than that of I_b, appear to arise from exciton complexes in which the intrinsic exciton is much more energetically bound than is the exciton in the complex which gives rise to the line I_b; in particular, note that the lines I_h and I_g lie about 0.035 eV below the energy of exciton A (3653.42 Å), while the line I_b lies only 0.015 eV below the intrinsic-exciton energy.

From the intrinsic-exciton energies, a number of relevant parameters directly related to the band structure can be determined. Using the ground-state exciton energy $(A, n = 1)$ and the first-excited-state energy $(A, n = 2)$, and assuming the exciton has a hydrogen-like set of energy levels, we can calculate the exciton binding energy as follows:

$$h\nu_n = E_{\text{gap}} - E_B/n^2 \tag{10.10}$$

where E_B is the exciton binding energy

$$E_B = \tfrac{4}{3}(0.0312) = 0.042 \text{ eV} \tag{10.11}$$

At $1.2\,^\circ$K the energy gap is

$$E_{\text{gap}} = 3.435 \text{ eV} \tag{10.12}$$

The reduced exciton spherical mass can be estimated from the relation

$$E_B = e^4 \mu/2\hbar^2 \varepsilon^2 = 13.6\,\mu/\varepsilon^2 \text{ eV} \tag{10.13}$$

where ε is the low-frequency dielectric constant. The dielectric constants of ZnO are as follows [21]:

$$\varepsilon_x = 8.47 \qquad \varepsilon_z = 8.84 \tag{10.14}$$

Therefore the reduced effective mass of the exciton is

$$\mu_x = 0.20\, m_0 \tag{10.15}$$

where m_0 is the free-electron mass.

The effective Bohr radius is given by

$$a_0{}^1 = a_0(\varepsilon_x/\mu_x)\, n^{\frac{1}{2}} = 22 \text{ Å} \tag{10.16}$$

The spin-orbit interaction energy has been calculated by Herman et al. [22] for a number of the Group IV, III–V, and II–VI compounds, and fairly good agreement with experimental values was obtained. In the case of ZnO, one finds a value of $+0.035$ eV for the spin-orbit energy calculated from this method. However a difficulty arises if one tries to obtain the spin-orbit interaction energy from the quasi-cubic model [23-25], which is given by the formula

$$\delta = \tfrac{1}{2}(2E_1 + E_2) + \tfrac{1}{2}[E_2{}^2 - 2E_1(E_1 + E_2)]^{\tfrac{1}{2}} \qquad (10.17)$$

where E_1 is the energy difference between excitons A and B, and E_2 is the energy difference between excitons B and C. In fact, one finds that δ is a complex number upon substitution of these energy differences into equation (10.17). The difference between the model of Thomas and that of Park *et al.* is twofold. First, there is a difference in the interpretation of the I_b (3669.66 Å) line. Thomas interprets this line as an intrinsic exciton; Park *et al.* interpret it as an extrinsic line. Second, Thomas assigns a Γ_7 symmetry to the top valence band, whereas Park *et al.* assign a Γ_9 symmetry to this band. More experiments are necessary in order to resolve these issues.

E. Exciton Structure in Photoconductivity of CdS

The fundamental absorption bands for a number of wurtzite compounds have recently been shown to possess a fine structure at low temperature, and this fine structure was explained in terms of exciton formation [4,8,15]. The observation and interpretation of the exciton spectra give detailed information concerning the electronic band structure of the material. To this end various optical methods, such as transmission, reflection, and luminescence measurements, have been employed. However, very few attempts have been made to interpret the photoconductivity structure near the fundamental absorption edge in terms of the energy band structure.

The existence of fine structure in photoconductivity spectra of CdS and CdSe at 77°K has been observed by several workers [26-28]. Gross and Novikov [27] have established the coincidence of photocurrent peaks and absorption maxima from their observation of the photoconductivity spectra of CdS, and they have indicated possible participation of excitons in photoconductivity. However, they found two types of crystals which they classified as Group I and Group II crystals. In Group I crystals the exciton absorption lines coincide with photocurrent maxima, whereas in Group II crystals the exciton lines coincide in position with photoconductivity minima.

Gross *et al.* [29] have recently reported the photoconductivity spectra of CdS at 4.2°K, but without identification of the peaks. Park and Reynolds [30] have investigated the photoconductivity spectra near the absorption edge in CdS, CdSe, and CdS: Se single crystals. In every case the photoconductivity maxima at the absorption edge coincided with the

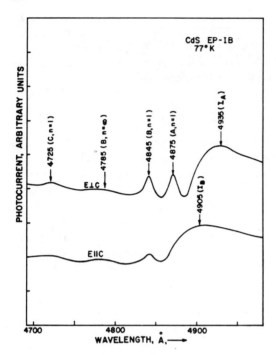

Fig. 9. Photoconductivity spectra of CdS at 77 °K in polarized light (after Park and Reynolds).

absorption maxima. At 77 °K, six photoconductivity peaks are observed in CdS as seen in Fig. 9. Among these peaks, those at 4935 Å (I_A) and 4875 Å $(A_n = 1)$ are active for $E \perp c$, while the distinct peaks at 4845 Å $(B_{n=1})$, 4785 Å $(B_{n=\infty})$, and 4725 Å $(C_{n=1})$ are active in both modes of polarization.

There is good agreement with Dutton [1] and Thomas and Hopfield [8] in the observation of three sharp reflection peaks—A at 4873 Å active only for $E \perp c$ and B and C at 4844.5 and 4738.7 Å, respectively, active in both modes of polarization.

On cooling the crystal from 77 to 4.2 °K, the number of peaks is increased and they appear more distinct as shown in Fig. 10.

The photoconductivity peak corresponding to the $n = 1$ state of exciton A is only active in $E \perp c$ and the peaks corresponding to the $n = 2, 3, \infty$ states of exciton A are seen in both modes of polarization. Peaks corresponding to the $n = 1$ state of exciton B and the $n = 1$ state of exciton C are also observed.

There are two peaks, 4800 Å and 4776 Å, that are active in both modes of polarization. From their position, they may correspond to the series limits of exciton A and exciton B. The series limits of exciton A and exciton B, assuming a hydrogen-like series, are 4801 and 4772 Å, respectively. In transmission measurements, excited states higher than $n = 4$ have not been reported, partly because of weak line strength at higher order and partly due to thickness of the crystal.

In addition to the photoconductivity peaks at energies corresponding to the position of the free excitons, some crystals show an oscillatory photoconductivity spectrum toward higher energies. Periodic structure in the spectral dependence of photoconductivity (PC) with a period corresponding to the longitudinal optical (LO) phonon energy has been observed in the intrinsic region of InSb and GaSb by Harbegger and Fan [31], by Stocker *et al.* [32], and more recently by Nasledov *et al.* [33] and Mczurczyk *et al.* [34]; Park and Langer also reported similar observations for CdS [35]. The phenomenon is illustrated in Fig. 11, which shows the photocurrent of a CdS platelet at 4.2 °K with the E vector of the incident light polarized (a) perpendicular and (b) parallel with respect to the crystalline c axis. The structure below the · absorption edge (2.582 eV for $E \perp c$) is due to a contribution of current carriers from dissociated excitons [30, 35]. The initial absorption of light at the exciton energy leads to the formation of direct excitons, which in part may decay radiatively; but such excitons may also dissociate in part

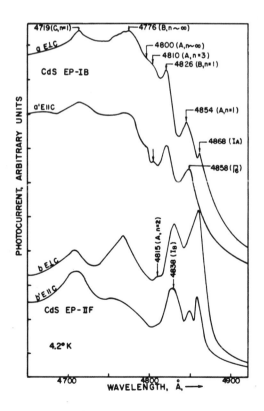

Fig. 10. Photoconductivity spectra of CdS at 4.2 °K in polarized light for two crystal samples. In the first crystal (curves a and a′) the largest part of sensitivity is concentrated in the intrinsic-exciton peaks, while in the second crystal (curves b and b′) the impurity peaks are more pronounced than intrinsic-exciton peaks (after Park and Reynolds).

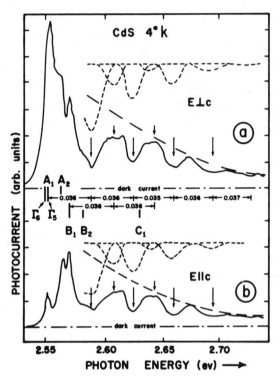

Fig. 11. Photoconductivity *versus* energy of exciting photons with (a) E vector perpendicular to the c axis and (b) E vector parallel to the c axis of a CdS platelet at 4.2 °K. Indicated by bars are the positions of excitons A, B, and C. Long arrows indicate A minima, short arrows B minima, and dashed curves show the separated contributions of the processes causing A and B minima (after Park and Langer).

into current carriers and thus appear as maxima in the spectral response curves of the photoconductivity. The dissociation might occur either by the interaction with impurity centers, or other excitons, or by the absorption of a phonon.

Toward higher energy, the structure marked by arrows is separated by equal energy differences which correspond approximately to the energy of the LO phonon in CdS. There are two series of minima indicated by long and short arrows. The first, let us call it the I series, is displaced by a LO phonon from the $A(n = 1)$ exciton peak ($\Gamma_5 = 4854$ Å for $E \perp c$); the second, the II series, is displaced by the same amount from the $B(n = 1)$ exciton peak at 4826 Å. In other words, the two sets of minima are displaced by an amount equal to the separation of the two upper valence bands. The position of the II minima is less conspicuous because of the composite nature of the two series; their positions become clearer when the composite spectral-response curve of the intrinsic region is decomposed as shown by the dashed curves. The spectral-response curve can be considered, in principle, as a superposition of an intrinsic unmodulated response (long dashes) and the energy-dependent oscillations indicated by short dashes with respect to a displaced base line.

An explanation for the above observation has been put forward by several other investigators [32, 34]; they propose that the oscillatory behavior of the photoconductivity is due to an oscillatory value of the electron lifetime in the conduction band as a function of its energy. The lifetime of the conduction electron at a certain energy value is smaller than that at adjacent energy values whenever this energy coincides with the sum of the energies of a shallow impurity [32], or band gap [31], plus an integral multiple m of the longitudinal optical phonon energy. The explanation of the data in Fig. 11 requires that the ground state of the excitons play the role of the "shallow impurity."

Figure 12 shows a model to explain the observations in Fig. 11. In an arbitrary direction of k-space, excitation may occur from the top two valence bands; this is indicated by dashed and solid vertical arrows, respectively. The positions of the transitions along k are chosen such that their corresponding energies are equal to the sum of an A or B exciton, respectively, plus a multiple of LO phonons, i.e., they are equal to the energies at which minima were observed. The dashed vertical transitions correspond to the I minima, indicated by long arrows in Fig. 11; the solid vertical transitions correspond to the II minima, indicated by short arrows. From these points in the conduction band, electrons might easily be scattered to the ground-state exciton level via LO phonons. The exciton in turn may be an A or B exciton depending on the valence band from which the hole has been captured. In both cases the formation of an A exciton will probably

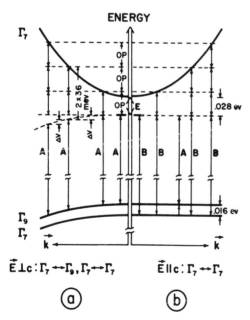

Fig. 12. Band model showing upper valence bands and conduction bands in CdS. E_{OP} stands for the energy of an optical phonon and A and B for the energies of excitons A and B, respectively. The binding energy of both excitons is identical (0.028 eV). Allowed transitions are $\Gamma_9-\Gamma_7$ for the E vector perpendicular to the crystal c axis and $\Gamma_7-\Gamma_7$ for the E vector parallel or perpendicular. (a) Illustrates the proposed origin of A minima. ΔV reflects the curvature of the valence band. (b) Here the curvature of the valence band is neglected for simplicity. The origins of A and B minima corresponding to the case realized in Fig. 11(b) are indicated. The case shown in Fig. 11 (a) corresponds to the sum of the mechanisms illustrated in Fig. 12 (a) and 12 (b) (after Park and Langer).

be favored because of the thermalization which will favor a higher concentration of holes in the upper valence band, but these details of the recombination process were not observable in the above experiments.

III. BOUND-EXCITON COMPLEXES

A. Theory

The general model adopted is that in which an exciton forms a complex which is held together by forces analogous to those that exist in the hydrogen molecule or the hydrogen-molecule ion. This effect was first predicted by Lampert [36] and demonstrated for the case of the wurtzite structure by Thomas and Hopfield [39] which will be discussed here.

Applying group theory of weakly bound localized states in crystals having a direct band gap at $k = 0$, one finds, for crystals having only one molecule per unit cell, a substitutional impurity with the bound states belonging to an irreducible representation of the point group of the crystal. However, in a crystal of several molecules per unit cell, the symmetry group which leaves the impurity fixed is smaller than the point group, and the symmetries of the states around an impurity will correspond to this smaller group.

The wurtzite structure has two molecules per unit cell, and the symmetry group of a point substitutional impurity is C_{3v}. The first three representations refer to an even number of particles (singular group), and the second three representations belong to the double group. The decomposition of the wurtzite point group into the representations of the impurity point group reveals

$$\Gamma_1 \rightarrow \Lambda_1 \qquad \Gamma_4 \rightarrow \Lambda_1 \qquad \Gamma_7 \rightarrow \Lambda_6$$

$$\Gamma_2 \rightarrow \Lambda_2 \qquad \Gamma_5 \rightarrow \Lambda_3 \qquad \Gamma_8 \rightarrow \Lambda_6 \qquad (10.18)$$

$$\Gamma_3 \rightarrow \Lambda_2 \qquad \Gamma_6 \rightarrow \Lambda_3 \qquad \Gamma_9 \rightarrow \Lambda_{45}$$

The representations Λ_4 and Λ_5 are degenerate by time reversal and are referred to as Λ_{45}. No nonaccidental degeneracy is lifted in this decomposition, since each representation of the point group decomposes into a unique representation of the impurity group.

The states Λ_3, Λ_{45}, and Λ_6 can have nonzero g values for a magnetic field parallel to the c axis, while only states Λ_3 and Λ_6 have a nonzero g value for a magnetic field perpendicular to the c axis. The corresponding possible g values of states for the point group are $g_{//}$ for $\Gamma_5, \Gamma_6, \Gamma_7, \Gamma_8$, and Γ_9 and $g_{\perp}$ for Γ_7 and Γ_8.

For light polarized parallel to the c axis, the optical dipole selection rules are $\Gamma_i \to \Gamma_i$ and $\Lambda_i \to \Lambda_i$. For light polarized perpendicular to the c axis, the optical selection rules are

$$
\begin{array}{lll}
\Lambda_1 \to \Lambda_3 & \Gamma_1 \to \Gamma_5 & \Gamma_4 \to \Gamma_6 \\
\Lambda_2 \to \Lambda_3 & \Gamma_2 \to \Gamma_5 & \Gamma_3 \to \Gamma_6 \\
\Lambda_3 \to \Lambda_1 + \Lambda_2 + \Lambda_3 & \Gamma_5 \to \Gamma_1 + \Gamma_2 + \Gamma_6 & \\
\Lambda_{45} \to \Lambda_6 & \Gamma_6 \to \Gamma_3 + \Gamma_4 + \Gamma_5 & \\
\Lambda_6 \to \Lambda_{45} + \Lambda_6 & \Gamma_7 \to \Gamma_7 + \Gamma_9 & \\
& \Gamma_8 \to \Gamma_8 + \Gamma_9 & \\
& \Gamma_9 \to \Gamma_7 + \Gamma_9 &
\end{array}
\tag{10.19}
$$

Some relaxation of selection rules occurs for the group of the impurity. For example, the transition $\Gamma_1 \to \Gamma_6$ for light polarized parallel to the c axis is forbidden, whereas the corresponding transition $\Lambda_1 \to \Lambda_3$ is allowed. Similarly, state Γ_5 has zero g values for $H \perp c$, but the corresponding state, Λ_3, can have a nonzero g value for $H \perp c$.

Consider any simple optical transition in which an electron bound to an impurity is taken from one band to another. Suppose the initial and final states can be assigned approximate effective-mass wave functions. The initial-state wave function can then be written as

$$
f_i(r)\, U_{i0}(r)
\tag{10.20}
$$

where $f_i(r)$ is a slowly varying function of r, and U_{i0} is the periodic part of the Bloch function of band i for wave vector zero. Similarly, the final-state wave function can be written

$$
f_f(r)\, U_{f0}(r)
\tag{10.21}
$$

Since f is a slowly varying function, the optical matrix element

$$
\int f_i(r)\, U_{i0}(r)\, P f_f^*(r)\, U_{f0}^*(r)\, d^3 r
\tag{10.22}
$$

can be approximately written as

$$
\left[\int f_i(r) f_f^*(r)\, d^3 r \right] \frac{1}{\Omega} \left[\int U_{i0}(r)\, P U_{if}^*(r)\, d^3 r \right]
\tag{10.23}
$$

where the second integration is carried out over the unit cell, whose volume is Ω. In this approximation, the only large optical matrix elements will arise when the analogous band-to-band transition is allowed.

A similar argument can be made to show that in this effective-mass approximation, large g values can be expected only when the parent energy-band wave function exhibits large g values.

Thomas and Hopfield concluded that, in this simple case of weakly bound states at substitutional impurities and energy bands at $k = 0$ in the wurtzite structure, it is reasonable to describe the states as though they belonged to the point group of the crystal rather than to the group of the impurity. Such a description gives the degeneracy of the states correctly. This description neglects certain optical transitions which are technically allowed, but are weak in the effective-mass approximation and will set equal to zero certain g values which should be much smaller than usual g values. The advantage of the description is that it neglects these small effects and thus permits the full use of group theory without the clutter of what should be small perturbations.

The electron g value g_e should be very nearly isotropic and, since the conduction band is simple and the g shift of the free electron is small, only weakly dependent on the state of binding of the electron. The hole g value g_h should be completely anisotropic with $g_{h\parallel}$ equal to zero (for the top Γ_9) for magnetic fields perpendicular to the hexagonal axis. It is to be expected that the hole g value will be sensitive to its state of binding, since the different valence bands will be strongly mixed in bound-hole states.

The binding energies representing the binding energy of an exciton to the center can be evaluated by scaling the known binding energies of states of hydrogen ions and molecules to exciton parameters. These values are only qualitative since the scaling ignores the complicated valence band structure, and the mass ratio m_e^*/m_h^* is treated as very small.

B. Sharp Line Spectra Near Absorption Edge in CdS

In this section optical effects observed on the low-energy side of the intrinsic excitons will be considered. These effects are attributed to bound-exciton complexes. Lampert [36] was the first to consider the problem from the standpoint of bound aggregates of two or more charged particles in a nonmetallic solid. He referred to such aggregates as "effective-mass-particle complexes." He described complexes analogous to H_2, H_2^+, and H^-. For structures analogous to H_2^+, the gross energy scheme is the electronic level scheme of H_2^+. Each electronic level has a fine structure similar to the vibration-rotation level scheme of H_2^+. The H_2 complex will behave in a similar manner, whereas the H^- complex will have only a single bound state. The first experimental observation of any of the above complexes was by Haynes [37] in silicon crystals. By observing the low-temperature emission from systematically doped crystals, he successfully identified the neutral donor and acceptor complexes. Thomas and Hopfield [38, 39] have observed a number of the bound complexes and have identified them with several absorption and emission lines in CdS.

1. *Analyses*

Thomas and Hopfield [39] have worked out the theory for bound excitons in CdS. The theory is based on the wurtzite structure of CdS with the salient features of the band structure such as band symmetries and selection rules being derived from group theory. Only the lowest states of the complexes were considered and the model was one of the complex being bound together by forces similar to those in the hydrogen molecule or the hydrogen-molecule ion. The perturbing effect of an applied magnetic field on the optical transitions from such complexes was also considered. Remembering the band structure at $k = 0$ for CdS, one recalls that three exciton series are present, one for each of the three valence bands. The above theory led to models representing the energy levels of the complexes that could be formed using holes from the two top valence bands. The energy levels corresponding to holes from the top valence band are shown in Fig. 13, whereas the energy levels corresponding to holes from the second valence

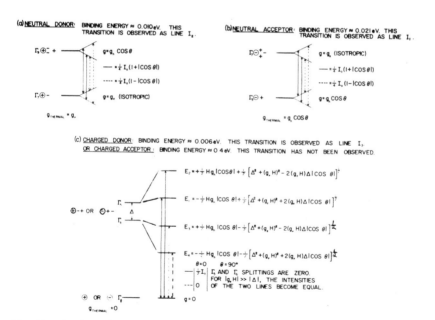

Fig. 13. A schematic representation of the energy levels of complexes which can be formed using holes from only the top valence band. In (a) and (b) linear splittings are observed, but in (c) the splittings are more complex and the energy levels are given. The expressions associated with the dashed or solid lines refer to the intensities of the transitions (after Thomas and Hopfield).

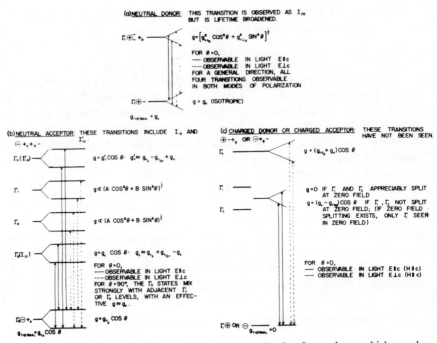

Fig. 14. A schematic representation of the energy levels of complexes which can be formed using holes from the second, or B, valence band as well as from the top, or A, valence band (after Thomas and Hopfield).

band are shown in Fig. 14. Thomas and Hopfield have successfully interpreted a number of the optical transitions in CdS on the basis of these models.

In considering transitions involving bound excitons formed from holes in the top valence band, it is seen that the g value of the electron is isotropic. The g value of the hole has the form $g = g_{h\parallel} \cos \theta$, where θ is the angle between the c axis of the crystal and the magnetic field direction. The symbols $\oplus$ and $\ominus$ refer to ionized donors and acceptors, respectively, and $-$ and $+$ refer to electrons and holes, respectively. Crystals containing imperfections such as neutral or ionized donors or acceptors can also bind excitons formed from holes in the second valence band. These transitions will be more energetic by the energy separation between the top two valence bands. The case involving the neutral donors will not be greatly different from the case when the exciton is formed from the hole in the top valence band with the exception that the transition will be unpolarized.

The remaining centers are distinctly unique as is shown in Fig. 14.

2. Optical Transitions

Thomas and Hopfield [39] have observed optical transitions in CdS associated with excitons bound to neutral donors and acceptors and to ionized donors. The ionized acceptor complex has not been observed. A strong line was observed at 4888.5 Å in both absorption and emission. This line was designated as I_1. The behavior of this in emission in a magnetic field is shown in Fig. 15(a). The splitting of the line as a function of the orientation of the crystal c axis with respect to the magnetic field is shown. With $c \perp H$ the line splits into a doublet, and from this splitting the electron g value was determined ($g_e = -1.76$). At an arbitrary orientation the line splits into a quartet demonstrating the anisotropic hole g value. The magnetic field splitting of this line is linear. One expects this if there is only one unpaired electronic particle. The transition must then arise from either a neutral donor or acceptor. Absorption measurements on the same line for the orientation $c \perp H$ showed no thermalization effects. This indicates a zero g value for the ground state from which one concludes that the complex is an exciton bound to a neutral acceptor site.

Another strong line at 4867.15 Å was observed in many CdS crystals. This line was identified as I_2. Similar to line I_1, it also showed a linear Zeeman effect. The splitting of this line as a function of magnetic field orientation is shown in Fig. 15(b). In the orientation $c \perp H$, I_2 also splits into a doublet. From the magnitude of the splitting an electron g value of

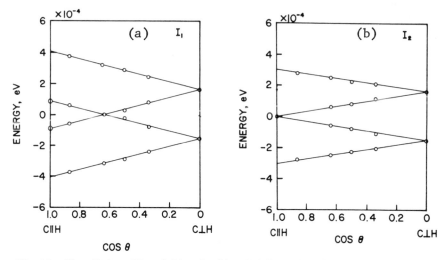

Fig. 15. The splitting of lines I_1 (a) and I_2 (b) at 31 kG as a function of cos θ at 1.6 °K as seen in fluorescence. The quantity θ is the angle between the c axis and the magnetic field. Notice that when $\theta \neq 0°$ or $\theta \neq 90°$ quartets occur. The zero of energy is taken to be at the center of each group of lines (after Thomas and Hopfield).

− 1.76 is again measured. In the orientation $c \parallel H$, no splitting is observed; therefore, the g values of the excited and ground states are equal. Unlike I_1, line I_2 shows thermalization effects in absorption for the orientation $c \perp H$. The high-energy component of the split doublet is the more intense. This confirms that there is an electron in the ground state and the complex is an exciton bound to a neutral donor site. In Fig. 15(b) the anisotropic hole g value is again observed.

Thomas and Hopfield observed a line at 4861.7 Å which they detected only in absorption. This line was designated I_3 and was found to behave quite differently in a magnetic field than I_1 and I_2. The splitting of this line as a function of magnetic field for the orientation $c \perp H$ is shown in Fig. 16. At zero field only the high-energy component is observed. When the field reaches approximately 10 kG, the low-energy component appears. The extrapolation of this component to zero field shows that the line is zero-field split. No thermalization was observed for absorption, indicating that the ground state was a singlet state. The zero-field splitting arises from an exchange interaction of an unpaired electron and an unpaired hole in the upper state. Such an interaction can occur in lines resulting from an exciton bound to ionized donors or acceptors.

The small binding energy of the exciton to the center strongly suggests that the complex is an exciton bound to an ionized donor. Further evidence supporting the identification of the complexes described above was provided by the infrared studies of Thomas and Hopfield. They found that, when the lines were observed in absorption using as little band-gap light as possible and simultaneously illuminating with infrared, I_1 and I_2 decreased in intensity

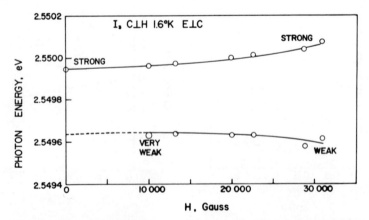

Fig. 16. The splitting of I_3 in a magnetic field with $c \perp H$, as seen in absorption. Notice that there is a zero-field splitting but that the lower state involves a forbidden transition and is only seen as it is mixed with the upper state by the magnetic field (after Thomas and Hopfield).

while I_3 increased. The infrared light was of an energy that preferentially ionized the acceptors. It was concluded that the free hole then ionized the donors. Thus, the neutral centers were decreased while the population of ionized centers increased.

Thomas and Hopfield conducted an investigation of several additional lines, a summary of which is given in Table VI. They were eminently successful in identifying all of these lines in terms of the models shown in Figs. 13 and 14.

Pedrotti and Reynolds [40] in a study of edge luminescence in CdS crystals described two distinct types of crystals. The distinction was made primarily on the fluorescence color, and, to a lesser extent, on presence, absence, and grouping of certain lines. The two types were classified as blue emission (blue crystal) and green emission (green crystal). Reynolds and Litton [41] observed a very intense line in the blue crystal at 4869.14 Å. This line was designated as I_5. The magnetic field splitting of I_5 as a function of the orientation of the c axis of the crystal with magnetic field direction is

Table VI. A Summary of the Properties of the Lines Described in the Text

Line	Active for	Energy (eV) [λ (Å)]	Approximate apparent width (10^3 eV)	Energy below exciton A (2.5537 eV) (eV)	Energy below exciton B (2.5687 eV) (eV)	Ground state
I_1	$E \perp c$	2.53595 (4888.5)	0.1	0.0177		Neutral acceptor
I_2 (many lines)	$E \perp c$	2.5471 (4867.15)	0.1	0.0066		Neutral donor
I_{1B}	$E \| c$	2.54887 (4863.7)	0.1		0.0198⎫	Neutral acceptor (trapped exciton
I_{1B}	$E \perp c$	2.54914	0.1		0.0196⎭	from band B)
I_3	$E \perp c$	2.5499 (4861.7)	0.1	0.0038		Ionized donor
I_{1B}'	$E \| c$	2.5504 (4860.8)	0.5		0.0183	Neutral acceptor (trapped exciton from band B)
Y	$E \| c$ $E \perp c$	2.55127 (4859.1)	0.1	0.0024		
X	$E \| c$ $E \perp c$	2.55206 (4857.6)	0.1	0.0016		
I_{2B}	$E \| c$ $E \perp c$	2.5626 (4837.7)	1.7		0.0061	Neutral donor (trapped exciton from band B)

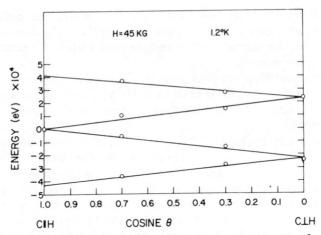

Fig. 17.   Splitting of the emission line I_5 as a function of cos θ in a field of 45 kG and at a temperature of 1.2 °K. Here θ is the angle between the crystalline c axis and the direction of the magnetic field. The zero of energy is taken to be the center of each group of lines (after Reynolds and Litton).

shown in Fig. 17. The g value is -1.76 for the orientation $c \perp H$. In the orientation $c \parallel H$ no splitting is observed, indicating that the g value of the hole is approximately equal to the g value of the electron. An examination of the I_5 multiplet intensities revealed that the intensity of each split component changed with magnetic field strength at 1.2 °K. The low-energy component increased in intensity as the high-energy component decreased with increasing magnetic field. It was concluded from these data that the line arises from a transition involving an exciton bound to a neutral acceptor site. A group of relatively long-wavelength emission lines was observed in some crystals ("blue" and "green"). The spectral positions of these sharp, narrow lines (< 0.1 Å half-width) are given in Table VII. The lines labeled I_6 and I_7 are particularly distinguished by their polarization characteristics: I_6 is polarized preferentially in the mode $E \parallel c$, while I_7 is practically unpolarized. Most of the CdS edge fluorescence is polarized in the mode $E \perp c$. This would follow from Fig. 2 where the optical selection rules allow a Γ_7–Γ_9 transition only for the mode $E \perp c$. Since the centers under investigation lie relatively close to a band, one might expect the transitions to reflect the band symmetry and obey the band-to-band optical selection rules.

Only one of the long-wavelength lines reported here showed a Zeeman effect, namely, I_6. The 5068.54- and 5069.18-Å lines are components of the zero-field-split I_6 line. The zero-field splitting is 3.1×10^{-4} eV. A plot of the splitting of the I_6 components as a function of magnetic field strength

Table VII.   A List of the CdS Emission Lines

Line	Wavelength (Å)	Energy (eV)	Active for
I_1	4888.47	2.53585	$E \perp c$
(doublet)	4888.47	2.53585	$E \perp c$
I_1	4888.18	2.53600	$E \perp c$
I_2	4867.17	2.54695	$E \perp c$
(doublet)	4867.17	2.54695	$E \perp c$
I_2	4866.98	2.54705	$E \perp c$
I_3	4861.66	2.54984	$E \perp c$
(zero-field split)	4862.25	2.54953	$E \perp c$
I_5	4869.14	2.54592	$E \perp c$
I_6	5068.54	2.44576	$E \| c$ (strong)
(zero-field split)	5069.18	2.44545	$E \| c$
I_7	5084.81	2.43793	Unpolarized

$(c \parallel H)$ is shown in Fig. 18. The high-energy component splits unsymmetrically in that it does not shift linearly with H; the low-energy component also seems to split, but its magnetically split high-energy component appears to be missing. In zero field, the I_6 components are of approximately equal

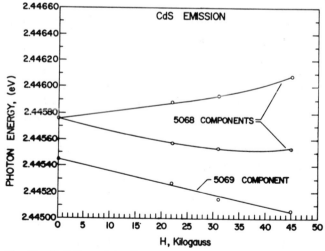

Fig. 18.   Splitting of the line I_6 as a function of magnetic field strength at 1.2 °K for the orientation $c \parallel H$. Based on the assertion that I_6 is a zero-field-split line, note that the high-energy component of the 5069-Å line is missing.

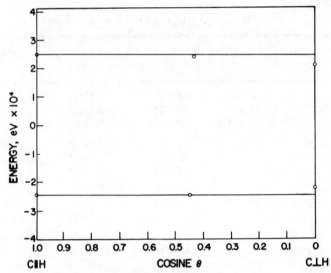

Fig. 19.   Splitting of the 5068-Å line (component of I_6) as a function of cos θ in a field of 45 kG at 1.2 °K (after Reynolds and Litton). Zero of energy is taken to be the center of the pair of lines.

intensity. With increasing magnetic field, the 5069.18-Å component shows an appreciable increase in intensity, whereas the magnetically split components of the 5068.54-Å line decrease with equal intensity as the field increases. The I_6 line is isotropically split in a magnetic field. The isotropic behavior of the split component of this line is observed in Fig. 19. Here is plotted the splitting of the 5068 component of I_6 as a function of field orientation (cos θ) at a constant field strength of 45 kG. The isotropic splitting of this line is not the usual behavior of CdS; equally unusual are the polarized emission characteristics of all these lines. From the data the I_6 line cannot be identified with any of the molecular complexes described by Thomas and Hopfield [39]. It should be pointed out that the exciton-to-center binding energy for I_6 (0.109 eV) is greater than the exciton binding energy (0.028 eV); hence, the exciton would dissociate before the complex could undergo a molecular-like dissociation. Such a situation might account for the observed polarization and magnetic field behavior of I_6.

C.   The ZnO Bound-Exciton Spectrum

A total of ten rather prominent, sharp and narrow, edge-emission lines have been observed in ZnO spectra, most of which showed Zeeman effects, similar to the effects which were previously reported for CdS [39]. Equally sharp and narrow (line half-widths $\lesssim$ 0.1 Å) were the eight absorption lines which were also observed over the same spectral range as the fluorescence lines; these lines resulted from the absorption of the exciting radiation as well as from the self-absorption of a continuum, background emission from the crystals. Most of the absorption lines also show Zeeman effects. In some cases, more than one line is observed to arise from a similar type of exciton complex (i.e., exciton bound to neutral donor, ionized donor, etc.); hence, in these cases, only the splitting of a representative line is discussed, while other lines that arise from a similar complex are simply indicated.

Figure 20 illustrates the general features of the ZnO fluorescence spectrum, showing several peaks of the broad-band, phonon-assisted, edge

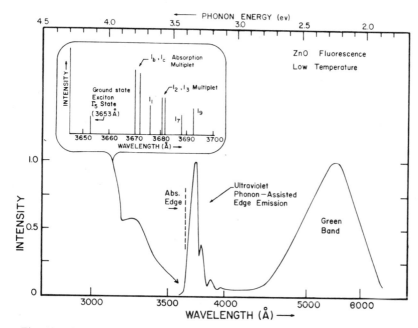

Fig. 20. A diagram of the ZnO fluorescence spectrum at low temperature. Shown in the spectrum is the well-known broad green band, as well as the characteristic phonon-assisted edge emission which appears in the near ultraviolet and whose peaks are separated by the longitudinal optical phonon energy, $\hbar\omega \approx 0.079$ eV. The inset diagram shows the highly resolved and detailed spectrum which appears between the absorption edge (ground-state exciton, Γ_5 state) and the phonon-assisted emission; this detailed spectrum is composed of many sharp lines, several of which form the basis of the present study. In the inset spectral diagram, lines are drawn to represent several of the fluorescence lines at the appropriate wavelength. While intensities are generally represented by line heights, intensities are not meant to be quantitatively comparable. The absorption multiplet I_b, I_c (3669.66, 3671.9 Å) is drawn as two fluorescence lines for the sake of comparison. Andress [43] previously studied the broad-band emission from ZnO and observed a spectral distribution similar to that shown in the diagram.

emission (near ultraviolet) and the well-known broad green band which peaks at about 5500 Å. Several investigators [42, 43] have studied the phonon-assisted peaks as well as the broad green peak in ZnO, and it has been shown that the relative intensities of these two peaks depend markedly on the electrical conductivity of the crystals, with the intensities of the two peaks becoming nearly equal at reasonably high conductivities; it has also been shown that the conductivity of ZnO is affected strongly by excess zinc in the crystals and that the conductivity increases rapidly with increasing zinc concentration. In the crystals used in the present experiments, the fluorescent intensities in both the ultraviolet and the green were rather strong, yet crystal resistivities were quite high, as determined from photoconductivity measurements with the same crystals. In addition to the platelets, we have also examined the edge fluorescence of several pale-green prism-type ZnO crystals, whose resistivities were quite low ($\rho \lesssim 1\,\Omega$-cm); however, the emission from these crystals, while fairly strong in both the ultraviolet and the green, was not characterized by the fluorescent line spectra of bound excitons. Also shown in Fig. 20 (inset diagram) is the highly resolved, "blown-up" picture of the spectral region between the absorption edge (A-exciton ground state) and the first peak ($n = 0$) of the phonon-assisted edge emission. It is this region of the spectrum that is of greatest interest here, since it contains the bound-exciton lines (emission and absorption). These exciton-complex lines, a few of which are pictured in the spectrum, are observable only in high-resolution spectra at low temperature. The emission and absorption lines are summarized in Tables VIII and IX, respectively. In addition to the lines listed in the tables, some sharp, relatively weak lines were observed at somewhat longer wavelengths; however, these lines are not discussed in the present work.

Table VIII.   A Summary of the ZnO Emission Lines Reported in the Text at 1.2°K with $H = 0$

Line	λ (Å)	Energy (eV)	$\bar{v}$ (cm^{-1})	Preferential polarization
I_1	3676.32	3.3720	27,201.1	$E\|c$
I_2	3680.63	3.3680	27,169.3	$E\perp c$
I_3	3681.59	3.3671	27,162.2	$E\|c$
I_4	3687.12	3.3621	27,121.4	$E\perp c$
I_5	3687.54	3.3617	27,118.3	$E\perp c$
I_6	3688.40	3.3609	27,112.0	$E\perp c$
I_7	3689.03	3.3604	27,107.4	$E\perp c$
I_8	3689.26	3.3601	27,105.7	$E\perp c$
I_9	3692.64	3.3571	27,080.9	$E\perp c$
I_{10}	3696.50	3.3536	27,052.6	$E\perp c$

Table IX.   A Summary of the ZnO Absorption Lines Reported in the Text at
1.2°K with $H = 0$*

Line	λ (Å)	Energy (eV)	$\bar{v}$ (cm^{-1})	Preferential polarization
I_a	3666.31	3.3812	27,275.5	$E\|c$
I_b	3669.66	3.3781	27,250.5	$E\perp c$
I_c	3671.99	3.3760	27,233.2	$E\|c$
I_d	3682.61	3.3662	27,154.6	$E\perp c$
I_e	3683.33	3.3655	27,149.3	$E\perp c$
I_f	3684.07	3.3649	27,143.9	$E\perp c$
I_g	3690.22	3.3593	27,098.6	$E\perp c$
I_h	3694.93	3.3550	27,064.1	$E\|c$

* Some of the absorption lines were observed as the self-absorption of a continuum emission from the crystals.

D.   Emission Line I_9

The emission line I_9 undergoes a linear splitting in a magnetic field. Plotted in Fig. 21 is the splitting of the I_9 line (photon energies of its split components in units of cm^{-1}) as a function of magnetic field strength with the crystalline c axis oriented perpendicular to the direction of the magnetic field ($c \perp H$); as can be seen in Fig. 21, I_9 splits into a doublet. In terms of the Thomas–Hopfield theory of bound excitons [39], one would expect a doublet splitting (with $c \perp H$) to arise from an exciton bound to a neutral donor or acceptor site in a $\Gamma_7 \rightarrow \Gamma_9$ optical transition, as shown in Fig. 13. In order to test further the theory as applied to the I_9 line, the Zeeman splitting of I_9 was examined, i.e., the line splitting was measured as a function of the crystal orientation (orientation of the c axis with respect to the

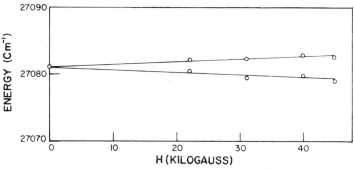

Fig. 21.   Splitting of the emission line I_9 (3692.64 Å) as a function of magnetic field strength at 1.2 °K for the orientation $c \perp H$.

magnetic field direction) in a constant magnetic field whose intensity was maintained at 45,000 G. The Zeeman splitting of I_9 is shown in Fig. 22. From the figure, it is obvious that I_9 splits into a quartet for all orientations except $c \perp H$ ($\theta = 90°$). Somewhat less obvious, however, is the fact that the high- and low-energy components of the quartet are not observed in the orientation $c \parallel H$ ($\theta = 0°$); these lines are missing because the transitions are not allowed by spin consideration. It was also observed that each of the four split components of I_9 were polarized in the mode $E \perp c$. In fact, all of the measurements concerning this line (especially the Zeeman splitting) tend to suggest that it arises from a bound-exciton complex of the neutral donor or acceptor type.

When emission and absorption lines, arising from bound-exciton complexes, undergo linear Zeeman splittings, Thomas and Hopfield have shown [39] that the hole and electron g values for such splittings obey the relation

$$g = g_e \pm g_h = g_{e0} \pm g_{h\parallel} \cos \theta \qquad (10.24)$$

where g_e and g_h are electron and hole g values, respectively; g_{e0} is the iso-tropic g value of the electron; $g_{h\parallel}$ is the g value of the hole when $c \parallel H$; and θ is, as usual, the angle between the c axis and the direction of H. Equation (10.24), implicit in Fig. 13(a) and (b), implies in the linearly split quartet of Fig. 22 that the outer pair of lines (high and low energy) splits as the sum of the g values ($g_e + g_h$) and that the inner pair of lines splits as the difference of the g values ($g_e - g_h$). Remembering that the hole in a neutral donor or acceptor complex is not spin-split when the crystal is oriented with $c \perp H$, we can calculate an electron g value from the usual relation

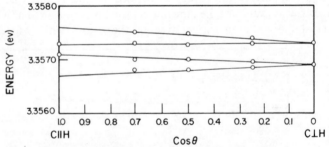

Fig. 22.   Splitting of the I_9 emission line as a function of $\cos \theta$ in a constant field of 45,000 G at 1.2 °K. Here, as in subsequent plots, θ is the angle between the crystalline c axis and the direction of the magnetic field. Note that this line splits as a doublet for the orientation $\theta = 0°$ and $\theta = 90°$, while for intermediate orientations ($0° < \theta < 90°$) quartets are observed.

$\Delta E = g\beta H$, where β is the Bohr magneton. Hence, from the data of Fig. 21, we calculate an electron g value of -1.9_3 for I_9. Recalling that the inner pair of lines in Fig. 22 splits as $g_e - g_h = \Delta E/\beta H$, we find $g_e - g_h = 0.69$, using ΔE at the orientation $c \parallel H$; hence, the g value for the hole of the I_9 complex is calculated to be $g_h = -1.93 + 0.69 = -1.24$. Thermalization effects were not observed in the split components of I_9. One would not expect to observe thermalization effects in the fluorescence from an exciton bound to a neutral donor site; on the other hand, this complex is not the only possibility in the present case. It is sometimes possible to observe thermalization effects in the emission from an exciton bound to a neutral acceptor site when the crystal is oriented with $c \perp H$; however, the observation of emission-line thermalization, in this case, requires that the exciton lifetime in the upper state be long enough for thermalization of the electron spins to occur. For example, consider the neutral acceptor complex of Fig. 13(b) in the orientation $c \perp H$, where the upper state $\ominus\,{}^+_+-$ is electron spin-split into a doublet, while the lower state $\ominus +$ is not split, since hole splitting is not allowed. Now, if the complex is sufficiently long-lived in its upper state, the electron spins can thermalize with either increasing magnetic field or decreasing temperature, so that the lower-energy component of the spin splitting will become more densely populated, giving rise to a low-energy emission line that is more intense than its high-energy counterpart. Hence, the I_9 line might arise from a neutral acceptor complex and still not show thermalization effects as a consequence of a short-lived upper state. Interestingly, in CdS thermalization has been observed in the magnetically split components of the 4888-Å emission line, a line which has been attributed to a neutral acceptor complex [39]. In the case of I_9 a lack of knowledge of the exciton-complex lifetime in the upper state precludes the possibility of distinguishing between a neutral donor and a neutral acceptor complex. The lines I_6, I_7, and I_8 have also been examined and have been found to be of the same type as line I_9, i.e., they are derived from an exciton bound to either a neutral donor or a neutral acceptor.

E. Emission-Line Multiplet, I_2-I_3

The emission-line multiplet I_2-I_3 is a zero-field split pair, as is revealed in the line splittings of Fig. 23, where the splittings of the high-energy component I_2 (3681.59 Å) are plotted as a function of magnetic field strength with the crystal oriented at $c \parallel H$. A zero-field splitting of 9.0×10^{-4} eV is measured, and a nonlinear splitting is indicated for the I_3 component. According to the observations of Thomas and Hopfield [39], the zero-field splitting of the multiplet suggests a bound-exciton complex composed of an intrinsic exciton molecularly bound to an ionized center; moreover, the small exciton-to-center binding energy (approximately 0.02 eV) makes it

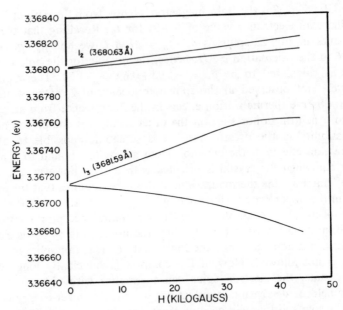

Fig. 23. Splitting of the $I_2 - I_3$ (3680.63 and 3681.59 Å) emission-line multiplet as a function of magnetic field strength for ZnO at 1.2 °K. The crystal is oriented with $c \| H$. Note the zero-field splitting at $H = 0$. Also note the nonlinear splitting of the I_3 component. The I_2 component is split but not resolved at the lower fields.

likely that the center is an ionized donor. As Thomas and Hopfield have shown [39], it can be argued that such a zero-field splitting of lines arises from a hole–electron, spin–spin exchange interaction in the upper state.

The Zeeman splitting of the $I_2 - I_3$ complex has been measured and is shown in Fig. 24, where the magnetic splittings of the two components (I_2 and I_3, circles) have been plotted as a function of orientation of the crystalline c axis with respect to the magnetic field direction ($\cos \theta$) while the field intensity was held constant at 45,000 G. A comparison of the data of Figs. 23 and 24 with the fourfold splitting scheme of Fig. 13(c) reveals that the I_3 line of the multiplet (low-energy component) corresponds to a transition from the Γ_6 exciton state. We have observed that the I_3 line is polarized in the mode $E \| c$, as expected for a Γ_6 exciton transition; similarly, the I_2 component of the multiplet corresponds to a Γ_5 exciton transition and is polarized in the mode $E \perp c$ also as expected. In the orientation $c \| H$ ($\cos \theta = 1$), the I_3 or Γ_6 component splits as the sum of the g values ($g_e + g_h$), whereas the I_2 or Γ_5 component splits as the difference of the g values ($g_e - g_h$), as shown by the large splitting for I_3 and the small splitting for I_2

in Fig. 24. In fact, on the basis of predicted splittings and polarization for an ionized complex in a $\Gamma_7 \leftrightarrow \Gamma_9$ optical transition, as shown in Fig. 13(c), one may reasonably conclude that the I_2–I_3 multiplet conforms to the specification of an exciton bound to an ionized donor or acceptor. Using the value of zero-field splitting quoted above ($\Delta = 9.0 \times 10^4$ eV) and a least-squares fit to the experimental splitting of Fig. 24 (circles), the energy level expressions in Fig. 13(c) (i.e., the levels E_1, E_2, E_3, and E_4) are plotted as the solid curves in Fig. 24. Rather good agreement (excellent for the I_2 component) is obtained between the calculated and experimental splittings of the I_2–I_3 complex; moreover, from these "weighted" line splittings, the following g values are obtained: $g_e = 1.9_5$ and $g_h = 1.5_1$. Of the several emission lines listed in Table VIII, the line I_1 also appears to arise from an ionized exciton complex; however, the intensities of its split components are too weak to permit a meaningful study in terms of the ionized complex model.

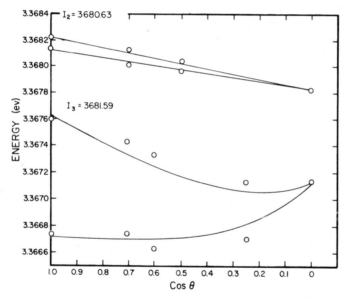

Fig. 24. Splitting of the I_2–I_3 (3680.63 and 3681.59 Å) emission-line multiplet at 1.2 °K as a function of cos θ in a constant magnetic field of 45,000 G. The circles represent the experimentally determined splittings for both components (I_2 and I_3), while the solid curves are a plot of theoretical splittings determined from the expressions of Fig. 13 (ionized complex). When the crystal assumes the orientation $c \perp H$ (cos $\theta = 0$), note that neither I_2 nor I_3 split at maximum field.

F.   Absorption-Line Multiplet, I_b–I_c

As is pointed out above, absorption spectra containing many sharp lines are often observed in selected ZnO crystals, primarily as a self-absorption of continuum fluorescence from the crystals. The most intense absorption line in these spectra is the line I_b (3669.66 Å); also, associated with this line is another weaker line which we have labeled I_c (3671.99 Å). Hence, as we shall show in the present measurements, the two absorption lines appear as a zero-field split pair and form the absorption multiplet, I_b–I_c. Thomas has previously observed these two lines and has labeled the line I_b as line A and the line I_c as line A_m; however, he has interpreted the line I_b as an intrinsic-exciton line, arising from the Γ_5 ground state of the A exciton (i.e., the ground-state exciton from the top valence band). Thomas has also examined the line I_c, but, like I_b, he has interpreted this line as an intrinsic-exciton line arising, in this case, from an exciton in the Γ_1 state and of top valence band origin. It is interesting that, if the Γ_1 exciton assignment is to hold for the line I_c, the exciton symmetry arguments require, in general, that the top valence band be of Γ_7 symmetry.

We have examined the splitting of the absorption-line multiplet I_b–I_c in a magnetic field. The experimentally determined Zeeman splittings are shown in Fig. 25 (circles) where, as usual, the line splittings are plotted as a

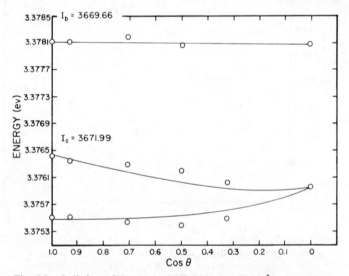

Fig. 25.   Splitting of the I_b–I_c (3669.66 and 3671.99 Å) absorption multiplet at 1.2 °K as a function of cos θ in a constant field of 45,000 G. As in Fig. 24, the circles represent experimental energies and splittings, while the solid curves are the calculated splittings (calculated from the energy level expressions of Fig. 13). The I_b component does not appear to split.

function of crystal orientation (splitting *versus* cos θ) in a magnetic field of constant field strength. As can be seen from Fig. 25, the I_c component of the multiplet shows a large splitting in the orientation $c \parallel H$ (cos $\theta = 1$); also, this line is polarized in the mode $E \parallel c$. The polarization and splitting of I_c suggest that this line derives from a Γ_6 exciton state, in direct analogy to the I_3 component of the emission-line multiplet I_2–I_3, which is shown to arise from an ionized donor or acceptor complex. If we treat the absorption line multiplet I_b–I_c as arising from a bound-exciton complex, in which the chemical center is ionized, we can again calculate the expected splittings for such a complex from the energy-level expressions (E_1, E_2, E_3, and E_4) of Fig. 13(c). Substitution of the measured value of zero-field splitting ($\Delta = 2 \times 10^{-3}$ eV) into the expressions of Fig. 13(c) yielded the following electron and hole g values obtained from a least-squares fit of the equations to the experimental data; $g_e = 1.9_5$ and $g_h = 1.7_4$. Using these parameters in the equations of Fig. 13(c), we have calculated the Zeeman splittings for the I_b–I_c multiplet. These splittings are plotted as the solid curves in Fig. 25. The line I_b is the high-energy component of the multiplet and corresponds to the Γ_5 exciton transition, whose splitting is proportional to the difference between the electron and hole g values ($g_e - g_h$). Since I_b components split as the difference between g values and since the electron and hole g values are of nearly the same magnitude in the complex (1.9$_5$ as compared to 1.7$_4$), the expected splitting for I_b is rather small. This prediction is borne out experimentally, since a splitting is not observed for the I_b components, as shown in Fig. 25. The circles represent the measured energies of the I_b line at the various orientations, and the single solid line drawn through these points represents the calculated splitting of I_b. As shown in Fig. 25, the experimental splitting of the I_c component at the intermediate angles ($0° \leqslant \theta \leqslant 90°$) is somewhat larger than that predicted by theory; however, a similar trend (deviation) is also observed for the I_3 component of the emission-line multiplet I_2–I_3. In spite of small deviations, there is good agreement between the experimental data and the general features of the theoretical model.

G. Vibrational Spectrum of a Bound-Exciton Complex

As previously mentioned, Lampert [36] pointed out that, in addition to the gross energy level or electronic level scheme of the H_2^+ complex, there will exist a fine structure for each electronic level similar to the vibration–rotation level scheme of H_2^+. This type spectrum was observed in selected CdS platelets by Reynolds *et al.* [44] and Collins *et al.* [45]. The densitometer trace of the emission series of one of these crystals is shown in Fig. 26. Evidence that would lend support to vibrational–electronic interpretations for this series is as follows: (1) a series of converging levels

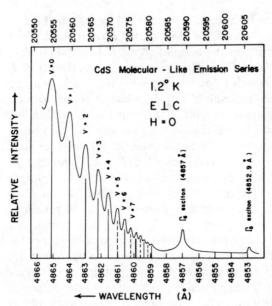

Fig. 26. Densitometer trace showing the line positions of the molecular-like series in CdS at 1.2 °K. The free-exciton lines Γ_6 and Γ_5 are also shown. Note that the Γ_6 is stronger in emission than the Γ_5. The quantities $v = 0, 1, 2, \ldots$ are the vibrational quantum numbers of the proposed exciton complex (after Reynolds, Litton, and Wheeler).

fitting a standard vibrational–electronic term scheme

$$E(v) = G(v) = D_0 + \omega_e(v+\tfrac{1}{2}) - \omega_e X_e(v+\tfrac{1}{2})^2 + \omega_e Y_e(v+\tfrac{1}{2})^3 - \ldots + \quad (10.25)$$

where v is the vibrational quantum number; the constants D_0, ω_e, X_e, Y_e are a measure of the potential describing the motion; the factor $\tfrac{1}{2}$ relates to the zero-point energy, $\hbar\omega_0$; and in our model D_0 is the electronic energy; (2) line-intensity distributions conforming to the Boltzmann factor; (3) level schemes extrapolating to the known exciton states; and (4) a series of rotational lines.

As can be seen from Fig. 26, the band of sharp, narrow (< 0.5 Å half-width) emission lines, whose intensities decay exponentially with increasing frequency, has a band head at 4865.08 Å and converges with increasing frequency toward a limit. More than 17 lines have been observed in the band, 15 of which are given in the data of Table X, and the line frequencies are reproducible from crystal to crystal.

The "series" (Table X) has been fitted to a standard vibrational–electronic scheme, the term values of which are given by

$$E(v) = [20{,}551.86 + 5.72(v+\tfrac{1}{2}) - 0.45(v+\tfrac{1}{2})^2 + 0.014(v+\tfrac{1}{2})^3] \text{ cm}^{-1} \quad (10.26)$$

where the (molecular) constants are $\omega_e = 5.72 \text{ cm}^{-1}$, $\omega_e X_e = 0.45 \text{ cm}^{-1}$, $\omega_e Y_e = 0.014 \text{ cm}^{-1}$, and $\hbar\omega_0 \approx 2.75 \text{ cm}^{-1}$ (the zero-point energy of complex). The fit is rather good except for the last several lines. If a quartic term $-\omega_e Z_e(v+\tfrac{1}{2})^4$ is added to equation (10.26), a better fit is obtained

Table X. A List of the Molecular-Like "Series" of CdS Emission Lines Reported in the Text*

| Vibrational quantum number v | Line position | | | Energy (cm^{-1}) E_0 [the zero of E_0 is referred to the first line of series (20,554.63 cm^{-1})] | Normalized intensity | ln (I/I_0) |
	λ (Å)	E (eV)	E (cm^{-1})			
0	4865.08	2.54804	20,554.63	0	100	0
1	4865.93	2.54760	20,559.50	4.87	37.5	−0.98
2	4862.95	2.54916	20,563.63	9.00	24.4	−1.41
3	4862.17	2.54957	20,566.95	12.32	12.4	−2.09
4	4861.53	2.54990	20,569.67	15.04	8.0	−2.53
5	4860.99	2.55019	20,571.93	17.30	5.2	−2.96
6	4860.57	2.55041	20,573.74	19.11	3.8	−3.27
7	4860.18	2.55061	20,575.35	20.72	3.1	−3.48
8	4859.86	2.55078	20,576.71	22.08	2.6	−3.64
9	4859.57	2.55093	20,577.94	23.31	2.2	−3.82
10	4859.34	·2.55105	20,578.91	24.28	1.9	−3.96
11	4859.11	2.55117	20,579.88	25.25	1.8	−4.02
12	4858.94	2.55126	20,580.62	—	—	—
13	4858.76	2.55136	20,581.39	—	—	—
14	4858.60	2.55144	20,582.08	—	—	—

* All of the lines in the "series" are strongly polarized in the mode $E \perp c$.

for the last three lines, but this discrepancy does not appreciably change the calculated constants of the vibrating complex.

The corrected and normalized intensities for the first 12 lines of the series are given in the data of Table VIII. From these measured intensities, ln I/I_0 is plotted as a function of line energy (E_0) in Fig. 27. With the use of all but the last two lines in this plot, a linear relationship is obtained (curve a), indicating that the line intensities do indeed conform to the Boltzmann factor; moreover, the slope of the line gives a temperature of 8.5°K, in fair agreement with the actual bath temperature of the experiment, 1.2°K. This temperature discrepancy may have been due to the fact that the intensities were determined from peak heights, without regard for line profiles and widths; also, it may have been that the exciton complex temperature was actually somewhat higher than that of the lattice.

Curve b of Fig. 27 is a plot of frequency difference of successive lines as a function of line frequency (ΔE versus E). As shown, the curve extrapolates very nearly to the Γ_6 emission line in the convergence limit. This limit implies a complex dissociation energy $D_e \sim 35 \text{ cm}^{-1}$. The dissociation or binding energy is shown graphically on the potential well of Fig. 28(b).

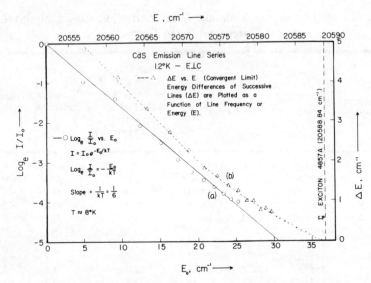

Fig. 27. Curve a is plot of the logarithm of the corrected series line intensities as a function of line energies, E_0. E_0 is the energy scale for the "series" referred to the lowest level or line (20,554.63 cm^{-1}) as zero. The experimental temperature was 1.2 °K. The intensities of the last two lines are thought to be in error, mainly because of convergence and line overlap. Curve b is the frequency or energy difference of successive lines. ΔE is never exactly zero in a vibrational spectrum comprised of a finite number of states; in this case it is thought to be close enough to zero in the limit to make the extrapolation valid. The curve and its extrapolation show the convergence limit for the vibrational spectrum (after Reynolds, Litton, and Wheeler).

A similar result is obtained when one plots the usual convergence, ΔE *versus* V, in conjunction with equation (10.25) and the quartic term. In the limit, one finds that there are some 25 vibrational states in the vibrating-complex potential well [Fig. 28(b)].

The CdS energy-band extrema, together with the potential well and energy scheme of the proposed vibrating complex, are given in the models of Fig. 28(a) and (b). Some details of the band model are shown, including several of the free-exciton states: the Γ_6 state is associated with the dissociation limit of the vibrating complex and is drawn at the same energy in both parts (a) and (b). In terms of the model, one can imagine the effective-mass particles, say, a hole from the A valence band and an electron from the Γ_6 state, bound together to form the free exciton. Now, if the free exciton is bound to an ionized donor or acceptor, a molecular-like complex is formed ($\oplus \mp$ or $\ominus \mp$). The bound-exciton complexes are created by the optical excitation of the crystals. The collapse of the excitons from the various vibrational levels in the potential will generate the observed emission spectrum. It is interesting to compare the observed binding energy (35 cm^{-1}) with some theoretical estimate of exciton-complex binding energies in CdS after Thomas and Hopfield [39]. On the basis of such a comparison, the ionized donor complex appears to be the most likely candidate for the bound exciton observed. If one assumes a functional form for the potential of Fig. 28(b), a calculated fit to the experimental data can be attempted. The assumed potential is the Morse function defined as

$$U(r-r_e) = D_e \left[1 - e^{-\beta(r-r_e)}\right]^2 \tag{10.27}$$

where D_e is the dissociation energy and β is a constant whose value is obtained from the experimental data. The solution to the Schrödinger wave equation with use of equation (10.27) for the potential gives the following term values

$$G(v) = \beta \sqrt{\frac{D_e h}{2\pi^2 c\mu}} (v+\tfrac{1}{2}) \frac{\hbar\beta^2}{-8\pi^2 c\mu} (v+\tfrac{1}{2})^2 \tag{10.28}$$

where μ is the effective mass and v is the quantum number which takes

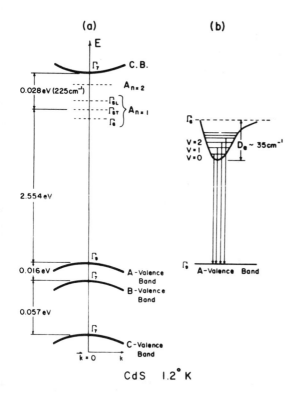

Fig. 28. (a) CdS band extrema at $k = 0$. The Γ_6 (4857 Å), Γ_{5T} (4854.5 Å), and Γ_{5L} (4853 Å) free-exciton states are indicated. The free-exciton binding energy (0.028 eV) is also indicated. (b) Energy scheme and vibrational potential well of the proposed bound-exciton complex, showing the Γ_6 exciton state as the dissociation limit (after Collins, Litton, and Reynolds).

integral values from zero until $G(v)$ is larger than the dissociation energy. Thus equation (10.28) gives us a finite number of levels associated with the vibrational energy states.

If one tries to fit the experimental data using only one series of lines, it is found that the value for the dissociation energy would have to lie above the first three term values which of course is not reasonable. We therefore conclude that there must be more than one series of lines in the observed spectrum. The evaluation of D_e and β/μ from experiment and the first few values of v substituted into equation (10.28) give a series of lines which fits more than half of the lines from the experimental data. The dissociation energy associated with these lines is 36.5 cm^{-1}, and the series extrapolates to the (free) Γ_6 exciton line (4857 Å). Investigating the remaining lines, one finds the first energy level (omitted) from the first series is 17.5 cm^{-1} or 4 Å above the ground state of the original series. Furthermore, this series of lines can be fitted using 36.5 cm^{-1} as a dissociation energy, and the second series appears to converge to the Γ_5 (free) exciton (4853 Å). However the lines in the upper end of the calculated series are missing from the experi-

Table XI.   (a) Term Values for the Series Which Converges to the Γ_6 Exciton Obtained from Equation (10.28) with $D_2 = 36.5$ cm^{-1} and $D_0 = 20,552.34$ cm^{-1} and (b) Term Values for the Series Which Converges to the Γ_5 Exciton Obtained from Equation (10.28) with $D_e = 36.5$ cm^{-1} and $D_0 = 20,570.04$ cm^{-1}

	(a)		(b)
v	$G(v) + D_0$ (cm^{-1})	v	$G(v) + D_0$ (cm^{-1})
0	20,554.60	0	20,571.71
1	20,558.92	1	20,574.87
2	20,562.94	2	20,578.00
3	20,566.67	3	20,580.91
4	20,570.12	4	20,583.66
5	20,573.25	5	20,586.26
6	20,576.14	6	20,588.70
7	20,578.72	7	20,590.98
8	20,581.01	8	20,593.11
9	20,583.01	9	20,595.07
10	20,584.72	10	20,596.89
11	20,586.15		
12	20,587.28		
13	20,588.12		

mental data. These missing lines would appear in a wavelength region where the crystal is highly absorbant; thus, it is not too surprising that they are not observed. The vertical solid lines drawn in Fig. 26 represent the semi-empirical calculation of the term values associated with the series which converges to the (free) Γ_6 exciton and the term values are given in Table XI(a). The vertical dashed lines in Fig. 26 represent calculations which are associated with the series converging to the Γ_5 (free) exciton. The values of the dashed lines are given in Table XI(b), and the curve in Fig. 26 is the plot of the intensity *versus* the wave number from the experiment performed by Reynolds *et al.* [44]. The band head for the series that converges to the Γ_5 free exciton is the 4861.7-Å I_3 line described by Thomas and Hopfield [39]. The band head for the series that converges to the Γ_6 free exciton is at 4865.08 Å. The energy difference between the band heads of the two series is the same as the energy difference between the convergence limits of the two series. The 4865.08-Å line might be the Γ_6 exciton associated with the I_3 complex.

IV. REFERENCES

1. D. Dutton, *Phys. Rev.* **112**:785 (1958).
2. H. Gobrecht and A. Bartschat, *Z. Physik* **136**:224 (1953).
3. L. R. Furlong and C. F. Ravilious, *Phys. Rev.* **98**:954 (1955).
4. E. F. Gross, *Nuovo Cimento Suppl.* **3**:672 (1956).
5. E. F. Gross, B. S. Razbirin, and M. Jakobson, *Zh. Tekhn. Fiz.* **27**:1149 (1957); *Soviet Phys. Tech. Phys. (English Transl.)* **2**:1043 (1957).
6. E. F. Gross and B. S. Razbirin, *Zh. Tekhn. Fiz.* **27**:2173 (1957); *Soviet Phys. Tech. Phys. (English Transl.)* **2**:2014 (1957).
7. J. L. Birman, *Phys. Rev. Letters* **2**:157 (1959); *J. Phys. Chem. Solids* **8**:35 (1959); *Phys. Rev.* **114**:1490 (1959).
8. D. G. Thomas and J. J. Hopfield, *Phys. Rev.* **116**:573 (1959).
9. R. J. Elliott, *Phys. Rev.* **108**:1384 (1957).
10. J. J. Hopfield and D. G. Thomas, *Phys. Rev.* **122**:35 (1961).
11. R. G. Wheeler and J. O. Dimmock, *Phys. Rev.* **125**:1805 (1962).
12. J. J. Hopfield and D. G. Thomas, *J. Phys. Chem. Solids* **12**:276 (1960).
13. J. Lambe and C. Kikuchi, *J. Phys. Chem. Solids* **8**:492 (1959).
14. J. J. Hopfield and D. G. Thomas, *Phys. Rev. Letters* **4**:357 (1960).
15. D. G. Thomas, *J. Phys. Chem. Solids* **15**:86 (1960).
16. J. J. Hopfield, *J. Phys. Chem. Solids* **15**:97 (1960).
17. Y. S. Park, C. W. Litton, T. C. Collins, and D. C. Reynolds, *Phys. Rev.* **143**:512 (1966).
18. E. I. Rashba, *Opt. i Spektroskopiya* **2**:508 (1957).
19. W. V. Lovitt, *Linear Integral Equations*, first edition, Dover Publications, Inc. (New York), 1950.
20. E. I. Rashba and G. E. Gurgenishvili, *Fiz. Tverd. Tela* **4**:1029 (1962); *Soviet Phys. Solid State (English Transl.)* **4**:759 (1962).
21. D. Berlincourt, Clevite Corporation (private communication).

22. F. Herman, C. D. Kuglin, K. F. Cutt, and R. L. Kortum, *Phys. Rev. Letters* **11**:541 (1963).
23. M. Balkanski and J. des Cloizeaux, *J. Phys. Radium* **21**:825 (1960).
24. J. J. Hopfield, *J. Phys. Chem. Solids* **15**:97 (1960).
25. S. L. Adler, *Phys. Rev.* **126**:118 (1962).
26. K. W. Boer and H. Gutjahr, *Z. Physik* **152**:203 (1958).
27. E. F. Gross and B. V. Novikov, *Soviet Phys. Solid State* **1**:321 (1959).
28. V. V. Eremenko, *Soviet Phys. Solid State* **2**:2315 (1961).
29. E. F. Gross, K. F. Lider, and B. V. Novikov, *Soviet Phys. Solid State* **4**:836 (1962).
30. Y. S. Park and D. C. Reynolds, *Phys. Rev.* **132**:2450 (1963).
31. M. H. Harbegger and H. Y. Fan, *Phys. Rev. Letters* **12**:99 (1964).
32. H. J. Stocker, C. Stannard, Jr., H. Kaplan, and H. Levinstein, *Phys. Rev. Letters* **12**:163 (1964).
33. D. N. Nasledov, Yu. G. Popof, and Yu. S. Smetanikova, *Fiz. Tverd. Tela* **6**:3728 (1964); *Soviet Phys. Solid State (English Transl.)* **6**:2989 (1965).
34. V. J. Mczurczyk, G. V. Tlemkov, and H. Y. Fan, *Phys. Rev. Letters* **21**:250 (1966).
35. Y. S. Park and D. W. Langer, *Phys. Rev. Letters* **13**:99(1964).
36. M. A. Lampert, *Phys. Rev. Letters* **1**:450 (1958).
37. J. R. Haynes, *Phys. Rev. Letters* **4**:361 (1960).
38. D. G. Thomas and J. J. Hopfield, *Phys. Rev. Letters* **7**:316 (1961).
39. D. G. Thomas and J. J. Hopfield, *Phys. Rev.* **128**:2135 (1962).
40. L. S. Pedrotti and D. C. Reynolds, *Phys. Rev.* **120**:1664 (1960).
41. D. C. Reynolds and C. W. Litton, *Phys. Rev.* **132**:1023 (1963).
42. B. Andress and E. Mollwo, *Naturwissenschaften* **46**:623 (1959).
43. B. Andress, *Z. Physik* **170**:1 (1962).
44. D. C. Reynolds, C. W. Litton, and R. G. Wheeler, *Proc. Intern. Conf. Phys. Semicond., Paris*, 1964.
45. T. C. Collins, C. W. Litton, and D. C. Reynolds, *Proc. Intern. Conf. Phys. Semicond., Paris*, 1964.

CHAPTER 11

Luminescence

F. Matossi

Physikalisch-Chemisches Institut
Freiburg im Breisgau, W. Germany

I. INTRODUCTION

One of the outstanding properties of most electronic semiconductors is their ability to luminesce, that is, to emit electromagnetic radiation under proper excitation and under conditions of thermal nonequilibrium.

As in the case of atoms, luminescence research should give information about energies of the stationary states of the electrons in the system considered. There are, however, several kinds of complications if the system is a solid. First, the levels are arranged in energy bands with or without localized levels in the forbidden band gaps. Second, the energy positions of the levels are spatially distributed; they are, for instance, different in the bulk and at the surface. Third, the levels may be different according to whether they are occupied or not. In addition, another complication concerns the approximative character of the theoretical models, which usually are one-electron models.

It should, nevertheless, be emphasized that even if we cannot construct rigorous and all-inclusive models, we do have simplified models that furnish an acceptable framework for the discussion of experiments. The main facts about photoluminescence and electroluminescence can indeed be interpreted, at least qualitatively, with models assuming simply the existence of localized and discrete donor and acceptor levels serving as activators or as traps for electrons and holes. In the usual Schön–Klasens model, the luminescence transition goes from the conduction band to an activator level, and electrons trapped in donor levels can be liberated thermally or otherwise to the conduction band. In the Lambe–Klick model, the luminescent transition goes from a high-lying activator level to the valence band, the electrons recombining with holes liberated from hole traps. Neither of these models can explain *all* the facts; both kinds of transitions may come into play in a specific semiconducting phosphor.

The above-mentioned models have proved to be suitable also for electroluminescence phenomena if the motion of electrons in electric fields and also the possibility of emptying electron or hole traps by electric fields are taken into account.

The mechanism of the excitation of electroluminescence still has not been determined unambiguously. Impact ionization would require very high fields, which may be present near phase boundaries, and sufficiently long electron paths in order to gain the necessary energy. In many instances, injection of minority carriers by suitable fields, which need not be very strong, may lead to luminescence by the recombination of the majority carriers with the minority carriers. This may happen particularly in p–n junctions. Since in such cases the carriers with the higher energy state are the majority carriers, one necessary condition of laser action— the overpopulation of the higher energy state—is satisfied.

This chapter is concerned mainly with electroluminescence and electrophotoluminescence (the modification of photoluminescence by electric fields) and related phenomena. So far, research has been concentrated on ZnS used as powdered material or as single crystals. It was possible to grow crystals of sufficient size that contained iron mainly as an impurity and could be doped with Al, Zn, or S. These crystals grown by sublimation were cubic, as revealed by electron spin resonance measurements, but contained stacking faults. The stacking faults have a definite influence on electroluminescence (injection luminescence only parallel to the fault planes) and on electronic and optical properties (anisotropy of conductivity and internal reflection at the fault planes). Crystals grown by transport reaction were purely cubic; they contained, however, large amounts of halogen.

We found it very profitable to conduct, in addition to the usual optical and electrical measurements, electron spin resonance experiments, which gave information about the chemical and crystallographic structure of the luminescence centers and the photocapacitive effect [change of capacitance by irradiation caused either by free electrons (photoconductive effect) or by displacement of trapped electrons (photodielectric effect)]. If the effect is due to the free electrons, the photocapacitive effect allows us to obtain conductivities without having to use contacts on the sample, thus avoiding the difficulties arising from disturbing contact properties.

II. RESULTS

Some of the main results obtained by a combination of the above-mentioned methods follow.

An important impurity in our "self-activated" crystals was Fe^{2+} (Cu was not present). By irradiation with ultraviolet light Fe^{2+} can be converted to Fe^{3+}, which can be measured by electron spin resonance experiments. This conversion is not quantitatively complete. By adding Cu in excess, however, complete conversion can be obtained. Irradiation with 910 mμ leads to charge transfer from "A-centers" of self-activated ZnS phosphors to Fe^{3+}; subsequent irradiation with 570 mμ reverses this process. Fe^{3+} is also quenched by 1200 mμ. The quenching of the A-centers is due to hole excitation [4].

Electron spin resonance revealed the A-center in cubic ZnS to be a halogen at a sulfur site neighboring on a Zn vacancy. One of the S atoms may trap a hole. This hole is fixed at low temperatures, but may hop among the different S sites with an activation of 0.057 eV [7].

The photocapacitive effect proved to be a powerful tool in observing conductivity in phosphors if the criteria for a photoconductive effect are met, viz., fast decay of photocapacitance after removal of the irradiation at low temperatures and strong dispersion effects at low frequencies with distinct maxima of photo-induced loss angle changes.

The photodielectric effect can be observed in powders. It can be modified by additional infrared irradiation. Combination of luminescence measurements with capacitance glow curves leads to the conclusion that reversible charge transfer among different traps may take place [8].

The photocapacitive effect in ZnS with stacking faults is anisotropic, equivalent to an anisotropic conductivity due to an extra scattering process of electrons passing the stacking fault planes. The existence of stacking faults can also be observed optically through incoherent internal reflection. From the amount of this reflected light, the average number of reflecting planes can be estimated to about 1000 cm^{-1}. The reflecting planes are, however, not necessarily the stacking fault planes as such, but planes where the average density of the stacking faults proper changes [9].

The "Gudden–Pohl flash," which is observed after applying an electric field to an excited ZnS phosphor, originates at the anode of the ZnS capacitor through recombination of electrons drifting from the cathode with a fixed positive space charge at the anode. The flash must then be delayed with respect to the time of application of the field. The delay time is interpreted as being the drift time of the electrons from cathode to anode. From this, the drift mobility can be directly computed as 88 cm^2/V-sec, while the Hall mobility is 110 cm^2/V-sec [11]. The assertion of considering the Gudden–Pohl flash as a surface phenomenon is proved directly by observing the site of the flash through a microscope, and indirectly by the dependence of the flash intensity on the polarity of electric fields acting during optical excitation superimposed on the field exciting the Gudden–Pohl flash [3].

The following results from as yet unpublished works may be mentioned. First, since the photocapacitive effect depends on the Debye length, this length can be determined from the measurements. Because of excess trapping of electrons at the stacking fault planes, the Debye lengths are larger for fields perpendicular to the planes than for the parallel direction. Second, hexacyanobenzene, $C_6(CN)_6$, is a semiconductor with an excitation energy of 2 eV (from dark conductivity *versus* temperature) corresponding to an observed absorption edge near this energy. Photoconductivity is observed with an activation energy of 0.26 eV.

III. REFERENCES

1. F. Matossi and H. Gutjahr, *Phys. Status Solidi* **3**:167 (1963).
2. A. Räuber, J. Schneider, and F. Matossi, *Z. Naturforsch.* **17a**:654 (1962).
3. H. Gutjahr, *Z. Physik* **168**:199 (1962).
4. A. Räuber and J. Schneider, *Phys. Letters* **3**:230 (1962).
5. J. Schneider *et al.*, *Phys. Letters* **5**:312 (1963).
6. D. Siebert and F. Matossi, *Physik Kondensierten Materie* **2**:334 (1964).
7. B. Dischler, A. Räuber, and J. Schneider, *Phys. Status Solidi* **6**:507 (1964).
8. F. Matossi and D. Siebert, *Z. Naturforsch.* **19a**:454 (1964).
9. D. Siebert and H. Teitge, *Z. Naturforsch.* **20a**:838 (1965).
10. D. Siebert, H. Teitge, and F. Matossi, *Z. Naturforsch.* **20a**:1309 (1965).
11. F. Matossi, K. Leutwein, and G. Schmid, *Z. Naturforsch.* **21a**:461 (1966).

CHAPTER 12

Lattice Vibrations

H. Bilz

Institut für Theoretische Physik
Frankfurt, Germany

I. GENERAL PROPERTIES OF LATTICE VIBRATIONS

A. Concept of Lattice Vibrations

In a crystal with N atoms or ions, there exist $3(N-1)$ lattice vibrations. For an insulator they can be derived, in principle, from the full Hamiltonian by using a Born–Oppenheimer treatment which separates these low-energy excitations from the comparatively high-lying electronic excitations. For a metal an even more complicated treatment which describes the dynamics of electrons and ions in a self-consistent way has to be used. Until now, no calculations existed from first principles, only some semi-empirical methods starting, e.g., from plausible lattice potentials (screened coulomb potential, Born–Mayer potential, etc.) with some parameters adjustable to match experimental data. In a rigorous treatment, one uses a formal development of the lattice potential in powers of the ion displacements. The expansion coefficients are restricted only by conservation laws and by the symmetry of the crystal, and they must be determined from experimental information. The harmonic approximation leads to the picture of free quasi-particles, called "phonons," each of which is characterized by its energy $\hbar\omega$, crystal momentum $\hbar q$, and branch index j. As can be seen from inelastic neutron spectroscopy, this picture is usually well satisfied since the lifetime of phonons is of the order of 100 vibration periods. The far-reaching analogy between quantum-mechanical and classical harmonic oscillators allows for a classical description of plane waves progressing in the direction of the wave vector $\mathbf{q}$. Thus, each lattice ion moves with the frequency ω and has an elliptic polarization which is uniquely related to the branch index j, but which has a simple form only in certain symmetry directions (transverse or longitudinal). The cubic and higher-order parts of the lattice potential cause complicated phonon–phonon interactions. In the following, we separate the *static* effects from the *dynamical* ones (phonon lifetime, infrared absorption, etc.). We therefore use a quasi-harmonic approximation [2] in which all single phonon quantities (frequencies, force constants, etc.) are considered at a fixed temperature and depend on it as a parameter.

B. Dispersion Relation

Let us expand the lattice potential φ in terms of displacement vectors $\mathbf{u}\begin{pmatrix} l \\ k \end{pmatrix} \equiv \mathbf{u}(i)$ of the ith ion in the lattice cell l with the basis index k and mass M_k:

$$\phi \approx \phi^{(2)} = \tfrac{1}{2} \sum_{lk\alpha} \sum_{l'k'\alpha'} \phi_{\alpha\beta} \begin{pmatrix} ll' \\ kk' \end{pmatrix} u_\alpha \begin{pmatrix} l \\ k \end{pmatrix} u_\beta \begin{pmatrix} l' \\ k' \end{pmatrix} \tag{12.1}$$

Using periodic boundary conditions, we have a finite volume V with N cells ($l = 1, \ldots, N$) and n ions in each cell ($k = 1, \ldots, n$), so there are $3nN$ different normal modes. Looking for stationary solutions with frequency ω

$$u_\alpha(i) = V_\alpha(i)\, e^{-i\omega t} \tag{12.2}$$

we obtain from the classical equations of motion

$$M_k\, \ddot{u}_\alpha(i) = - \sum_{i',\beta} \phi_{\alpha\beta}(ii')\, u_\beta(i') \tag{12.3a}$$

$$\omega^2 M_k\, V_\alpha(i) = \sum \phi_{\alpha\beta}(ii')\, V_\beta(i') \tag{12.3b}$$

For the solution of this system of $3nN$ homogeneous equations, it is convenient to introduce an affine transformation:

$$w_\alpha(i) \equiv w_\alpha \begin{pmatrix} l \\ k \end{pmatrix} \equiv (M_k)^{\frac{1}{2}} V_\alpha(i)$$

$$V_{\alpha\beta}(ii') = (M_k M_{k'})^{-\frac{1}{2}} \phi_{\alpha\beta}(ii') \tag{12.4}$$

We then obtain

$$\omega^2 w_\alpha(i) = \sum V_{\alpha\beta}(ii')\, w_\beta(i') \tag{12.5}$$

or in matrix notation

$$\omega^2 \mathbf{w} = \mathbf{Vw} \tag{12.6}$$

Here $\mathbf{w}$ is a $3nN$-dimensional vector

$$\mathbf{w} = \begin{bmatrix} w_x \begin{pmatrix} 1 \\ 1 \end{pmatrix} \\ \vdots \\ w_t \begin{pmatrix} N \\ n \end{pmatrix} \end{bmatrix}$$

and $\mathbf{V}$ is a corresponding $3nN \times 3nN$ matrix $(V_{\alpha\beta})$. We note that the force constants $\phi_{\alpha\beta}$ and $V_{\alpha\beta}$ depend only on the difference $l - l'$ due to the crystal periodicity. The eigenvalues of equation (12.6) are obtained from the secular equation

$$\det |\mathbf{V} - \omega^2 \, \mathbf{I}_1| = 0 \tag{12.7}$$

where $\mathbf{I}_1$ denotes a $3nN$ unit matrix.

In solving equation (12.7) we make use of the periodicity of the crystal. If $\mathbf{a}_1$, $\mathbf{a}_2$, $\mathbf{a}_3$ is a triple vector, defining the elementary cell in ordinary space, then the center point of the lth cell is described by the vector

$$\mathbf{X}(l) = l_1 \, \mathbf{a}_1 + l_2 \, \mathbf{a}_2 + l_3 \, \mathbf{a}_3 \tag{12.8}$$

and a particular atom with index k is described by

$$\mathbf{X}(l, k) = \mathbf{X}(l) + \mathbf{X}(k) \tag{12.8a}$$

It is known from Floquet's theorem for periodic potentials that there exists a representation

$$w_\alpha(i) = w_\alpha \begin{pmatrix} l \\ k \end{pmatrix} = y_\alpha(k) \, e^{i\mathbf{q}\mathbf{X}(l,k)} \tag{12.9}$$

where the wave vector $\mathbf{q}$ is restricted to the values

$$q_i = \frac{2 \, \pi m_i}{N_i} \qquad \begin{matrix} m_i = 1, 2, ..., N_i \\ N_1 \, N_2 \, N_3 = N \end{matrix} \tag{12.10}$$

For exactly these values of $\mathbf{q}$ we have

$$w_\alpha \begin{pmatrix} l_1 + N_1, \, l_2 + N_2, \, l_3 + N_3 \\ k \end{pmatrix} = w_\alpha \begin{pmatrix} l_1 \, l_2 \, l_3 \\ k \end{pmatrix} \tag{12.11}$$

This means that the periodic boundary conditions are fulfilled. The remaining problem to solve consists of the $3n$ equations

$$\omega^2 \, y_\alpha(k) = \sum_{k'\beta} D_{\alpha\beta} \begin{pmatrix} \mathbf{q} \\ kk' \end{pmatrix} y_\beta(k') \tag{12.12}$$

with

$$D_{\alpha\beta} \begin{pmatrix} \mathbf{q} \\ kk' \end{pmatrix} = \sum_l V_{\alpha\beta} \begin{pmatrix} l \\ kk' \end{pmatrix} e^{-i\mathbf{q}[\mathbf{X}(l) + \mathbf{X}(k) - \mathbf{X}(k')]} \tag{12.13}$$

This corresponds to the secular equation

$$\det (\mathbf{D} - \mathbf{I}_2 \, \omega^2) = 0 \tag{12.14}$$

where $\mathbf{D} = (D_{\alpha\beta})$ is the dynamical matrix and $\mathbf{I}_2$ is a $3n$ unit matrix.

Solution of equation (12.14) gives $3n$ solutions $\omega_j(\mathbf{q})$ for each wave vector $\mathbf{q}$, namely, the $3n$ different acoustic and optic branches j.

Introducing the reciprocal lattice with the three basis vectors $\mathbf{b}_j$ defined by

$$\mathbf{a}_i \mathbf{b}_j = \delta_{ij} \tag{12.15}$$

we can restrict the representation of $\omega_j(\mathbf{q})$ to the first elementary cell or Brillouin zone in $\mathbf{q}$-space, which is equal to 2π times the reciprocal lattice spacing. This exhibits the periodicity of $\omega(\mathbf{q})$ in $\mathbf{q}$-space.

Without proof, we note some important properties of the dispersion relation $\omega(\mathbf{q})$. First,

$$\omega_j(\mathbf{q}) = \omega_j(-\mathbf{q}) \tag{12.16}$$

Second, three acoustic branches always exist:

$$\lim_{\mathbf{q}\to 0} \omega_j(\mathbf{q}) = 0 \qquad \text{for} \qquad j = 1, 2, 3 \tag{12.17}$$

Third, $\omega_j^2(\mathbf{q}) > 0$ for $\mathbf{q} \neq 0$ or $j \neq 1, 2, 3$. This reality condition for $\omega(\mathbf{q})$ comes from the stability of the lattice and conversely could be regarded as $3nN$ conditions for the force constants $\phi_{\alpha\beta}\begin{pmatrix} l & l' \\ k & k' \end{pmatrix}$ in ordinary space.

C. Normal Coordinates

The special vectors $\mathbf{y}(k)$ belonging to the eigenvalues $\omega_j(\mathbf{q})$ of the dynamical matrix $\mathbf{D}$ are the generally complex eigenvectors $\mathbf{e}\left(k \Big| \begin{matrix} q \\ j \end{matrix}\right)$ which determine the (generally elliptic) polarization of each lattice mode. They can be assumed to be an orthonormal system in the following way:

$$\mathbf{e}^*\begin{pmatrix} \mathbf{q} \\ j \end{pmatrix} \mathbf{e}\begin{pmatrix} \mathbf{q} \\ j' \end{pmatrix} = \delta_{jj'} \tag{12.18}$$

and

$$\sum_j e_\alpha^*\left(k \Big| \begin{matrix} \mathbf{q} \\ j \end{matrix}\right) e_\beta\left(k' \Big| \begin{matrix} \mathbf{q} \\ j \end{matrix}\right) = \delta_{\alpha\beta}\, \delta_{kk'} \tag{12.19}$$

Here $\mathbf{e}(q)$ is a $3n$-component vector:

$$\begin{bmatrix} e_\alpha\left(1 \Big| \begin{matrix} \mathbf{q} \\ j \end{matrix}\right) \\ \vdots \\ e_\gamma\left(n \Big| \begin{matrix} \mathbf{q} \\ j \end{matrix}\right) \end{bmatrix}$$

From

$$D^*_{\alpha\beta}\begin{pmatrix} \mathbf{q} \\ kk' \end{pmatrix} = D_{\alpha\beta}\begin{pmatrix} -\mathbf{q} \\ kk' \end{pmatrix} \tag{12.20}$$

it follows that $\mathbf{e}^*\begin{pmatrix} \mathbf{q} \\ j \end{pmatrix} = \pm e\begin{pmatrix} -\mathbf{q} \\ j \end{pmatrix}$. Here the positive sign is used.

In a normal vibration with frequency $\omega_j(\mathbf{q})$, each lattice ion $\begin{pmatrix} l \\ k \end{pmatrix}$ takes part in the displacement:

$$u_\alpha \begin{pmatrix} l \\ k \end{pmatrix} \begin{vmatrix} \mathbf{q} \\ j \end{vmatrix} = (NM_k)^{-\frac{1}{2}} Q\begin{pmatrix} \mathbf{q} \\ j \end{pmatrix} e_\alpha \begin{pmatrix} k \begin{vmatrix} \mathbf{q} \\ j \end{vmatrix} \end{pmatrix} \times \exp\left[i\mathbf{q}\mathbf{X}(l,k) - i\omega_j(\mathbf{q})\,t\right] \tag{12.21}$$

The amplitudes $Q\begin{pmatrix} \mathbf{q} \\ j \end{pmatrix}$ of these normal vibrations are the so-called normal coordinates with the reality condition

$$Q^*\begin{pmatrix} \mathbf{q} \\ j \end{pmatrix} = Q\begin{pmatrix} -\mathbf{q} \\ j \end{pmatrix}$$

D. Density of States and Critical Points

We denote the number of lattice vibrations per interval $\omega \ldots \omega + d\omega$ by $n(\omega)$ and the corresponding number per interval $\omega^2 \ldots \omega^2 + d\omega^2$ by $N(\omega^2)$. Furthermore, we assume that the volume V_a of the elementary cell is small compared with the volume $V = NV_a$ of the crystal, so that the eigenvalues $\omega_j(\mathbf{q})$ form a quasi-continuous set for all practical purposes. The normalization of $n(\omega)$ leads to

$$\int_0^{\omega_{max}} n(\omega)\,d\omega = 1 \tag{12.22}$$

We can express $n(\omega)$ in the following way:

$$n(\omega) = \lim_{\substack{\Delta\omega \to 0 \\ N \to \infty}} \frac{1}{\Delta\omega} \cdot \frac{1}{3nN} \sum_{\mathbf{q}j(\omega)}^{(\omega + \Delta\omega)} \text{(number of eigenvalues)} \tag{12.23a}$$

$$= \lim_{\Delta\omega \to 0} \frac{1}{\Delta\omega} \frac{V_a}{(2\pi)^3} \frac{1}{3n} \iiint_\omega^{\omega + \Delta\omega} d^3\mathbf{q} \tag{12.23b}$$

$$= \frac{V_a}{3n} \iint \frac{dS_\mathbf{q}}{|\nabla \omega_j(\mathbf{q})|} \qquad \omega = \omega_j(\mathbf{q}) \tag{12.23c}$$

We have made use of the fact that to each discrete eigenvalue there belongs a volume $2\pi^3/NV_a$ in $\mathbf{q}$-space. A surface element in this space is denoted by $dS_\mathbf{q}$.

The corresponding relations for $N(\omega^2)$ follow simply from the equality

$$n(\omega) = 2\,\omega N(\omega^2) \qquad (12.24)$$

Anomalous behavior of the density $n(\omega)$ is expected at frequencies for which $\nabla\omega_j(\mathbf{q})$ equals zero or changes sign discontinuously. At these "critical points" (usually at the Brillouin zone boundary), characteristic features appear which are reflected in thermal and optical properties of the crystals. The theory of these critical points has been developed mainly by van Hove [3], Phillips [4], and Rosenstock [5] and can be summarized as follows:

1. An analytical critical point P_i with frequency $\omega_c = \omega(\mathbf{q}_c)$ and index i, in an s-dimensional $\mathbf{q}$-space, has a Taylor expansion in its neighborhood of the form

$$\omega^2(\mathbf{q}) = \omega^2(\mathbf{q}_c) + \sum_{r=1}^{s} \varepsilon_r\,\tilde{\mathbf{q}}_r^{\,2} \qquad (12.25)$$

with

$$\varepsilon_r = \pm 1$$

where i is the number of negative ε_r in this expansion and $\tilde{\mathbf{q}}$ is defined in a locally transformed system.

2. There exist in an s-dimensional Brillouin zone for each branch at least $\binom{s}{i}$ analytical critical points P_i. Generally there are topological conditions (Morse relations) which relate the different $P_i's$ (topological set).

3. From the cubic or lower symmetry of the Brillouin zone, there follow also some critical points (symmetry set).

4. The smallest topological set of critical points which contains the symmetry set is called the minimal set. Apparently an analysis of the density of states of a single branch cannot be carried out with less than the minimal set.

The foregoing statements are not restricted to analytical critical points. They can be extended to nonanalytical, but topologically equivalent critical points.

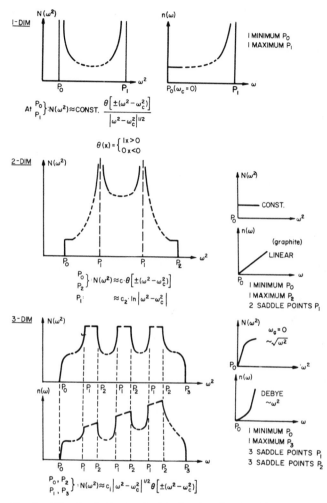

Fig. 1. The simplest sets of critical points for densities $n(\omega)$ and $N(\omega)$ for a single branch in one, two, and three dimensions.

Figure 1 shows the simplest sets of critical points for densities $n(\omega)$ and $N(\omega)$ for a single branch in one, two, and three dimensions, which are consistent with the requirements above. From this figure, it turns out that density maxima [for $n(\omega)$!] in the three-dimensional **q**-space arise only at saddle points P_2 or more generally if a P_1 and a P_2 have nearly the same frequency.

II. LATTICE VIBRATIONS IN DIATOMIC CUBIC CRYSTALS

A. General Discussion

Diatomic cubic crystals such as the alkali halides, the homopolar semiconductors germanium, silicon, and diamond, and the III–V compounds (GaAs, etc.) are by far the most investigated nonmetallic crystals, both theoretically and experimentally. There are only a few other substances, such as solid argon, CaF_2, and $SrTiO_3$, where the lattice vibrations are rather well-known. We shall therefore focus our attention on the diatomic crystals.

Until 10 years ago all theoretical calculations of dispersion curves have used the "rigid ion model". In this model, force constants $\phi_{\alpha\beta}$ between the few (main, first, and second) nearest neighbors are used without taking into account the deformation of the valence electron "clouds" which lead to long-range forces between the ions. Only the coulomb forces of ions in ionic crystals, such as KCl, were properly taken into account by using Ewald's method of summing up the coulomb forces and including the splitting of longitudinal and transverse optical modes with very long wavelengths due to the different boundary conditions of these waves at the crystal surface (Kellermann model).

After some earlier indications from specific heat and infrared spectra (e.g., the Lyddane–Sachs–Teller relation), the measuring of dispersion curves with inelastic neutron scattering (starting in the 1950's) clearly showed the breakdown of the force constant models. Herman [6] in 1959 showed that the measured dispersion curves of germanium could not be described satisfactorily without taking into account force constants to fifth or sixth nearest neighbors. This indicated that long-range forces have to be considered which are not related to the short-range repulsive part of the potential. The natural step was an extension of the model to include the dipole forces induced during the vibrations, due to the polarizability of the atoms or ions. First Tolpygo and Mashkevich [7] and later Cochran [8] and Hardy [9] introduced dipole approximations. The most pictorial form is the "shell model" introduced by Cochran, following Dick and Overhauser, which has been used with remarkable success for alkali halides [10], germanium, silicon and diamond [11], and III–V compounds [12]. To understand the essential points of this model, let us first discuss a one-dimensional example which can be regarded as representative for the longitudinal model of a diatomic cubic crystal in one of the main symmetry directions in **q**-space.

B. Diatomic Linear Chain

Consider a one-dimensional crystal with alternating masses m_1 and m_2 and only one spring constant f (Fig. 2).

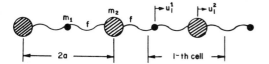

Fig. 2. Diatomic linear chain with alternating masses.

The displacements in the lth cell are $u\begin{pmatrix} l \\ 1,2 \end{pmatrix}$ for the ions with masses $m_{1,2}$. The spring constant is

$$\phi\begin{pmatrix} 0 \\ 12 \end{pmatrix} = \phi\begin{pmatrix} 1 \\ 21 \end{pmatrix} = -f$$

With

$$u\begin{pmatrix} l \\ k \end{pmatrix} = (m_k)^{-\frac{1}{2}}\, y(k)\, e^{-i\{\omega t - qa[l+(k-1)\frac{1}{2}]\}} \tag{12.26}$$

we easily find the dynamical matrix

$$D\begin{pmatrix} q \\ kk' \end{pmatrix} = \begin{bmatrix} \dfrac{2f}{m_1} & -\dfrac{2f}{(m_1 m_2)^{\frac{1}{2}}}\cos\dfrac{qa}{2} \\[3ex] -\dfrac{2f}{(m_1 m_2)^{\frac{1}{2}}}\cos\dfrac{qa}{2} & \dfrac{2f}{m_2} \end{bmatrix} \tag{12.27}$$

The solution of the secular equation (12.14) gives

$$\omega^2 = \frac{f}{m_1 m_2}\,[m_1 + m_2 \pm (m_1^2 + m_2^2 + 2m_1 m_2 \cos qa)^{\frac{1}{2}}] \tag{12.28}$$

Two branches, one optical and one acoustical, are obtained (Fig. 3).

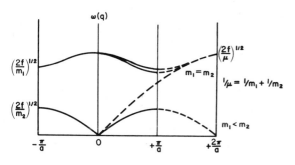

Fig. 3. Dispersion curve of a diatomic chain.

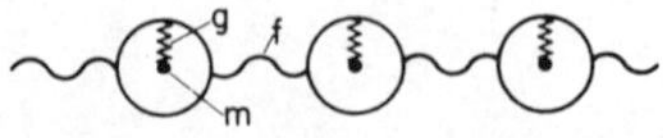

Fig. 4. Simple linear monatomic shell model.

If $m_1 = m_2$, the simple monatomic lattice with the dispersion relation prevails

$$\omega^2(q) = \frac{4f}{m}\sin^2\frac{aq}{2} \tag{12.29}$$

where the first Brillouin zone is doubled (Fig. 3).

Consider now a simple linear shell model. The monatomic case is shown in Fig. 4, where f is now the repulsive shell–shell force constant and g is the repulsive shell–core spring constant. No coulomb forces are taken into account. If u_l and v_l are the displacements of the core and the shell in the lth cell, the equations of motion are

$$m\ddot{u}_l = -mu_l\omega^2 = g(u_l - v_l)$$

$$0 = g(v_l - u_l) + f(v_l - v_{l+1} - v_{l-1}) \tag{12.30}$$

The second equation is only a subsidiary condition due to the adiabatic approximation in which no resultant force is acting on the electrons.

The solution of equation (12.30) is easily obtained and can be written in the following form:

$$\omega^2 = 4\frac{f(q)}{m}\sin^2\frac{aq}{2} \tag{12.31}$$

with

$$f(q) \equiv f \cdot \frac{1}{1 + 4\frac{f}{g}\sin^2\frac{aq}{2}}$$

Equation (12.31) shows that the result can be compared with the rigid ion model [equation (12.29)] by introducing a q-dependent spring constant $f(q)$.

If the valence electrons are strongly coupled to the ion cores, that is, if $g \gg f$, then $f(q) \approx f$ and we obtain again the result of the rigid ion model. If on the other hand the electrons are only weakly bound, i.e., $g \ll f$, then for values $aq \gtrsim 1$ it follows that

$$\omega^2 \propto g/m = \text{const.} \tag{12.32}$$

Thus, there is a large region in the q-space with very flat dispersion, which generally shows the effect of high polarizability of lattice ions on the dispersion curves (Fig. 5).

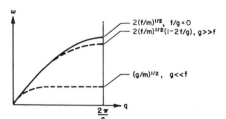

Fig. 5. Dispersion curves in the monatomic simple shell model.

The discussion will now be extended to a diatomic chain with *one* polarizable ion (Fig. 6). This model applies to alkali halides, such as NaI, KBr, etc. Without showing the detailed calculations, we will merely state the results which are

$$\omega_{1,2}^2 = \frac{f^*}{m_1 m_2}\bigg([m_1 + m_2 + 2f\sin^2(qa/2)\,m_2/g]$$

$$\pm \{[m_1 + m_2 + 2f\sin^2(qa/2)\,m_2/g]^2 - 4f/f^*\,m_1 m_2 \sin^2(qa/2)\}^{\frac{1}{2}}\bigg) \quad (12.33)$$

with

$$f = f^*(1 + 2f/g)$$

For $g \gg f$, equation (12.28) is rederived. If $g \gtrsim f$, one must distinguish between the case in which the light ion is polarizable ($m_2 < m_1$) and the case in which the heavy ion is polarizable ($m_1 < m_2$) (which is usually the case). The dispersion curves for both cases are shown in Fig. 7.

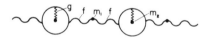

Fig. 6. Diatomic chain with one polarizable ion.

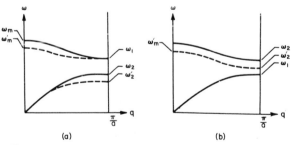

(a)   (b)

Fig. 7. Dispersion curves for the diatomic chain with one polarizable ion (— — —) compared with the rigid ion model. Equation (12.34) describes the parameters present. (a) Heavy ion polarizable ($m_2 > m_1$). (b) Light ion polarizable ($m_2 < m_1$).

The shape of the dispersion curves for the rigid ion model and for the shell model as shown in Fig. 7(a) is very similar to those of alkali halides (NaI and BKr), especially for the longitudinal modes (Fig. 1). Equations (12.34) describe the parameters present in Fig. 7:

$$\omega_m^2 = 2f/\mu \qquad \omega_m'^2 = 2f^*/\mu \qquad 1/\mu = 1/m_1 + 1/m_2 \qquad (12.34)$$
$$\omega_1^2 = 2f/m_1 \qquad \omega_2^2 = 2f/m_2 \qquad \omega_2'^2 = 2f^*/m_2$$

We shall next discuss the three-dimensional case.

C. Diatomic Cubic Crystals

For the calculation of three-dimensional crystals, there is an additional problem compared with the one-dimensional case, namely, to take into account properly the point group of the crystal. This leads to a restriction of the number of independent force constants $\phi_{\alpha\beta}$, which are restricted moreover by some conservation laws.

Let us consider the case of alkali halides in more detail. Start with the slightly modified equation (12.12)*

$$m_k \omega^2 U_\alpha(k) = \sum_{k'\beta} M_{\alpha\beta} \begin{pmatrix} q \\ kk' \end{pmatrix} U_\beta(k') \qquad (12.35)$$

or in matrix notation

$$\mathbf{m}\omega^2 \, \mathbf{U} = \mathbf{M}\mathbf{U} \qquad (12.36)$$

with

$$U_\alpha(k) = (m_k)^{\frac{1}{2}} y_\alpha(k) \qquad (12.37)$$

$$M_{\alpha\beta} \begin{pmatrix} q \\ kk' \end{pmatrix} = (m_k \, m_{k'})^{\frac{1}{2}} D_{\alpha\beta} \begin{pmatrix} q \\ kk' \end{pmatrix}$$

Here $\mathbf{m}$ is a diagonal matrix:

$$\mathbf{m} = \begin{pmatrix} \mathbf{m}_1 & \\ & \mathbf{m}_2 \end{pmatrix} \qquad \mathbf{m}_i = m_i \, \mathbf{I}_3 \qquad (12.38)$$

$$\mathbf{I}_3 = \begin{pmatrix} 1 & & \\ & 1 & \\ & & 1 \end{pmatrix}$$

The term $\mathbf{M}$ is the dynamical matrix [apart from the factor $(m_k \, m_{k'})^{\frac{1}{2}}$] and $\mathbf{U}_\alpha$ is a column matrix with the six elements $U_x(1), U_x(2), \ldots, U_z(2)$. The

* The notation of Cowley et al. [13] is followed here.

next task is to solve the secular equation

$$\det |\mathbf{M} - \omega^2\, \mathbf{m}| = 0 \tag{12.39}$$

The solution depends on the number of force constants $\Phi_{\alpha\beta}$ involved in the calculations. For example, if only nearest neighbor interactions are taken into account, then matrices of the following form prevail:

$$\left[\phi_{\alpha\beta}\begin{pmatrix}0\\12\end{pmatrix}\right] = \begin{bmatrix} \phi_{xx}\begin{pmatrix}0\\12\end{pmatrix} & \phi_{xy}\begin{pmatrix}0\\12\end{pmatrix} & \cdots \\ \vdots & \ddots & \\ \vdots & & \phi_{zz}\begin{pmatrix}0\\12\end{pmatrix} \end{bmatrix} \tag{12.40}$$

A symmetry operation S of the point group (for alkali halides: O_h) has the property of leaving one ion at rest, say, at $x(l, k)$, but bringing all others to equivalent lattice sites. This means that the matrix $\phi_{\alpha\beta}$ transforms as

$$\phi\begin{pmatrix}ll''\\kk'\end{pmatrix} = \mathbf{S}\phi\begin{pmatrix}ll'\\kk'\end{pmatrix}\mathbf{S}^T \tag{12.41}$$

where S is a matrix representing a symmetry operation. For example, the inversion matrix is simply

$$\mathbf{S} = -\mathbf{I}_z \tag{12.42}$$

Making use of the symmetry operations of the point group O_h, we obtain the result that the nine elements of the matrix $\phi_{\alpha\beta}$ are reduced to where only two are independent:

$$\left[\phi_{\alpha\beta}\begin{pmatrix}0\\12\end{pmatrix}\right] = \begin{pmatrix} a & b & b \\ b & a & b \\ b & b & a \end{pmatrix} \tag{12.43}$$

Similar results are obtained for second nearest neighbors, etc.
In ionic crystals, lattice potentials are of the form

$$\Phi(r) = \tfrac{1}{2}\sum_{\substack{ll'\\kk'}} \left[\frac{Z_k Z_{k'}}{r} + \omega(r)\right] \tag{12.44}$$

The first part is the coulomb interaction between ions with charges Z_k and distance γ. The second part denotes the repulsive interaction and can be approximated by a Born–Mayer potential:

$$\omega(r) \sim \omega_0\, e^{-r/\rho} \tag{12.45}$$

Therefore, the force constants and the elements of the dynamical matrix can be split into two parts, a coulomb part and a repulsive one. It follows that

$$
\left.\begin{aligned}
\phi\left({l \atop kk'}\right) &= \phi^{(c)}\left({l \atop kk'}\right) + \phi^{(R)}\left({l \atop kk'}\right) \\[4pt]
\mathbf{M} &= \mathbf{M}^{(c)} + \mathbf{M}^{(R)} \\[4pt]
M_{\alpha\beta}^{(c)}\left({\mathbf{q} \atop kk'}\right) &\equiv Z_k Z_{k'} C_{\alpha\beta}\left({\mathbf{q} \atop kk'}\right) \\[4pt]
M_{\alpha\beta}^{(R)}\left({\mathbf{q} \atop kk'}\right) &\equiv R_{\alpha\beta}\left({\mathbf{q} \atop kk'}\right)
\end{aligned}\right\} \tag{12.46}
$$

Equation (12.36) then reads

$$
\omega^2\, m\mathbf{U} = (\mathbf{R} + \mathbf{ZCZ})\,\mathbf{U} \tag{12.47}
$$

where $\mathbf{Z}$ is a diagonal matrix of the same form as $\mathbf{m}$. The evaluation of the coulomb lattice sums

$$
C_{\alpha\beta}\left({\mathbf{q} \atop kk'}\right) = -\sum_l \left\{ \frac{\partial^2}{\partial x_\alpha\,\partial x_\beta}\frac{1}{|\mathbf{r}|}\, e^{i\mathbf{q}[\mathbf{x}(l)+\mathbf{x}(k)-\mathbf{x}(k')]} \right\} \tag{12.48}
$$

has been done by Kellermann [14] for one thousand points in the first Brillouin zone. In this calculation of KCl, he considered only nearest neighbor interactions for repulsive potential.

Then two parameters were retained [equation (12.43)], one of which is given by the equilibrium condition (invariance against a uniform infinitesimal translation of the crystal)

$$
\sum_{l,k'} \varphi_{\alpha\beta}\left({l \atop kk'}\right) = 0 \tag{12.49}
$$

which leads to

$$
\sum_{k'} M_{\alpha\beta}\left({0 \atop kk'}\right) = 0 \tag{12.50}
$$

Fitting the only adjustable parameter to the measured compressibility and assuming $Z_1 = -Z_2 = e$, Kellermann calculated the dispersion curves for NaCl. The results of analogous calculations for NaI and KBr, respectively, are shown in Figs. 8 and 9 (dotted lines). The result agrees rather well with the experimental curves measured by neutron spectroscopy with the exception of the longitudinal modes. In particular, the Lyddane–Sachs–Teller relation

$$
\omega_{LO}^2(0)/\omega_{TO}^2(0) = \varepsilon_0/\varepsilon_\infty
$$

is not fulfilled ($\varepsilon_\infty = 1$ in the Kellermann model).

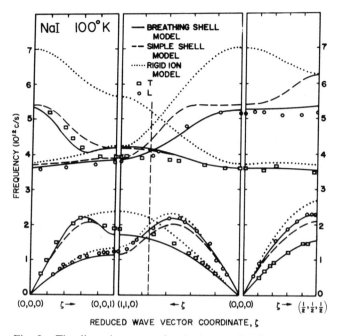

Fig. 8. The dispersion curves for the lattice vibrations of sodium iodide. Comparison of the experimental curves (dotted lines) with the calculated ones is based on the different models.

As mentioned above, the rigid ion model can be improved by taking into account the polarization of the ions. With use of the description of the shell model, dipole parts $\phi^{(s)}$ and $\mathbf{M}^{(s)}$ must be added to the force constant and the dynamical matrix. The equations of motion for the electron shells degenerate to subsidiary conditions as discussed in section II B. We denote the relative displacement of a shell and a core of the same ion by

$$W_x\begin{pmatrix} l \\ k \end{pmatrix} = W_x\begin{pmatrix} \mathbf{q} \\ k \end{pmatrix} e^{i\mathbf{q}[x(l,\,k)-x(l'k')]} \tag{12.51}$$

The latter causes a dipole moment at the lattice site $\begin{pmatrix} l \\ k \end{pmatrix}$:

$$\mathbf{P}\begin{pmatrix} l \\ k \end{pmatrix} = Y_k\,\mathbf{W}_x\begin{pmatrix} l \\ k \end{pmatrix} \tag{12.52}$$

where Y_k is the shell charge. Then, the extended equations of motion

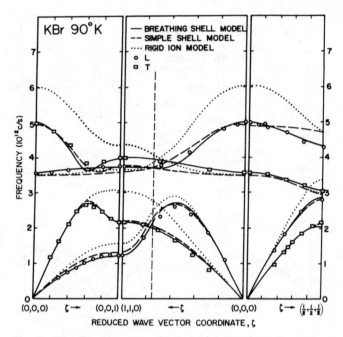

Fig. 9.   The dispersion curves for the lattice vibrations of potassium
bromide. Comparison of the experimental curves (dotted lines) with
the calculated ones is based on the different models.

are

$$\omega^2\, mU = (R + ZCZ)\, U + (T + ZCY)\, W$$

$$0 = (T^+ + YCZ)\, U + (S + YCY)\, W \qquad (12.53)$$

$$\equiv \tau^+\, U + \sigma W$$

Here T and S are the repulsive parts of the shell–core and the shell–shell
interactions, analogous to R. In addition, S contains a diagonal matrix
$T_6(Y_k{}^2/\alpha_k)$, where the $\alpha_k{}'s$ are the polarizabilities of the ions.

Finally the dynamical matrix becomes

$$M = (R + ZCZ) - \tau\sigma^{-1}\, \tau^+ \qquad (12.54)$$

Figures 8 and 9 show dispersion curves for NaI and KBr (dashed lines),
where a simple shell model analogous to that in section II B has been used
for the diatomic linear chain. That means only one ion (I^-, Br^-) is treated
as polarizable. Only three parameters are adjustable: the anion = shell —
cation = core repulsive constant, the shell–core spring constant, and the

shell charge Y_k. The three fitted empirical quantities are the static dielectric constant ε_0, the high-frequency constant ε_∞, and the compressibility. Inspection of the figures reveals that the agreement is remarkably improved for the optical branches, especially for the longitudinal optical mode. Nevertheless there remain some discrepancies between measured and calculated curves. At the point $(\frac{1}{2}, \frac{1}{2}, \frac{1}{2})$ the same result as for the shell model is obtained. At this point the shell model doesn't work, since the anions are at rest while the cations are moving symmetrically. This can be overcome by introducing additional spring constants between second nearest neighbors and formal polarizabilities of the cations [13]. One needs altogether nine parameters to fit the measured dispersion curves within the limits of experimental error. This does not seem to be very satisfactory from a physical point of view.

Schröder [15] has shown that the compressibilities of the ions must be taken into account. Quantum mechanically, this can be justified by the fact that the wave functions of excited states are not only of p-type (corresponding to dipole excitations), but also of s-type. This leads to a spherical extension or compression of the electron shell ("breathing shell model"). Without introducing any additional parameters, a modified system of equations of motion follows, where

$$\mathbf{m}\,\omega^2\,\mathbf{U} = (\mathbf{R}+\mathbf{ZCZ})\,\mathbf{U}+\tau\mathbf{W}+\mathbf{QV}$$

$$0 = \tau^+\,\mathbf{U}+\sigma\mathbf{W}+\mathbf{QV} \qquad (12.55)$$

$$0 = \mathbf{Q}^+(\mathbf{U}+\mathbf{W})+\mathbf{HV}$$

where $\mathbf{Q}$ denotes a modified isotropic interaction, $\mathbf{V}$ is the change of the radius of the shell and $\mathbf{H}$ denotes the compression spring matrix which depends on the other springs.

Calculations are shown in Figs. 8 and 9 where the three elastic constants $(c_{11}, c_{12}, \text{ and } c_{44})$, the two dielectric constants ε_0 and ε_∞, and the transverse optic ω_0 are used to determine the six free parameters of the model (it corresponds to model II of Cowley et al. [13], but with $\alpha_1 = 0$). Calculated and measured dispersion curves agree quite satisfactorily. The proposed model can be used to calculate the dispersion curves of diatomic crystals without information from inelastic neutron scattering. A similar investigation is in progress for homopolar crystals.

III. REFERENCES

1. M. Born and K. Huang, *Dynamical Theory of Crystal Lattices*, Oxford University Press (London), 1954; A. A. Maradudin, E. W. Montroll, and G. H. Weiss, "Theory of Lattice Dynamics in the Harmonic Approximation," *Solid State Phys. Suppl.* 3 (1963); W. Cochran, "Lattice Vibrations," *Rept. Progr. Phys.* **26**:1 (1963).

2. G. Leibfried and W. Ludwig, *Solid State Phys.* **12**:276 (1961); R. A. Cowley, *Advan. Phys.* **12**:421 (1963).
3. L. van Hove, *Phys. Rev.* **89**:1189 (1953).
4. J. C. Phillips, *Phys. Rev.* **104**:1263 (1956).
5. H. B. Rosenstock, *Phys. Rev.* **97**:290 (1955); *J. Phys. Chem. Solids* **2**:44 (1957).
6. F. Herman, *J. Phys. Chem. Solids* **8**:405 (1959).
7. K. P. Tolpygo and U. S. Mashkevich, *Soviet Phys. JETP (English Transl.)* **5**:435 (1957).
8. W. Cohran, *Proc. Roy. Soc. (London)* **A253**:260 (1959).
9. J. R. Hardy, *Phil. Mag.* **4**:1278 (1959); J. R. Hardy and A. M. Karo, *Phil. Mag.* **5**:859 (1960).
10. A. D. B. Woods, B. N. Brockhouse, R. A. Cowley, and W. Cochran, *Phys. Rev.* **131**:1025 (1963).
11. B. N. Brockhouse and P. K. Iyengar, *Phys. Rev.* **111**:747 (1958); G. Dolling, in: *Inelastic Scattering of Neutrons in Solids and Liquids*, Vol. *II*, International Atomic Energy Agency (Vienna), 1965, p. 37; J. L. Warren, G. R. Wenzel, and J. L. Yarnell, in: *Inelastic Neutron Scattering*, Vol. *I*, International Atomic Energy Agency (Vienna), 1965, p. 361.
12. G. Dolling and J. L. T. Waugh, *Proc. Intern. Conf. Lattice Dynamics*, R. F. Wallis (ed.), Copenhagen, 1963, p. 19.
13. R. A. Cowley, W. Cochran, B. N. Brockhouse, and A. D. B. Woods, *Phys. Rev.* **131**:1030 (1963).
14. E. W. Kellermann, *Phil. Trans. Roy. Soc.* **A238**:S13 (1940).
15. U. Schröder, *Solid State Comm.* **4**:347 (1966).

CHAPTER 13

New Spectral and Atomistic Relations in Physics and Chemistry of Solids

Johannes N. Plendl

Air Force Cambridge Research Laboratories
Office of Aerospace Research
Bedford, Massachusetts

I. INTRODUCTION

Equations that describe relationships between diverse characteristics of solids and atomic parameters have always been of much interest for the physics and chemistry of solids. Their practical value depends on their range of validity with regard to such characteristics as crystalline structures, valences, bond strengths, and types of bonding. It is well known that relations of general validity are difficult to achieve, particularly when based on models. A substantial improvement can be expected, however, when we deduce them from experimental data of a great variety of solids.

An attempt in this direction is shown here. It is based on a number of relations derived by the author within the last years. They deal with more than 100 solids and treat the following characteristics: infrared spectra, specific heat, bulk modulus, change of compressibility with pressure, lattice anharmonicity, exponent of repulsion, lattice cohesive energy, zero-point energy, bond strength, valence, and hardness.

Ample evidence has been given that the various characteristics are tightly interwoven and that they can be reasonably explained by simple atomistic concepts.

In reviewing formerly and newly derived relations, each one is graphically confirmed by experimental data of a great number of solids which differ widely in their structures, bond strengths, valences, and types of bonding. This verification includes many new data obtained from very recent experiments.

In summary, each characteristic is presented in multiple expressions in relation to other characteristics as well as diverse atomic parameters, indicating a wide interchangeability of them. By taking advantage of this fact, additional relations have been concluded, and new data have been determined which were inaccessible to measurements.

II. SPECTRAL AND ATOMISTIC RELATIONSHIPS

The search for a single characteristic frequency, which represents the vibrational spectrum of a crystalline substance as a whole, is a very old one and is familiar to most physicists and chemists. In the early stage, however, the spectral information available from experimental data was rare and incomplete. Hence the early investigators [1-4] had to rely on certain hypotheses, as Fig. 1 indicates in historical order. Nevertheless, results describing a few simple characteristics of solids were remarkably good. Today, having available a wealth of infrared spectra of solids from experimental data, a more realistic approach to the study of solids, from an analytical viewpoint, may be appropriate. It may lead to a comprehensive understanding of the properties of solids from an atomistic point of view.

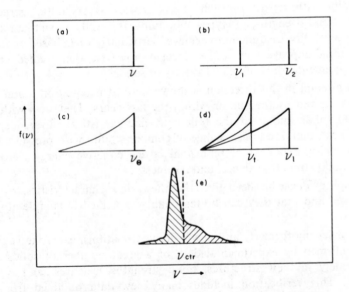

Fig. 1. Graphic presentation of the various characteristic frequencies as proposed by (a) Einstein [1] in 1907, (b) Nernst and Lindemann [2] in 1911, (c) Debye [3] in 1912, (d) Born and v. Karman [4] in 1913 and 1923, and (e) Plendl [5] in 1960.

A. Introduction of a "Centro-Frequency" [5]

The characteristics of a solid, in all their complexity, are expressed in the vibrational (far infrared) spectrum, which discloses the lattice cohesive forces. Hence the far infrared spectra are an indispensable means for the study of solids. As a first step, one has to find an efficient and true representation of the experimental lattice vibration spectrum. For this purpose, we defined a "frequency of the center of gravity" or simply "centro-frequency", formulated as

$$v_{ctr} = \int_{v_1}^{v_2} v f(v)\, dv \Big/ \int_{v_1}^{v_2} f(v)\, dv \qquad (\text{sec}^{-1}) \qquad (13.1)$$

The integration averages $f(v)$ ($=$ reflectivity or absorption coefficient) over the entire frequency range of the experimental lattice vibrational spectrum. In this manner, the centro-frequency unambiguously characterizes this spectrum in its entirety. It is illustrated in Figs. 2a and 2b for diamond and LiF, respectively. These figures clearly indicate that the resonance lines may occur at both sides of v_{ctr} and that the centro-frequency may or may not coincide with any of the lattice resonance frequencies of the solid.

B. Centro-Frequency Versus Debye Frequency [5]

An interesting observation may be made when we compare Debye's characteristic frequency (from specific heat data)

$$v_{\Theta} = (k/h)\, \Theta \qquad (\text{sec}^{-1}) \qquad (13.2)$$

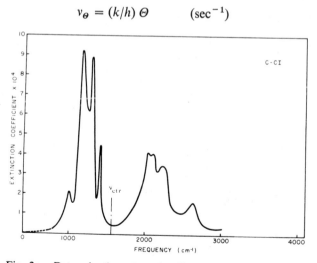

Fig. 2a. Determination of v_{ctr} for diamond (I) from the lattice vibrational spectrum.

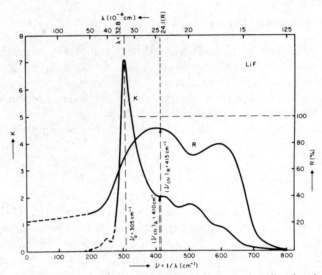

Fig. 2b.   Determination of ν_{ctr} for LiF from lattice vibrational
spectra.

with the centro-frequency derived from the infrared spectrum. It appears
that ν_Θ and ν_{ctr} are very closely related to one another, as is shown in Fig. 3,
The examples of Fig. 3 cover various valences, bond strengths, bonding
types, and crystalline structures. The diagonal completely averages the data

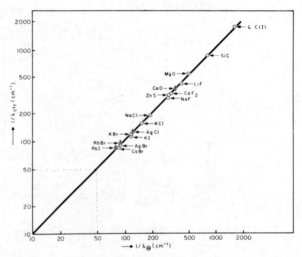

Fig. 3.   Equality of Debye's characteristic frequency from
specific heat with the centro-frequency from infrared spectra.

and agreement is within the limits of error. We thus conclude that

$$v_{ctr} \equiv v_{\Theta} \quad \text{or} \quad v_{ctr} = (k/h)\,\Theta \qquad (\text{sec}^{-1}) \qquad (13.3)$$

A complete confirmation of the new relations is shown by plotting the calculated values of a great number and variety of solids *versus* the experimental data. In every case, the diagonal averages the plotted points and their deviations remain within the limits of error, as is shown in Fig. 3. This method affords instant evaluation of accuracy and validity.

C. Centro-Frequency from Elastic Data [11]

Inasmuch as the concept of a centro-frequency, derived from the optical spectrum, has a thermodynamical analogy in equation (13.3), it also has a mechanical analogy which may be derived from the elastic (acoustic) spectrum as

$$v_c = v_c/2r_0 = \sqrt{K/\rho}/2r_0 = (2r_0\sqrt{\varkappa\rho})^{-1} \qquad (\text{sec}^{-1}) \qquad (13.4)$$

where $v_c = \sqrt{K/\rho}$ describes the integrated value of the sound velocity in the lattice.

In comparing v_{ctr} with v_c (from cubic compressibility), we obtain Fig. 4, which covers 28 dielectric compounds and elements of various crystalline

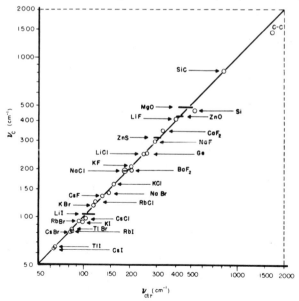

Fig. 4. Equality of centro-frequency from cubic compressibility v_c with centro-frequency from infrared spectra v_{ctr} for dielectric solids.

structures, valences, bond strengths, and bonding types. Again the diagonal completely averages the data and agreement is within the limits of error. We conclude that

$$v_{ctr} \equiv v_c \quad \text{or} \quad v_{ctr} = (2r_0 \sqrt{\varkappa\rho})^{-1} \qquad (\text{sec}^{-1}) \qquad (13.5)$$

D. Debye Temperature from Compressibility [11]

Equating the relations (13.3) and (13.5), we obtain

$$v_\Theta \equiv v_c \quad \text{or} \quad (k/h)\,\Theta = (2r_0 \sqrt{\varkappa\rho})^{-1} \qquad (\text{sec}^{-1}) \qquad (13.6)$$

and infer the following general relation for calculating Debye temperatures:

$$\Theta = h/2r_0\,k\,\sqrt{\varkappa\rho} \qquad (°K) \qquad (13.7)$$

The above relations are further tested in the graph of Fig. 5 for 34 metallic solids (mainly elements), having various structures, valences, and bond strengths. In this manner, the relations (13.4) to (13.7) are shown valid for a total of 62 solids, having either ionic or covalent or metallic bonding.

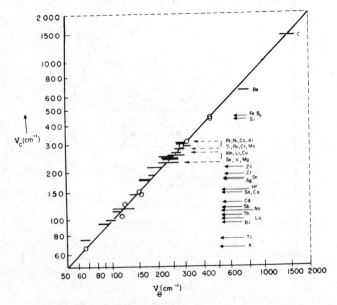

Fig. 5. Centro-frequency from cubic compressibility v_c versus Debye characteristic frequency for metallic solids v_Θ.

E.   Compressibility and Lattice Cohesive Energy [12]

The bulk modulus K may be expressed as a function of the lattice cohesive energy by

$$K = (U/VN)\,(q/m)\,(1/Z) \qquad (\text{dynes/cm}^2) \qquad (13.8)$$

or

$$\varkappa = (VN/U)\,(m/q)\,Z \qquad (\text{dynes/cm}^2)^{-1} \qquad (13.9)$$

The relationship (12.9) is tested in the graph of Fig. 6 for 45 solids, covering binary and ternary compounds of various valences, bond strengths, and crystalline structures. Once more the diagonal averages the data and agreement is within the limits of error. Since in many cases the cubic compressibility is experimentally more readily determinable than the lattice cohesive energy, solving equation (13.9) for

$$U = (V/\varkappa)\,(m/q)\,Z \qquad (\text{ergs/molecule}) \qquad (13.10)$$

may become a helpful formulation, when a Born–Haber cycle determination of U is difficult to achieve.

Combining equations (13.4), (13.5), and (13.8), we obtain the relation

$$v_{\text{ctr}} = \left(\frac{1}{2\,r_0}\right)\sqrt{(1/\rho\,Z)\,(U/VN)\,(q/m)} \qquad (\text{sec}^{-1}) \qquad (13.11)$$

which will facilitate discussion of section II-L.

F.   Harmonic and Anharmonic Lattice Vibrations [6]

In assuming the ideal case of purely harmonic lattice vibrations, we derive a "definite frequency" from the lattice cohesive energy U of a solid as

$$v_d = \left(\frac{1}{2\,\pi r_0}\right)\sqrt{2\,U/Zm_r} \qquad (\text{sec}^{-1}) \qquad (13.12)$$

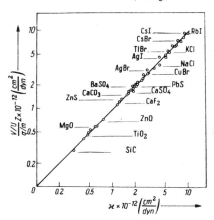

Fig. 6.   Correlation between experimental and calculated compressibility data.

In comparing the definite frequency with the centro-frequency, representing the more common case of anharmonic vibrations, we derive a "factor of anharmonicity" defined by

$$F(A) = v_{ctr}/v_d \qquad (13.13)$$

The "unique case", where $F(A)$ = unity, corresponds to the argon configuration which, for a compound, is closely realized in KCl. For all other solids, $F(A)$ deviates from unity, and the deviation is a measure of the degree of anharmonicity (nonlinearity) of the cohesive forces. It may be conveniently expressed by

$$1 - F(A) = 1 - (v_{ctr}/v_d) \qquad (13.14)$$

where $1 - F(A) = 0$ is realized for the purely linear or harmonic vibration.

G. Analytical Expression of Lattice Anharmonicity [6]

The factor of anharmonicity is a function of the reduced mass, as is shown for a variety of 27 solids in curve a of Fig. 7. The reciprocal function

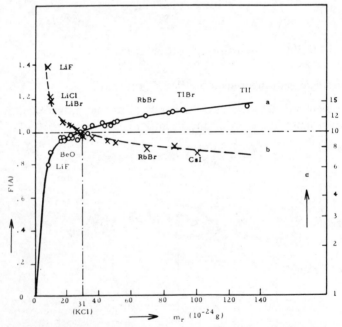

Fig. 7. Anharmonicity data of lattice vibrations, derived from spectral data as a function of the reduced mass m_r. Curve a: plot of $X_m{}^{1/p}$ (calculated data) and also of n (calculated data); circles plot values of $F(A)$ (experimental data). Curve b: plot of $X_m{}^{-1/p}$ (calculated data); crosses plot values of $(\psi_0/\varkappa_0)/Z$ CN (experimental data).

$1/F(A)$ (see curve b) will be discussed below. Curve a allows a simple formulation of the factor of anharmonicity if we introduce the concept of a related "reduced mass"

$$X_m = m_r/m_{r(A)} \qquad (13.15)$$

where the reduced mass m_r of a solid is compared to the reduced mass of KCl, of which $m_{(r)A}$ is $31 \times 10^{-24} g$ (argon configuration). In this case, we obtain the expression

$$F(A) = X_m^{1/p} \qquad (13.16)$$

where $p = 9 = $ constant (in most cases).

H. The Two Types of Lattice Anharmonicity [6]

Lattice anharmonicity or nonlinearity of the cohesive forces in individual solids can produce one of two different types of force characteristic, depending on whether the stiffness of the restoring force increases or decreases with the displacement of the atoms from the state of equilibrium. The first case is called a "hard force characteristic" and is here defined by $F(A) > 1$. The other case is called a "soft force characteristic" and is here defined by $F(A) < 1$. The linear characteristic, $F(A) = 1$, an exceptional case in solids, defines the change-over point between the two types of anharmonic force characteristics [6]. Figure 8 graphically represents the three different types of force characteristic, linear and anharmonic "hard" and

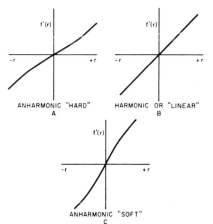

Fig. 8.   The different types of anharmonic force characteristic (soft and hard) and the linear characteristic schematically represented.

"soft," as they may exist in individual solids, with slope and bending of the characteristic varying from solid to solid.

From an atomistic point of view, a soft force characteristic results from a shielding effect, caused by mutual interpenetration of the electron clouds of adjacent atoms. It occurs when $m_r < m_{r(A)}$, thus representing a low electron density. By contrast, the hard force characteristic may be caused by mutual deformation of the electron clouds without interpenetration. It occurs when $m_r > m_{r(A)}$, thus indicating high electron density. This atomistic concept calls for a classification of the solids which is discussed in section II-L.

I. Lattice Cohesive Energy and Zero-Point Energy [6,9]

Combining equations (13.12), (13.13), and (13.16), we obtain

$$v_{ctr} = (X_m^{1/p}/2\,\pi r_0)\,\sqrt{2\,U/Zm_r} \qquad (\sec^{-1}) \qquad (13.17)$$

This relationship is plotted *versus* experimental data in Fig. 9 and shows excellent agreement. Hence, solving equation (13.17) for U, we obtain a useful relation for calculating lattice cohesive energy data:

$$U = 2ZNm_r\,(\pi v_{ctr}\,r_0\,X_m^{-1/p})^2 \qquad (\text{ergs/molecule}) \qquad (13.18)$$

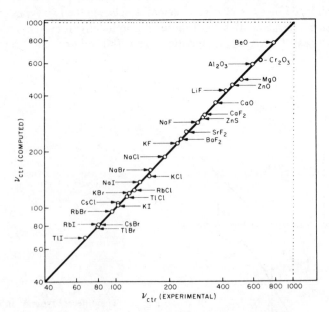

Fig. 9. Equality of centro-frequencies computed from lattice cohesive energy data with centro-frequencies from infrared spectra.

Furthermore, based on $v_{ctr} \equiv v_\Theta$, we obtain an additional relation for calculating Debye temperatures:

$$\Theta = (hX_m^{1/P}/2\pi kr_0)\sqrt{2\,U/ZNm_r} \qquad (^\circ K) \qquad (13.19)$$

Over and above, the concept of centro-frequency may be used for the computation of zero-point energy (E_0) resulting in

$$E_0 = [(3\,Nh/4\pi r_0)\sqrt{2\,U/ZNm_r} \times X_m^{1/P}]_{T=0} \qquad (\text{ergs/mole}) \qquad (13.20)$$

or

$$E_0 = (3\,Nh/4\,r_0\,\sqrt{K/\rho})_{T=0} \qquad (\text{ergs/mole}) \qquad (13.21)$$

J. Lattice Anharmonicity and Exponent of Repulsion [6]

The anharmonicity of lattice vibrations is assumed to be caused by the forces of repulsion. Hence the exponent of repulsion n should have a very close relationship to the factor of anharmonicity. We found that

$$F(A) = X_m^{1/P} = \log n \qquad (13.22)$$

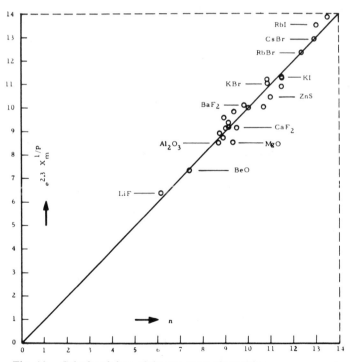

Fig. 10. Calculated data of the exponent of repulsion n *versus* experimental data.

or

$$n = \exp{(2.30\ X_m^{1/p})} \tag{13.23}$$

Relation (13.23) is tested by the graph of Fig. 10, where the calculated data of n are plotted *versus* the generally accepted values of n for 25 solids. Again, the diagonal completely averages the data and agreement is within the limits of error.

Combining equations (13.11) and (13.17), based on $v_c \equiv v_{ctr}$, we obtain the relation

$$F(A) = \pi \sqrt{(m_r/2)\ (q/m)\ (N/\rho V)} = \pi \sqrt{(m_r/2)\ (N/M)\ (q/m)} \tag{13.24}$$

where $\rho V = M$. It relates the factor of anharmonicity to the packing of the atoms within the lattice.

K. Lattice Anharmonicity and Change of Compressibility with Pressure[6]

Obviously the compressibility of a solid is strongly influenced by the repulsive forces within the lattice which also cause the anharmonicity. Hence we may expect a close relationship between the cubic compressibility and the factor of anharmonicity $F(A)$. If the valence Z is taken into account, the change of compressibility with pressure (ψ_0) divided by the compressibility at absolute zero $(\varkappa_0)_{T=0}$ (coined "relative compressibility") is related to $F(A)$ as follows

$$(\psi_0/\varkappa_0)_{T=0} = Z\ CN/F(A) = Z\ CN\ X_m^{-1/p} \tag{13.25}$$

where CN is the coordination number of the lattice structure. This relation is tested in Fig. 11 by plotting values of the function $Z\ CN/F(A)$ *versus* the experimental data of $(\psi_0/\varkappa_0)_{T=0}$ for 17 solids of various structures, bond strengths, and valences. Considering small uncertainties due to the need for reduction to absolute zero. we find that Fig. 11 confirms equation (13.25). An additional confirmation is shown in Fig. 7, where data of the term

$$1/F(A) = \frac{(\psi_0/\varkappa_0)_{T=0}}{Z\ CN} \tag{13.25a}$$

are plotted *versus* m_r (as crosses). They fit (within the limits of error) curve b which represents

$$1/F(A) = X_m^{-1/p} \tag{13.16a}$$

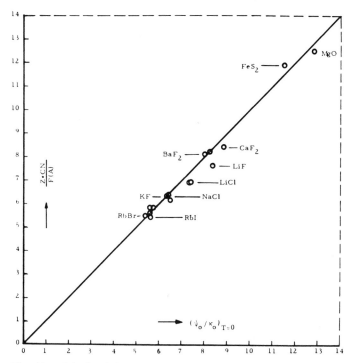

Fig. 11. Calculated data of effective lattice anharmonicity $Z \, CN/F(A)$ *versus* experimental data of relative compressibility $(\psi_0/\varkappa_0)_{T=0}$.

L. Hardness and Lattice Anharmonicity [7,8]

Hardness determined by scratch or abrasion tests* may be defined as the energy required to remove a certain volume of material during testing. We found such hardness data proportional to the lattice cohesive energy divided by the molecular volume

$$H = \text{constant} \times (U/VN) \qquad (\text{ergs/cm}^3) \qquad (13.26)$$

In examining relation (13.26) using the Moss hardness data of about 80 solids, we obtained the graph of Fig. 12. It clearly indicates a split into two different slopes when $H > 4$. In explaining this result, we refer to the anharmonicity curve b of Fig. 7 which in a different fashion is represented in Fig. 13. This graph indicates two definitive conditions: (1) the cross-over point between the two types of anharmonic force characteristic at KCl (argon configuration, X_m = unity) and (2) the change-over point between

---

* In contrast to hardness measured by indentation techniques.

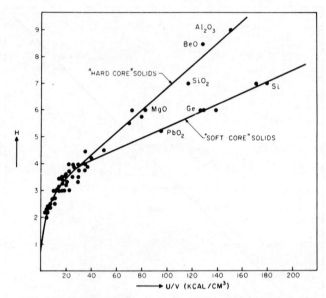

Fig. 12. Relation between hardness H and volumetric lattice cohestive energy (U/V) for soft-core and hard-core solids in the range below $H = 9$.

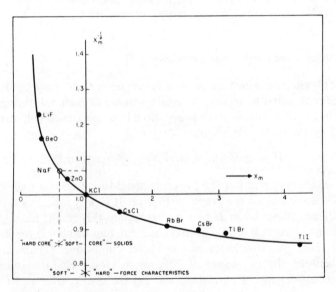

Fig. 13. Anharmonicity function $1/F(A) = X_m^{-1/p}$ versus the related mass X_m for a number of binary compounds.

the two hardness branches at NaF (neon configuration). Here the mutual interpenetration of the atoms reaches $X_m = \frac{2}{3}$, and the solids become divided into two types, that is, the "hard-core" solids when the electron configuration corresponds to the neon configuration ($X_m \leqslant \frac{2}{3}$), and the "soft-core" solids for all other solids ($X_m > \frac{2}{3}$).* This concept agrees with the data of Fig. 12 and so should explain the split into the two branches. For very small values of U/V (< 30 kcal/mole), the two slopes unite and form a single curve.

The hard-core solids (characterized by a deep mutual interpenetration of the atoms) can reach and exceed $H = 9$, in contrast to the soft-core solids. When referred to the combined Moss–Wooddell scale, the hard-core solids may extend toward $H = 42$ which is the average value for diamond. This is depicted in Fig. 14, which shows that the hardness range of the hard-core solids is many times wider, their maximum hardness is many times higher, but their total number is many times smaller than the corresponding data of the soft-core solids.

---

* The terms "hard-core" and "soft-core" solids were coined in an earlier paper [8] to correspond to the hardness properties. This is in contrast to the traditional terms of "hard" and "soft" characteristic of the restoring forces.

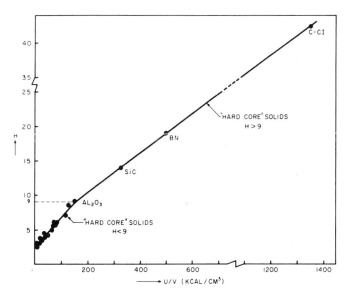

Fig. 14. Relationship between hardness H and volumetric lattice cohesive energy (U/V) for hard-core solids, over the entire hardness range.

The introduction of equation (13.26) into equation (13.11) establishes a relationship between v_{ctr} and hardness:

$$U/V = 4 r_0{}^2 v_{ctr}^2 \rho Z N (m/q) \qquad (\text{ergs/cm}^3) \qquad (13.27)$$

M. Atomistic Expressions for Hardness and Compressibility [8,12]

Combining relations (13.26) and (13.18) and introducing

$$V = \text{constant} \times r_0{}^3 \qquad (\text{cm}^3) \qquad (13.28)$$

we found

$$H = C_{str} Z(f/r_0) \qquad (\text{ergs/cm}^3) \qquad (13.29)$$

where

$$f \propto U/Z r_0{}^2 \qquad (\text{dynes/cm}^3) \qquad (13.30)$$

and

$$C_{str} = N \rho r_0{}^3 / 2 M \qquad (13.31)$$

Relation (13.29) presents an atomistic expression of hardness. A similar atomistic expression for the cubic compressibility was obtained by combining relations (13.8), (13.26), and (13.29), resulting in

$$K = 1/\varkappa = C_{str} (f/r_0) (q/m) \qquad (\text{dynes/cm}^2) \qquad (13.32)$$

Comparing both the atomistic relation for (abrasion) hardness [equation (13.29)] and that for compressibility [equation (13.32)], we find almost indentity except for the factor Z (valence). It is present in the expression of abrasion hardness, which is proportional to both the bond strength and the number of bonds per molecule. However, it is missing in the case of compressibility which is proportional to the bond strength only. Furthermore, it is interesting to note that the two expressions, in a way, correspond to the two types of hardness testing, scratch or abrasion *versus* pressure indentation. This may help to explain why data obtained from the two representative types of hardness testing are not directly comparable for many solids.

N. Valence from Hardness *Versus* Bulk Modulus

When we combine the relations (13.29) and (13.32), we obtain

$$Z = F(H/K) \qquad (13.33)$$

This represents a new method for determining the valence in solids.

O. Determination of Inaccessible Data

As an example of the usefulness of the relations given in this study, we may select equation (13.18). Based on new spectral data, the lattice cohesive energy has been determind for such solids as SiC, BN (cubic), and diamond (I) (for which a Born–Haber cycle cannot be achieved because of the very high sublimation temperatures). The data thus obtained are listed in Table I, together with the measured data of Al_2O_3 (corundum). The U/M and U/V values correspond to the mechanical qualities of these materials.

III. SUMMARY OF MUTUAL RELATIONSHIPS

As is well known, the bonding forces, which hold matter together in the solid state, determine the atomic vibrations within the lattice. Hence the study of the lattice vibration spectrum discloses the much wanted insight into the characteristics of solids from an atomistic point of view, since the material properties are decisively determined by the bonding forces. However, for effectively dealing with an experimental vibrational spectrum as a whole, we need a true characterization by one single frequency which best represents all the particular characteristics of a solid. Such a representative frequency was shown to be the frequency of the center of gravity of the entire vibrational spectrum. For a great variety of solids, it was shown that other characteristic frequencies obtained from thermodynamical data (specific heat) and from mechanical data (cubic compressibility) are identical (within the limits of error) to the optical (infrared) centro-frequency. The unique advantage of the centro-frequency concept is that it could be developed into a succesful formulation of the nonlinearity of the bonding forces. This preponderant lattice anharmonicity was found to have a decisive influence on the distinguishing properties of individual solids.

Based on the two leading concepts, that of centro-frequency and that of lattice anharmonicity, many useful equations were derived which successfully interrelate the diverse characteristics of solids. These equations have been verified for a great variety of solids differing widely in bond strengths, valences, crystalline structures, and bonding types. When presenting them as multiple expressions between various characteristics, atomic parameters, and physical constants, their mutual relationships become apparent. Now we can determine characteristic data of solids which are experimentally inaccessible, by using other characteristic data accessible to measurements. In the process of lining up the mutual relationships between diverse characteristics as given below, some new interrelations were also disclosed.

Table I. Determination of Lattice Cohesive Energy from Infrared Spectra for Some Solids of Extremely High Sublimation Temperatures

No.	Substance	Structure	Z	M (g)	V (cm³)	r_0 (10⁻⁸ cm)	m_r (10⁻²⁴ g)	$X_m^{1/p}$	ν_{etr} (cm⁻¹)	U (kcal/mole)	U/M (kcal/g)	U/V (kcal/cm³)
0	Al_2O_3	Corundum	3.2	102	24.5	1.91	16.8	0.93	590	3610 *	35.5	148
1	SiC	Zinc blende	4	40.1	12.6	1.89	14.1	0.92	830	4200 †	105	330
2	BN (cubic)	Zinc blende	3	24.8	7.12	1.57	10.2	0.86	1175	3420 †	138	480
3	C–C(I)	Diamond	4	24.0	6.90	1.54	10.1	0.86	1600	8300 †	350	1200

* Measured data.
† Calculated data.

A. Centro-Frequency

$$\nu_{ctr} = \int_{\nu_1}^{\nu_2} \nu f(\nu)\, d\nu \Big/ \int_{\nu_1}^{\nu_2} f(\nu)\, d\nu \qquad (\text{sec}^{-1})$$

$$= (k/h)\, \Theta \qquad (\text{sec}^{-1})$$

$$= 1/2\, r_0 \sqrt{\varkappa\rho} \qquad (\text{sec}^{-1})$$

$$= \left(\frac{1}{2\, r_0}\right) \sqrt{(1/\rho Z)\,(U/VN)\,(q/m)} \qquad (\text{sec}^{-1})$$

B. Anharmonicity

$$F(A) = X_m^{1/p} \qquad X_m = m_r/m_{r(A)}$$

$$= 2\pi r_0\, \nu_{ctr}/\sqrt{2\, U/ZNm_r}$$

$$= \pi \sqrt{(Nm_r/2\,M)\,(q/m)}$$

C. Exponent of Repulsion

$$n = \exp\,(2.303\, X_m^{1/p})$$

D. Debye Temperature

$$\Theta = (h/k)\, \nu_{ctr} \qquad (^{\circ}\text{K})$$

$$= h/2\, r_0\, k \sqrt{\varkappa\rho} \qquad (^{\circ}\text{K})$$

$$= (hX_m^{1/p}/2\pi r_0\, k) \sqrt{2\, U/ZNm_r} \qquad (^{\circ}\text{K})$$

E. Lattice Cohesive Energy

$$U = (V/\varkappa)\,(m/q)\, ZN \qquad (\text{kcal/mole})$$

$$= 2ZNm_r\,(\pi\nu_{ctr}\, r_0\, X_m^{-1/p})^2 \qquad (\text{kcal/mole})$$

$$= F(HV) \qquad (\text{kcal/mole})$$

F. Zero-Point Energy

$$E_0 = [(\tfrac{3}{2})\, Nh\nu_{ctr}]_{T=0} \qquad (\text{kcal/mole})$$

$$= [(\tfrac{3}{4})\, Nh/r_0 \sqrt{\varkappa\rho}]_{T=0} \qquad (\text{kcal/mole})$$

$$= [(\tfrac{3}{4})\, NhX_m^{1/p} \sqrt{2\, U/Zm_r\, N}/\pi r_0]_{T=0} \qquad (\text{kcal/mole})$$

$$= [(\tfrac{3}{4})\, Nh \sqrt{(1/\rho Z)\,(U/VN)\,(q/m)}/\pi r_0]_{T=0} \qquad (\text{kcal/mole})$$

G. Compressibility $\varkappa_0$ and Change of Compressibility with Pressure ψ_0

$$(\psi_0/\varkappa_0)_{T=0} = Z \, CN \, X_m^{-1/p}$$

$$= [(CN/\pi r_0 \, v_{ctr}) \sqrt{ZU/2 \, m_r \, N}]_{T=0}$$

$$(\varkappa_0)_{T=0} = [(V/U)(m/q) \, Z]_{T=0} \qquad\qquad (\text{dynes/cm}^2)^{-1}$$

$$(\psi_0)_{T=0} = [CN \, Z^2 \, X_m^{-1/p} \, (V/U)(m/q)]_{T=0} \qquad (\text{dynes/cm}^2)^{-1}$$

$$= \{[(m/q) \, CN/\pi r_0 \, v_{ctr}] \times \sqrt{Z^3 \, NV^2/2 \, m_r \, U}\}_{T=0} \quad (\text{dynes/cm}^2)^{-1}$$

H. Hardness ($H = U/V$), Bulk Modulus (K), and Bond Strength (f)

$$H = C_{str} \, Z(f/r_0) \qquad\qquad (\text{ergs/cm}^3)$$

$$C_{str} = N\rho r_0^3/2 \, M$$

$$K = C_{str}(f/r_0)(q/m) \qquad\qquad (\text{dynes/cm}^2)$$

I. Valence

$$Z = F(H/K)$$

IV. CONCLUSIONS

This chapter has given a number of atomistic interrelations between diverse characteristics (properties) of solids. In applying these relations to a great variety of solids, general validity has been shown.

In expressing each characteristic by several relations, the interchangeability of various characteristics and atomic parameters has been indicated. By taking advantage of this fact, new relations have been concluded, and certain data have been determined which experimentally were inaccessible.

Most relations in this paper were derived from infrared spectra and so treat the dielectric solids. For the metallic solids, where infrared spectra are not available, different approaches are needed, some of which are indicated here, and others are under study.

Ample evidence has been given that diverse characteristics of simple compounds and elements can be reasonably explained by simple atomistic concepts.

Further research in this direction, now under way, will substantially increase the number of characteristics to be treated and relationships to be derived. This too, will enhance the interchangeability of characteristics and atomic parameters. More emphasis will be given to the metallic solids.

The type of research presented here may ultimately lead to the determinative principles that govern material properties, based entirely on natural constants. Such principles will enable us, in advance of experimental trial, either to foresee serious failures in materials or to predict succesful utilisation of solids, when exposed to extreme environmental conditions. In addition, they may help us to develop new materials of superior qualities to withstand extremes in temperature, high pressure, high voltage, or radiation.

V. APPENDIX: EXPERIMENTAL DETERMINATION OF THE CENTRO-FREQUENCY

So far the centro-frequency has been defined as the center of gravity of the vibrational spectrum of a solid. In order to determine v_{ctr} correctly, we need to know the *entire* spectrum of lattice vibrations which, as far as the infrared spectra are concerned, constitutes the entire spectrum in either reflection or absorption. It should be readily apparent that the following points should be carefully considered before any experimental evaluation is attempted:

1. The spectra should represent a plot of either intensity of reflection or absorption *versus* frequency (wave numbers in cm^{-1}). Many spectra are published on a wavelength scale (linear or logarithmic) and require replotting before any integration can be performed. As for reflection spectra, $f(v)$ of equation (13.1) may be directly equated with R, the reflectivity or percent reflection as compared to a standard highly reflective surface (gold–silver or aluminized mirror), whereas $f(v) = \ell$ is used for the absorption-type spectrum. Here ℓ denotes the extinction coefficient which is defined as follows:

From Beer's law we have $I_1 = I_0 e^{-\alpha d_1}$, where I_0 denotes the intensity of the light incident (normal) on a crystal plate of thickness d_1 and I_1 denotes the intensity of the light after passage through the crystal. In this way, the *absorption* coefficient α_1 is defined; it is related to the *extinction* coefficient ℓ by $\ell = \alpha\lambda/4\pi$. The extinction coefficient is a dimensionless quantity; therefore it is here preferred over the absorption coefficient (cm^{-1}). The absorption spectrum (extinction coefficient *versus* frequency) may be derived from transmission measurements on specimen of varying thickness or, where experimental setup or high absorbance of the materials do not permit, by means of Simon's or Robinson's method [10].

2. The limiting frequencies of the absorption-type spectrum can be conveniently defined as those frequencies at which $f(v) = \ell$ becomes very small and remains so there after. The *high*-frequency limit of the reflection spectrum may be equally defined, while the low-frequency reflection

limit does not necessarily approach a low value. In any case, the spectrum should include the entire region of anomalous dispersion due to lattice vibration, to which may be added the condition that only that part of the spectrum representing the vibrations that contribute to the specific heat of the solid should be included. In the case of simple compounds, such as the alkali halides, this entails the entire spectrum; but more complex substances, especially those containing distinct units or complex ionic groups, may need a different treatment. As an example, consider the vibrational spectrum of SiO_2. The centro-frequency as determined from the reflection spectrum of SiO_2 had always been found to be too high and not in accordance with the values of the characteristic frequency determined from heat capacity data. If it is recalled that the high-frequency part of the spectrum may be referred to the "internal vibration" of the SiO_2 tetrahedra and that these tetrahedra are retained on melting, it may be safe to assume that the centro-frequency should be taken over that part of the spectrum which reflects interactions directly determining the cohesion, e.g., the low-frequency part [9].

VI. ACKNOWLEDGMENT

The author is particularly grateful to P. J. Gielisse, S. S. Mitra, and L. C. Mansur for their unfailing collaboration over several years, and many valuable discussions as well.

VII. NOTATION

C_{str}	=	Structure constant
CN	=	Coordination number of crystalline structure
E_0	=	Lattice zero-point energy
f	=	Force constant
$F(A)$	=	Factor of anharmonicity
h	=	Planck's constant
H	=	Hardness
k	=	Boltzmann's constant
K	=	Bulk modulus
m	=	Number of component atoms
m_r	=	Reduced mass
$m_{r(A)}$	=	Reduced mass of KCl (argon configuration)
M	=	Molecular weight
n	=	Exponent of repulsion
N	=	Avogadro's number
q	=	Number of atoms per molecule
r_0	=	Interatomic distance
R	=	Reflectivity

T = Temperature
U = Lattice cohesive energy
v_c = Bulk velocity of sound
V = Molecular volume
X_m = Related reduced mass $[= m_r/m_{r(A)}]$
Z = Valence
Θ = Debye temperature from specific heat
$\varkappa$ = Cubic compressibility
v_c = Analog frequency to v_{ctr} from elastic data
v_{ctr} = Frequency of the center of gravity of the infrared spectrum
v_d = Definite frequency
v_θ = Debye frequency from specific heat
ρ = Density
ψ_0 = Change of compressibility with pressure

VIII. REFERENCES *

A. Direct References

1. Early History

1. A. Einstein, "Die Planck'sche Theorie der Strahlung und die Theorie der spezifischen Waerme," *Ann. Phys.* **22**:180 (1907).
2. W. Nernst and F. Lindemann, "Spezifische Waerme und Quantentheorie," *Ann. Phys.* **39**: 789 (1921); *Z. Elektrochem.* **17**: 817 (1911).
3. P. Debye, "Zur Theorie der spezifischen Waermen," *Ann. Phys.* **39**(4): 789 (1912).
4. M. Born and Th. v. Karman, "Zur Theorie der spezifischen Waerme," *Physik. Z.* **14**: 15 (1913); M. Born, *Atomtheorie des festen Zustandes*, Teubner Verlag (Leipzig), 1923.

2. Background Papers †

5. J. N. Plendl, "Center Law of Lattice Vibration Spectra," *Phys. Rev.* **119**:1598 (1960).
6. J. N. Plendl, "Some New Interrelations in the Properties of Solids Based on Anharmonic Cohesive Forces," *Phys. Rev.* **123**:1172 (1961).
7. J. N. Plendl and P. J. Gielisse, "Hardness of Nonmetallic Solids on an Atomic Basis," *Phys. Rev.* **125**:828 (1962).
8. J. N. Plendl and P. J. Gielisse, "Atomistic Expression of Hardness," *Z. Krist.* **118**:404 (1963).
9. J. N. Plendl and P. J. Gielisse, "An Analytical Approach to the Study of Solids," Monograph, Air Force Cambridge Research Laboratories, 1963.
10. J. N. Plendl and P. J. Gielisse, "Infrared Spectra of Inorganic Dielectric Solids," *Appl. Opt.* **3**:943 (1964).
11. J. N. Plendl and P. J. Gielisse, "Characteristic Frequencies from Infrared and Elastic Data," *Appl. Opt.* **4**:853 (1965).
12. J. N. Plendl, S. S. Mitra, and P. J. Gielisse, "Compressibility, Cohesive Energy, and Hardness of Non-Metallic Solids," *Phys. Status Solidi* **12**:367 (1965).

---

* The titles of the papers are given in order to indicate which papers discuss certain aspects in detail.
† The background papers contain many detailed references pertinent to this presentation.

B. General References

1. Reference Tables

International Critical Tables, Vol. III, McGraw-Hill Book Co. (New York), 1928.

J. D'Ans and E. Lax, *Taschenbuch fuer Chemiker und Physiker*, Springer-Verlag (Berlin), 1943.

W. E. Forsythe, *Smithsonian Physical Tables*, Smithsonian Institute (Washington D.C.), 1954.

H. H. Landolt and R. Boernstein, *Zahlenwerte und Funktionen, Vol. 1/4, Kristalle* (and earlier volumes), Springer-Verlag (Berlin), 1955.

Handbook of Chemistry and Physics, 45th edition, Chemical Rubber Co. (Cleveland, Ohio), 1965.

American Institute of Physics Handbook, 2nd edition, McGraw-Hill Book Co. (New York), 1963.

2. Review Articles

J. Sherman, "Crystal Energies of Ionic Compounds and Thermodynamical Applications," *Chem. Rev.* **11**:93 (1932).

M. Born and M. Goeppert-Mayer, "Dynamische Gittertheorie der Kristalle," *Handbuch der Physik, Vol. 24/11*, Springer-Verlag (Berlin), 1933, p. 623.

M. Blackman, "The Theory of Specific Heat of Solids," *Rept. Progr. Phys.* **8**:11 (1941).

M. Blackman, "The Specific Heat of Solids," *Handbuch der Physik, Vol. VII/1*, Springer-Verlag (Berlin), 1955, p. 325.

G. Leibfried, "Gittertheorie der mechanischen und thermischen Eigenschaften der Kristalle," *Handbuch der Physik, Vol. VII/1*, Springer-Verlag (Berlin), 1955, p. 104.

J. de Launey, "The Theory of Specific Heat and Lattice Vibrations," *Solid State Physics, Vol. 2*, Academic Press (New York), 1956, p. 219.

S. S. Mitra, "Vibration Spectra of Solids," *Solid State Physics, Vol. 13*, Academic Press (New York), 1961, p. 1.

S. S. Mitra, "Debye Characteristic Temperatures of Solids," *J. Sci. Ind. Res. (India)* **21A**:76 (1962).

J. N. Plendl and P. J. Gielisse, "An Analytical Approach to the Study of Solids," Monograph, Air Force Cambridge Research Laboratories, Bedford, Massachusetts, 1963.

S. S. Mitra and P. J. Gielisse, "Infrared Spectra of Crystals," in: H. A. Szymanski (ed.), *Progress in Infrared Spectroscopy*, Plenum Press (New York), 1964.

J. N. Plendl, "New Concepts in the Physics of Solids," Monograph, Air Force Cambridge Research Laboratories, Bedford, Massachusetts, 1966.

3. Textbooks

C. Schaefer and F. Matossi, *Das Ultrarote Spektrum*, Springer-Verlag (Berlin), 1930.

J. C. Slater, *Introduction to Chemical Physics*, McGraw-Hill Book Co. (New York), 1939.

F. Seitz, *The Modern Theory of Solids*, McGraw-Hill Book Co. (New York), 1940.

K. Rice, *Electronic Structure and Chemical Binding*, McGraw-Hill Book Co. (New York), 1940.

J. J. Stoker, *Nonlinear Vibrations*, Interscience (New York), 1950.

C. Zwikker, *Physical Properties of Solid Materials*, Pergamon Press (London), 1954.

M. Born and K. Huang, *Dynamical Theory of Crystal Lattices*, Clarendon Press (Oxford), 1954.

C. Kittel, *Introduction to Solid State Physics*, John Wiley & Sons (New York), 1956.

A. J. Dekker, *Solid State Physics*, Prentice-Hall (Englewood Cliffs, N.J.), 1957.

F. O. Rice and Teller, *The Structure of Matter*, John Wiley & Sons (New York), 1959.

W. Kleber, *Angewandte Gitterphysik*, W. de Gruyter Co. (Berlin), 1960.

CHAPTER 14

Infrared and Raman Spectra Due to Lattice Vibrations

Shashanka S. Mitra

Department of Electrical Engineering
University of Rhode Island
Kingston, Rhode Island

I. INTRODUCTION

The infrared and Raman spectra of a solid may arise from a number of causes. We shall, however, limit our discussion here to the infrared absorption and reflection and Raman scattering in a crystal due to the interaction of the incident electromagnetic fields with its vibrational modes. Related topics such as infrared spectra due to the transitions among the low-lying electronic levels of a solid, spectra due to impurities or free carriers in a semiconductor, the absorption of radiation due to the excitation of a local mode around an impurity center in a solid, the various magneto-optic effects in semiconductors observed usually in the far infrared, and electronic Raman spectra of solids shall not be included in the present article.

In recent years a number of reviews [1–3] have been published on the vibrational spectra of solids. However, their emphasis has been laid chiefly on the long-wavelength vibrations of the internal modes of molecular crystals. The purpose of writing another article is twofold: (1) to present a unified account of the optically active vibrational transitions in solids, molecular as well as ionic, semi-ionic, and covalent crystals and (2) to point out the role of $k \neq 0$ ($|k| = 2\pi/\lambda$) modes in the interaction of more than one vibrational mode with the incident photon resulting in the combination and overtone bands. We shall especially provide a comprehensive review of the multiphonon infrared absorption in the II–VI and III–V semiconducting crystals, which has aroused considerable interest in recent years.

No attempt will be made to describe the experimental methods of obtaining infrared transmission and reflection spectra of single crystal specimens. It is assumed that the reader is familiar with elements of crystallography and the application of group theory to molecular vibration problems.

These factors excluded, this paper attempts to present a complete and self-contained account. Although no special effort has been made to present a comprehensive account of the Raman spectra of solids, frequent reference to Raman measurements is made because of its complementary nature to the infrared data.

II. NORMAL VIBRATIONS IN A CRYSTAL

A crystal may be regarded as a mechanical system of nN particles, where n is the number of particles per unit cell and N is the number of unit cells contained in the whole crystal. Such a crystal will have $3nN$ degrees of freedom of which $3Nn - 3$ are the linearly independent normal modes of oscillation of the crystal and three are pure translations. The very large number ($\sim 10^{24}$) of modes belonging to a macroscopic piece of a crystal necessitates the description of the frequency spectrum in terms of a frequency distribution function.

The frequency spectrum of the nuclear motions in a solid can be determined by constructing the classical equations of motion for the lattice points and obtaining the solutions of the normal modes as plane waves. The vibration frequencies occur as the $3n$ roots of the secular equation, involving the wave vector $|\mathbf{k}| (= 2\pi/\lambda)$, which may take N values.

Three of these roots approach zero as the wave vector tends to zero and designate the *acoustic branches*. The remaining $3n - 3$ branches are termed *optical branches* and approach finite limits as the wave vector vanishes. These constitute the fundamental vibration spectrum of the crystal. The frequency distribution $g(\nu)$ in an individual branch is obtained by calculating the number of roots in each small interval of frequency. As an elementary but illustrative example of the foregoing observations, a one-dimensional crystal consisting of two dissimilar atoms is considered.

A. The Linear Diatomic Chain

Let the unit cell consist of two particles of mass M and m (as shown in Fig. 1) located at the even- and odd-numbered lattice points $2n$ and $2n+1$, respectively. Let a be the nearest-neighbor distance. Only nearest-neighbor interactions are assumed, which can be represented by a single force constant f. The displacements u_{2n} and u_{2n+1} of the even and odd particles are given

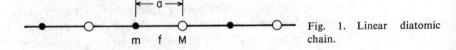

Fig. 1. Linear diatomic chain.

by the following equations of motion:

$$M\ddot{u}_{2n} = f(u_{2n+1} + u_{2n-1} - 2u_{2n})$$
$$m\ddot{u}_{2n+1} = f(u_{2n+2} + u_{2n} - 2u_{2n+1})$$

(14.1)

Assuming solutions of the form

$$u_{2n} = y_1 \exp i(2\pi vt + 2nka)$$
$$u_{2n+1} = y_2 \exp i[2\pi vt + (2n+1)ka]$$

(14.2)

and substituting them in equation (14.1), one obtains two equations for the amplitudes y_1 and y_2. The compatibility condition for these equations is

$$\begin{vmatrix} 2f - 4\pi^2 v^2 M & -2f\cos ka \\ -2f\cos ka & 2f - 4\pi^2 v^2 m \end{vmatrix} = 0$$

(14.3)

which gives

$$v^2 = \frac{1}{4\pi^2}\left[\frac{f}{\mu} \pm \left(\frac{f^2}{\mu^2} - \frac{4f^2\sin^2 ka}{Mm}\right)^{\frac{1}{2}}\right]$$

(14.4)

where μ is the reduced mass per unit cell. This relation between v and the wave vector k is known as the *dispersion relation*. The finite length $2Na$ of the lattice restricts the possible values of $\mathbf{k}$ in the range of $-\pi/2a \leqslant \mathbf{k} \leqslant \pi/2a$). The region between these limits of $\mathbf{k}$ is termed the *first Brillouin zone*.

Equation (14.4) has two sets of solutions depending on the positive or the negative signs, which are the optical and acoustic branches. In the long-wave limit (small k), the two roots are

$$v = \frac{1}{2\pi}\left(\frac{2f}{\mu}\right)^{\frac{1}{2}} \quad \text{optical}$$

(14.5)

$$v = \frac{1}{2\pi}\left(\frac{2f}{M+m}\right)^{\frac{1}{2}} ka \quad \text{acoustic}$$

(14.6)

As $\mathbf{k} \to 0$, the frequency of the acoustical branch tends to zero, as expected, since it then represents the motion of the crystal as a whole. It can be shown that the frequency distribution function $g(v)$, defined such that $g(v)\,dv$ gives the number of frequencies between v and $v+dv$, has maxima at $(1/2\pi)(2f/m)^{\frac{1}{2}}$ and $(1/2\pi)(2f/M)^{\frac{1}{2}}$ corresponding to the boundary of the Brillouin zone, $\mathbf{k} = \pi/2a$. The dispersion relations for the two branches are shown in

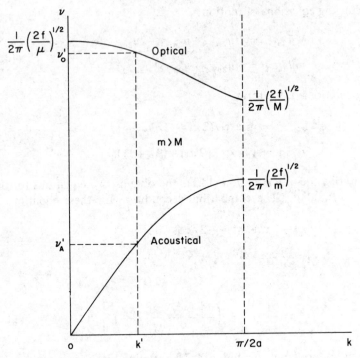

Fig. 2. The dispersion relation for a linear diatomic chain.

Fig. 2. The frequency distribution for the diatomic linear lattice is given in Fig. 3.

When electromagnetic radiation interacts with a crystal lattice, a photon may be absorbed provided the wave vector and the energy are conserved. For infrared radiation of wavelength, say, around $50\,\mu$, the wave vector $\mathbf{k} = (2\pi/\lambda) \approx 10^3\,\text{cm}^{-1}$. This is extremely small compared with the wave vector corresponding to the edge of the Brillouin zone, $\pi/2a \simeq 10^8\,\text{cm}^{-1}$. Thus, one may consider the wave vector associated with the electromagnetic field in the infrared region to be essentially zero. The infrared spectrum for a diatomic linear lattice in the long-wave limit ($\mathbf{k} \sim 0$) corresponding to the center of the Brillouin zone will then consist of a single line at a frequency equal to $(1/2\pi)\,(2f/\mu)^{\frac{1}{2}}$, provided it is associated with a changing transition moment. This fundamental infrared absorption originates from the relative motion of the M lattice as a whole against the m lattice. For a homopolar diatomic linear lattice, this fundamental mode is forbidden in the infrared and is active only in Raman scattering.

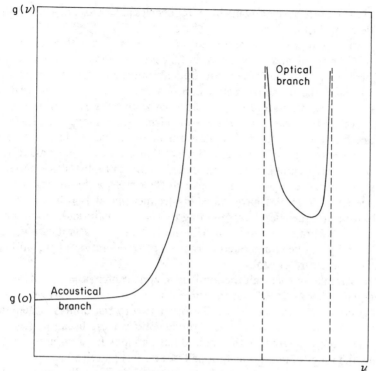

Fig. 3. The frequency distribution function for a linear diatomic chain.

B. Vibration of Three-Dimensional Lattices

Analogous equations of motion can be constructed for the two- and three-dimensional lattices also. Such equations were first set up by Born and von Karman [4]. The solutions obtained are in terms of plane waves of the form $\exp i(2\pi t - \mathbf{k} \cdot \mathbf{r}_j)$. For a monatomic lattice, there are N independent values of $\mathbf{k}$, each of which is associated with three different modes having different and orthogonal polarization directions. Thus, there are $3N$ independent lattice modes. The energy of each mode is quantized, and the term *phonon* is used to describe a quantized lattice vibration. For a simple Bravais lattice (one particle per unit cell) such as sodium or copper, all the branches have zero frequencies in the long-wave limit. These, therefore, do not have any optical vibration spectra. In a lattice with basis, i.e., one in which the entire crystal is obtained by the primitive translations of a unit cell containing more than one particle, the optical branches are separated from the acoustic ones and the former have nonvanishing frequencies at $k = 0$. These constitute the optical spectrum.

To sum up, a crystal composed of N primitive cells with n atoms per cell has $3Nn$ normal modes of vibration. These $3Nn$ modes are distributed on $3n$ branches, $3n-3$ of which are optical and 3 are acoustic branches. In each branch, the mode frequency depends upon the wave vector, which can assume all values in and on the Brillouin zone. However, all modes of vibration other than those in which equivalent atoms move identically in phase are forbidden as fundamentals both in the infrared absorption and the Raman scattering. This is due to the fact that the wave vector associated with a photon is essentially zero compared with those of most of the phonons of the same energy, except for the ones near the center of the Brillouin zone. The three acoustic modes have vanishing frequencies at $k = 0$. The $3n-3$ optical branch frequencies corresponding to $k = 0$ constitute the fundamental modes of vibration, whose distribution, symmetry classification, and optical activity can be enumerated only from a consideration of the unit cell instead of the entire crystal.

For example, in rock salt the number of zone-center ($k = 0$) fundamentals is equal to $3 \times 2 - 3 = 3$. Of the three fundamentals, one corresponds to the longitudinal optical mode and the other two to the doubly degenerate transverse optical models. The electromagnetic waves, because they are transverse, cannot interact with longitudinal phonons in an infinite crystal, and thus the longitudinal optical mode frequency is inactive in the infrared. The transverse optical mode, on the other hand, is optically active and is observed as a resonant frequency by measuring the transmittance of thin films for infrared radiation or as the reststrahlen band in the reflection spectrum. For homopolar crystals such as Ge or Si, the transverse optical mode is only active in the Raman scattering.

C. Symmetry and Unit Cell Modes

In this section, attention is confined to the modes of zero wave vector, i.e., the ones corresponding to the center of the Brillouin zone. It was mentioned earlier that these are the only modes that may be optically active as fundamentals. They are $3n-3$ in number, where n is the number of particles per Bravais unit cell. The equivalent particles in the crystal lattice move in identical phase for these modes of oscillation. In other words, all the unit cells of the lattice are in the same phase throughout the period of execution of the fundamental optical modes. The symmetry classification of the $k = 0$ modes of vibration can therefore be accomplished by taking into consideration the factor group of the crystal, instead of the entire space group. The factor group represents the group formed by the symmetry elements contained in the smallest unit cell (Bravais cell). A factor group is always isomorphous with one of the 32 crystallographic point groups. The $3n-3$ long-wavelength fundamental modes of a crystal are also known as the *factor group fundamentals*.

In the case of crystals consisting of a number of structural units, it may be assumed that the forces holding together the atoms within a group are much stronger than those keeping the various groups together. For example, in a single crystal of naphthalene, the carbon and hydrogen atoms in a naphthelene molecule are held together by chemical bonds, whereas the individual molecules in the crystal are held chiefly by van der Waals type of binding. The former is about two orders of magnitude larger than the latter. The modes arising chiefly from the oscillation of atoms within a group are termed *internal* modes. The motions of the groups relative to one another give rise to the *external* modes, which are also known as *lattice* modes. Since the forces among the groups are usually weaker than those among the atoms within a group, the external vibrations are expected to occur at lower frequencies than the internal ones. The external modes can be further subdivided into translatory and rotatory types which, in the limit of vanishing forces among the groups, correspond to pure translations and rotations, respectively. The rotatory lattice modes are often referred to as *librational* modes and may be associated with any polyatomic group in a crystal. In a crystal like rock salt, the only possible vibrational modes are lattice modes. No internal mode occurs because of the lack of any molecular group in such a crystal.

The determination of selection rules and the symmetry classifications of the unit cell fundamentals of a crystal can be accomplished by two methods. Bhagavantam and Venkatarayudu [5] consider the atoms in the unit cell a large molecule, and the usual method [6] for deriving the optical activities and the symmetry classification of the normal modes of a molecule is applied. Halford [7], on the other hand, suggested use of the site approximation in which the local symmetry of a molecular group in the unit cell is considered. Hornig [8] has shown that, proceeding from the site approximation one may obtain the same results as Bhagavantam and Venkatarayudu. Winston and Halford [9] have derived both methods by considering the motions of a crystal segment composed of an arbitrary number of unit cells and subject to the Born–von Kármán boundary conditions. The factor group analysis of Bhagavantam and Venkatarayudu is most suitable in classifying the lattice vibrations. Mitra [10] has indicated an extension of the method suitable for crystals containing linear polyatomic groups. A brief summary of the method is presented here.

The procedure consists in writing down the character table and irreducible representations of the isomorphous point group corresponding to the space group of the crystal. N_k, the number of times a particular irreducible representation Γ_k is contained in another representation Γ, is given by

$$N_k = \frac{1}{N} \sum_j h_j \chi_k(R) \chi_j'(R) \tag{14.7}$$

where N is the order of the group; and h_j, the number of group operations contained in the jth class. Respectively, $\chi_k(R)$ and $\chi_j'(R)$ are the characters of the group operation R in the representations Γ_k and Γ. The normal modes are classified by making specific selections of the representation Γ and its appropriate characters $\chi_j'(R)$. One needs to consider the characters for the electric dipole moment vector (τ) and the polarizability or the symmetric tensor (α) representations, respectively, in order to obtain the infrared- and Raman-active fundamentals. The selection rules can be summed up as

$$\frac{1}{N} \sum_j h_j \chi_k(R) \chi_j'(T) \quad \begin{array}{l} = 0 \text{ IR forbidden} \\ \neq 0 \text{ IR permitted} \end{array} \tag{14.8}$$

and

$$\frac{1}{N} \sum_j h_j \chi_k(R) \chi_j'(\alpha) \quad \begin{array}{l} = 0 \text{ Raman forbidden} \\ \neq 0 \text{ Raman permitted} \end{array} \tag{14.9}$$

The group characters $\chi_j'(R)$ for the various representations are given in Table I. Two especial cases of selection rules applicable to the factor group

Table I.   Group Characters $\chi_j'(R)$ for Various Representations *

Representation	Group character
All unit cell modes (3n Cartesian coordinates)	$\chi_j'(n_i) = \omega_R(\pm 1 + 2 \cos \phi_R)$
Acoustic modes (dipole moment vector)	$\chi_j'(T) = \pm 1 + 2 \cos \phi_R$
Translatory lattice modes	$\chi_j'(T') = [\omega_R(s) - 1](\pm 1 + 2 \cos \phi_R)$
Rotatory lattice modes	$\chi_j'(R') = [\omega_R(s-p)]\chi_j'(P)$
Symmetric tensor	$\chi_j'(\alpha) = (\pm 1 + 2 \cos \phi_R) 2 \cos \phi_R$

* Notation: ω_R is the number of atoms invariant under the symmetry operation R; $\omega_R(s)$ is the number of structural groups remaining invariant under symmetry operation R; $\omega_R(s-p)$ is the number of polyatomic groups remaining invariant under an operation R; p is the number of monatomic groups; ϕ_R is the angle of rotation corresponding to the symmetry operation R. Plus and minus signs stand, respectively, for proper and improper rotations; $\chi_j'(P)$ is $(1 \pm 2 \cos \phi_R)$ for nonlinear polyatomic groups; $\pm 2 \cos \phi_R$ for operations $C(\phi_R)$ and $S(\phi_R)$ in a linear polyatomic group; and 0 for operations $C_2(\theta)$ and σ_v in a linear polyatomic group.

fundamentals are worth remembering. These are: (1) A totally symmetric mode is always Raman active. (2) In a crystal with a point of inversion, a normal mode symmetric with respect to the center of symmetry is Raman active and forbidden in the infrared, whereas an antisymmetric mode with respect to the center is infrared active but forbidden in the Raman scattering.

D. Normal Modes of Oscillation in Gypsum

1. Factor Group Analysis

The foregoing considerations will now be used in the elucidation of the group fundamentals of gypsum, $CaSO_4 \cdot 2H_2O$. The crystal belongs to the space group $C_{2h}^6 (C2/c)$ with two molecules per Bravais unit cell. Gypsum is thus composed of three types of structural groups: (1) monatomic ions Ca^{++}, (2) polyatomic ions, $SO_4^=$, and (3) H_2O molecules. The structure is shown in Figs. 4 and 5. Exclusive of the long-wavelength accoustic modes, there are a total of 69 factor group fundamentals. The internal modes belong to the sulfate ions and the water molecules only. They are expected to occur in the crystal at nearly the same frequencies as for the isolated systems $SO_4^=$ and H_2O. The lattice modes are caused by the motions of the structural groups (eight in number in the case of gypsum) relative to each other. The classification of the various modes of vibrations has been carried out by means of the factor group analysis shown in Table II. This readily gives the symmetry species and the activities of the external modes. The number

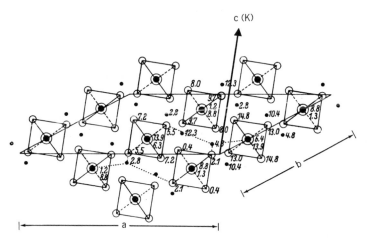

Fig. 4. Projection of the crystal structure of gypsum on the (010) plane. Heights are in 10^{-8} cm. Squares indicate SO_4 tetrahedra. The upper of the two numbers beside the symbol ⊙ gives the height of the S. [Reprinted by permission from W. A. Wooster, Z. Krist. **94**:375 (1936).]

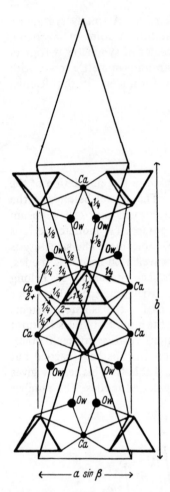

Fig. 5. Projection of the gypsum structure with the Wooster c axis normal to the paper. Numbers indicate the disturbance of valence bonds. [Reprinted by permission from W. A. Wooster, *Z. Krist.* **94**:375 (1936).] The diamond-shaped outline (drawn over Wooster's diagram) shows the Bravais unit cell.

of internal modes belonging to the $SO_4^=$ and H_2O groups is obtained by subtracting the number of external modes from the total number of modes belonging to any particular irreducible representation. In order to ascertain the number of internal modes belonging to $SO_4^=$ and H_2O separately, an artifice due to Hass and Sutherland [11] is most useful. The lattice is first considered composed only of $CaSO_4$ to obtain the sulfate fundamentals. Next, the lattice is regarded as consisting only of water molecules, ignoring the $CaSO_4$. This procedure is justified since the number of internal modes of each kind of molecule depends only on the number of that kind of molecule in the unit cell.

Table II
Character Table and Distribution of Unit Cell Modes for Gypsum

C_{2h}^6	E	C_2	i	σ_h	n_i	T	T'	R'	n_i' $SO_4^=$	n_i' H_2O	IR	Raman
A_g	1	1	1	1	17	0	5	4	5	3	f	$p(\alpha_{xx}, \alpha_{yy}, \alpha_{zz}, \alpha_{xz})$
A_u	1	1	-1	-1	17	1	4	4	5	3	$p\,M_y$	f
B_g	1	-1	1	-1	19	0	7	5	4	3	f	$p(\alpha_{xy}, \alpha_{yz})$
B_u	1	-1	-1	1	19	2	5	5	4	3	$p\,M_x, M_y$	f
ϕ_R	0°	180°	180°	0°								
$\omega_R(s)$	24	4	0	0								
$\omega_R(s-p)$	8	4	0	0								
$\omega_R(s-p)$	6	2	0	0								
h_j	1	1	1	1								
$\chi_j{}'(n_i)$	72	-4	0	0								
$\chi_j{}'(T)$	3	-1	-3	1								
$\chi_j{}'(T')$	21	-3	3	-1								
$\chi_j{}'(R')$	18	-2	3	0								
$\chi_j{}'(\alpha)$	6	2	6	2								

2.   Internal Modes of the Sulfate Ions

The internal vibrations of the sulfate ions in gypsum will now be discussed by the site group method of Halford [7]. The free sulfate ion has a tetrahedral symmetry. The symmetry species and the designation of the normal modes of vibration of a XY_4 molecule belonging to the T_d point group are given in Table III. There are four distinct modes for the free molecule. The nondegenerate $v_1(A_1)$ and the doubly degenerate $v_2(E)$ modes are Raman active and are forbidden in the infrared.

The two triply degenerate modes $v_3(F_2)$ and $v_4(F_2)$ are active both in Raman and infrared. The normal modes of vibration are shown in Fig. 6.

In the gypsum crystal, the C_2 crystal axis coincides with a C_2 axis of each $SO_4^=$ group. The motions of the sulfate ions in the crystal thus have to be classified as those of a molecule with C_2 symmetry only. However, the intermolecular forces giving rise to the reduction in symmetry in the

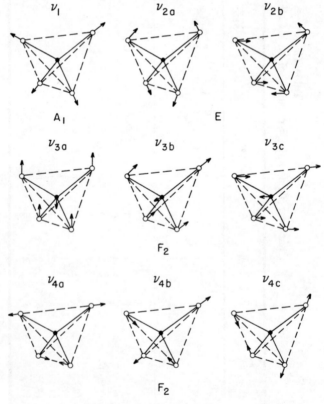

Fig. 6.   Normal vibrations of a tetrahedral XY_4 molecule.

Table III
Character Table and Distribution of Normal Modes of an XY_4 Molecule in the T_d Point Group

T_d	E	$8C_3$	$6\sigma_d$	$6S_4$	$3S_4^2 = 3C_2$	n	Designation	Activity IR	Activity Raman
A_1	1	1	1	1	1	1	ν_1	f	$p(\alpha_{xx}, \alpha_{yy}, \alpha_{zz})$
A_2	1	1	-1	-1	1	0		f	f
E	2	-1	0	0	2	1	ν_2	f	$p(\alpha_{xx}, \alpha_{yy}, \alpha_{zz})$
F_1	3	0	-1	1	-1	0		f	f
F_2	3	0	1	-1	-1	2	ν_3, ν_4	pTx, Ty, Tz	$p(\alpha_{xy}, \alpha_{yz}, \alpha_{zx})$

Table IV. Character Table and Distribution of Normal Modes of an XY_4 Molecule in the C_2 Point Group

C_2	E	$C_2(v)$	n	Activity IR	Activity Raman
A	1	1	5	pTy	$p(\alpha_{xx}, \alpha_{yy}, \alpha_{zz}, \alpha_{zx})$
B	1	-1	4	pTx, Tz	$p(\alpha_{xy}, \alpha_{yz})$

crystal are weak compared with the intramolecular forces within an $SO_4^=$ group. The internal modes of the ions may therefore be considered not very different from those of a tetrahedral molecule under a small perturbation having the symmetry of the lattice, causing a breakdown of the degeneracies. The symmetry species of the C_2 point group are given in Table IV. The fundamental modes of the free molecule are distributed between the two irreducible representations of the C_2 point group, as shown in the correlation chart of Fig. 7. Since the $v_1(A_1)$ and $v_2(E)$ fundamentals of the free molecule are completely symmetric with respect to the twofold rotation operation, they correlate with the A species of the C_2 point group. They thus become permitted fundamentals in the infrared with the transition moment parallel to the $C_2(y)$ axis. If the coupling forces are weak, these modes will be only weakly active in the infrared and may not appear at all in the case of vanishing coupling. The triply degenerate F_2 modes v_3 and v_4 of the free molecule now split into three single frequencies each. For each set, one of the modes is symmetric (A) and the other two are antisymmetric with respect to the twofold axis. Correspondingly, the former transition moment lies along the $C_2(y)$ axis and the latter lies perpendicular (x and z) to the C_2 axis.

In the above considerations, only one SO_4 ion was subjected to the perturbation having the symmetry of the lattice. It is, however, necessary to investigate the motions of all the molecules in the unit cell. In the present case, there are two $SO_4^=$ ions per Bravais cell, each capable of undergoing nine fundamental modes of oscillation. Each mode, therefore, is further split into two depending on whether symmetric (g) or antisymmetric (u) to the point of inversion in the unit cell. This is shown in the last column of the correlation diagram, which is identical with the distribution of the internal

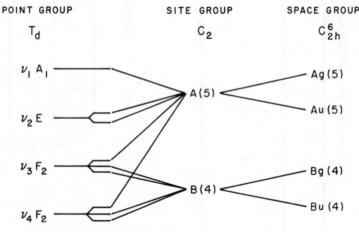

Fig. 7.   Correlation chart for the SO₄ vibrations.

modes (n') of the $SO_4^=$ ions in Table II. Thus, although there are only four normal modes possible in the free tetrahedral $SO_4^=$ ion, there may be as many as eighteen in the crystal, corresponding to the motions of all unit cells in phase ($\mathbf{k} = 0$).

3. Internal Modes of the Water Molecules

The three fundamental modes of vibration due to an isolated water molecule are shown in Fig. 8. All of them are active in both the infrared and the Raman spectrum. The modes v_1 and v_2 are symmetric with respect to the molecular symmetry axis with the transition moments parallel to this direction, whereas the v_3 mode is antisymmetric with the transition moment normal to the molecular symmetry axis.

There are four molecules of water of crystallization per unit cell of gypsum. The crystal internal modes may now be constructed by the superposition of the molecular modes of the four water molecules in the unit cell in such a way that they satisfy the symmetry requirements of the crystal. In the absence of any interaction among the molecules, each molecular mode becomes fourfold degenerate in the crystal. Introduction of a small amount of coupling leads to a breakdown of degeneracy. Each molecular fundamental splits into four distinct crystal modes belonging one each to the four symmetry species of the C_{2h} factor group.

The four crystal modes, say, corresponding to the v_1 frequency of H_2O should occur at slightly different frequencies. Since the extent of the coupling

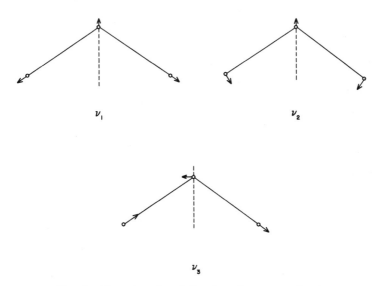

Fig. 8. Normal modes of vibration of a water molecule.

among the water molecules is not known, the crystal modes may be regarded as arising from the limiting motions of an array of molecules oriented in the lattice positions but with vanishing interactions among them. Such a model is known as the *oriented gas model* first proposed by Ambrose, Elliot, and Temple [12] and Pimentel and McClellan [13]. Though this model is incapable of predicting the extent of splitting of the molecular modes in the crystal, it is useful in deducing the relative intensities of the various crystal fundamentals. Let angle θ denote the orientation of the water symmetry axis to the b axis of the crystal, as indicated in Fig. 9. The ratio of the transition moments of the A_u and B_u bands arising from the v_1 and v_2 molecular modes will be given by cot θ. Hence, the ratio of the extinction coefficients of the $A_u(\|)$ to $B_u(\perp)$ bands known as the *dichroic ratio* [14] is given by $\cot^2 \theta$. The angle θ in gypsum, as determined from the X-ray diffraction data, is 52°, which predict a dichroic ratio of 0.613 for the A_u and the B_u crystal

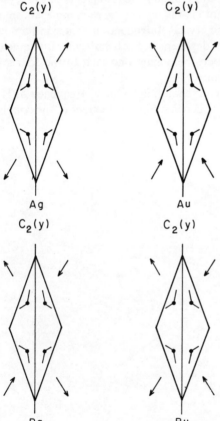

Fig. 9. v_1 and v_2 Modes of water molecules in gypsum.

modes corresponding to the symmetric (v_1 and v_2) molecular vibrations. A similar analysis can be carried out for the antisymmetric v_3 molecular vibration as well. There, the pertinent angle is that between the line joining the hydrogen nuclei in a water molecule and the b axis, which is 40° in gypsum. Hence, the predicted dichroic ratio for the v_3 water fundamental in gypsum is 1.40.

E. The Effect of Crystal Field on the Internal Modes

When compared with the spectrum of an isolated molecule, the infrared spectrum of a solid has the following distinctive characteristics: (1) appearance of entirely new bands in the low-frequency region ($< 800 \text{ cm}^{-1}$) due to the external modes; (2) changes in vibration frequencies, occasionally accompanied by changes of intensities; and (3) splitting of single bands into multiplets. In the present section it is intended to furnish a qualitative explanation of the last two features.

The concept of an isolated molecule may nearly be achieved in a gas at low pressure and, to a certain extent, in a dilute solution with an inert solvent. The interactions with other molecules may be entirely neglected in such a situation, and the observed spectrum may be explained from the consideration of the potential energy for the internal coordinates of a single molecule. In the liquid and the solid state, the interactions with the environment in general and the immediate neighbors in particular are expected to have perturbing effects causing changes in the spectrum. Thus, by investigating the modifications occurring in the spectrum while going from the vapor to the liquid and the solid phases, it is possible in principle to obtain some information regarding the nature of the intermolecular forces in the condensed states.

The vibrational potential energy of a molecular crystal in general may be represented [15] by

$$V = \sum_n V_n + \sum_n \sum_k V_{nk} + V_l + V_{ln} \qquad (14.10)$$

where V_n is the potential energy due to all the internal coordinates of the nth molecular group in the crystal; $\sum_n V_n$, the sum of the internal potential energy for all the molecules contained in the crystal; V_{nk}, the potential-energy cross terms between the internal coordinates of the nth and the kth molecules; $\sum_n \sum_k$, the sum over all pairs of molecules; V_l, the potential arising from the external coordinates, i.e., due to the displacements of the centers of gravity and changes in orientations of the various structural units in the crystal; and V_{ln}, the cross terms involving the internal and the lattice coordinates.

In the oriented gas approximation, only the first term need be considered. However, the potential energy for a molecule in the crystal will be different from that for the corresponding gas molecule owing, what is called the *static field effect*. This results in the usual difference in frequencies noted between the corresponding gas and condensed phase modes and includes the influence of the surrounding lattice in its equilibrium configuration on the molecule in question. Thus, V_n incorporates the site symmetry of this molecule. If the site group has a lower symmetry than the molecular point group, intrinsic degeneracy, if any, of the molecular modes no longer exists in the crystal as was observed in the case of the sulfate vibrations in gypsum. Splitting will result in the presence of a nonvanishing static field effect in addition to the shift of frequencies relative to those of the free molecule. It should be borne in mind that a comparison of spectra in the solid and gas phases may be somewhat hypothetical in nature. In many instances, the isolated molecule cannot be studied, as is the case with ions in a solid.

An approximate theory explaining the condensed-phase frequency shifts was given by West and Edwards [16] and Bauer and Magat [17]. Considering the overall effect of the environment, they obtained the following relation:

$$\frac{\Delta v}{v_0} = C \ \frac{\varepsilon - 1}{2\varepsilon + 1} \tag{14.11}$$

where v_0 is the gas-phase frequency; Δv, the frequency shift; ε, the dielectric constant of the medium; and C, a property of the medium determined empirically. Since only the bulk properties of the medium are included in the relation (equation (14.11)], departures from it are very frequent [18]. Pullin [19] has made some advances by considering the influence of the medium involving the anharmonic terms in vibrational potential energy and higher-order terms in the electric moment. However, the theory is of limited applicability because the expression for the frequency shift contains far too many undeterminable parameters. Buckingham [20] has given a quantum mechanical theory for a diatomic molecule in a solvent. The interaction potential of the solvent–solute system is expressed as a power series of the internuclear displacement of the atoms of the solute molecule. It is introduced as a perturbation on the anharmonic-oscillator Hamiltonian. An analogous theory for the solid is possible which replaces the solvent by the lattice itself. However, no complete theory has yet been advanced explaining the static field splitting satisfactorily. This is partly because the parameters involved usually far outnumber the experimentally determinable quantities.

The cross terms $\sum_n \sum_k V_{nk}$ represent the *dynamical crystal effects* arising from the intermolecular interactions. The most important of these is the *correlation field splitting* associated with the site-group to the factor-group transformation.* For example, the single OH⁻ ion has only one internal mode due to the hydroxyl stretching vibration. However, in the brucite crystal $Mg(OH)_2$, there are two internal modes due to OH stretching which, in turn, results from two OH⁻ ions per unit cell. In fact, one expects two different crystal modes to exist for each **k** value. The magnitude of the splitting depends upon the extent of interaction between one OH⁻ ion in a unit cell with all other nonequivalent OH⁻ ions in the entire crystal. For the **k** = 0 internal modes due to the water molecules in gypsum, such a splitting was noted earlier.

The static and the dynamic field effects are usually present simultaneously. Hexter [21], e.g., has examined methyl chloride, CH_3Cl, for this condition. He has predicted that a nondegenerate (A_1) internal fundamental will be split into two infrared-active frequencies by the intermolecular coupling (correlation) and a doubly degenerate (E) fundamental will be split in two by the site symmetry C_S (static field effect) in addition to a further doubling by intermolecular coupling. However, only three of the four crystal modes of the latter type are infrared active under the space group C_{2y}. Dows [22] has observed the correlation field splitting for the nondegenerate modes v_1 and v_2 of CH_3Cl and for v_2 and v_3 of CH_3Br and CH_3I. The v_3 of CH_3Cl shows an isotope shift of 6 cm⁻¹ in the gas phase due to the sizable abundance of Cl^{35} and Cl^{37}. In the crystal, three lines are observed, presumably due to a superposition of the isotope effect and the correlation field shift, with the lower of the Cl^{35} components coinciding with the upper of the Cl^{37} components. The absorption bands due to the crystal modes derived from the doubly degenerate E modes are more difficult to interpret. The v_6 of CH_3Cl is not split, while in CH_3Br and CH_3I the observed splitting amounts to 5 and 7 cm⁻¹, respectively. The site splitting is expected to be more prominent in CH_3Cl than in CH_3Br and CH_3I because of the large dipole moment and closer packing factor in the first. The splittings, however, increase in the reverse order, which indicates that the intermolecular coupling is the predominant source of splitting. Except for the carbon–halogen stretching mode, the splittings cannot be accounted for by dipole–dipole interactions alone. Dows has shown [23] that an intermolecular hydrogen–hydrogen repulsion potential satisfactorily explains the observed effects in the modes involving the hydrogen motions.

---

* In analogy with the splitting of electronic levels, some authors prefer the term *Davydov splitting*.

III. INFRARED DISPERSION BY IONIC CRYSTALS

It may be recalled that, in the long-wave limit, the optical vibrations of a diatomic lattice correspond to the motion of one type of atoms, all in phase, relative to the other kind. In ionic crystals, such a motion is associated with strong electric moments and hence can directly interact with the electric field of proper polarization from incident electromagnetic radiation. In the vicinity of the resonance frequency, one thus expects drastic changes in the optical properties of such a crystal. In this section, it is intended to discuss briefly the dispersion of infrared radiation by cubic diatomic ionic crystals with optical isotropy.

A. Interaction with the Radiation Field

Huang [24] has given a phenomenological theory of infrared dispersion in ionic crystals, salient features of which are described below. If $\mathbf{u}$ represents the displacement of the positive ions relative to the negative ions, a reduced displacement vector $\mathbf{w}$ may be expressed as $\mathbf{w} = \mathbf{u}(\mu/v_a)^{1/2}$, where μ/v_a is the reduced mass per Bravais unit cell. The macroscopic equations describing the polar motions are then

$$\ddot{\mathbf{w}} = b_{11}\mathbf{w} + b_{12}\mathbf{E} \tag{14.12}$$

and

$$\mathbf{P} = b_{21}\mathbf{w} + b_{22}\mathbf{E} \tag{14.13}$$

where $\mathbf{E}$ is the electric field and $\mathbf{P}$ the dielectric polarization defined by

$$\mathbf{E} + 4\pi\mathbf{P} = \mathbf{D} = \varepsilon\mathbf{E} \tag{14.14}$$

The b's are constants characteristic of the solid, the nature of which is to be ascertained. It can be shown [25] that $b_{12} = b_{21}$ as a consequence of the principle of the conservation of energy. The linearity of the equations (14.12) and (14.13) implies that anharmonicity and higher-order terms in the electric moment are neglected.

Considering the periodic solutions,

$$(\mathbf{w}, \mathbf{E}, \mathbf{P}) = (\mathbf{w}_0, \mathbf{E}_0, \mathbf{P}_0)\, e^{2\pi i v t} \tag{14.15}$$

equations (14.12) and (14.13) are reduced to

$$-4\pi^2 v^2\, \mathbf{w} = b_{11}\mathbf{w} + b_{12}\mathbf{E} \tag{14.16}$$

and

$$\mathbf{P} = b_{21}\mathbf{w} + b_{22}\mathbf{E} \tag{14.17}$$

Elimination of $\mathbf{w}$ from equations (14.16) and (14.17) yields

$$\mathbf{P} = \left(b_{22} + \frac{b_{12} b_{21}}{-b_{11} - 4\pi^2 v^2} \right) \mathbf{E} \tag{14.18}$$

Substitution of equation (14.18) in equation (14.14) readily gives the dielectric constant ε in terms of the b coefficients

$$\varepsilon = 1 + 4\pi b_{22} + \frac{4 b_{12} b_{21}}{-b_{11} - 4\pi^2 v^2} \tag{14.19}$$

The similarity of equation (14.19) to the infrared dispersion formula

$$\varepsilon = \varepsilon_\infty + \frac{(\varepsilon_0 - \varepsilon_\infty) v_0^2}{v_0^2 - v^2} \tag{14.20}$$

is obvious. In equation (14.20), v_0 is the dispersion frequency; $\varepsilon_\infty = n^2$; n is the refractive index for light of $v \gg v_0$ in a nondispersive region; and ε_0 is the DC or low-frequency dielectric constant.
The b coefficients may now be obtained by comparing equation (14.19) with equation (14.20)

$$b_{11} = -4\pi^2 v_0^2 \tag{14.21}$$

$$b_{12} = b_{21} = (\varepsilon_0 - \varepsilon_\infty)^{\frac{1}{2}} v_0 \tag{14.22}$$

and

$$b_{22} = \frac{\varepsilon_\infty - 1}{4\pi} \tag{14.23}$$

We have, however, yet to identify the dispersion frequency v_0.
Since, macroscopically, the crystal is electrically neutral, one can apply the Gauss law.

$$\nabla \cdot \mathbf{D} = \nabla \cdot (\mathbf{E} + 4\pi \mathbf{P}) = 0 \tag{14.24}$$

$\mathbf{P}$ can then be eliminated from equation (14.13), which gives

$$\nabla \cdot \mathbf{E} = \frac{-4\pi b_{21}}{1 + 4\pi b_{22}} \nabla \cdot \mathbf{w} \tag{14.25}$$

The solenoidal and irrotational solutions of this equation correspond to the transverse and longitudinal waves, respectively, for which

$$\nabla \cdot \mathbf{w}_t = 0 \qquad \text{solenoidal} \tag{14.26}$$

and

$$\nabla \times \mathbf{w}_l = 0 \qquad \text{irrotational} \tag{14.27}$$

where

$$\mathbf{w} = \mathbf{w}_t + \mathbf{w}_l \tag{14.28}$$

Consequently the equation of motion (equation (14.12)] can be split into two parts

$$\mathbf{w}_t = b_{11}\ddot{\mathbf{w}}_t = -4\pi^2 v_0^2 \mathbf{w}_t \tag{14.29}$$

and

$$\mathbf{w}_l = \left(b_{11} - \frac{4\pi b_{12} b_{21}}{1+4\pi b_{22}}\right)\mathbf{w}_l = -\left(\frac{\varepsilon_0}{\varepsilon_\infty}\right)4\pi^2 v_0^2 \mathbf{w}_l \tag{14.30}$$

Since $\mathbf{w}_t$ and $\mathbf{w}_l$ are periodic with the transverse and longitudinal frequencies v_t and v_1, it follows that

$$v_t = v_0 \tag{14.31}$$

and

$$v_l = \left(\frac{\varepsilon_0}{\varepsilon_\infty}\right)^{\frac{1}{2}} v_0 = \left(\frac{\varepsilon_0}{\varepsilon_\infty}\right)^{\frac{1}{2}} v_t \tag{14.32}$$

Thus, the dispersion frequency is identical with the transverse optical mode frequency, and the longitudinal frequency v_l is given by the Lyddane–Sachs–Teller [26] formula [equation (14.32)].

In a nonionic crystal such as diamond or Ge, in the absence of polar interactions, the atomic motions are determined only the local elastic restoring forces. Therefore, the second term of equation (14.12) vanishes. But, since macroscopically $E \neq 0$, it follows that

$$b_{12} = 0$$

and, consequently,

$$v_l = v_t = v_0 \tag{14.32'}$$

Since the macroscopic equations of motions are only valid for the long-wave limit, the above considerations apply only to the zone center ($\mathbf{k} = 0$) or the factor group modes.

B. Reststrahlen Spectrum

As a consequence of the infrared dispersion by ionic crystals, electromagnetic radiation with frequencies in the vicinity of the dispersion frequency undergoes selective reflection. For normally incident radiation in the case of an ideally ionic crystal, the reflectivity may be 100%, for which equation (14.20) holds good. This selective reflection of radiation in the neighborhood of the optical lattice mode frequencies in ionic crystals is known as the *reststrahlen phenomenon*. It can be understood by means of equation

(14.20) along with the Fresnel formula

$$R = \left|\frac{n-1}{n+1}\right|^2 \tag{14.33}$$

where R is the fraction of light intensity reflected by an optically isotropic medium when light is incident perpendicular to its surface, and $n = \sqrt{\varepsilon(v)}$ is the refractive index and may assume complex values. It will be noticed from the dispersion relation [equation (14.20)] that, at $v = 0$, $\varepsilon(v) = \varepsilon_0$, which is its static value. As v increases, n increases steadily above the value $n = \sqrt{\varepsilon_0}$. When v reaches the dispersion frequency v_0, ε and hence n become infinite and the crystal becomes perfectly reflecting with $R = 1$. When v is further increased by an infinitesimal amount, $v_0{}^2 - v^2$ takes a negative but infinitesimally small, value which makes $\varepsilon = -\infty$. With increasing v, ε remains negative until becoming zero again for a frequency satisfying the relation

$$0 = \varepsilon_\infty + \frac{(\varepsilon_0 - \varepsilon_\infty)}{v_0{}^2 - v^2} v_0{}^2$$

the solution of which is $v = (\varepsilon_0/\varepsilon_\infty)^{\frac{1}{2}} v_0$. The refractive index $n = \sqrt{\varepsilon}$ is therefore imaginary between the values v_0 and $v = (\varepsilon_0/\varepsilon_\infty)^{\frac{1}{2}} v_0$, and $R = 1$ over this range. However, these two values of v are precisely the transverse and the longitudinal optical modes, as may be seen from equations (14.31) and (14.32). The reststrahlen band for an ideally ionic crystal is shown in Fig. 10 (top curve), for which a band of perfect reflection exists between the frequencies v_t and v_l.

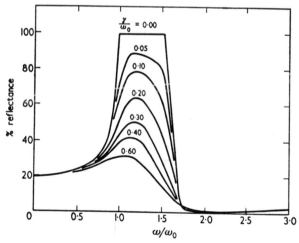

Fig. 10.   Reflection spectra of a damped oscillator. (Reprinted by permission from S. S. Mitra [122].)

In all real diatomic cubic ionic crystals, the observed reflectivity shows characteristic reststrahlen bands with high reflectivity between the frequencies v_1 and v_t. However, in shape or intensity the observed reflectivity does not agree quantitatively with that of an ideal ionic crystal. This is evident in Fig. 11, where the reflection spectrum of AlSb is shown. This discrepancy is due to the fact that equation (14.20), though capable of representing the dispersion of a real crystal at frequencies away from v_0, is inadequate in the immediate vicinity of the dispersion frequency. In real crystals, the strong reflection in the reststrahlen region is also bound to be associated with strong absorption, whereas equation (14.20) predicts no such selective absorption.

A relation of the form of equation (14.20) was obtained from the phenomenological equations of motion [equations (14.12) and (14.13)]. The last two equations neglected all but linear terms, with the consequence being mutually independent lattice waves. However, in real crystals, they are coupled by anharmonic and higher-order electric-moment terms, which play an important role in the dissipation of energy. Huang [24] has shown that the energy density of the lattice waves is predominantly mechanical, manifested in the oscillations of the particles, and only a small portion is associated with the electromagnetic field. Thus, in the steady state, a small amount of energy dissipated by the lattice waves owing to a small amount

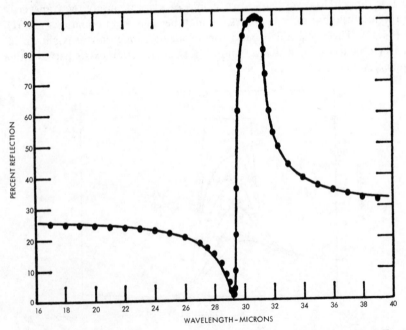

Fig. 11.   Reststrahlen spectrum of AlSb. (Reprinted by permission from W. J. Turner and W. E. Reese [28].)

of coupling between the modes results in a drastic reduction of the electromagnetic energy flux, with the consequent absorption of radiation by the medium.

A more realistic dispersion formula may be obtained by the phenomenological introduction of a damping term that represents a force always opposed to the motion and proportional to the velocity. This procedure provides a method for including the effect of energy dissipation in the neighborhood of v_0. Equation (14.12) now takes the form

$$\ddot{\mathbf{w}} = b_{11}\mathbf{w} - \gamma\dot{\mathbf{w}} + b_{12}\mathbf{E} \tag{14.34}$$

Introducing the periodic solutions $(\mathbf{w}, \mathbf{E}) = (\mathbf{w}_0, \mathbf{E}_0)e^{2\pi i v t}$, one obtains

$$-4\pi^2 v^2 \mathbf{w} = (b_{11} + 2\pi i v\gamma)\mathbf{w} + b_{12}\mathbf{E} \tag{14.35}$$

Comparing equation (14.35) with equation (14.16), one notes that the addition of the damping term is equivalent to replacing the coefficient b_{11} with $b_{11} + i2\pi v\gamma$, with the corresponding change in the dispersion formula [equation (14.19)]. Therefore, equation (14.20) may be written as

$$\varepsilon(v) = \varepsilon_\infty + \frac{\varepsilon_0 - \varepsilon_\infty}{1 - (v/v_0)^2 - i(\gamma/v_0)(v/v_0)} \tag{14.36}$$

which now includes absorption.

Now, a plane electromagnetic wave of phase velocity $c/\sqrt{\varepsilon} = 2\pi v/k$ and frequency v may be represented by

$$\mathbf{E} = \mathbf{E}_0 \exp 2\pi i v\left(\mathbf{e}\cdot\mathbf{r}\frac{\sqrt{\varepsilon}}{c} - t\right) \tag{14.37}$$

where $\mathbf{e}$ is a unit vector in the direction of the Poynting vector. In an absorbing medium, the dielectric constant $\varepsilon(v)$ and hence the refractive index represent complex quantities given by

$$\sqrt{\varepsilon(v)} = n + i\varkappa \quad \text{and} \quad \varepsilon(v) = \varepsilon' + i\varepsilon'' = (n^2 - \varkappa^2) + 2in\varkappa \tag{14.38}$$

Equation (14.37) thus takes the form

$$\mathbf{E} = \mathbf{E}_0 \exp\left(-2\pi v\varkappa e\cdot\frac{\mathbf{r}}{c}\right)\exp\left[2\pi i v\left(\mathbf{e}\cdot\mathbf{r}\frac{n}{c} - t\right)\right] \tag{14.39}$$

and simultaneously describes the effects of refraction and attenuation. The real quantities n and $\varkappa$ are the refractive index and the extinction coefficient, respectively. The first term in equation (14.39) represents attenuation. In terms of the absorption coefficient α, the attenuation is given by

$$|E^2| = |E_0^2| \exp(-\alpha \mathbf{e} \cdot \mathbf{r}) \qquad (14.40)$$

Therefore, the relationship between α and $\varkappa$ is

$$\alpha = \frac{4\pi\varkappa}{\lambda}$$

By expanding equation (14.36) in terms of its real and imaginary components, it follows that

$$\varepsilon' = n^2 - \varkappa^2 = \varepsilon_\infty + \frac{(\varepsilon_0 - \varepsilon_\infty)\left[1 - (v/v_0)^2\right]}{[1 - (v/v_0)^2]^2 + (\gamma/v_0)^2 (v/v_0)^2} \qquad (14.41)$$

and

$$\varepsilon'' = 2n\varkappa = \frac{(\varepsilon_0 - \varepsilon_\infty)(\gamma/v_0)(v/v_0)}{[1 - (v/v_0)^2]^2 + (\gamma/v_0)^2 (v/v_0)^2} \qquad (14.42)$$

As a conseqence of equation (14.38), the reflectance R of an absorbing medium is given by

$$R = \frac{(n-1)^2 + \varkappa^2}{(n+1)^2 + \varkappa^2} \qquad (14.43)$$

The reflection spectrum of a damped oscillator for several values of the damping factor γ/ω_0 (where $\omega_0 = 2\pi v_0$) is shown in Fig. 10.

C. Determination of the Optical Constants in the Reststrahlen Region

The refractive index n and the extinction coefficient $\varkappa$ are known as the optical constants of an absorbing medium. These can be determined as functions of frequency from the reflection spectrum by using equations (14.41) to (14.43). If the values of static or low-frequency ($\sim$ 1 kcps) ε_0 and $n = \sqrt{\varepsilon_\infty}$ in the visible or ultraviolet region free from any electronic transition are known, R may be calculated as a function of v/v_0 for several values of γ/ω_0. The inverse of the peak reflectivity is approximately linear with the damping constant [27]. Figure 12 shows a plot of $1/R_{max}$ *versus* γ/ω_0 for the damped oscillator of Fig. 10. From the observed value of the peak reflectivity, a value may be obtained for γ/ω_0. Values of n and κ may then be calculated from equations (14.41) and (14.42). For a more accurate evaluation, equations (14.41) to (14.43) are fitted to the observed reststrahlen band by machine programming. The best values of v_0, ε_0, ε_∞, and $\gamma/2\pi v_0$ thus obtained are used to calculate n and $\varkappa$. Figure 13 gives the calculated optical constants [28] for AlSb from the reststrahlen spectrum shown in Fig. 11. In this figure, the best fit is given for the calculated values of R by using

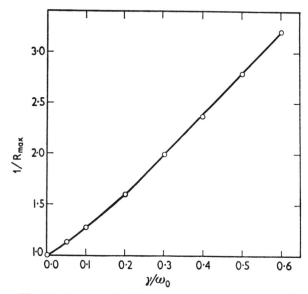

Fig. 12. Inverse of peak reflectivity *versus* damping constant. (Reprinted by permission from S. S. Mitra [122].)

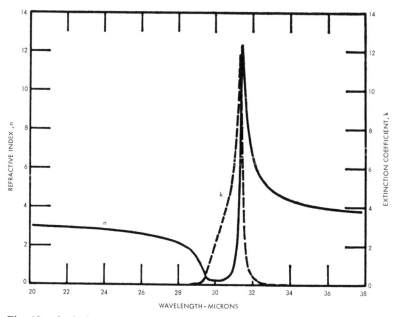

Fig. 13. Optical constants of AlSb. (Reprinted by permission from W. J. Turner and W. E. Reese [28].)

the empirical parameters. It may be mentioned here that, although the maximum of $\varkappa$ occurs very near $\nu = \nu_0$, it is $n\varkappa\nu$ that undergoes the maximum precisely at $\nu/\nu_0 = 1$.

The dispersion frequency ν_0 may also be determined directly by transmission measurements. However, for a polar crystal, the strong absorption in the immediate vicinity of ν_0 calls for the use of very thin samples of the crystal. For extremely thin ($< 1\ \mu$) samples, it may be shown that the minimum transmission occurs exactly at the dispersion frequency.

Very often, an elaborate determination of the optical constants from the analysis of the reststrahlen band may not be necessary for the evaluation of ν_0. It may be obtained from ν_m, the frequency at which the maximum reflection occurs, provided ε_0 and ε_∞ are known. For this purpose, Havelock [29] has shown that, if γ/ω_0 is small, the ratio ν_m/ν_0 is approximately independent of the damping constant and is given by

$$\frac{\nu_m}{\nu_0} = \left(1 + \frac{\varepsilon_0 - \varepsilon_\infty}{6\varepsilon_\infty - 4}\right)^{\frac{1}{2}} \tag{14.44}$$

Since, for most ionic crystals, the damping constant γ/ω_0 is relatively small, the above formula may be used for the determination of the dispersion frequency. This is also equal to the zone-center transverse optical (TO) mode frequency.

Under higher resolution, the reststrahlen band is occasionally accompanied by some structure, usually a shoulder, in the high-frequency side. This side band corresponds to a second resonance frequency arising from the combination of an acoustic phonon with an optical phonon in accordance with the conservation principles discussed in Section VII. The analysis of such a reststrahlen band may be accomplished [30] by assuming two sets of parameters, one each for the two resonance frequencies, in equations (14.41) and (14.42). In the case of ionic crystals with more than one infrared-active fundamental giving rise to as many reststrahlen bands (e.g., quartz [31] and sapphire [32], multiresonance damped-oscillator calculations may be performed for the determination of the optical constants. One such spectrum for Al_2O_3 is shown in Fig. 14.

D. Determination of the Longitudinal Optical-Mode Frequency at $k \sim 0$

The infrared-inactive zone-center longitudinal optical (LO) mode frequency ν_l of ionic crystals may be evaluated by several methods. Usually ν is computed from ν_t by using the Lyddane–Sachs–Teller (LST) relationship (equation (14.32)] provided ε_0 and ε_∞, the values of the dielectric constant at very low and high (visible) frequencies, are available. However, there may be considerable uncertainties as to the values of ε_0 and ε_∞ to

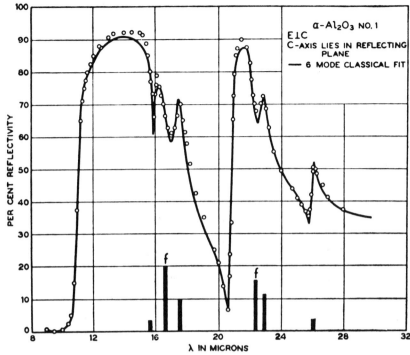

Fig. 14. Reflection spectrum of a single crystal of sapphire with the electric field perpendicular to the c axis of the crystal. Solid line, a multiresonance damped-oscillator fit; circles, experimental points. (Reprinted from A. S. Barker, Jr. [32].)

be used in many substances. This applies especially to crystals that are not good insulators and those with more than one reststrahlen band or with low-energy electronic transitions. A second method of obtaining v_l is to fit the observed reststrahlen spectrum with the use of a dispersion relation in which v_l is a variable parameter instead of v_t. This method is not very accurate either because dielectric dispersion relations need several undetermined parameters for a good fitting and the data may be fitted quite accurately with some range of values of v_l. Thirdly, v_l may be determined by Drude's method, which consists in finding the frequency at which the real part of the dielectric constant goes through zero, i.e., where $n = \varkappa$. This method is often used for determining v_l from infrared reflection due to a single damped oscillator. For strongly ionic crystals with a vanishingly low value of the damping constant, for which equation (14.20) may be used as an approximate dispersion relation, v_l corresponds to the frequency at which the reflectivity attains a minimum on the high-frequency side of the reststrahlen band.

For infrared dispersion involving multiple oscillators, although a generalized LST relation,

$$\prod_i \frac{TO_i}{LO_i} = \left(\frac{\varepsilon_0}{\varepsilon_\infty}\right)^{\frac{1}{2}} \tag{14.45}$$

involving all the long-wavelength longitudinal and transverse optical modes (LO_i and TO_i) holds good [33]. It does not, however, yield the LO mode frequencies individually. Often, Drude's rule has been invoked [34] for the multiresonance oscillators, also, although without any theoretical justification. Furthermore, when some of the oscillators are much weaker than the others, the n and κ curves may not cross at as many points as there are expected longitudinal branches which makes the application of Drude's rule impractical. Chang and Mitra [35] have shown that the Drude rule as stated above is not correct for multiresonance damped oscillators although approximately true when only a few oscillators, all of appreciable strength, are present. A rigorous procedure suitable for all cases is shown to be to identify the minima of the modulus of the complex dielectric constant ($|\varepsilon|$) with LO modes. This can be easily done by just performing a Kramers–Kronig analysis (see section III G) of the reflection spectrum to obtain ε' and ε''. They have also shown that it is possible to obtain the LO frequencies individually as functions of oscillator parameters.

In some crystals, the selection rules permit the $k \sim 0$ LO modes to be active in the Raman scattering. This is the case with the cubic zinc-blende and the hexagonal wurtzite structures. For such crystals, the determination of the LO modes is thus most direct. The Raman spectrum of ZnSe, which crystallizes in the zinc-blende structure, is shown in Fig. 15, which indicates both the LO and TO peaks.

Inelastic neutron scattering studies may yield dispersion curves for the LO mode frequency as a function of the wave vector k. A fairly accurate value of ν_l can be obtained from the extrapolation of these curves to $k = 0$. Brockhouse and associates have obtained ν_l for several alkali halide crystals by this method [36].

Finally, Berreman [37] has recently shown that thin films of cubic ionic crystals have sharp, strong infrared absorption and reflection at frequencies characteristic of the LO mode of long wavelength when the radiation is p polarized and incident obliquely to the surface.

E.   Polaritons, the Photon–Phonon Mixed Modes in Ionic Crystals at Long Wavelengths

We should like to examine in some detail the photon–phonon interactions in the vicinity of $k = 0$ in ionic crystals and the validity of the LST relation given in equation (14.32). In an ionic crystal, the out-of-phase

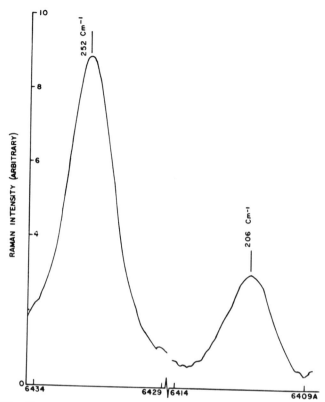

Fig. 15. First-order Raman spectrum of ZnSe showing both the longitudinal (252 cm⁻¹) and the transverse (206 cm⁻¹) optic modes (Mitra, unpublished).

motion of the ions produces a long-range electromagnetic field in addition to the short-range forces. The short-range forces may be lumped together as an isotropic restoring force, while the long-range forces are represented by an electromagnetic field satisfying Maxwell's equations

$$\nabla \cdot \mathbf{D} = 0 \qquad \nabla \cdot \mathbf{H} = 0$$

$$\nabla \times \mathbf{H} = \frac{1}{c}\frac{\partial \mathbf{D}}{\partial t} \qquad \nabla \times \mathbf{E} = -\frac{1}{c}\frac{\partial \mathbf{H}}{\partial t} \tag{14.46}$$

and the equations (14.12) and (14.13) describing the motion of the ions. In equation (14.46), $\mathbf{D}$ is the electric displacement; $\mathbf{E}$ is the electric field; and $\mathbf{H}$ is the magnetic field.

It is in the constant b_{11} of equation (14.12) that all the short-range forces have been lumped together as a restoring force. This theory assumes that the ions can be treated as a continuum, and, therefore, it is only valid for wavelengths much longer than a lattice spacing.

The dispersion relations which can be derived for the optical lattice modes are

$$\omega = \omega_l = \omega_0 \left(\frac{\varepsilon_0}{\varepsilon_\infty}\right)^{\frac{1}{2}} \tag{14.47}$$

for the longitudinally polarized mode and

$$\left(\frac{kc}{\omega}\right)^2 = \varepsilon_\infty + \frac{(\varepsilon_0 - \varepsilon_\infty)}{\omega_0{}^2 - \omega^2} \tag{14.48}$$

for the transversely polarized modes. These relations were first obtained by Huang [24]. The frequency ω $(= 2\pi\nu)$ as a function of wave number is shown in Fig. 16. In an ionic crystal when the ions with opposite charges are vibrating out of phase, we have five different modes present. These modes are a longitudinal lattice mode, two transverse lattice modes, and two transverse electromagnetic modes. The four transverse modes are strongly coupled in ionic crystals. We shall call these modes *polarization modes* as they result from a long-range polarization of the crystal by the ionic motion. Thus, these elementary excitations can no longer be described as pure phonons or pure photons. For such a coupled excitation, the term *polariton* has also been used [38]. It is convenient to divide Fig. 16 into three regions. In the very short wavelength region C, we can separate the polarization

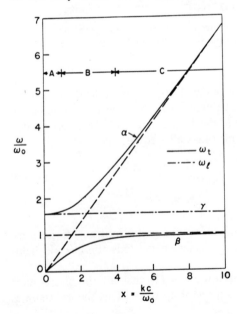

Fig. 16. The dispersion curves for the two doubly degenerate transverse polarization modes α and β and the longitudinal polarization mode γ are shown for NaCl. The dashed lines represent the dispersion curves for the uncoupled electromagnetic and lattice modes. (Reprinted from T. N. Casselman, S. S. Mitra, and H. N. Spector [42].)

modes into a doubly degenerate lattice mode (optical phononlike) β of frequency ω_0 and a doubly degenerate electromagnetic mode (photonlike) α of phase velocity $c/\sqrt{(\varepsilon_\infty)}$. In this region, the coupling between the photon modes and phonon modes is very weak as the transverse electromagnetic field generated by the out-of-phase motion of the ions is negligible. We note that, in this region, the LST relation between the frequencies of the LO and TO phonons is reasonably valid. In the intermediate region B, the coupling between the electromagnetic and lattice waves has become quite strong. In fact, the β mode, which was phononlike in C, starts becoming photonlike, and the α mode, which was photonlike in C, becomes phononlike. Here, the modes are so mixed that one cannot say that one mode is electromagnetic and the other is a lattice vibration. In the long-wavelength region A, the modes have reversed their roles. The uppermost branch α, which was originally photonlike in C, has become predominantly phononlike in A with a limiting frequency at $\mathbf{k} = 0$ of ω_l. The lower branch β, which was originally phononlike in C, has now become predominantly photonlike in A with a phase velocity $c/\sqrt{(\varepsilon_0)}$. The upper branch, which we identify as primarily a TO phonon mode, is now degenerate with respect to the LO phonon mode γ. This degeneracy agrees with the result of Rosenstock [39], who has shown that, in general for an infinite crystal with cyclic boundary condition, the optical modes will be degenerate at $\mathbf{k} = 0$.

The LST relation holds in region C but not in regions A and B. The limits of validity of the LST relation that follow from Huang's theory have been discussed by Barron [40], who showed that it should hold for wavelengths of 10^{-4} cm $> \lambda > 10^{-6}$ cm. The longest wavelength for which the LST relation is valid is at the boundary of region C, while the short-wavelength limit arises because, for shorter wavelengths, the Huang theory is no longer valid.

Implicit in the Huang theory is the assumption of an infinite crystal. Hardy [41] has shown that, for a finite crystal with long-range forces, the Huang theory is valid for those modes in which $\mathbf{k} > 2\pi/l$, where l is a sample dimension. Thus, the Huang theory is valid for a finite crystal as long as its dimensions are an order of magnitude greater than the wavelength of the polarization mode of interest. The essential point is that, for a finite crystal of dimensions l, there is a region

$$\frac{2\pi}{l} < \mathbf{k} \leqslant 10^4 \text{ cm}^{-1}$$

where the Huang theory clearly indicates that the polarization modes do not obey the LST relation.

Casselman, Mitra, and Spector [42] have pointed out that, if neutron scattering data could be taken with sufficient accuracy in this very low range of $\mathbf{k}$, departures from the LST relation would be found. They have also considered the absorption and reflection of infrared radiation by ionic crystals. The infrared absorption is maximum at a frequency of ω_0. A wave inside the crystal having this frequency will have a wave number given by $|\mathbf{k}| = \omega_0 n(\omega_0)/c$ where $n(\omega_0)$ is the real part of the index of refraction, $\sqrt{\varepsilon}$. In NaCl, e.g., $\omega_0 = 3 \times 10^{13}$ sec^{-1} and the measured value of $n(\omega_0)$ is 7. Therefore, inside the crystal, the wave has a wave number $7(\omega_0/c)$, which yields $x = 7$ for the dimensionless abscissa of Fig. 16. This puts the infrared radiation of frequency ω_0 in region C, where the transverse phonon of the same wave number also has frequency ω_0. The maximum absorption of infrared occurs at ω_0 because, at this frequency, the infrared photon can be resonantly absorbed and an optical phonon of the same frequency emitted. As the infrared frequency ω is increased from ω_0, the theory predicts that incoming infrared radiation will be strongly reflected. This reflection will not be perfect as there are some damping effects which have not been taken account of in the theory. Experimentally, the frequency ω_l, at which the reflection coefficient has its minimum, is sometimes identified as the LO phonon frequency. In fact, it is the limiting TO phonon frequency as $\mathbf{k} \to 0$ (i.e., the limiting frequency of the upper transverse branch in Fig. 16). This can be seen because the incoming radiation has only transverse polarization and therefore cannot interact with longitudinal modes except in thin films [37]. Also, if we take the experimental value of the index of refraction at ω_l, $n(\omega_l)$, we see that the wave number of the incoming radiation inside the crystal falls close to $\mathbf{k} = 0$ and it is the upper transverse branch of the optical phonon modes which are interacting with the electromagnetic radiation at this frequency. Thus, the infrared reflection measurements are consistent with the theory of Huang in the long-wavelength region as long as the frequency ω_l is interpreted as the frequency of the TO phonon mode as $\mathbf{k} \to 0$.

Recently, Henry and Hopfield [43] have investigated the Raman scattering by polaritons. They have experimentally mapped the β and γ branches (Fig. 16) for GaP by a forward-scattering experiment in which the angle between the scattered Stokes radiation and incident laser beam (used for excitation) was varied from 0 to 7°.

GaP belongs to the zinc-blende structure, in which the TO is both infrared and Raman active and the LO is also Raman active at $\mathbf{k} \sim 0$. Figure 17 shows the scattering geometry used by Henry and Hopfield, where $\mathbf{k}_L$, $\mathbf{k}_S$, and $\mathbf{q}$ are the wave vectors of the incident laser photon, scattered Stokes photon, and polariton inside the crystal. In the usual perpendicular-mode Raman scattering, $\mathbf{q} \sim \sqrt{2}\,\mathbf{k}_L$, and, thus, the polariton is very phononlike. However, for $\theta < 3°$ inside the crystal, the energy of the Stokes-Raman component of the scattered radiation shifts to energies significantly below the TO energy.

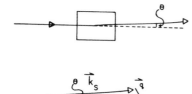

Fig. 17. Scattering geometry for photon–phonon mixed modes. Notation: k_L and k_S are the wave vectors of the laser light and the Stokes shifted light, respectively; q is the wave vector of the scattered polariton; and θ is the scattering angle.

The polariton energy $\hbar\omega_q$ is obtained from the conservation of energy

$$\hbar\omega_q = \hbar\omega_L - \hbar\omega_S \qquad (14.49)$$

where ω_L and ω_S are the frequencies of the laser and Raman-scattered radiation. Conservation of wave vectors similarly determines the polariton wave vector. With the aid of Fig. 17, one may write

$$|\mathbf{q}| = (k_L^2 + k_S^2 - 2k_L k_S \cos\theta)^{\frac{1}{2}}$$
$$\simeq \left[\left(\frac{\partial k}{\partial\omega}\right)_{\omega=\omega_L}^2 \omega_q^2 + k_L k_S \theta^2\right]^{\frac{1}{2}} \qquad (14.50)$$

where $k_L = n(\omega_L)\,\omega_L/c$ and $k_S = n(\omega_S)\omega_S/c$. The experimental data of Henry and Hopfield on GaP are compared with the theoretical dispersion curve constructed from the infrared data of Kleinman and Spitzer [44] in Fig. 18. It may be noted that a 20% shift in the polariton frequency in GaP occurs as the scattering angle is varied from the near-forward to forward direction.

The shift of the polariton frequency with θ is much more dramatic for certain uniaxial crystals [45]. The frequency of the polariton that conserves energy and wave vector can be made small if one makes the incident light an ordinary ray and the Stokes–Raman light an extraordinary ray with the maximum index. The polariton frequency increases as θ increases. Porto, Tell, and Damen [46] have found that the long-wavelength TO phonon of ZnO that occurs at $407\ \mathrm{cm}^{-1}$ (e.g., in a perpendicular-scattering mode) becomes a polariton resonance at $160\ \mathrm{cm}^{-1}$ in the forward-scattering mode.

F. Frequency-Dependent Damping and Dielectric Dispersion Theories

According to the classical dispersion theory discussed in Section III, A and B, it is assumed that the dielectric constant of an ionic crystal may be represented by

$$\varepsilon = \varepsilon_\infty + \sum_{j=1}^{n} \frac{4\pi\rho_j}{1 - (v/v_j)^2 - i(\gamma_j/v_j)(v/v_j)} \qquad (14.36')$$

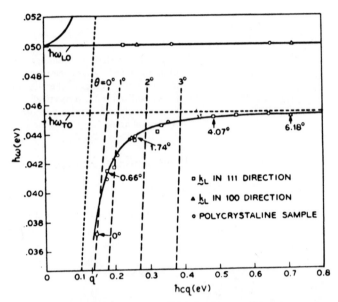

Fig. 18. Observed polariton and LO phonon dispersion curves.
○, □, and △, experimental points; solid lines, theoretical dispersion
curves. (Reprinted from C. H. Henry and J. J. Hopfield [43].)

where $\varepsilon_0 = \varepsilon_\infty + \sum_{j=1}^{n} 4\pi\rho_j$. The sum is intended to include all resonances
necessary to represent the dielectric constant. This equation is basically
phenomenological and is useful because of its simple form and because it
represents experimental data extremely well.

A proper understanding of the dispersion of far-infrared radiation in
ionic dielectrics can only be achieved by treating the problem quantum
mechanically. The expression for the electric-susceptibility tensor

$$\chi_{ij} = \frac{1}{4\pi}(\varepsilon_{ij} - \delta_{ij})$$

varies in detail depending both upon the choice of the Hamiltonian representing the crystal and the incident electromagnetic field and upon the particular approximate method used for obtaining a solution. Born and
Huang [24], Maradudin and Wallis [47], Mitskevich [48], and Neuberger and
Hatcher [49] among others have obtained the optical constants of a dielectric
crystal in this manner. These results, although qualitatively similar in
essence to the classical one, differ in detail from it and also among one
another.

In the theory of Born and Huang, the Hamiltonian contains quadratic and cubic terms in the crystalline potential and an interaction term through which the electric field couples to the dipole moment of the vibrating lattice. The electric susceptibility obtained by them contained a damping factor γ which was a function of both frequency and temperature. When it is evaluated at the resonance frequency v_0, one finds

$$\text{Im } \chi(v, T)|_{v=v_0} \sim [\bar{\gamma}(v_0, T)]^{-1} \tag{14.51}$$

Here, the susceptibility tensor reduces to a scalar for cubic crystals and is dominated by its imaginary part. In equation (14.51), $\bar{\gamma}$ represents an effective or averaged damping constant. Born and Huang have also shown that, in the high-temperature limit ($kT > hv_0$), the imaginary part of χ decreases inversely as T^3 in the center of the dispersion region.

The Hamiltonian used by Mitskevich contains several Born–Mayer type of potentials with adjustable parameters. This treatment includes contributions from electric multipole moments in addition to those from the third- and fourth-order mechanical anharmonic terms. Mitskevich obtains the following expression for the dielectric constant of a cubic ionic crystal:

$$\varepsilon(\omega, T) = \varepsilon_\infty + \frac{(\varepsilon_0 - \varepsilon_\infty)\,\Omega_0{}^2}{\Omega_\omega{}^2 - \omega^2 + i\gamma(\omega, T)\,\omega} \tag{14.52}$$

where Ω_ω is the dispersion frequency; $\gamma(\omega, T)$ is the damping constant and Ω_0 is a constant. The form of ε is close to the classical expression (14.36) except that the damping constant is now a function of both frequency and temperature. Wehner [50] has recently obtained the expression (14.52) by the powerful Green's function technique. He has also used a frequency-dependent damping constant in the analysis of the reststrahlen spectrum of LiF. It is worth noting that the peaks in the γ *versus* ω curves of Wehner actually correspond to the various resonances assumed in the application of equation (14.36') in a classical, frequency-independent damping-constant analysis of the same reststrahlen spectrum performed by Jasperse, Kahan, Plendl, and Mitra [51].

Mitskevich has computed the cubic and fourth-order contributions of the crystalline potential to γ in the high-temperature limit and obtained

$$\gamma_3(\omega, T) \sim \frac{T}{\omega_0{}^4}\,\phi_3(\omega) \tag{14.53}$$

and

$$\gamma_4(\omega, T) \sim \frac{T^2}{\omega_0{}^6}\,\phi_4(\omega) \tag{14.54}$$

The $\phi_3(\omega)$ and $\phi_4(\omega)$ are functions which depend on sums over the density of phonon states.

Maradudin and Wallis have treated the problem in a rigorous manner. They have used the formalism of quantum-field theory and the density-matrix approach in treating quantum statistical aspects of the problem. They develop the operator equations of motion of a suitably transformed set of normal coordinates, and the associated equations for the damping constant. Approximate solutions to these operator equations are obtained, and the general formalism of Kubo [52] is used to calculate the electric-susceptibility tensor. They obtain

$$\chi_{lm} = (2V_0)^{-1} \sum_j \frac{M_{lj}M_{mj}}{\omega(0j)} \left[\frac{1}{\omega + \omega(0j) + \Delta\omega(0j) + i\gamma(0j)} \right.$$

$$\left. - \frac{1}{\omega - \omega(0j) - \Delta\omega(0j) - i\gamma(0j)} \right] \quad (14.55)$$

where the γ is given by

$$\gamma(0j) = \frac{h}{4} \sum_{k'j'k''j''} \frac{|V(0j, k'j, k''j'')|^2}{\omega(0j)\,|\omega(k'j')|\,\omega(k''j'')}$$

$$\times (n_{k'j'} + \tfrac{1}{2})\,\delta[\omega - \omega(k'j') - \omega(k''j'')] \quad (14.56)$$

and the occupation number $n_{k'j'}$ for phonons (a boson) is given by

$$n_{k'j'} = \left\{ \exp\left[\frac{\hbar\,|\omega(k'j')|}{kT} \right] - 1 \right\}^{-1} \quad (14.57)$$

Here, j refers to the jth optical branch; $\mathbf{k}$ is a wave vector; the V's are the cubic parts of the Hamiltonian when it is expressed in terms of the normal coordinates; and the $\Delta\omega$'s are the frequency shifts. The form of γ shows that the phonons are coupled through the cubic terms in the Hamiltonian, which produces damping of the main resonance. A complete calculation of γ as a function of ω would require the knowledge of the phonon dispersion relations.

Although the expressions for the electric susceptibility obtained from the classical or the various quantum mechanical theories differ from each other in detail, at $v = v_0$ one obtains essentially the same result. At the center of the dispersion region ($v = v_0$) the Maradudin–Wallis result reduces to

$$\chi|_{v=v_0} \sim v_0^{-1}(T) \left[\frac{1}{2v_0(T) + i\gamma} - \frac{i}{\gamma} \right] \quad (14.58)$$

Since $v_0 \gg \gamma(v, T)$, we find that the imaginary part of χ is the same as that given by equation (14.51) obtained from the Born and Huang theory to within a phase factor. At $v = v_0$, the classical damped oscillator yields

$$1 + 4\pi\chi|_{v=v_0} = \varepsilon_\infty + i \left(\frac{4\pi\rho_0}{v_0} \right) \gamma^{-1} \tag{14.59}$$

We note that all the representations of χ give essentially the same results when evaluated at $v = v_0$. This is significant. It is thus valid to use the classical dispersion theory in the vicinity of $v = v_0$. In fact, if a quantitative estimate is made, it is found that the $\mathrm{Im}\chi$ from various theories agree within 1% when evaluated at $v = v_0$, although they may differ from one another in the wings of the fundamental lattice absorption. Thus, it may not be valid to draw conclusions about the behavior of the optical constants derived from a classical damped-oscillator fit of reststrahelen spectrum in regions away from $v = v_0$.

G. Analysis of Reflection Spectrum Due to Internal Modes by Kramers–Kronig Dispersion Relations

Certain infrared-active internal fundamentals of some crystals may also be very intense. This makes it necessary to use extremely thin ($\sim 1\,\mu$) samples in order to obtain satisfactory transmission data. With available techniques, however, it is well-nigh impossible to prepare single crystals of this thickness in various orientations for many crystals. Auxiliary information on the absorption in such situations may be available from the reflectivity, which is necessarily large because of anomalous dispersion associated with the large extinction coefficient. It is possible to relate the observed reflection spectrum to the optical constants near the dispersion frequency by means of Maxwell's equations.

The maximum reflection associated with an infrared-active internal mode seldom exceeds 50% and is usually within 20%. The method described in the earlier section for the analysis of the lattice reststrahlen band is therefore not suitable for the determination of the optical constants in the vicinity of an internal mode frequency of a molecular crystal. In such cases, n and κ may be obtained from the measurements of reflection spectra at two widely separated angles of incidence. Šimon [53] has reviewed the experimental procedure and the method of calculation. This method, however, works only for isotropic solids. Since the extinction coefficient is a function of the angle of incidence for anisotropic crystals, this method is rendered unsuitable. Robinson and Price [54] have modified Šimon's method to permit the evaluation of the optical constants in the neighborhood of a strong absorption band from the measurement at normal incidence.

Thus, n and $\varkappa$ are given by

$$n = \frac{1-r^2}{1+r^2-2r\cos\theta} \qquad \varkappa = \frac{-2r\sin\theta}{1+r^2-2r\cos\theta} \tag{14.60}$$

The quantity $R = r^2$ is the experimentally observed reflectivity. The phase difference θ between the incident and the reflected waves is obtained from the Kramers–Kronig [55] dispersion relations

$$\theta_c = \frac{1}{\pi}\int_0^\infty \frac{d\ln r}{dv}\ln\left|\frac{v+v_c}{v-v_c}\right|dv \tag{14.61}$$

Or in its equivalent form

$$\theta_c = 2\frac{v_c}{\pi}\int_0^\infty \frac{\ln r(v)-\ln r(v_c)}{v^2-v_c^2} \tag{14.62}$$

Here, θ_c is the value of the phase difference at a frequency v_c. The integral is evaluated numerically. The limits of integration are, in practice, the two extremities of the band where the reflection approaches constant values.

The infrared transmission in the hydroxyl stretching region of single-crystal brucite, $Mg(OH)_2$, is very low for certain orientation of the crystal and polarization of incident radiation. This is especially true for the absorption near $3700\ cm^{-1}$ for radiation polarized parallel to the c axis. The absorption is too high, which makes the accurate determination of its band

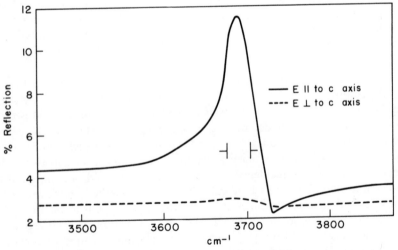

Fig. 19. Reflection spectrum of brucite from 3450 to 3850 cm⁻¹. (Reprinted by permission from S. S. Mitra [1].)

position very difficult. The associated high reflectivity in this region is shown in Fig. 19 for near normal incidence on a plane cut perpendicular to the cleavage plane. The only reflection band has a maximum at 3690 cm^{-1}. It is completely dichroic, with the infrared transition moment vector oriented along the c axis of the crystal. The optical constants obtained from an analysis of this band are given as functions of frequency in Fig. 20. The peak of the extinction coefficient curve occurs at 3705 cm^{-1}, and its value is 0.57. Thus, for a sample of 1-μ thickness, the maximum absorption at this frequency will be 94%, which explains the difficulty encountered in the measurement of single-crystal transmission spectrum. The transmission spectrum of a powdered sample gives a value of 3700 cm^{-1} for the position of this absorption band, in excellent agreement with the value obtained from the analysis of the reflection spectrum.

The Robinson and Price method of analysis is rigorously applicable only to crystals of orthorhombic or higher symmetry. Under certain conditions, however, the method may also be applicable to crystals of lower symmetry. This method may give erroneous results for ferroelectric solids. The use and applicability of the Robinson–Price method has recently

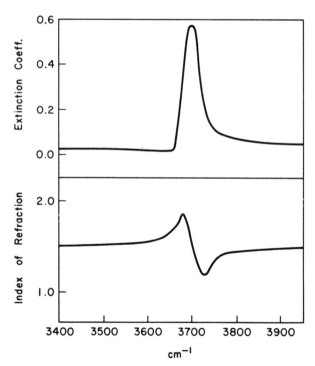

Fig. 20.   Optical constants of brucite in the region 3400–3900 cm^{-1}. (Reprinted by permission from S. S. Mitra [1].)

been reviewed and critically examined by Schatz, Maeda, and Kozima [56] and by Bowlden and Wilmshurst [57]. It is worth mentioning here that, for the analysis of lattice infrared spectrum of ionic solids, it is often advantageous to perform both Kramers–Kronig analysis and a damped-oscillator-model calculation.

H. Attenuated Total Reflection

It has been noted in the foregoing that the reflection spectral measurements are, in general, useful for two reasons. Firstly, the analysis of the reflection spectrum affords a means of obtaining the optical constants as a function of frequency, and, secondly, it permits spectral measurements on an otherwise intractable sample such as a highly absorbing crystal for which it is impractical to prepare a specimen of required thinness. The reflection technique, however, is limited to only the regions for which a substance has a high value of the extinction coefficient. For example, crystals with extinction maxima in the range of zero to 0.2 will show reflection spectrum of low overall intensity. They will be devoid of contrast, which renders the determination of the optical constants with any accuracy impossible. Since a large number of absorption bands associated with the vibrational modes of a great number of organic and nonpolar solids are weak, the ordinary reflection technique gives very little useful information for such materials.

The above difficulties can be overcome by a novel reflection technique developed by Fahrenfort [58], known as the *attenuated total reflection* (ATR). This technique utilizes the interface between a dielectric of high refractive index and the specimen as the reflecting surface, instead of that between air and the sample as is used in the conventional technique. The refractive index n at the interface will thus be less than 1 in the new method. If the angle of incidence θ and the dielectric are selected in such a way that $n \leqslant \sin \theta$, total reflection ensues for wavelengths where the sample is non-absorbing. If, for the second medium however, $\kappa \neq 0$, part of the incident radiation will be absorbed by the surface layers, which will reducing the reflected energy. It may be shown that significant reflection attenuation takes place even for very small values of the extinction coefficient. The attenuated reflection spectrum is in many ways similar to a transmission spectrum because, with increasing extinction coefficient, the attenuation increases progressively and thus decreases the light intensity leaving the sample. Optical constants may also be determined from such spectra with a slightly different Kramers–Kronig analysis than described earlier.

Single crystals of KRS5, AgCl, or Ge seem to be suitable dielectrics for ATR measurements. These crystals are transparent in different regions of the infrared radiation and have refractive indices between 2 and 4 over a considerable part of the region. Thus, they may be used for producing inter-

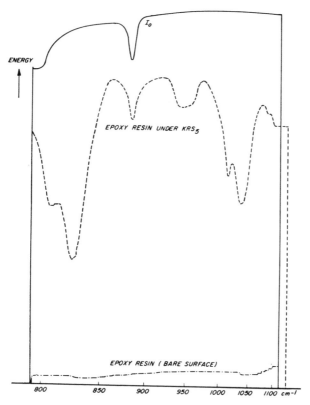

Fig. 21. Comparison between conventional reflection spectrum and ATR spectrum of a solid epoxy resin. (Reprinted from J. Fahrenfort [58].)

faces of $n < 1$ with any absorbing specimen with refractive index less than 2 Figure 21 shows the comparison between the conventional reflection spectrum and the ATR spectrum of a solid epoxy resin.

I. Emission Spectrum in the Infrared

Electromagnetic radiation incident on a solid is partly reflected, partly absorbed, and the remainder transmitted. If all the radiation is absorbed, the solid is termed a *black body*. If the solid is at a higher temperature than the environment, it also possesses an emission spectrum. The thermal radiation emitted by a black body follows Planck's law. The emission from a solid which is not a perfect absorber is less than that of an ideal black body at the same temperature. The two quantities are related by Kirchhoff's law.

For partially absorbing solids, the emittance is related to the reflectance and the transmittance. By considering multiple reflections in a plane-parallel sample, McMahon [59] has derived expressions for the apparent reflectivity, the apparent transmissivity, and the emissivity E in terms of the true reflectivity and the true transmittance. The true reflectivity is the fraction of incident radiation that is reflected from the first surface and is given by the Fresnel formula [equation (14.33)]. The true transmittance is the fraction of light entering the solid that reaches the second surface, i.e., without any internal reflection, and is given by

$$T = e^{-\alpha t} \tag{14.63}$$

where $\alpha = (4\pi\kappa)/\lambda$ is the absorption coefficient, and t is the thickness of the sample. McMahon's relationships for the apparent or the observed quantities are

$$R^* = R\left[1 + \frac{T^2(1-R)^2}{1-R^2\,T^2}\right] \tag{14.64}$$

$$T^* = T\,\frac{(1-R)^2}{1-R^2\,T^2} \tag{14.65}$$

and

$$E = \frac{(1-R)\,(1-T)}{1-RT} \tag{14.66}$$

By adding equation (14.64) to (14.66), one obtains

$$R^* + T^* + E = 1 \tag{14.67}$$

Haas [60] has measured R^*, T^*, and E as functions of frequency for $CaCO_3$ and $NaNO_3$ and has found that equation (14.67) is strictly obeyed. At first glance, it appears from equation (14.67) that no new information is available from emission measurements which may not be obtained from reflection or transmission measurements. However, a closer scrutiny reveals that, for certain spectral regions, emissivity measurements can be very effective in acquiring optical data which are difficult to obtain by transmission or reflection measurements. Substitution of equation (14.63) in equation (14.66) gives

$$E = \frac{(1-R)\,(1-e^{-\alpha t})}{1-Re^{-\alpha t}} \tag{14.68}$$

in the case of an almost transparent sample with very low absorption coefficient, $\alpha t \ll 1$, and equation (14.68) reduces to

$$E \simeq \alpha t \tag{14.69}$$

The direct measurement of the emittance thus offers a distinct advantage over the reflectance and transmittance measurements. For low values of αt, the emissivity measurement is a sensitive method of determining α. Using a nominal sample size, Stierwalt and Potter [61] have measured extremely low absorption coefficients of Si, Ge, and CdS. The determination of the absorption coefficient of these materials by transmission method would have required very thick samples in the regions where the absorption coefficient is as low as $0.2\ \mathrm{cm}^{-1}$. The emission technique therefore seems eminently suitable for the study of absorptance in the overtone and the combination regions. Figure 22 gives the spectral emittance of a 1.7-mm-thick single crystal of silicon at several temperatures.

For almost opaque solids, on the other hand, the absorption coefficient and hence αt for nominal sample thickness are large, so that $T \simeq 0$ and $R^* \simeq R$. The emissivity is therefore given by

$$E = 1 - R \qquad (14.70)$$

and the direct measurement of E permits an accurate determination of small changes of the reflectance.

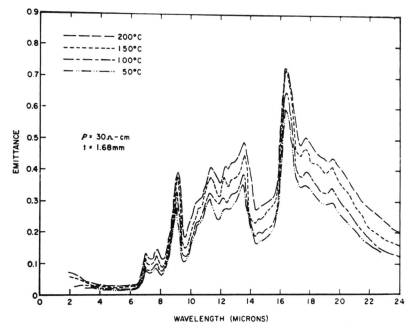

Fig. 22. Spectral emittance of single-crystal silicon as a function of temperature. (Reprinted by permission from D. L. Stierwalt and R. F. Potter [61].)

IV. RAMAN SCATTERING

The inelastic scattering of photons by phonons is known as the *Raman effect* in solids. The energy lost (Stokes) or gained (anti-Stokes) by the photon in such a process is accomplished by the creation or annihilation of a phonon. Brillouin scattering is a special case of Raman scattering, involving the low-frequency acoustic phonons.

The incident electromagnetic field couples with the phonon field through induced dipole moment. This is accomplished by the variation of the electronic polarizability tensor α_{lm} with the lattice configuration during a normal vibration. The tensor α_{lm} may be expanded as

$$\alpha_{lm} = \alpha_{lm}^{(0)} + \alpha_{lm}^{(1)} u + \alpha_{lm}^{(2)} u^2 + \ldots \tag{14.71}$$

where u is a nuclear displacement during a normal vibration. Thus, for the jth normal mode,

$$u_j = u_{j0}\, e^{i\omega_j t} \tag{14.72}$$

and

$$\alpha_{kl}^{(1)} = \alpha_{kl,j} = \left(\frac{\partial \alpha_{kl}}{\partial u_j}\right)_{u=0} \qquad \text{etc.} \tag{14.73}$$

If

$$E = E_0\, e^{i\omega t} \tag{14.74}$$

denotes the incident electric field of the radiation of frequency ω, the induced dipole moment

$$\mathbf{M} = \alpha \mathbf{E} \tag{14.75}$$

becomes, with the aid of equations (14.71), (14.72), and (14.74),

$$M = \alpha^{(0)} E_0\, e^{i\omega t} + \alpha^{(1)} u_{j0} E_0\, e^{i(\omega \pm \omega_j)t} + \alpha^{(2)} u_{j0}^2 E_0\, e^{i(\omega \pm 2\omega_j)t} \ldots \tag{14.76}$$

The first term of equation (14.76) gives rise to Raleigh scattering in which the frequency of the radiation remains unchanged during the scattering process. The second term, involving the derivative of the electronic polarizability, constitutes the first-order Raman scattering. The incident photon $(\omega, \mathbf{k})$ is absorbed, and the system makes a transition from an initial state (n, v), where n and v represent the electronic and vibrational quantum numbers, to an intermediate state (n', v'). The system eventually retreats to the final state $(n, v \pm 1)$. In the process, a phonon $(\omega_j, \mathbf{q})$ is created or destroyed and a photon of different frequency and wave vector $(\omega_s, \mathbf{k}_s)$ is emitted such that the energy and the momentum are conserved

$$\omega = \omega_s \pm \omega_j \tag{14.77}$$

$$\mathbf{k} = \mathbf{k}_s \pm \mathbf{q}_j \tag{14.78}$$

A. Scattering Angle and Phonon Wave Vector

The frequency of the exciting radiation is usually chosen in a region where there is negligible dispersion, i.e., in a region where $\omega \gg \omega_j$. Thus, the relative difference between ω and ω_s is small. Therefore, to a good approximation $|\mathbf{k}| \simeq |\mathbf{k}_s|$. For the scattering geometry of Fig. 23, the conservation of momentum reduces to the Bragg condition

$$q = 2k \sin \frac{\phi}{2} = \frac{2\omega n(\omega)}{c} \sin \frac{\phi}{2} \qquad (14.79)$$

where Φ is the angle between the incident and scattering directions; and $n(\omega)$, the refractive index of the medium at ω. Thus, the wave vector of the phonon involved in the scattering process depends on the angle $\varnothing$. Depending on the nature of the problem, three different scattering geometries are generally employed: (1) forward scattering with $\phi = 0$, (2) perpendicular scattering with $\phi = 90°$, and (3) backscattering with $\phi = 180°$. The phonon wave vector involved in each case is, respectively, $q \simeq 0$, $q \simeq \sqrt{2}\,k$ and $q \simeq 2k$. For visible and ultraviolet light usually employed for exciting Raman effect, k is between 10^4 and 10^5 cm^{-1}, which is about 10^{-4} to 10^{-3} of q_{max}. Thus in the first-order Raman or Brillouin scattering, only the phonons from near the center of the first Brillouin zone are involved.

For a crystal of diamond structure, the long-wavelength optic mode is triply degenerate and is active in the first-order Raman effect. Since the frequency *versus* wave-vector curve for this optic phonon is essentially flat near $\mathbf{q} = 0$, one would not expect any shift of the Raman frequency of the diamond (1332 cm^{-1}) as a function of ϕ. However, this is not the case with ionic crystals that exhibit first-order Raman spectra (e.g., zinc-blende, wurtzite, or fluorite structures). Such an angle dependence of the Raman frequency of GaP has already been discussed under E of Section III. The

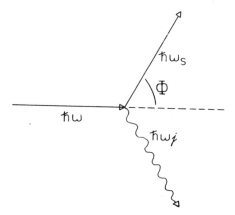

Fig. 23. Raman scattering geometry.

acoustic mode frequencies vary linearly with $\mathbf{q}$ near $\mathbf{q} \simeq 0$. Thus, Brillouin scattering depends strongly on the scattering angle.

B. Intensity of First-Order Raman Processes

The intensity of a Raman band can be defined in terms of the scattering efficiency S defined as the ratio of the number of scattered phonons produced per ũnit of time per unit of cross-sectional area of the crystal in solid angle $d\Omega$ about the direction of observation to the number of incident photons crossing the unit area in the unit time. For right-angle scattering geometry and unpolarized light, Smith [62] finds

$$S = \frac{3\hbar\,\omega_s^4\,L\,d\Omega}{\rho c^4\,\omega_j}\,|\alpha^{(1)}|^2\,(n_j+1) \tag{14.80}$$

for the Stokes component. For the anti-Stokes component, the last term is n_j instead of (n_j+1). Here L is the length of the crystal in the incident direction; ρ, its density; and n_j, the phonon occupation number as given in equation (14.57). An examination of equation (14.80) reveals a number of interesting points: (1) The Raman band intensity strongly depends (fourth power) on the frequency of the exciting light; (2) the closer the Raman line is to the exciting line, the stronger it is; and (3) unlike the infrared spectrum, the first-order Raman intensity has a temperature dependence through the terms (n_j+1) and n_j for the Stokes and anti-Stokes components, respectively. This is so because the first-order Raman scattering is really a second-order (three-particle) process, as can be seen from Fig. 23.

As in the case of infrared absorption, the polarization of radiation and the orientation of crystal samples affect the intensity of Raman lines. The situation is more complicated in this case because the incident and scattered lights may be differently polarized. However, the study of the intensity of Raman lines in varying polarization and sample orientations is a valuable guide when assigning the observed frequencies to specific modes of oscillation. The polarization of the scattered radiation is often expressed as the degree of depolarization ρ. The following example will elucidate the significance of ρ.

In the case of highly symmetric molecules such as methane, the polarizability ellipsoid is a sphere and the induced electric moment is developed only along the direction of the electric field. When a gas consisting of such molecules is irradiated by polarized or unpolarized light and the scattered radiation is examined at right angles to the direction of incidence, it is found to be completely polarized perpendicular to the scattering plane containing the incident and scattered beams. However, in the general case of a nonspherical polarizability ellipsoid, $\mathbf{M}$ and $\mathbf{E}$ have different directions except along the axes of the ellipsoid where they coincide. For such a solid, M is no longer restricted to the plane at right angles to the beam but is distributed in all directions with varying probability. Consequently, the scattered light observed perpendicular to the direction of incidence is no longer totally polarized. The deviation from the complete polarization of the scattered beam is quantitatively measured as the degree of depolarization. It is defined in terms of scattering efficiencies as

$$\rho = \frac{S_{\parallel}}{S_{\perp}}$$ (14.81)

where the $\parallel$ and $\perp$ signs refer to polarizations of the scattered beam with respect to the plane of scattering.

The degree of depolarization is obviously a function of the scattering angle and the orientation of the crystal. For a uniaxial crystal, Nedungadi[63] has shown that 27 different types of Raman spectra may be obtained with transverse scattering ($\phi = 90°$), of which only 12 are relatively important. Using some simplifying assumptions regarding the components of the polarizability tensor, Bhagavantam [64] has calculated the intensity of certain Raman-active modes of calcite for the 12 different cases arising from the variation of orientation and polarization. For both transverse and longitudinal scattering, Saksena [65] has given explicit results for the depolarization and the intensities of the different Raman-active modes of crystals belonging to the various symmetry classes, with the restriction that light propagates along the principal axes of a crystal. Chandrasekharan [66] has obtained Raman scattering matrices for cubic crystals in the most general configuration. For the calculation of Raman scattering efficiencies and extensive tables of the symmetric tensors of the polarization quotients, $\alpha^{(1)}$, needed for the evaluation of ρ, the reader is referred to the review by Loudon [67].

The intensity and depolarization of a Raman line measured by conventional excitation (by helical Toronto arcs) and detection systems are quite often different from the predicted values because of the modifications caused by convergence error, optical density of the sample, reflection loss, spectral sensitivity of the photomultiplier tube, and other factors. With the advent of the laser-light-excited Raman spectroscopy and detection by phase-sensitive or photon-counting methods, many of these errors can now be completely eliminated. Furthermore, some of the scattering configurations unobtainable by conventional excitation methods can now be easily achieved by using laser beams. Porto and co-workers [68] have recently demonstrated the ease and versatility of laser Raman spectroscopy in the analysis of vibrational spectra of solids.

V. PRESSURE AND TEMPERATURE DEPENDENCE OF LONG-WAVELENGTH OPTICAL PHONONS

The infrared absorption spectrum of a diatomic ionic cubic crystal such as NaCl or ZnS is expected, from elementary theory, to consist of a single line associated with the optical lattice mode of an essentially zero-propagation vector. For diatomic homopolar crystals such as diamond or silicon, this factor group fundamental should be a single line active in the Raman scattering only. The resonance absorption in reality, however, is not infinitely sharp but shows the natural line width due to radiation damping. The excited state may also, by some coupling mechanism, transform into a state in which two or more different phonons replace the optical one. This process also limits the lifetime of the optical phonon, which causes it to broaden. Since the probability of emission or absorption of a phonon is temperature dependent, the widths of the long-wavelength optical phonons are also temperature dependent. The one-phonon peaks due to fundamental infrared absorption or due to first-order Raman scattering also shift with temperature and pressure. Since the change in the width and the position of the fundamental infrared and Raman bands due to the effect of temperature and pressure are the same as the phonon width and shift, such measurements provide important information regarding the dynamics of the lattice. Similar information may also be available from neutron scattering and vibronic fine structure measurements.

A. Pressure and Temperature Dependence of One-Phonon Optic Frequencies

When the anharmonic contributions are included, the vibrational potential energy of a solid has the form

$$V = \sum_i^{\mathbf{q}} f_j^{\mathbf{q}}(Q_j^{\mathbf{q}})^2 + \sum_{ijk}^{\mathbf{q}_1\mathbf{q}_2\mathbf{q}_3} b_{ijk}^{\mathbf{q}_1\mathbf{q}_2\mathbf{q}_3} Q_i^{\mathbf{q}_1} Q_j^{\mathbf{q}_2} Q_k^{\mathbf{q}_3} +$$

$$+ \sum_{ijkl}^{\mathbf{q}_1\mathbf{q}_2\mathbf{q}_3\mathbf{q}_4} c_{ijkl}^{\mathbf{q}_1\mathbf{q}_2\mathbf{q}_3\mathbf{q}_4} Q_i^{\mathbf{q}_1} Q_j^{\mathbf{q}_2} Q_k^{\mathbf{q}_3} Q_l^{\mathbf{q}_4} + \cdots \qquad (14.82)$$

Here, f, b, c, etc., are harmonic, cubic, quartic, etc., force constants in suitable units, and Q_i is a normal coordinate. The superscripts refer to wave vectors, and the subscripts refer to branches.

The harmonic part (the first term) of the potential can be solved exactly, and the cubic and higher-order terms may be treated as perturbations. The unperturbed solution of the wave equation may be denoted by $\psi^0(\mathbf{n})$, where

$$\mathbf{n} = n_1^{\mathbf{q}_1}, n_2^{\mathbf{q}_2}, \ldots \qquad (14.83)$$

where the occupation number n_i^q is given by equation (14.57). The eigenvalues are given by

$$E^0 = \sum_{i,q} (n_i^q + \tfrac{1}{2}) \hbar \omega_i(\mathbf{q}) \tag{14.84}$$

where the eigenfrequencies are

$$\omega_i^2(\mathbf{q}) = 2f_i(\mathbf{q}) \tag{14.85}$$

If the perturbation by the higher-order terms is small compared with the harmonic term (which is usually the case), it may be shown that, to a good approximation, the eigenvalue of the perturbed state $\psi'(\mathbf{n})$ may still be expressed as a sum of oscillator energies

$$E' = \sum_{i,q} (n_i^q + \tfrac{1}{2}) \hbar \omega_i'(\mathbf{q}) \tag{14.86}$$

where the "quasi-normal mode" frequencies $\omega'(\mathbf{q})$ are different from the harmonic oscillator frequencies. The shift thus depends on the anharmonic-potential constants and the occupation number $\mathbf{n}$

$$\omega_i'(\mathbf{q}) = \omega_i(\mathbf{q}) + F(b_{ijk}^{q_1 q_2 q_3}, c_{ijkl}^{q_1 q_2 q_3 q_4}, n) \tag{14.87}$$

Since ω' depends on $\mathbf{n}$, it implies that

$$\omega_i'(\mathbf{q}) = f(T) \tag{14.88}$$

where T denotes temperature.

The change of the one-phonon frequency with temperature is, however, not entirely due to the anharmonic coupling of the phonon in question with other phonons but is also due to the thermal expansion of the crystal. Although the phenomenon of thermal expansion itself occurs as a result of anharmonicity, it affects the "quasi-harmonic" frequencies through the harmonic-force constant $f_i(\mathbf{q})$, which may change with volume. The purely anharmonic contribution to the shift, known as the *self-energy* shift, will be explicitly manifested if spectra are taken as functions of T at constant volume, instead of at constant pressure.

1. *Pressure Dependence of Long-Wavelength Optical Phonons in Ionic Crystals*

Since the harmonic force constant may be a function of volume, there is a volume dependence of the lattice frequencies over and above the anharmonic effects which are manifested through change of temperatures. The volume dependence of $\omega_j(\mathbf{q})$ is given by

$$\gamma_j(\mathbf{q}) = - \frac{d \ln \omega_j(\mathbf{q})}{d \ln V} \tag{14.89}$$

Within the limits of a quasi-harmonic oscillator model, the Grüneisen parameter $\gamma_j(\mathbf{q})$ is expected to be a constant. The Grüneisen parameter γ_j for the long-wavelength optic modes may be experimentally determined from a study of the pressure dependence of fundamental infrared absorption in ionic crystals at constant temperature to avoid the "self-energy" shift.

Mitra, Postmus, and Ferraro [69] have recently measured the pressure dependence of the $\mathbf{q} \sim 0$ optical phonon frequencies of a large number of ionic crystals. Their results for a few selected crystals are shown in Fig. 24. The Grüneisen parameter may be determined by using the appropriate $P - V$ data. Plots of $\ln \nu$ versus $\ln V$ are shown in Fig. 25. For NaF, slight deviation from linearity at low pressures may be noted, which may be indicative of a P dependence of $\gamma_j(\mathbf{q})$. The Grüneisen parameters obtained from the slopes of these lines are given in Table V. One may also obtain an approximate value of γ from isothermal compressibility χ, using the relation

$$\gamma_j(\mathbf{q}) = \frac{1}{\chi \omega_j(\mathbf{q})} \left[\frac{\partial \omega_j(\mathbf{q})}{\partial P} \right]_T \tag{14.90}$$

which is especially suitable when extensive $P - V$ data are not available. The γ's obtained from relation (14.90) are also given in Table V.

An estimate of γ is possible from the Born and Huang theory of the long-wavelength optical modes in ionic crystals. The expression for the $\mathbf{k} \sim 0$ TO mode is given [70] by

$$-\omega_t^2 = -\frac{f}{\mu} + \frac{(4\pi/3)\,(e^2/\mu V_a)}{1 - (4\pi/3)\,(\alpha_+ + \alpha_-/V_a)} \tag{14.91}$$

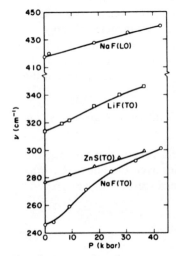

Fig. 24. Effect of pressure on the $\mathbf{k} \sim 0$ optic mode frequencies of a few ionic crystals. (Reprinted from S. S. Mitra, C. Postmus, and J. R. Ferraro [69].)

where f is the nearest-neighbor force constant; μ, the reduced mass per Bravais unit cell of volume $V_a = 2r_0{}^3$, where r_0 is the nearest-neighbor distance; and α_+ and α_-, the ionic polarizabilities. Assuming the rigid-ion

Table V
Observed and Calculated Grüneisen Parameters for the Long-Wavelength Optic Modes

	Experimental				Calculated			
	From equation (14.89) and Fig. 25		From equation (14.90)*		Born–Mayer[†]		r^{-n} repulsion[‡]	
Crystal	γ_{TO}	γ_{LO}	γ_{TO}	γ_{LO}	γ_{TO}	γ_{LO}	γ_{TO}	γ_{LO}
LiF	2.15	...	2.59	...	2.44	0.88	4.36	0.84
NaF	2.80	0.74	2.95	0.64	2.43	0.86	3.00	0.95
ZnS	1.75	...	1.85	...	...	...	2.43	0.98

* Estimated approximately from equation (14.90). Isothermal compressibility for LiF and NaF are from Born and Huang [70], p. 52; for ZnS from C. F. Cline and D. R. Stephens, *J. Appl. Phys.* **36**:2869 (1965).
† Potential constants are from Born and Huang [70], p. 26.
‡ Potential constants are from F. Seitz, *Modern Theory of Solids*, McGraw-Hill Book Company (New York), 1940, pp. 80–83.

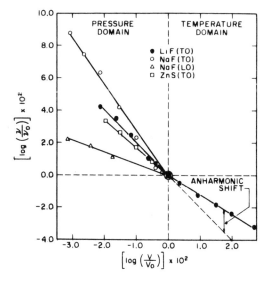

Fig. 25. Plot of $\ln(v/v_0)$ *versus* $\ln(V/V_0)$ for a number of optic modes, where v_0 and V_0 are, respectively, the phonon frequency and volume at 298 °K and 1-atm pressure. Data on the left-hand side of the figure are from pressure measurements; on the right-hand side, data are from temperature-dependence measurements. (Reprinted from S. S. Mitra, C. Postmus, and J. R. Ferraro [69].)

model of Kellerman [71], where $\alpha_+ = \alpha_- = 0$, one gets

$$\omega_t^2 = \frac{f}{\mu} - \frac{2\pi}{3} \frac{e^2}{\mu r_0^3} \tag{14.92}$$

For the long-wavelength longitudinal mode, by using the LST relation [equation (14.32)], the rigid-ion approximation yields

$$\omega_l^2 = \frac{f}{\mu} + \frac{4\pi}{3} \frac{e^2}{\mu r_0^3} \tag{14.93}$$

where it has been assumed that the high-frequency dielectric constant $\varepsilon_\infty = 1$.

For the average potential energy per ion pair, one may now use the Born–Mayer potential

$$u = -\frac{\alpha e^2}{r} + Mbe^{-\sigma(r/r_0)} \tag{14.94}$$

where α is the Madelung constant; M, the coordination number; and b and σ are potential constants. By using equation (14.94), f may be shown to be [72]

$$f = \frac{2b\sigma(\sigma-2)}{r_0^2} e^{-\sigma} \tag{14.95}$$

Hence, TO and LO modes at $\mathbf{k} \sim 0$ are given by

$$\omega_t^2 = \frac{2b(r_0/\rho)\,[(r_0/\rho)-2]}{\mu r_0^2} e^{-r_0/\rho} - \frac{2\pi}{3} \frac{e^2}{\mu r_0^3} \tag{14.96}$$

and

$$\omega_l^2 = \frac{2b(r_0/\rho)\,[(r_0/\rho)-2]}{\mu r_0^2} e^{-r_0/\rho} + \frac{4\pi}{3} \frac{e^2}{\mu r_0^3} \tag{14.97}$$

in terms of the volume-independent parameters b and ρ. For the Grüneisen parameters, one therefore obtains [73]

$$\gamma_t = \frac{(f/6)\,(\sigma^2 - 2\sigma - 2)/(\sigma-2) - (\pi/3)\,(e^2/r_0^3)}{f - (2\pi/3)\,(e^2/r_0^3)} \tag{14.98}$$

and

$$\gamma_l = \frac{(f/6)\,(\sigma^2 - 2\sigma - 2)/(\sigma-2) + (2\pi/3)\,(e^2/r_0^3)}{f + (4\pi/3)\,(e^2/r_0^3)} \tag{14.99}$$

where f may be evaluated from

$$f = \frac{\sigma-2}{3} \frac{\alpha e^2}{r_0^3} \tag{14.100}$$

Expressions for γ_t and γ_l may also be obtained from equations (14.92) and (14.93) by using a potential function involving an inverse power type of repulsive energy of the form

$$u = -\frac{\alpha e^2}{r} + \frac{Mb}{r^n} \qquad (14.101)$$

γ_t and γ_l are given by

$$\gamma_t = \frac{(f/6)\,(n+2) - (\pi/3)\,(e^2/r_0^3)}{f - (2\pi/3)\,(e^2/r_0^3)} \qquad (14.102)$$

and

$$\gamma_l = \frac{(f/6)\,(n+2) + (2\pi/3)\,(e^2/r_0^3)}{f + (4\pi/3)\,(e^2/r_0^3)} \qquad (14.103)$$

where

$$f = \frac{(n-1)\,\alpha e^2}{3\,r_0^3} \qquad (14.104)$$

Similar expressions for the zinc-blende structure have been given by Bienenstock and Burley [74] and for the CsCl structure by Postmus, Ferraro, and Mitra [75].

In the above expressions, it had been assumed that the ions are rigid. For the more realistic case of polarizable ions, one may use the Clausius–Mossoti relation [76]

$$\frac{4\pi}{3}\,(\alpha_+ + \alpha_-)/V_a = \frac{\varepsilon_\infty - 1}{\varepsilon_\infty + 2} \qquad (14.105)$$

to obtain [51]

$$\gamma_t = \frac{(f/6)\,[(\sigma^2 - 2\sigma - 2)/(\sigma - 2)] - \tfrac{1}{3}\,\pi\,[(\varepsilon_\infty + 2)/3]\,(e^2/r_0^3)}{f - (2\pi/3)\,[(\varepsilon_\infty + 2)/3]\,(e^2/r_0^3)} \qquad (14.106)$$

and

$$\gamma_l = \frac{(f/6)\,[(\sigma^2 - 2\sigma - 2)/(\sigma - 2)] + \tfrac{2}{3}\,\pi\,[(\varepsilon_\infty + 2)/3]\,(e^2/r_0^3)}{f + (4\pi/3)\,[(\varepsilon_\infty + 2)/3]\,(e^2/r_0^3)} \qquad (14.107)$$

using the Born–Mayer potential [equation (14.94)]. The potential function with an inverse power repulsive term [equation (14.101)], on the other hand, gives

$$\gamma_t = \frac{(f/6)\,(n+2) - \tfrac{1}{3}\,\pi\,[(\varepsilon_\infty + 2)/3]\,(e^2/r_0^3)}{f - \tfrac{2}{3}\,\pi\,[(\varepsilon_\infty + 2)/3]\,(e^2/r_0^3)} \qquad (14.108)$$

and

$$\gamma_l = \frac{(f/6)\,(n+2)+(2\,\pi/3)\,[(\varepsilon_\infty+2)/3]\,(e^2/r_0^3)}{f+(4\,\pi/3)\,[(\varepsilon_\infty+2)/3]\,(e^2/r_0^3)} \tag{14.109}$$

The experimental values of γ_t and γ_l obtained from the pressure-dependent measurements are compared with the ones calculated from relations (14.98), (14.99), (14.102) and (14.103) in Table V. Except for ZnS, the agreement is as good as can be expected. The discrepancy in the case of ZnS may be due to the use of a potential which is only valid for an ideally ionic crystal. The Grüneisen parameters for the long-wavelength optic modes may depend on the ionic character of a crystal. For a homopolar covalent crystal, one may not only expect that $\gamma_{TO} = \gamma_{LO}$, but perhaps that their value is somewhere in between those for an ionic crystal such as LiF. For ZnS with a structure closely related to that of diamond and some covalent character in its binding, it is thus not surprising to find its γ_{TO} somewhat smaller than that of LiF or NaF.

2. Temperature Dependence of Long-Wavelength Optical Phonons in Ionic Crystals

For the understanding of the temperature dependence of $v_j(\mathbf{q})$, equation (14.89) is not sufficient because, in addition to the purely volume-dependent part obtainable from the Grüneisen equation of state, one needs to consider the contribution from various anharmonic (cubic and higher) terms in the potential energy of the lattice, described earlier as the "self-energy" shift. We define the observed shift in the frequency as

$$\Delta v_{i(\text{obs})} = v_i(0) - v_i(T) \tag{14.110}$$

where $v_i(0)$ and $v_i(T)$ are the observed one-phonon resonance frequencies at $0\,°K$ and $T°K$, respectively. In practice, $v_i(0)$ may be obtained, from the extrapolation of low-temperature data. The third law of thermodynamics requires that the thermal expansion 3α vanish as $T \to 0$; consequently, v should approach a constant value as $T \to 0$ by virtue of equation (14.89), which may be rewritten as

$$\gamma_i = -\frac{1}{3\alpha}\frac{1}{v_i}\left(\frac{\partial v_i}{\partial T}\right)_P \tag{14.111}$$

This is more so because, at very low temperatures, the anharmonic effects also tend to diminish. To separate out the contribution of thermal expansion to observed Δv, we define

$$\Delta v_G = v(0) - v_G \tag{14.112}$$

v_G, the $\mathbf{q} \sim 0$ one-phonon optic mode in the Grüneisen approximation, is given by

$$v_{iG} = v_i(0) \exp\left(-3\,\gamma_i \int_0^T \alpha\, dT \right) \tag{14.113}$$

The anharmonic part of Δv may be obtained as

$$\Delta v_{iAN} = \Delta v_{i(\text{obs})} - \Delta v_{iG} \tag{14.114}$$

Often the directions of shifts arising from the two effects, i.e., anharmonicity and thermal expansion, may be opposite, which results in a smaller apparent temperature dependence. The temperature dependence of the TO frequency of LiF has recently been reported by Jasperse, Kahan, Plendl, and Mitra [51]. Their high-temperature data are plotted on a volume scale in the right-hand side of Fig. 25. It is evident that the straight line extrapolated from the pressure domain does not coincide with the line obtained from temperature measurements. The difference may thus be attributed to a purely anharmonic contribution to the frequency shift, steadily increasing with increasing temperature.

Maradudin [77] and Maradudin and Fein [78] have treated the problem of scattering of neutrons by an anharmonic crystal. They obtained the one-phonon scattering cross section for the coherent scattering of thermal neutrons by retaining the cubic and quartic anharmonic terms in the Hamiltonian for the crystal. These treatments, although derived for neutron scattering, may also be applied to other experiments involving phonon lifetimes, as has been indicated by Loudon [67] for the Raman scattering. The expressions for the frequency shift and the width of one-phonon lines given by Maradudin and Fein are too complicated to be evaluated explicitly except for very idealized Bravais lattices. However, from the expressions they have given, it is clear that the magnitudes of the shift should increase with temperature, the dependence being linear for $T \geqslant \Theta$, where Θ is the debye temperature of the crystal. In Tables VI and VII, are presented Δv_G and Δv_{AN} for LiF and MgO obtained [51] in the fashion outlined in the foregoing. In the high-temperature limit $(T \geqslant \Theta)$, it may be seen that, for both LiF and MgO, the anharmonic contribution is in the opposite direction to the volume effect, as predicted by the Maradudin and Fein theory. The quartic term in the anharmonic potential makes a much larger contribution to the shift than the cubic term does. Whereas the Grüneisen contribution and the small cubic anharmonic contribution are in one direction (decreasing frequency with increasing temperature), the quartic anharmonic contribution is in the opposite direction. The first two contributions to the shift are in the direction of the observed shift, which is smaller than the value predicted by these two contributions alone. Compensation occurs through the third contribution. As predicted by the Maradudin and

Table VI
Grüneisen and Anharmonic Contributions to the Frequency Shift of the k ≃ 0 TO Mode in LiF* †

T °K	ν_G cm^{-1}	$\Delta\nu_G$ cm^{-1}	ν_{obs} cm^{-1}	$\Delta\nu_{obs}$ cm^{-1}	$\Delta\nu_{AN}$ cm^{-1}
100	319	1	315	5	4
200	312	8	310	10	2
300	301	19	306	14	− 5
500	276	44	298	22	−22
700	249	71	289	31	−40
900	219	101	279	41	−60
1100	188	132	269	51	−81

* Thermal-expansion data used in the calculation of ν_G are from *American Institute of Physics Handbook*, 2nd ed., McGraw-Hill Book Company (New York), 1957, pp. 4–73.
† γ used is the average of that from columns 8 and 10 of Table 3 of Jasperse *et al.* [51].

Fein theory, in the high-temperature limit, $\Delta\nu_{AN}$ indeed varies linearly with temperature.

Cowley [79] and Ipatova, Maradudin, and Wallis [80] have given semi-empirical treatment of the shift of the long-wavelength optic phonons with temperature. In essence, these treatments agree with the predictions of the Maradudin and Fein theory and with the experimental results on LiF and

Table VII
Grüneisen and Anharmonic Contributions to the Frequency Shift of the k ≃ 0 TO Mode in MgO* †

T °K	ν_G cm^{-1}	$\Delta\nu_G$ cm^{-1}	ν_{obs} cm^{-1}	$\Delta\nu_{obs}$ cm^{-1}	$\Delta\nu_{AN}$ cm^{-1}
100	408	0	405	3	3
200	406	2	402	6	4
300	402	6	400	8	2
600	385	23	392	16	− 7
900	367	41	383	25	−16
1400	335	73	368	40	−33
1900	304	104	356	52	−52

* Thermal-expansion data used in the calculation of ν_G are from A. Goldsmith, H. J. Hirschhorn, and T. E. Waterman, Wright Air Development Center, WADC Tech. Rept. No. 58-476, Vol. III, 1960 (unpublished).
† The γ used is the average of that from columns 8 and 10 of Table 3 of Jasperse *et al.* [51].

Table VIII
Temperature Dependence of k ~ 0 TO and LO Phonon Frequencies of KBr; the Calculated Values are from Cowley [79] and the Experimental Values are from Jones et al. [81]

T, °K	Thermal expansion, cm^{-1}	Fourth-order anharmonic, cm^{-1}	Third-order anharmonic, cm^{-1}	ν (calc.) cm^{-1}	ν (exptl.) cm^{-1}
		Transverse Optic			
5	− 3.2	2.4	−2.0	121.7	
90	− 4.8	3.5	−3.2	120.0	120
200	− 9.2	6.6	−5.9	116.0	117
290	−12.8	9.4	−7.9	113.0	114
400	−17.8	12.9	−9.5	110.3	111
		Longitudinal Optic			
5	− 2.4	1.8	−0.4	168.7	170
90	− 3.4	2.6	−2.0	166.7	167
200	− 6.5	4.8	−6.2	161.7	163
290	− 9.2	6.8	−8.7	158.3	

MgO described earlier. Such calculations need complete phonon dispersion curves. Cowley has used the neutron scattering data of Woods et al. [36] on KBr and NaI. In Table VIII, we compare the calculated TO and LO frequencies of KBr with those observed by Jones et al. [81].

In the case of RbI, the infrared data exist only over a small temperature range (up to 300°K). By using the thermal-expansion data of Schuele and

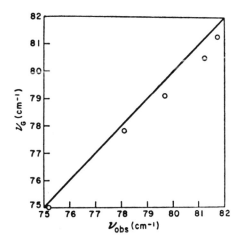

Fig. 26. TO frequency of RbI. Comparison of observed value with that calculated from equation (14.113) at various temperatures (4 to 300°K). (Reprinted from J. R. Jasperse, A. Kahan, J. N. Plendl, and S. S. Mitra [51].)

Table IX
Shift of Long-Wavelength Optic Phonon of Diamond

T, °K	$\Delta \nu$ (obs),* cm^{-1}	$\Delta \nu$ (calc),† cm^{-1}
10	0	3.27
300	1	3.30
500	2	3.60
1000	13	5.13

* Unpublished data, Mitra and Namjoshi.
† From Cowley [83].

Smith [82], ν_G for the crystal has been calculated [73]. Within the experimental error, they agree with the observed frequency shown in Fig. 26, which indicates that the volume dependence of the frequency almost entirely accounts for the observed shift over the range of temperature considered and that the anharmonic contributions are negligibly small or cancel out.

The theory of the Raman scattering of light by crystals of the diamond structure has recently been discussed by Cowley [83], who has made detailed calculations of the Raman spectra of germanium, silicon, and diamond. The anharmonicity introduces a coupling between the one- and two-phonon scattering, which in part shifts the one-phonon frequency. By fitting the experimental phonon dispersion curves with shell-model calculations, Cowley obtained numerical values for the shift of the one-phonon Raman peak of diamond. His calculated values are compared with our recent unpublished data in Table IX. The comparison is rather poor.

B. Temperature Dependence of the Width of Long-Wavelength Optic Phonons

The finite lifetimes of the phonons result in broadening of the associated spectral lines. The decay of a quasi-normal mode may take place through many channels involving phonons of different branches and wave vectors coupled through anharmonic terms in the crystal Hamiltonian. In the case of covalent crystals, the higher-order electric moments are also responsible for such coupling. Szigeti [84] has given an elementary perturbation treatment in which he obtains an expression for the total oscillator strength associated with a one-phonon transition. To gain an understanding of the behavior of the half-width (inverse lifetime) of individual one-phonon transitions requires a many-body theory, as has been used by Cowley [79]. We shall, however, give a qualitative explanation based on the treatment of Maradudin and Wallis [47] of a frequency-dependent damping factor.

If one examines equation (14.56) for the damping constant, it suggests that, when evaluated at $v = v_1$, one finds [51] the damping constant to have the form

$$\frac{\gamma(v, T)}{v_1}\bigg|_{v=v_1} = \frac{\text{constant}}{v_1^4} \left[\left(\exp \frac{hv_1}{kT} - 1\right)^{-1} + \tfrac{1}{2}\right] \qquad (14.115)$$

where the constant represents an average value from the summation, and v_1 stands for TO at $\mathbf{k} \sim 0$. The v_1 varies with temperature and, as we have seen earlier, linearly for $T \geqslant \Theta$. It is difficult to know whether there is any temperature dependence buried in the constant factor shown in equation (14.115). However, other calculations of γ in the high-temperature limit ($kT > hv_1$) suggest that γ/v_1, is consistent with equation (14.115). A graph of the damping constant for the main band in LiF and MgO as a function of temperature is shown [51] in Fig. 27. For LiF, kT is about three times hv_1 at 1060 °K, and, for MgO, kT is about four times hv_1 at 1950 °K. The effective temperature dependence of γ_1/v_1 for LiF is about $T^{3/2}$ and for MgO, about T. Now we notice that equation (14.115), suggested by the Maradudin–Wallis theory (also by Mitskevich [48] and Neuberger [49] in the high-temperature limit), appears to contain a $v_1(T)^{-4}$ multiplicative term which would

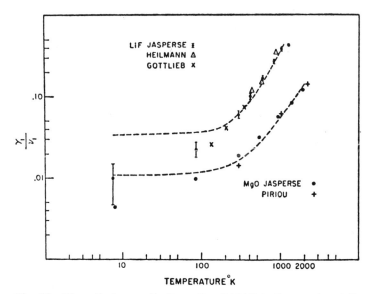

Fig. 27. Theoretical curve fit of equation (14.115) to the experimentally determined damping constant. (Reprinted from J. R. Jasperse, A. Kahan, J. N. Plendl, and S. S. Mitra [51].)

tend to make the temperature dependence of γ_1/ν_1 more than the first power of T when $kT > h\nu_1$. A similar result would hold for MgO only to a lesser extent because the coefficient $d\nu_1/dT$ for MgO is about one-half of that for LiF. The dotted line in Fig. 27 represents a curve fit of the experimental data to equation (14.115). Data reported by Heilmann [85] and Gottlieb [86] for LiF and by Piriou [87,88] for MgO are also included in this graph. Equation (14.115) fits the data reasonably well at high temperatures but deviates at the lower temperatures. It is true that the experimental error associated with γ_1/ν_1 increases rapidly at low temperatures, but the theoretical expression (curve-fitted at the higher temperatures) appears to lie well outside of the estimated error. The Born–Huang theory predicts that, at the center of the resonance, γ_1/ν_1 varies in proportion to the cube of the temperature in the high-temperature limit. This dependence seems to be much stronger than is indicated by experiment. The high-temperature limit of the Maradudin–Wallis results and the treatments by Neuberger [49] and Blackman [89], in which each mode of vibration in the crystal is assigned an energy kT, produce similar results and agree reasonably well with equation (14.115). We note that Hass found that the damping constant of NaCl in the high-temperature limit was proportional to a power of T slightly less than 2 [90]. It is plausible that his results could also be explained by equation (14.115) and a temperature dependence in ν_1 even stronger than the one we observe for LiF.

The behavior of $\gamma(\nu_1, T)$ at low temperatures ($kT \ll h\nu_1$), as suggested by equation (14.115), is dominated by the constant $\frac{1}{2}$ that is inside the brackets. The $\frac{1}{2}$ is there because quantum oscillators are believed to have zero-point energy in their ground state. If the $\frac{1}{2}$ were not present, the damping constant would go to zero exponentially with decreasing temperature. In Fig. 27, we see that, if equation (14.115) is curve fitted at high temperatures (because experimental errors in γ are less there), the theoretical curve deviates from the experimental values at low temperatures. It seems that the experimentally determined damping constant evaluated at ν_1 has a definite bend at lower temperatures, which suggests that γ does not vanish as the temperature becomes arbitrarily small. More recently, Ipatova, Maradudin, and Wallis [80] have calculated the damping factors for the TO of LiF as a function of temperature, using the calculated phonon dispersion curves of Karo and Hardy [91]. These calculated values are compared with the experimental values of Jasperse, Kahan, Plendl, and Mitra [51] in Table X. Considering the various errors involved in such a calculation, the agreement should be judged qualitatively good.

Table X
Theoretical and Experimental Values of γ/ν for LiF at Several Elevated Temperatures

T, °K	γ/ν_t (theor.)*	γ/ν_t (exptl.)†
293	0.0943	0.060
443	0.1569	0.110
593	0.2292	0.168
893	0.4030	0.305

* See Ippatova, Maradudin, and Wallis [80].
† See Jasperse et al. [51].

VI. EFFECT OF ELECTRIC FIELD ON OPTICAL PHONONS

Homopolar molecules, such as H_2, which do not exhibit any fundamental infrared vibration spectrum are known to become infrared active in the presence of an electric field. Such an effect was predicted by Condon[92] and was experimentally observed for H_2 molecules by Crawford and Dagg[93] at fields of 10^5 V/cm. An analogous effect may also be observed in the solids.

In homopolar crystals such as diamond or silicon, the effective ionic charge on an atom is zero. These crystals, therefore, do not absorb infrared radiation owing to a one-phonon process involving the long-wavelength optic phonons. However, an externally applied static electric field may induce an effective ionic charge and the $\mathbf{k} \sim 0$ optic phonons, which are normally Raman active, may also become infrared active. The theory is similar to that of Raman scattering discussed in Section IV. The electric field induces a dipole moment [as given by equation (14.76)], which, in the case of a static field $\mathbf{E}(0)$, may be written as

$$\mathbf{M} = \alpha^{(0)} \cdot \mathbf{E}(0) + u_j \alpha^{(1)} \cdot \mathbf{E}(0) + \dots \qquad (14.116)$$

Since $u_j = u_{j0} e^{i\omega_j t}$, the induced dipole moment varies in amplitude and orientation at the frequency of the normal mode in question. The induced effective ionic charge is given by

$$q^* = \frac{\partial M}{\partial u} = \alpha^{(1)} E(0) = \left(\frac{\partial \alpha}{\partial u}\right) E(0) \qquad (14.117)$$

The intensity of the field-induced absorption will thus be given by

$$I \propto q^{*2} = \left| \left(\frac{\partial \alpha}{\partial u}\right) \mathbf{E}(0) \right|^2 \qquad (14.118)$$

Such an electric-field-induced infrared absorption in diamond was indeed observed by Anastassakis, Iwasa, and Burstein [94]. They made transmission measurements at room temperature on a type IIa diamond plate. In the absence of an electric field, there was no measurable absorption in the region of $1332 \, \text{cm}^{-1}$, the frequency of the $k \sim 0$ optical vibration modes of diamond. In the presence of a field of 1.2×10^5 V/cm, a transmission minimum was observed at $1336 \, \text{cm}^{-1}$. The shift of $4 \, \text{cm}^{-1}$ was larger than the experimental error and was tentatively attributed to an electric-field-induced shift in frequency [95]. From the intensity of this line, $\partial \alpha / \partial u$ was determined to be $4 \times 10^{-16} \, \text{cm}^2$ per unit cell.

In Section IV, an expression for the Raman scattering efficiency S was given by equation (14.80). Using the above value of $\alpha^{(1)} = \partial \alpha / \partial u$, Anastassakis et al. [94] obtain an experimental value of $S = 3 \times 10^{-7}$ in agreement with Loudon's theoretical estimate of 10^{-6} to 10^{-7} based on a deformation-potential calculation [96].

A related but opposite effect may be expected for the NaCl structure. Ionic crystals of the NaCl type possess an inversion symmetry at each atomic site. They have, thus, fundamental ($k \sim 0$) infrared spectrum but no first-order Raman effect because the long-wavelength optic phonons have odd parity. However, an applied electric field may remove the inversion symmetry of the crystal, and a field-induced first-order Raman spectrum of NaCl is thus possible, although not observed so far.

Fleury and Worlock [97] have observed such an AC electric-field-induced Raman effect in the cubic perovskite $KTaO_3$. Under the influence of an electric field (10^4 V/cm and 210 cps), an entirely new Raman band was observed. At $300 \, ^\circ K$, this Raman line occurred at $85 \, \text{cm}^{-1}$ in agreement with that found by infrared studies of Miller and Spitzer [98] for the low-frequency ferroelectric "soft" TO mode. As the temperature was decreased, the frequency of the field-induced scattering decreased as expected for the soft mode [99].

The intensity of the scattered radiation may be obtained from a simple perturbation theory. The matrix element x for induced scattering from a phonon of odd parity is given by

$$x \propto \sum_{u',u''} \frac{\langle u''' | \mathbf{E}_s \cdot \mathbf{r} | u'' \rangle \langle u'' | \mathbf{E}_0 \cdot \mathbf{r} | u' \rangle \langle u' | \mathbf{E}_{ac} \cdot \mathbf{r} | u \rangle}{(\omega'' \pm \omega_{ac} \pm \omega_0)(\omega' \pm \omega_{ac})} \qquad (14.119)$$

where $|u\rangle$ is the initial state of the crystal; $|u'\rangle$ and $|u''\rangle$ are virtual excited states of odd and even parity, respectively; $|u'''\rangle$ is a one-phonon state associated with $|u\rangle$; $\mathbf{E}_0$ and $\mathbf{E}_s$ and ω_0 and ω_s are the field strengths and frequencies of the incident and scattered optical fields, respectively; and $\mathbf{E}_{ac}$ and ω_{ac} are the field strength and frequency of the applied AC field. The scattered intensity which is proportional to x^2 is thus proportional to

E_{ac}^2. Such a quadratic variation of the integrated intensity was indeed observed [100] for the electric-field-enhanced Raman scattering of the soft mode of $SrTiO_3$.

An applied electric field can also alter the optic mode frequencies in crystals through anharmonic terms of the short-range forces. For ferroelectric perovskite structures, the lowest TO branch at $\mathbf{k} \sim 0$ is the ferroelectric soft mode and is expected to be strongly shifted. For a cubic crystal, a frequency shift proportional to the square of the applied field is expected. However, the presence of a noncentrosymmetric electronic-charge distribution about each ion can give rise to a contribution linear in the applied field. Such a linear shift was indeed noted by Schaufele, Weber, and Silverman [100] for the soft mode of $SrTiO_3$ at $77\,°K$ in the field range of 0 to $30\,kV/cm$.

At lower temperatures, the effect of an electric field on the soft mode of ferroelectric crystals is much enhanced. Worlock and Fleury [101] report: (1) At small applied fields, the soft mode frequency of $SrTiO_3$ varies from $11\,cm^{-1}$ at $8\,°K$ to $85\,cm^{-1}$ at $250\,°K$ and may be explained by the temperature dependence of the dielectric constant by using the LST relation. (2) At temperatures $< 55\,°K$, the spectrum of the field-induced scattering exhibits a field-dependent structure. At $8\,°K$, the soft mode frequency shifts from 11 to $45\,cm^{-1}$ as the field is changed from $400\,V/cm$ to $12\,kV/cm$. (3) The efficiency of the induced Raman scattering and the shifting and splitting of phonon frequencies as well become increasingly effective as the temperature is lowered.

VII. COMBINATIONS AND OVERTONES

In the harmonic approximation, only the fundamental modes of vibration in a solid are possible. Combinations and overtones do not arise. Moreover, in the infrared absorption or Raman scattering, the fundamentals should appear as sharp lines. Their optical activity depends on whether the equivalent atoms move identically in phase. If, for a normal mode, θ_{ij} is the phase difference between the ith and jth unit cells, the condition for the mode to be optically active is that $\theta_{ij} = 0$, which corresponds to the modes that belong to the center of the Brillouin zone with $\mathbf{k} = 0$. The appearance of combinations and overtones in the spectra, the line broadening in the condensed phases, and the phenomenon of thermal expansion of solids, however, indicate the presence of cubic or higher-order terms in the vibrational-potential-energy function of a solid. Inclusion of anharmonicity in the general theory of the vibrational spectra of crystals not only allows combinations and overtones but puts very little restriction on the selection rules applicable to such transitions.

A. Multiphonon Absorption in Simple Crystals

For polar crystals, the mechanism of the interaction of the electromagnetic field with phonons that causes the side bands has been explained [102] as due to the anharmonic part of the potential energy associated with the lattice vibrations. The interaction takes place through the dipole moment associated with the lattice mode. In homopolar crystals, the factor group fundamental is inactive because the dipole moment is zero, and, thus, the above mechanism does not apply. For such systems, Lax and Burstein [103] have shown that two or more phonons can interact directly with the radiation through terms in the electric moment of second or higher order in the atomic displacements. Such combinations, of either origin, will be optically active if the energy and wave vectors are conserved according to

$$hv = \sum_i (\pm hv_i) \qquad (14.120)$$

and

$$\mathbf{q} + n\mathbf{K} = \sum_i (\pm \mathbf{k}_i) \qquad (14.121)$$

where hv and $\mathbf{q}$ are the energy and the wave vector of the absorbed photon; hv_i and $\mathbf{k}_i$ are the energy and the wave vector of the ith phonon; $\mathbf{K}$ is a reciprocal lattice vector; and n is an integer. The positive and negative signs indicate emission and absorption, respectively. Since the wave vector of the photon in the infrared region is small compared to the phonon wave vector, the second condition becomes

$$n\mathbf{K} = \sum_i (\pm \mathbf{k}_i) \qquad (14.122)$$

For one- and two-phonon processes, $n = 0$, and, for a three-phonon process, $n = 0, \pm 1$. In the event of interaction of the photon with a single phonon, the latter must be a zero-wave-vector phonon, i.e., the factor group fundamental that belongs to the zone center. A two-phonon-sum band requires $k_1 = -k_2$ as a consequence of equation (14.122), while, for a difference band, the requirement is $k_1 = k_2$.

Strictly speaking, the combinations should give rise to continuous absorption. Absorption maxima, however, occur because of singularities in the phonon frequency distribution. Any structure in the infrared or Raman spectra thus reflects structure in the frequency dependence of the combined density of states of the participating phonons or of the interaction which gives rise to the transition. The regions, or points of high phonon concentration, in the phonon dispersion are known as critical points (CP). The singularities corresponding to the CP's occur where the dispersion curves for the individual branches are flat. A CP in a phonon branch is then a point where $\nabla_k \omega(k)$ either is zero or changes sign discontinuously. In general, if there are p number of CP in the first Brillouin zone of a crystal, then $3np$ individual frequencies, where $3n$ is the number of branches, may participate in the second- or higher-order spectra. In the case of the cubic diamond and zinc-blende structures, the relevant CP, e.g., are

Γ, the center of the Brillouin zone $(0, 0, 0)$

L, the center of a face with coordinates $(\frac{1}{2}, \frac{1}{2}, \frac{1}{2})$

X, the center of a face with coordinates $(1, 0, 0)$

W, the peak of coordinates $(1, \frac{1}{2}, 0)$

K, the edge of the zone with coordinates $(\frac{3}{4}, \frac{3}{4}, 0)$.

Each of the CP phonons are related to one of the irreducible representations of the crystal space group. With the use of space group theory, one may, in principle, enumerate the selection rules governing the infrared and Raman spectra of the crystal, allowing for the possibility of creation and annihilation of one or more of these CP phonons. Birman [104] has worked out the space-group selection rules in the diamond and zinc-blende structures that are relevant to the analysis of infrared-absorption and Raman scattering processes. Mitra [105] has applied Birman's selection rules to the understanding of the infrared spectrum of GaAs. Loudon and Johnson [106] have applied them to the interpretation of the spectra of diamond Si and Ge. Burstein, Johnson, and Loudon [107] have obtained the selection rules for the NaCl structure, whereas Ganesan and Burstein[108] have done so for the fluorite structure. Nusimovici and Birman [109] have worked out some of the selection rules for the wurtzite structure, details of which have also been considered by Ziomek and Mitra [110].

The observed infrared and Raman spectra of crystal may be completely understood in terms of the CP phonons if these are known either from neutron scattering data or from a reliable theoretical calculation, e.g., the shell model. The reverse, i.e., an unambiguous assignment of the CP phonons from the observed infrared and Raman spectra alone, may not always be possible. In the absence of detailed dispersion curves, it is still, however, possible to obtain broad zone-boundary phonon frequencies consistent with the space-group selection rules. Thus, for the diatomic cubic crystals (NaCl, CsCl, diamond, and zinc-blende types), one sometimes assumes four characteristic phonons: LO, TO, LA, and TA, corresponding to the frequencies of the longitudinal and transverse optic and acoustic phonon branches at the edge of an approximate spherical Brillouin zone.

The temperature dependence of the absorption spectrum at any frequency of incident radiation depends on the number of phonons available to take part in a proposed multiphonon process at a given temperature. The temperature dependence of the number n_i of phonons of frequency v_i is given by equation (14.57)

$$n_i = [\exp{(hv_i/kT)} - 1]^{-1} \tag{14.57a}$$

The probability of absorption or emission of one of these phonons is proportional to the square of the matrix element of the phonon creation a^+ or annihilation a^- operator, respectively. These matrix elements are given by [111]

$$\langle \psi(n_i) \, | a^+ | \, \psi(n_i + 1) \rangle = (n_i + 1)^{\frac{1}{2}} \quad \text{emission} \tag{14.123}$$

$$\langle \psi(n_i) \, | a^- | \, \psi(n_i - 1) \rangle = n_i^{\frac{1}{2}} \quad \text{absorption} \tag{14.124}$$

The probability of absorption of a phonon is thus proportional to n_i, whereas the probability of emission of a phonon is proportional to $1 + n_i$. Let us suppose an absorption band is assigned to a two-phonon process such that the absorption of a photon is accompanied by the emission of two phonons ($i = 1, 2$) in accordance with the conservation principles (14.120) and (14.121). The probability of the process is proportional to $(1 + n_1)(1 + n_2)$. The net absorption is obtained by correcting for spontaneous emission of the photon by the reverse process in which two phonons are absorbed. Then the temperature dependence of the net absorption is proportional to $(1 + n_1)(1 + n_2) - n_1 n_2 = 1 + n_1 + n_2$. Similarly, for an assigned three-phonon process such as $LO(v_1) + TO(v_2) - LA(v_3)$, the absorption coefficient will be proportional to $(1 + n_1)(1 + n_2)n_3 - n_1 n_2 (1 + n_3)$. Similar expressions can be obtained for a Raman spectrum also. Temperature-dependence factors for second-order infrared and Raman spectra are given in Table XI. The two conservation laws plus the temperature dependence of the absorption intensity subject the phenomenological analysis of the spectral data to rather stringent self-consistency requirements.

Table XI
Temperature Dependence Factors for Two-Phonon Infrared and Raman Processes

Combination	Infrared	Raman	
		Stokes	Anti-Stokes
$2v$	$1 + 2n$	$1 + 2n + n^2$	n^2
$v_1 + v_2$	$1 + n_1 + n_2$	$1 + n_1 + n_2 + n_1 n_2$	$n_1 n_2$
$v_1 - v_2$	$n_2 - n_1$	$n_1 n_2 + n_2$	$n_2 n_1 + n_1$

B. Combinations and Overtones in Crystals Containing MolecularGroups[112]

It was observed that the fundamental modes of vibration that may be infrared or Raman active belong to the center of the Brillouin zone. They may be subdivided into internal, translatory external, and rotatory external modes in the case of crystals containing molecular groups. Their symmetry classifications and selection rules are derived by factor group analysis, and these constitute the $k = 0$ modes of the optical branches. The acoustic modes have zero frequencies in the factor group approximation and hence are excluded. However, the latter will have nonvanishing values at the zone boundary and thus may take part in combination tones.

In general, there may be the following types of combinations in a molecular crystal:

$$\text{Internal } + \text{internal} \qquad\qquad (14.125)$$

$$\text{Internal } + \text{external} \qquad\qquad (14.126)$$

$$\text{Internal } + \text{acoustic} \qquad\qquad (14.127)$$

$$\left.\begin{array}{l}\text{External } + \text{external}\\ \text{External } + \text{acoustic}\\ \text{Acoustic} + \text{acoustic}\end{array}\right\} \quad \text{Multiphonon} \qquad (14.128)$$

The only selection rules that may be operative are the two conservation laws (14.120) and (14.121). Whereas the interactions (14.125) to (14.127) involve molecular modes, the last three interactions denoted by equation (14.128) are formally equivalent to the multiphonon combinations discussed in the earlier section.

For a crystal such as brucite, $Mg(OH)_2$, with five atoms per unit cell, there are 15 distinct branches. Ten of these branches correspond to optical lattice modes, three to the acoustic modes, and two to the internal modes[113]. In the simplest case of only one CP corresponding to $k/k_{max} = 1$, in addition to the multiphonon processes, one needs to consider the combination of 2 internal mode frequencies with the 13 external mode frequencies. It may be emphasized that these combining frequencies are not the observed fundamentals (which are for $k \sim 0$) but their values at the zone boundary. Furthermore, any degeneracy (accidental or otherwise) will probably break down at the zone boundary, which will make all these combining frequencies distinct. In a hexagonal crystal such as $Mg(OH)_2$ (space group $P\bar{3}m$), many CP are expected to occur. Thus, in reality, it may be necessary to consider the combination of 4 internal modes with 26 external modes, or 6 with 39, etc. It is needless to say that such a spectrum is expected to be very complex. The envelope shape of the combination bands will depend on the dispersion relations and may be worked out only if the latter are known. However, to our knowledge, the lattice dynamics has not yet been completely worked out for any molecular crystal, except for a recent attempt with anthracene [114].

For real molecular crystals with strong chemical bonds within a molecular group and with loose bindings such as the van der Waals type between the groups, there may be some simplification possible. This applies to combinations of only the internal modes of the molecular groups. Usually in such crystals, the interactions between the identical groups within a unit cell give rise to the so-called correlation field splitting discussed earlier, but interaction between groups in different unit cells may be neglected. Consequently, the **k** dependence of the internal mode frequencies may be limited, with the result that the zone-boundary frequencies may not be very different from the zone-center fundamentals. In such a situation, combination tones may be expected of the type observed in molecular spectra in the gaseous phase. In addition, if the molecular groups are heavy and have very loose bindings between them, the phonon frequencies may be expected to be quite small and may be manifest only in the envelope shape of the internal fundamentals and their combinations. A situation like this may be expected in a crystal such as naphthalene or anthracene.

In the electronic spectrum of solid N_2, Vegard [115] observed spectral spacings of 40 and 69 cm^{-1}, which were assigned to the excitation of librational modes. The infrared absorption spectrum of crystalline N_2 also shows this structure [116] and has been explained as arising from a libration–vibration combination. However, Ewing and Pimentel [117], from a consideration of the heat of sublimation (using an r^{-6}, r^{-9} potential function), have shown that the broad absorptions of solid N_2 may, in part at least, be caused also by combinations involving translations. Similar combinations have also been observed by them in solid carbon monoxide. In the infrared absorption spectra of solid N_2O and CO_2, using thick samples, Dows [118] has observed several combination bands. These he assigned as originating from the mixing of the libration modes with the bending mode of an XY_2 molecule. Hydrogen is one of the few molecules that undoubtedly shows free rotation in the solid state because of the extremely weak intermolecular forces. It is interesting to note, however, that the same intermolecular forces, along with quadrupolar interactions, are able to induce fundamental infrared absorption in the solid and the liquid states, in spite of the fact that such absorption is forbidden for the free molecule. The fine structure of the vibration spectrum of crystalline H_2 investigated by Gush et al. [119] can only be explained by the presence of a more or less unhindered rotation. Additional broad absorption features are assigned to combinations of a stretching mode with translational modes. The nature of molecular rotation in solids and the interaction of librational modes with the stretching type of internal modes have been discussed [120] elsewhere by the author and will not be further elaborated on here.

It has been a frequent practice to use the factor-group selection rules also in the analysis of the combination spectra of the molecular modes. Unfortunately, theoretical justification exists only for application to the fundamentals and not to such an extension. We shall use the gypsum crystal to illustrate this point. Consider the band at $2112 \, \text{cm}^{-1}$, which seems numerically to be arising from the combination of the two sulfate internal modes $v_3(B_g) = 1117 \, \text{cm}^{-1}$ and $v_1(A_u) = 1000 \, \text{cm}^{-1}$. Now, from the character table given in Table II, one finds the product

$$A_u \cdot B_g = B_u \qquad (14.129)$$

It is then argued that the combination band should manifest the symmetry properties of the B_u species, which are the infrared activity and transition moments along the $b(C_2)$ axis, i.e., in the (010) plane. However, the conclusions may not necessarily be correct because of the following reasons. Firstly, the factor-group selection rules may not be operative. A combination band judged inactive by the factor-group selection rules may, as a matter of fact, be active. Secondly, the numerical agreement is no guarantee because such an agreement also suggests a harmonic force field. However, the occurrence of a combination mode requires an anharmonic field. Finally, the combining frequencies may correspond to any k value and may be different from the observed $k = 0$ fundamentals.

If the lattice modes in a crystal are substantially large, as in the case of light molecular groups with relatively strong binding, the analysis of the combination modes arising from interactions with the internal modes is not straightforward. For example, in the alkali and alkaline earth hydroxides, which are predominantly ionic with relatively light groups, the lattice modes range from 100 to $600 \, \text{cm}^{-1}$. The combination of the internal with the lattice modes does indeed give an immensely complicated near-infrared spectrum [121] which can hardly be understood from a factor group standpoint.

To our knowledge, brucite, $Mg(OH)_2$, is the only ionic molecular crystal for which the multiphonon combination bands and the importance of the zone-boundary acoustic modes have been pointed out [122]. For a complete understanding of the internal–external combination spectra of crystals containing polyatomic groups, it may be necessary to know the dispersion of individual branches (from elastic constant, specific heat, neutron scattering data, etc.) and the location of the CP in the Brillouin zone. In addition, high-resolution experimental data at several temperatures, in the near- as well as in the far-infrared regions and Raman scattering studies, may be necessary. In principle, the selection rules for combination modes can be worked out from group theoretical considerations involving the entire space group. It appears to us, however, that the quantitative study of vibration spectra of solids is still in its infancy and the present knowledge of it does not yet permit a thorough analysis for crystals of relatively complex structures containing even the simplest of the polyatomic groups. Even the shape of the combination bands of a simple crystal like Si is not predictable with much success [123]. This situation exists with the lattice dynamics and experimental dispersion curves being fairly well known. Thus, it is still a far way before the envelope shape of the combination bands of a crystal like $Mg(OH)_2$ (whose elastic constants are not even known) can be predicted on a realistic basis. It still remains to advance specific mechanisms which make the interactions of the internal and external modes in the hydroxide crystals possible. Hexter and Dows' [124] model and the modifications proposed by Mitra [120] of a librating vibrator may, however, be regarded as attempts to construct such a mechanism. However, they incorporate only a portion of the possible interactions, namely, the coupling of the internal with the rotatory lattice modes. The interaction of the internal with the external modes should be fairly universal and should be observed in varying degrees in most crystals containing polyatomic groups. High-resolution spectroscopy in transmission and in emission may be the key to revealing the interaction.

Buchanan [125] has reported complex near-infrared spectra for NaOH and KOH under certain conditions, whereas Snyder *et al.* [126] have reported rather simple spectra. The importance of impurities and point defects in accentuating forbidden transitions in Si and Ge is fairly well known [127]. One therefore suspects that the lattice–internal and phonon–phonon combinations in molecular crystals may also be affected by impurities and defects. This probably explains the difference between the spectra of Buchanan and those of Snyder *et al.* for NaOH and KOH.

The vibrational combination spectrum (whether among lattice modes, between internal and lattice modes, or among internal modes) of solids is a highly complex situation. The concepts related to an isolated molecule (in gas phase or dilute solution) or factor-group considerations should be applied with caution to such spectra and, at best, may be regarded as very poor approximations. An almost infinite number of frequencies are possible in a crystal corresponding to a normal mode for which the vibrations in successive unit cells are out of phase by varying amounts. Each mode becomes active in combination with another mode of the same or a different vibrational frequency having a corresponding compensating phase difference.

VIII. EXAMPLES

In this part, it is intended to describe in some detail the optical vibration spectra of a few typical classes of crystals with a view to illustrating some of the principles discussed earlier. We shall start with the example of a diatomic cubic crystal AlSb followed by GaAs. Next, gypsum, a semi-ionic molecular crystal will be discussed; and finally, the infrared spectra of truly molecular crystals cyclopropane and naphthalene with van der Waals type of binding will be presented.

Except for certain diatomic crystals, we shall not attempt to present a complete bibliography of all published experimental work on the infrared and Raman spectra of solids. In recent years, considerable interest has been evidenced in the study of the lattice infrared spectrum of semiconducting compounds of simple structure. In the following sections, a review of the multiphonon spectra of such crystals will be presented.

A. Multiphonon Infrared Absorption in AlSb

AlSb belongs to the zinc-blende type of structure with two atoms per Bravais unit cell. The lattice infrared spectrum of AlSb has been measured by Turner and Reese [28]. The reststrahlen spectrum and its analysis has been given earlier. The maximum of the extinction coefficient occurs at $31.37 \pm 0.5 \mu$ (319 cm^{-1}), which is identified with the zone-center TO mode v_t. The zone-center LO mode v_l is infrared inactive but may be identified with the short-wavelength minimum of the reflection spectrum at 29.4μ. A more precise value may be obtained from the LST relation. The values of ε_∞, ε_0, and v_t are obtained from the dispersion analysis of the reststrahlen band. The calculated value of v_l is 340 cm^{-1}.

The absorption coefficient of AlSb as a function of wavelength is shown in Fig. 28. It was obtained from several transmission spectral measurements. The absorption peaks are due to the lattice absorption involving one, two, three, and four phonons. The strongest absorption around 31 μ is due to the zone-center TO mode phonon v_t, the exact position of which was determined from the analysis of the reststrahlen band. The position of the zone-boundary characteristic phonons are estimated from the 7 two-phonon band assignment as indicated in Fig. 28. The triplet 2 LO, LO + TO, and 2 TO is better resolved at 77 °K. The peaks shift toward higher frequencies by 5 cm^{-1} at 77 °K. The assignment is confirmed by a comparison of the net power absorbed in a two-phonon summation at two temperatures.

A temperature factor f is defined as the ratio of the magnitudes of the lattice band at 300 and 77 °K. It may be seen from our discussion of multiphonon absorption in simple crystals that the calculated temperature factor f_{calc} is given by

$$f_{calc} = \frac{(1 + n_1 + n_2)_{300°K}}{(1 + n_1 + n_2)_{77°K}} \qquad (14.130)$$

The observed temperature factor f_{obs} was obtained after correction for the background due to free carrier absorption, impurity absorption, and the tail of the fundamental absorption at v_t. The assignment for the two-phonon

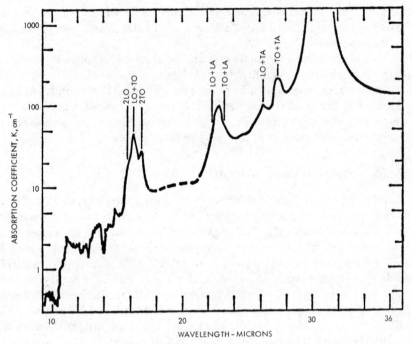

Fig. 28.   Absorption spectrum of AlSb. (Reprinted by permission of W. J. Turner and W. E. Reese [28].)

Table XII
Two-Phonon Combination Bands in AlSb As Calculated from the Four Characteristic Zone-Boundary Phonon Frequencies LO = 316 cm^{-1}, TO = 297 cm^{-1}, LA = 132 cm^{-1}, and TA = 65 cm^{-1}*

$\lambda(\mu)$	ν (obs), cm^{-1}	f_{obs}	Assignment	Calculated	
				ν, cm^{-1}	f_{calc}
27.50	363	2.83	TO + TA	362	2.84
26.30	380	2.86	LO + TA	381	2.82
23.29	429	2.27	TO + LA	429	2.23
22.33	448	1.91	LO + LA	448	2.20
16.81	595	1.70	2 TO	594	1.62
16.31	613	1.57	LO + TO	613	1.59
15.80	633	1.56	2 LO	632	1.55

* From W. J. Turner and W. E. Reese [28].

bands along with their temperature dependence is given in Table XII. Turner and Reese's assignment of the two optical phonons is quite arbitrary. As we shall note later, for a crystal of low ionicity such as AlSb, it is preferable to designate the highest zone-boundary phonon as TO and the next highest as LO in a spherical Brillouin zone scheme. The assignment of the combination bands, however, is not otherwise affected. Thirty-eight addition bands were observed in the short-wavelength side of the 2 LO band at 15.8 μ. These were assigned as three- and four-phonon combination bands, as shown in Table XIII. Many of the combinations violate the selection rules [104]. However, in the absence of calculated or experimental dispersion curves for AlSb, it is almost impossible to arrive at a unique assignment. A value of the effective ionic charge q^* may be obtained from the relation

$$\mathbf{q}^* = 2\pi v_t \left(\frac{\varepsilon_0 - \varepsilon_\infty}{4\pi}\right)^{\frac{1}{2}} \left(\frac{3}{\varepsilon_\infty + 2}\right) \left(\frac{\mu v_a}{Ze}\right)^{\frac{1}{2}} \qquad (14.131)$$

due to Szigeti [128], where μ is the reduced mass per unit cell of volume v_a, and Ze is the formal charge on an ion. For AlSb, $\mathbf{q}^* = 0.48$, indicative that it is only partially ionic with a considerable amount of covalency.

Based on coulomb attractive forces and nearest-neighbor repulsive forces, Brout [129] has given the sum rule

$$\sum_{i=1}^{6} \omega_i^2(\mathbf{k}) = \frac{18\, r_0}{\chi\mu} \qquad (14.132)$$

where $\hbar\omega_i(\mathbf{k})$ is the phonon energy of the ith vibrational branch at wave vector $\mathbf{k}$; r_0 is the nearest-neighbor distance; and χ is the compressibility. This relation is not expected to hold precisely correct for the zinc-blende type of crystal. However, it may be expected that the sum of the squares of the lattice frequencies should be a constant for each wave vector $\mathbf{k}$. A comparison of the Brout sum calculated at the zone boundary with its value at the zone center ($\mathbf{k} = 0$) thus should provide an additional check on the assignment of the zone-edge phonon frequencies. Such a comparison for AlSb is made below. At $\mathbf{k} = 0$,

$$\omega_l^2 + 2\omega_t^2 = 1.13 \times 10^{28}\ \sec^{-2} \qquad (14.133)$$

From the frequencies LO, TO, LA, and TA of Table XII, the value of the Brout sum for some $\mathbf{k}$ at or near the zone boundary is

$$LO^2 + 2\,TO^2 + LA^2 + 2\,TA^2 = 1.08 \times 10^{28}\ \sec^{-2} \qquad (14.134)$$

A value of $1.11 \times 10^{28}\ \sec^{-2}$ is obtained when LO and TO designations are reversed. In equations (14.133) and (14.134), the factor 2 arises from the assumed degeneracy of the transverse modes.

Table XIII
Summary of the Three- and Four-Phonon Combination Bands in AlSb*

Line	No.	Assignment	Expected		Observed	
			ν (cm^{-1})	$\lambda(\mu)$	ν (cm^{-1})	$\lambda(\mu)$
	1	3 LO + LA	1080	9.26	1081	9.26
	2	2 LO + TO + LA	1061	9.42	1059	9.44
	3	LO + 2 TO + LA	1042	9.60	1042	9.60
	4	3 TO + LA	1023	9.77	1023	9.77
	5	3 LO + TA	1013	9.87	1012	9.88
	6	2 LO + TO + TA	994	10.06	996	10.03
	7	2 TO + LO + TA	975	10.26	975	10.26
	8	3 TO + TA	956	10.46	959	10.43
A		3 LO	948	10.55	949	10.54
B		2 LO + TO	929	10.76	928	10.78
C		LO + 2 TO	910	10.99	911	10.98
	9	2 LO + 2 LA	896	11.16	896	11.16
D		3 TO	891	11.22	893	11.20
	10	3 LO − TA	883	11.32	883	11.33
	11	LO + TO + 2 LA	877	11.40	876	11.42
	12	2 LO + TO − TA	864	11.57	865	11.56
	13	2 TO + 2 LA	858	11.65	858	11.65
	14	LO + 2 TO − TA	845	11.83	846	11.82
	15	2 LO + LA + TA	829	12.06	828	12.07
	16	3 LO − LA	816	12.25	816	12.26
	17	LO + TO + LA + TA	810	12.34	809	12.36
	18	2 LO + TO − LA	797	12.55	795	12.57
	19	2 TO + LA + TA	791	12.64	792	12.63
	20	LO + 2 TO − LA	778	12.85	778	12.85
E		2 LO + LA	764	13.09	765	13.07
	21	3 TO − LA	759	13.17	760	13.15
F		LO + TO + LA	745	13.42	745	13.42
	22	LO + TO + 2 TA	743	13.46	741	13.50
G		2 TO + LA	726	13.77	728	13.73
	23	2 TO + 2 TA	724	13.81	725	13.80
	24	LO + 3 LA	712	14.04	712	14.05
	25	2 LO + LA − TA	699	14.31	700	14.29
H		2 LO + TA	697	14.35	696	14.36
	26	TO + 3 LA	693	14.43	689	14.52
	27	LO + TO + LA − TA	680	14.70	683	14.63
I		LO + TO + TA	678	14.75	675	14.83
	28	2 TO + LA − TA	661	15.13	661	15.12
J		2 TO + TA	659	15.17	660	15.16

* From W. J. Turner and W. E. Reese [28].

B. Space-Group Selection Rules and Infrared-Active Phonon Processes in GaAs [105]

In this section, it is intended to apply the space-group selection rules to the analysis of the infrared lattice spectrum [130] of GaAs. The CP in the zinc-blende-structure [131] phonon spectrum are at Γ, X, L, and W. Here, Γ corresponds to the Brillouin zone center at $\mathbf{k} = 0$. Values of the phonon modes near Γ for GaAs are available from the infrared reflection measurement of Hass and Henvis [132] and the transmission measurement of Iwasa et al. [133]. The assignment of the phonon frequencies belonging to the CP at X and L is straightforward from the dispersion relations given by Waugh and Dolling [134]. The CP at W presents some problems. The W is located at $(1, \frac{1}{2}, 0)$ and is not in a (110) plane, for which the experimental values are available. For reasons given by Phillips [135], it will be a good approximation to assume $v(W) = v(\Sigma)$, where Σ denotes a CP along the [110]

Table XIV
Critical Points, Coordinates, and Phonon Modes for GaAs
(Space Group T_d^2)

Critical point	Coordinate	Phonon mode	Frequency,* cm^{-1}
Γ	$(0, 0, 0)$ $(1/a)$	TO (Γ)	267 (273)†
		LO (Γ)	285 (297)†
X	$(2\pi, 0, 0)$ $(1/a)$	TO (X)	252
		LO (X)	241
		LA (X)	227
		TA (X)	79
L	(π, π, π) $(1/a)$	TO (L)	261
		LO (L)	238
		LA (L)	209
		TA (L)	62
$W(\Sigma)$‡	$(2\pi, \pi\ 0)$ $(1/a)$	O_1 (W)	263
		O_2 (W)	215
		O_3 (W)	250
		A_1 (W)	116
		A_2 (W)	188
		A_3 (W)	79

* Data from Waugh and Dolling [134] if not stated otherwise.
† The values obtained by Hass and Henvis [132] from infrared reflection measurements. A value of 290 cm^{-1} has been obtained by Iwasa et al. [133] for the LO (Γ) mode from infrared transmission measurements on thin films.
‡ See text for explanation.

Table XV
Infrared-Active Two-Phonon Processes in GaAs

Assignment	Calculated, cm^{-1}	Observed,* cm^{-1}
2 TO $(\Gamma)^\dagger$	534 (546)†	
2 LO (Γ)	570 (594)	578; 593
TO (Γ) + LO (Γ)	552 (570)	578
2 TO (X)	504	509
2 TA (X)	158	
2 TO (L)	522	523
2 TA (L)	124	
2 LO (L)	476	468
2 LA (L)	418	411
TO (X) + LO (X)	493	494
TO (X) + LA (X)	479	
TA (X) + LO (X)	320	321
TA (X) + LA (X)	306	307
TO (X) + TA (X)	331	333
LO (X) + LA (X)	468	468
TO (L) + LO (L)	499	494
TO (L) + LA (L)	470	468
TA (L) + LO (L)	300	307
TA (L) + LA (L)	271	
TO (L) + TA (L)	323	321
LO (L) + LA (L)	447	442
O_1 (W) + O_2 (W)	478	
$O_1 + O_3^\ddagger$	513	509
$O_1 + A_1$	379	387
$O_1 + A_2$	451	456
$O_1 + A_3$	342	
$O_2 + O_3$	465	468
$O_2 + A_1$	331	333
$O_2 + A_2$	403	411
$O_2 + A_3$	294	
$O_3 + A_1$	366	(360)
$O_3 + A_2$	438	442
$O_3 + A_3$	329	333
$A_1 + A_2$	304	307
$A_1 + A_3$	195	
$A_2 + A_3$	267	

* Data from Cochran *et al.* [130]. The values in the parentheses are not listed by Cochran *et al.* but estimated from their diagrams.

† Using Hass and Henvis [132] assignment.

‡ In the rest of this column, O_i and A_i (where $i = 1, 2, 3$) mean $O_i(W)$ and $A_i(W)$.

direction near $\mathbf{k}$ = (0.75, 0.75, 0). There is, however, a lack of clear separation at W of branches into transverse and longitudinal. The CP phonon assignment for GaAs is given in Table XIV. There are a total of 16 different phonon frequencies that may take part in infrared-active combinations.

All infrared-active two-phonon combinations are listed in Table XV. It is evident that the total number of possible combinations far exceeds the absorption maxima recorded by Cochran et al. [130]. However, all observed bands below 600 cm^{-1} can be assigned to one or the other of the multitude of binary combinations. The infrared-active three-phonon combinations are even more numerous. High-resolution spectra and extension of range may resolve some of the multiple assignments. In Table XVI, some possible assignments of the high-frequency bands are shown.

Although the observed infrared spectra of zinc-blende and diamond types of crystals may be completely understood in terms of the CP phonons known from the neutron scattering, the reverse, i.e., an unambiguous assignment of the CP phonons from the observed infrared spectrum, may

Table XVI
Some Infrared-Active Three-Phonon Processes in GaAs

Observed,* cm^{-1}	Assignment	Calculated, cm^{-1}
578	TA (X) + TO (Γ) + LA (X)	573 (579)†
	TA (X) + TO (X) + LO (X)	572
	TO (L) + LO (L) + TA (X)	578
593	TA (X) + LO (Γ) + LA (X)	591 (603)
	TA (X) + TO (Γ) + LO (X)	587 (593)
	2 TO (L) + TA (X)	601
694	TO (L) + LA (L) + LA (X)	697
	LO (L) + LA (L) + TO (X)	699
714	TO (L) + LA (L) + LO (X)	711
	TO (X) + LA (X) + LO (X)	720
(752)	TO (L) + LO (L) + TO (X)	751
	LO (X) + LA (X) + LO (Γ)	753 (765)
	TO (X) + TO (Γ) + LA (X)	746 (752)
	3 TO (X)	756
770	2 TO (L) + TO (X)	774
	TO (X) + LO (Γ) + LA (X)	764 (776)

* See footnote * of Table XV.
† See footnote † of Table XV.

not always be possible. In the absence of detailed dispersion curves, however, it is still possible to obtain broad zone-boundary values of LO, TO, LA, and TA, etc., consistent with the space-group selection rules.

C. Lattice Infrared Spectra of Diatomic Crystals

1. NaCl and CsCl Type of Crystals

Czerny [136], Barnes and Czerny [137], Barnes [138], Hohls [139], Klier [140], Heilman [141], Abeles and Mathieu [142], and Frohlich [143] have measured the absorption and reflection spectra of a large number of alkali halides. Hass [144] recently studied the reflection spectra of LiCl, LiBr, KF, RbF and CsF, which were not previously investigated. Randall, Fuller, and Montgomery [145] have measured the infrared absorption spectra of thin films of alkali halides, including that of LiI. Jones et al. [81] also have determined the infrared lattice vibration parameters for a large number of alkali halides and silver halides. In addition to the absorption maximum corresponding to the TO mode, a few subsidiary maxima on the short-wavelength side of the absorption and reflection spectra are also observed. However, the number of multiphonon absorption bands in the alkali halides is less than that in a III–V compound such as AlSb discussed in an earlier section. It appears that the extent of multiple-phonon interactions giving rise to absorption maxima in a crystal depends markedly on the effective ionic charge. For example, the spectra of the II–VI compounds, which are less ionic than the alkali halides or the alkaline earth halides [146], show more structure than the spectra of the last two types of solids. The multiphonon structure is even more pronounced in III–V compounds and in homopolar crystals such as diamond.

Hass [147] has studied the reflection spectrum of NaCl from 300 to 985 °K in the region of the fundamental lattice absorption. The damping constant, deduced from a single dispersion formula, varied approximately with the square of the temperature. This is shown to be consistent with theoretical expectation. Similar work by Jasperse, Kahan, Plendl, and Mitra [51] on LiF and MgO has already been discussed.

The alkali halides do not show first-order Raman spectra. The second-order spectrum has been reviewed by Mitra [148] and thus need not be repeated here. Stekhanov and Eliashberg [149] have reported the spectra of KBr and KBr-KCl mixed crystals. They find that, in addition to a second-order spectrum, there occurs a first- order spectrum in the mixed crystals owing to the lattice defects. They have also measured [150] the Raman spectrum due to the local mode of Li^+ in KCl.

Table XVII lists the zone-center phonon frequencies v_t and v_l (calculated from the LST formula [26]), the high- and low-frequency dielectric

Table XVII
Zone-Center Phonon Frequencies and the Effective Ionic Charge of Some NaCl and CsCl Type of Crystals

Crystal	ν_t cm^{-1}	ν_l cm^{-1}	ε_0	ε_∞	q*	Ref.
LiH	590	1120	12.9	3.6	0.52	151
LiF	306	659	8.81	1.9	0.87	51
LiCl	191^a	398	12.0	2.7	0.73	144
LiBr	159^b	325	13.2	3.2	0.68	144
LiI	144					145
NaF	244	418	5.1	1.7	0.93	69; 152
NaCl	164	264	5.9	2.25	0.74	81
NaBr	134	209	6.4	2.6	0.70	81
NaI	117	176	6.6	2.91	0.71	145
KF	190	326	5.5	1.5	0.88	144
KCl	142	214	4.85	2.1	0.81	81
KBr	113	165	4.9	2.3	0.76	81
KI	101	139	5.1	2.7	0.71	81
RbF	156	286	6.5	1.9	0.95	144
RbCl	116	173	4.9	2.2	0.84	81
RbBr	88	127	4.9	2.3	0.84	81
RbI	75	103	5.5	2.6	0.75	81
CsF	127					145
CsCl	99	165	7.2	2.6	0.85	81
CsBr	73	112	6.5	2.8	0.78	81
CsI	62	85	5.65	3.0	0.67	81
TlCl	63	158	31.9	5.1	0.80	81
TlBr	43	101	29.8	5.4	0.82	81
AgCl	106	196	12.3	4.0	0.71	81; 153
AgBr	79	138	13.1	4.6	0.70	81; 153
MgO	401	718	9.64	3.01	0.88	51
NiO	401	580	11.75	5.7	0.84	154
CoO	349	546	13.0	5.3	0.89	154
MnO	262	552	22.5	4.95	1.08	155

^{a} Randall et al. [145] give ν_t = 204 cm^{-1}.
^{b} Randall et al. [145] give ν_t = 171 cm^{-1}.

constants ε_∞ and ε_0, and the Szigeti effective ionic charge q*. Hanlon and Lawson [156] have noticed a correlation between the values of q* and the difference in atomic polarizability calculated from the Clausius–Mosotti formula [157]. The zone-boundary phonon frequencies of only a few alkali halides have so far been determined, using neutron spectrometry [36] and from the analysis of multiphonon Raman spectra [107].

2. II–VI Compounds

So far, infrared lattice vibration parameters have been reported for ZnO, ZnS, ZnSe, ZnTe, CdS, CdSe, and CdTe. ZnO, CdS, and CdSe crystallize in the hexagonal wurtzite structure, whereas ZnSe, ZnTe, and CdTe crystallize in the cubic zinc-blende structure. ZnS can occur in either structure. Small anisotropy splittings have been noted for all the wurtzite crystals. Table XVIII gives the values of the TO and LO phonon frequencies at $k \sim 0$, then high- and low-frequency dielectric constants, and the Szigeti effective ionic charge.

In Table XIX the $k = 0$ factor-group selection rules for the optic phonons of crystals of diamond, zinc-blende and wurtzite structures are given. Although the wurtzite structure is very closely related to the zinc-blende structure, it has a more complex first-order spectrum owing to its having more particles per unit cell. An unambiguous determination of these modes is possible from the study of polarized Raman spectrum.

The analysis of Raman spectrum of uniaxial crystals obtained by conventional techniques may be quite involved and may often predict inaccurate results. The use of lasers as the excitation source simplifies the matter; especially, extensive angular-dependence and polarization measurements make unambiguous identification of symmetry character of the Raman-active lattice modes possible. Recently, such measurements on CdS and ZnO have been made.

The Raman spectrum of CdS was measured by Poulet and Mathieu [165] by conventional excitation. Tell, Damen, and Porto [166], on the other hand,

Table XVIII
$k \sim 0$ Optic Phonon Frequencies
and the Effective Ionic Charge of Some II–VI Compounds

Crystal	ν_t cm^{-1}	ν_l cm^{-1}	ε_0	ε_∞	q^*	Ref.
CdTe	141	168	10.20	7.13	0.74	158
CdSe (∥)^a	166	211	10.16	6.30	0.86	159
(⊥)	172	210	9.29	6.20	0.80	
CdS (∥)	235	306	9.00	5.32	0.91	212
(⊥)	242	304	8.47	5.32	0.87	
ZnTe	179	206	10.38	7.8	0.62	161
ZnSe	207	253	9.20	6.10	0.72	162
ZnS	274	350	8.3	5.0	0.96	163
ZnO (∥)	377	575	8.75	3.75		164
(⊥)	406	589	7.8	3.70		

^{a} For wurtzite structures, ∥ and ⊥ notations relate to the c axis.

Table XIX
k = 0 Factor Group Fundamentals

Structure	Space group	Particles per unit cell	Symm. species, k = 0	No. of modes, degeneracy	Activity
Diamond	$O_h{}^7$	2	F_{2_g}	1 (3)	R
Zinc blende	$T_d{}^2$	2	F_2	1 (3)	R; IR
Wurtzite	C_{6v}^4	4	A_1	1	R; IR
			B_1	2	Inactive
			E_1	1 (2)	R; IR
			E_2	2 (2)	R

have recently measured the Raman spectrum of this crystal excited by an He–Ne laser (6328 Å) and an argon ion laser (5145 Å). Their data on angular-dependence and polarization measurements do not agree with Poulet and Mathieu's assignment. Tell et al.'s assignment of the Raman-active $k \sim 0$ modes are 44 cm^{-1} (E_2), 252 cm^{-1} (E_2), 235 cm^{-1} (E_1 transverse), 228 cm^{-1} (A_1 transverse), and 305 cm^{-1} (E_1, A_1 longitudinal). They have also observed seven multiphonon transitions in the Raman spectrum.

The lattice modes of CdS belonging to the edge of the first Brillouin zone were derived by Marshall and Mitra [167] from the analysis of the infrared spectrum of the crystal. Balkanski, Besson, and LeToullec [160] have also investigated the infrared spectrum of CdS. Their results are essentially in agreement with those of Marshall and Mitra. However, Balkanski et al.'s more extensive data make possible more accurate determination of the acoustic mode frequencies. The polarized infrared spectrum between 80 and 600 cm^{-1} consists of about 35 bands, which are satisfactorily explained as multiphonon combinations of 10 representative zone-edge modes (in wave numbers): $LO_1 = 297$, $TO_1 = 259$, $TO_{1'} = 238$, $LO_2 = 277$, $TO_2 = 199$, $TO_{2'} = 208$, $TO_3 = 69$, $LA_1 = 151$, $TA_1 = 56$, and $TA_2 = 51$. The Raman bands not assigned to allowed $k \sim 0$ fundamentals may also be explained as multiphonon combinations of the above zone-edge modes [162].

The Raman spectrum of single-crystal ZnO was first reported by Mitra and Bryant [168], who used excitation by an Hg arc source. Damen, Porto, and Tell [169] have subsequently obtained the Raman spectrum, using laser radiation. They give the following assignments for the $k \sim 0$ Raman-active fundamentals: 101 cm^{-1} (E_2), 437 cm^{-1} (E_2), 407 cm^{-1} (E_1 transverse), 380 cm^{-1} (A_1 transverse), 583 cm^{-1} (E_1 longitudinal), and 574 cm^{-1} (A_1 longitudinal). They have also observed a few multiphonon bands. Mitra and Marshall [170] have reported multiphonon infrared absorption in ZnO.

Multiphonon structure has been observed in cubic ZnS by Deutsch [171]. Marshall and Mitra [167] have shown that Deutsch's assignment is erroneous and have given a new assignment. They have also investigated the transmission spectrum of hexagonal ZnS, which differs slightly from the cubic form in the positions of the absorption bands.* Aven, Marple, and Segall [172] have reported multiphonon absorption bands of ZnSe, which were assigned by Mitra [173] using a four-phonon scheme in the spherical Brillouin zone approximation. Mitra [162] has subsequently measured a more detailed multiphonon structure in the infrared spectrum of ZnSe and has given a CP phonon assignment. Geick, Perry, and Mitra [159] have noted a few multiphonon structures in the infrared spectrum of CdSe and assigned them, using a four-phonon scheme. A multiphonon infrared spectrum of ZnTe has been reported by Nahory and Fan [174] and by Narita, Harada, and Nagasaka [161]. Their assignments of the four zone-boundary phonons agree well except for the LO frequency. Narita et al.'s assignment seems to be relatively more acceptable. Infrared-active multiphonon bands have been reported for CdTe by Mitshuishi [158], Stafsudd, Haak, and Radisavljevic [175], and Bottger and Geddes [176]. Mitshuishi has given a CP phonon assignment, using the X and L CP. Except for the lack of constancy of the Brout sum rule [equation [14.132]], his assignment seems more acceptable when compared with the four-phonon schemes used by Stafsudd et al. or Bottger and Geddes. Slack, Ham, and Chrenko [177] have given yet another rival set of four-phonon assignments from a study of the vibronic structure of F_e^{+2} in CdTe. In the absence of any experimental dispersion curves obtainable by neutron scattering measurements or of theoretical curves with the use of a shell model, it appears that it is a futile attempt to obtain unique CP phonon frequencies from infrared data. At best, one may arrive at broad zone-boundary values of LO, TO, LA, and TA phonons with due care that there are no gross violations of the space-group selection rules.

3. *III–V Compounds*

The reststrahlen spectra of InSb, InAs, GaSb, InP, and AlSb have been investigated by Hass and Henvis [178] and of GaP by Kleinman and Spitzer [179]. The reststrahlen spectra of BN and BP have been measured by Gielisse, Mitra, Plendl, Griffis, Mansur, Marshall, and Pascoe [180], and that of AlN by Collins, Lightowlers, and Dean [181]. Mooradian and Wright [182] have measured the first-order Raman spectra of AlSb, InP GaP, and GaAs and have obtained accurate values of the zone-center TO

---

* Note added in the proof: Brafman and Mitra (*Phys. Rev.*, July 15, 1968) have measured the Raman spectrum of hexagonal ZnS. Their assignment of the $k \sim 0$ Raman-active fundamentals: 72 cm^{-1} (E_2), 286 cm^{-1} (E_2), 273 cm^{-1} (E_1, A_1 transverse), and 351 cm^{-1} (E_1, A_1 longitudinal).

Table XX
k ~ 0 Optic Phonon Frequencies
and the Effective Ionic Charge of Some III–V Compounds

Crystal	ν_t cm^{-1}	ν_l cm^{-1}	ε_0	ε_∞	q*	Ref.
InSb	185	197	17.88	15.68	0.42	178
InAs	219	243	15.15	12.25	0.56	178
InP	304	345	12.61	9.61	0.68	182; 178
GaSb	231	240	15.69	14.44	0.33	178
GaAs	269	292	12.90	10.90	0.51	182; 178
GaP	367	403	10.18	8.46	0.58	182; 179
AlSb	319	340	11.21	9.88	0.48	182; 28
AlP	440	501				182a
AlN	667	916	9.14	4.84	1.2	181
BP	820	834			0.25^a	180
BN	1056	1304	7.1	4.5	1.14	180; 183

^{a} Estimated value.

and LO phonons. Brafman, Mitra, and Gielisse [183] have measured the Raman spectra of AlN, BN, and BP. In Table XX we present the long-wavelength optic phonon frequencies, the high- and low-frequency dielectric constants, and the Szigeti effective charge.

The multiphonon structure in the infrared spectra of III–V compounds has recently been reviewed by Spitzer [184] and Stierwalt and Potter [185]. Except for GaAs for which neutron scattering data [134] exist, many of the CP phonon frequencies reported for the other III–V compounds may not be too reliable. Multiphonon infrared absorption has also been reported [180, 181] for BN and AlN.

4. IV–IV Compounds

In homopolar crystals such as Ge and Si, the k ~ 0 optic phonons are infrared inactive because the effective ionic charge or the derivative of the dipole moment during such vibrations is zero. However, type I diamonds do exhibit an absorption band corresponding to this mode. The type II diamonds, on the other hand, do not show this fundamental absorption. The long-wavelength optic phonons of the diamond type of crystals are degenerate and Raman active. Both types I and II diamonds show strong first-order Raman scattering around 1332 cm^{-1}. The interesting spectroscopic difference between types I and II diamonds will not be further elaborated here. The reader is instead referred to an earlier review [186]. For Ge and Si, the zone-center optical modes have been determined by Brockhouse and Iyengar [187] and Brockhouse [188], respectively, from inelastic neutron scattering measurements. Their first-order Raman spectra excited by an argon laser have also been recently reported [189].

The multiphonon absorption in diamond has been described by Hardy and Smith [190] and in Ge and Si by Brockhouse and Iyengar [187] and Johnson [191], respectively. The lattice absorption in SiC has been reported by Spitzer, Kleinman, and Frosch [192] and Patrick and Choyke [193]. SiC occurs in both wurtzite and zinc-blende structures, but the infrared spectra of the two are only slightly different. SiC is partially ionic, and hence a separation of the TO and LO modes at the Brillouin zone center takes place. The relevant data on the IV–IV compounds including the phonon frequencies at X and L CP for diamond [194], Si, and Ge are given in Table XXI.

Table XXI
Critical-Point Phonon Frequencies per Centimeter for Diamond, Silicon, Germanium, and Silicon Carbide

	Γ*	X†	L†
Diamond			
TO	1332	1076	1210
LO	1332	1193	1242
LA	0	1193	1035
TA	0	803	552
Silicon			
TO	520	463	489
LO	520	411	420
LA	0	411	378
TA	0	150	114
Germanium			
TO	301	275	280
LO	301	230	247
LA	0	230	215
TA	0	82	65

	Silicon carbide	
	Zone center‡	Zone boundary§
LO	970	851
TO	793	770
LA	0	540
TA	0	363

* Raman scattering.
† Neutron scattering.
‡ Infrared measurements [193]
§ Infrared measurements [192]

5. Trends in the Characteristic Phonon Frequencies of the Zinc-Blende Type of Crystals

In the absence of a knowledge of CP phonons, the multiple structure observed in the lattice spectra of the zinc-blende type of crystals may be accounted for in terms of four characteristic phonon energies. These belong to the four branches at the Brillouin zone boundary: TA (transverse acoustic), LA (longitudinal acoustic), LO (longitudinal optical), and TO (transverse optical). In certain cases, however, some doubts exist concerning the identification of the optical branch phonons as transverse and longitudinal modes. No such difficulty exists for the acoustical branches. Here, the transverse frequency is smaller than the longitudinal one for all values of the propagation vector $\mathbf{k}$, except at the zone center where both are zero. Keyes [195] has noted some correlation between the zone-boundary phonon energies of the zinc-blende type of crystals and the mass ratio of their constituent atoms and their ionicity $\mathbf{q}^*$. The latter correlation enables one to assign the higher two optical phonon frequencies to definite longitudinal and transverse branches.

For homopolar crystals such as Si ($\mathbf{q}^* = 0$), the optical branches are degenerate at the zone center. However, the TO phonons have higher energies than the LO at or near the zone boundary. For crystals of low ionicity, the LO mode is at a higher frequency compared with the TO mode (LST rule) at the zone center. However, near the zone boundary, the situation is reversed with the TO modes having higher values than the LO modes. This is the case with InSb, GaAs, and GaSb. In the case of GaP and AlSb, however, the LO and TO assignments of Keyes are just the reverse of those of Kleinman and Spitzer [179] and Turner and Reese [28]. The LO phonons have higher energy than the TO phonons at the zone boundary as well as at the zone center for crystals of large effective ionic charge such as NaCl. Figure 29 shows the effect of ionicity on the dispersion curves as discussed above. Keyes has also predicted that the frequencies of the optical branches at the zone boundary coincide for a crystal with effective ionic charge $\mathbf{q}^* = 0.7$ (this value of $\mathbf{q}^*$ we term *intermediate*), and this has been noted by Mitra [173] for ZnSe.

Mitra and Marshall [196] have noted that, for both II–VI and III–V compounds, the $\mathbf{q}^*$ is a linear function of $(v_1 - v_t)/v_t$, the zone-center optical phonon separation. This indicates that, for these compounds, the product of the reduced mass μ per unit cell and the cell volume is a constant. They also note that the zone-center and the zone-boundary Brout sums $\sum_{i=1}^{6} \omega_i^2(\mathbf{k})$ not only are approximately equal, but they also are a linear function of the reciprocal of the reduced mass. This indicates that the interatomic force constants do not change appreciably from member to member in either

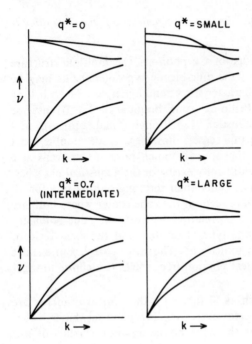

Fig. 29. The effect of ionicity on dispersion curves.

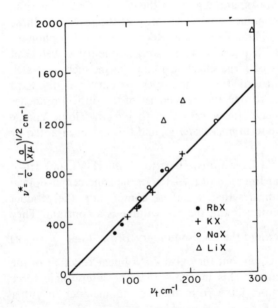

Fig. 30. Relation between ν^* and ν_t for the alkali halides.

II–VI or III–V compounds. Departure from this rule has been noted when lighter members are included [180]. For the alkali halides, on the other hand, such is not true. A plot of v_t versus $(a/\chi\mu)^{\frac{1}{2}}$ yields a straight line, where χ and a are, respectively, the compressibility and lattice constants. This is shown in Fig. 30.

The Brout sum is related to χ, μ, and r_0, as was noted in equation (14.132). For the zinc-blende and diamond type of crystals, the appropriate equation is

$$\sum_{i=1}^{6} \omega_i^2 (\mathbf{k}) = \frac{16 \sqrt{3} \, r_0}{\chi\mu} \tag{14.135}$$

Equation (14.135), contrary to the general belief [197], is obeyed by the III–V and the IV–IV compounds surprisingly well. The calculated and observed values of χ in a few cases are compared in Table XXII.

D. Long-Wavelength Optic Phonons of Mixed Crystals

In recent years, considerable interest has been evidenced in the study of mixed crystals. In this section, we present a brief review of this interesting topic. As far as the behavior of the long-wavelength optical phonons are concerned, there seem to exist two types of mixed crystals. In one class of mixed systems, termed here the one-mode behavior type, each of the $\mathbf{k} \sim 0$ optic mode frequencies (infrared-or Raman-active) varies continuously and approximately linearly with concentration from the frequency characteristic of one end member to that of the other end member. Furthermore, the strength of the mode remains approximately constant. Examples [198–205] of this behavior are $Na_xK_{1-x}Cl$, $Ni_xCo_{1-x}O$, KCl_xBr_{1-x}, $K_xRb_{1-x}Cl$, $(Ca, Ba)_xSr_{1-x}F_2$, $GaAs_xSb_{1-x}$, $Zn_xCd_{1-x}S$, and $ZnSe_xTe_{1-x}$. In the other class of mixed crystal systems, termed here the two-mode behavior type, two-phonon frequencies for each of the allowed optic modes of the pure crystal are observed to occur at frequencies close to those of the end members. In addition, the strength of each phonon mode of the mixed crystal is approximately proportional to the mole fraction of the component it represents. Examples [206–213] of the two-mode type of behavior are $InAs_xP_{1-x}$, Ge_xSi_{1-x}, $GaAs_xP_{1-x}$, CdS_xSe_{1-x}, and ZnS_xSe_{1-x}.

It is worth mentioning here two additional observations concerning the behavior of mixed crystals. The reststrahlen bands of the pure components are mostly broad (ionic nature) and overlap in the one-mode type of systems. The separation between the $\mathbf{k} \sim 0$ LO and TO frequencies of a diatomic crystal may be considered the spread of a reststrahlen band. In the two-mode type of systems, on the other hand, the bands are well separated. Balkanski et al. [210] have suggested that the long-range average crystal potential variation may be large enough for the one-mode systems to shift the eigen-

Table XXII
Comparison of Calculated and Observed Compressibilities

Compound	χ in 10^{-12} cm^2/dyne		
	Brout	Szigeti	Experimental*
C	0.227	0.227	0.226
Si	1.010	1.010	1.023
Ge	1.173	1.173	1.330
SiC	0.378	0.557	(0.473)
InSb	2.078	1.934	2.132
InAs	1.750	1.568	1.727
GaSb	1.716	1.644	1.855
GaAs	1.338	1.193	1.337
AlSb	1.751	1.624	1.694
CdTe	3.220	2.691	2.360
CdS	2.21	1.768	1.631
ZnS	1.47	1.309	1.196
LiF	1.87	1.17	1.43
LiCl	5.84	4.60	3.17
LiBr	3.40	2.28	3.91
NaF	2.05	1.75	2.06
NaCl	4.18	4.22	3.97
NaBr	6.23	5.74	4.74
NaI	7.98	7.42	6.21
KF	4.40	3.01	3.14
KCl	6.17	5.40	5.50
KBr	7.26	6.37	6.45
KI	9.73	8.45	8.07
RbF	5.43	3.44	3.66
RbCl	6.72	5.92	6.17
RbBr	7.87	6.59	7.24
RbI	9.64	6.32	9.02

* For source of experimental data see Mitra and Marshall [196].

frequencies of each constituent toward a unique value but is not enough for the two-mode systems. This seems to lead one to make a distinction between ionic and covalent crystals and to suggest this as a criterion for the one-mode and two-mode types of behavior. However, the fact that CdS_xSe_{1-x} and ZnS_xSe_{1-x} belong to the two-mode class whereas $Cd_xZn_{1-x}S$ and $ZnSe_xTe_{1-x}$ belong to the one-mode type does not lend support to such a criterion. Secondly, Lucovsky et al. [214] have pointed out that the conditions necessary for the existence of localized and gap modes [215] in mixed systems are fulfilled for two-mode systems and are not for the one-mode systems. Chang and Mitra [216] have derived a simple criterion for the prediction of one- or two-mode type of behavior of a mixed crystal. A two-mode (one-mode) type of mixed crystal of the type $AB_{1-x}C_x$ must (not) have one substituting element whose mass is smaller than the reduced mass of the compound formed by the other two elements.

The behavior of long-wavelength optic phonons as functions of mole fraction is shown in Fig. 31. The top part of the figure illustrates the behavior of $Ni_{1-x}Co_xO$, which belongs to the one-mode class [199], while the bottom part illustrates the behavior of $ZnS_{1-x}Se_x$, which belongs to the two-mode class [213]. The one-mode type of behavior of a crystal AB_xC_{1-x} is often treated [217] by the *virtual crystal* model, where one replaces the atoms B and C with a virtual atom such that the reduced mass μ of the primitive unit cell is given by

$$\mu^{-1} = M_A^{-1} + X M_B^{-1} + (1-x) M_C^{-1} \qquad (14.136)$$

It also assumes that the nearest-neighbor force constant varies linearly from one end member to the other. To explain the behavior of the TO modes of $GaAs_xP_{1-x}$, which is a two-mode type of mixed crystal, Chen, Shockley, and Pearson [208] have developed the *random-element-isodisplacement* (REI) model. It assumes that the cations as well as the anions of like species

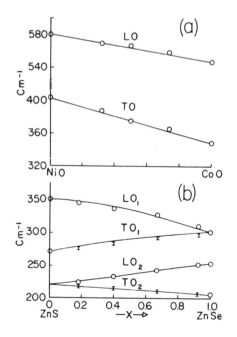

Fig. 31. Concentration dependence of long-wavelength optic modes of mixed crystals. Solid line, Chang and Mitra theory; circles and I, experimental points. (a) $Ni_{1-x}Co_xO$; (b) $ZnS_{1-x}Se_x$.

vibrate as rigid units, i.e., each unit vibrates with one phase and amplitude. One notices that this assumption of isodisplacement is exactly true for the reststrahlen frequency in an ordered diatomic crystal, since there is no phase shift from the unit cell for $k \sim 0$ modes. The second assumption of the REI model is randomness, which suggests that each atom is subjected to forces produced by a statistical average of the neighbors and no effects of order are actually present. The statistical average of the neighbors depends on the fraction parameter x only. Chang and Mitra[216] have modified the REI model to include the polarization field, which enables it to predict the long-wavelength LO frequencies as well. Under certain conditions, the modified REI model may also be made to work for mixed crystals with *one-mode* type of behavior, including the prediction of the virtual-crystal-model result. Given the properties of the "pure" end members AB and AC, the Chang and Mitra theory can then determine whether AB_xC_{1-x} is going to display one-mode or two-mode type of behavior and predict the concentration dependence of the TO and LO modes. A comparison of the prediction of this theory with experimental data on $Ni_xCo_{1-x}O$ and ZnS_xSe_{1-x} is shown in Fig. 31.

E.   The Infrared Spectrum of Gypsum, $CaSO_4 \cdot 2H_2O$

The structure and the vibrational modes of gypsum have been discussed previously. In this section, we describe briefly the infrared spectra of single crystals of gypsum studied by Hass and Sutherland[11] in the 450 to 3800 cm^{-1} region. The measurements consist of transmission and reflection spectra obtained with plane-polarized radiation on three different crystal sections (010, $\bar{1}$01, and $\bar{2}$01). The Raman spectrum of the crystal has been investigated by Krishnan[218], Rousset and Lochet[219], and Stekhanov[220].

Hass and Sutherland observed that, for plane-polarized light incident normal to the (010) plane, the maximum dichroism usually occurred when the E vector made an angle (ϕ) of either 9 or 99° with the a axis. For the ($\bar{1}$01) face, the E vector was either parallel to the C_2 axis and gave the A_u bands or in the (010) plane with $\phi = 33°$ and gave the B_u bands. For the ($\bar{2}$01) section, the E vector was either parallel to the C_2 axis (A_u bands) or in the (010) plane with $\phi = 99°$ (B_u bands). The transmission and reflection spectra of a (010) section of gypsum are shown in Fig. 32. The observed frequencies, intensities, and dichroism of the absorption bands of gypsum recorded by Hass and Sutherland are listed in Table XXIII.

The four fundamental vibration frequencies of the free sulfate ion were reported by Kohlrausch[221] from a study of the Raman spectrum in solution. These, along with their degeneracies and symmetry species, are given below:

$$v_1 = 981 \text{ cm}^{-1} A(1)$$
$$v_2 = 451 \text{ cm}^{-1} E(2)$$
$$v_3 = 1104 \text{ cm}^{-1} F_2(3)$$
$$v_4 = 613 \text{ cm}^{-1} F_2(3)$$

Table XXIII
Observed Frequencies, Intensities, and Dichroism of $CaSO_4 \cdot 2H_2O$ Absorption Bands*

Assignment	Species	Direction,[†] deg	Method [‡]	Frequency, cm^{-1}	Intensity,[§] cm^{-1}	Width,** cm^{-1}
$v_{R'}(H_2O)$	B_u	99	R	450	..	vb
$v_{R'}(H_2O)$	A_u	..	R	580	..	b
$v_4(SO_4)$	A_u	..	R	602	30	17
$v_4(SO_4)$	B_u	9	R	604	30	20
$v_4(SO_4)$	B_u	99	R	672	35	16
$v_1(SO_4)$	A_u	..	R	1000	..	..
$v_3(SO_4)$	B_u	9	R	1118	120	32
$v_3(SO_4)$	A_u	..	R	1131	100	34
$v_3(SO_4)$	B_u	99	R	1142	130	27
$v_4 + v_4(SO_4)$	B_u	9	R	1205	..	..
$v_2(H_2O)$	B_u	9	R	1623	12	16
$v_2(H_2O)$	A_u	..	R	1685	6.5	25
$v_3 + v_1(SO_4)$	B_u	19	T	2112	..	..
$v_3 + v_1(SO_4)$	B_u	124	T	2130		
$v_{R''} + v_2(H_2O)$	B_u	9	T	2198		
$v_{R''} + v_1(H_2O)$	B_u	9	T	2235		
$v_2 + v_2(H_2O)$	B_u	19	T	3248		
$v_2 + v_2(H_2O)$	B_u	19	T	3350		
$v_1(H_2O)$	B_u	19	R	3410	42	49
$v_1(H_2O)$	A_u	..	R	3430	8	b
$v_3(H_2O)$	B_u	..	R	3490	2	
$v_3(H_2O)$	B_u	Unp††	T	3495		
$v_3(H_2O)$	A_u	...	R	3537	60	67
$v_3(H_2O)$	B_u	19	T	3560		

* Taken from Hass and Sutherland [11].
† Direction of transition moment in (010) plane.
‡ R = data derived from reflection measurements; T = data derived from transmission measurements.
§ Where numerical values are given, the number is the integrated intensity defined as $K = \int k \, dv$, where k is the extinction coefficient and v is the frequency per centimeter.
** vb = very broad; b = broad.
†† Unpolarized.

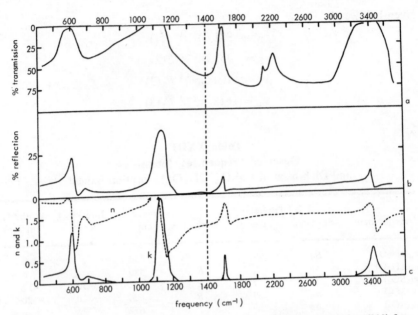

Fig. 32. Transmission (a) and reflection (b) spectra of gypsum normal to (010) face. n and k deduced from middle curve are plotted in (c). (Reprinted by permission of M. Hass and G. B. B. M. Sutherland [11].)

Thus, in the infrared spectrum of the crystal, one expects to find one band near 980 cm^{-1} polarized parallel to the C_2 axis, which is indeed the case. The v_1 frequency of the SO_4 ion in gypsum is found at 1000 cm^{-1} and shows A_u character. The v_2 mode of the sulfate ion is expected to be split into two infrared-active A_u bands near 450 cm^{-1}. However, no A_u bands were observed near 450 cm^{-1}. These bands (like the v_1 band) may be very weak or may be outside the range of observation of Hass and Sutherland. The v_3 mode, expected near 1104 cm^{-1}, indeed shows three components. The 1118 and 1142 cm^{-1} bands have B_u character, while the third one at 1131 cm^{-1} has A_u character in accordance with the predictions made in the section on the normal modes of gypsum. The v_4 mode also has three components in the infrared spectrum of the crystal, the 604 and 672 cm^{-1} bands belonging to the B_u species and the 602 cm^{-1} band belonging to the A_u species. Table XXIV lists 16 of the 18 fundamental vibrations assigned to the sulfate ion in gypsum. This includes also the Raman (*gerade* representations) assignments of Rousset and Lochet.

A comparison of the frequencies of the crystal modes with the corresponding free ion modes reveals that all the components of v_1 and v_3 vibra-

Table XXIV
Sulfate Ion Fundamentals*

Solution		Gypsum	K
$981\nu_1$ — A_g		1006	
— A_u		1000	
$451\nu_2$ — A_g		492	
A_u		–	
A_g		413	
A_u		–	
$1104\nu_3$ — A_g		1144	
A_u		1131	(100)
B_g		1138	
B_u		1142	(130)
B_g		1117	
B_u		1118	(120)
$613\nu_4$ — A_g		621	
A_u		602	(30)
B_g		669	
B_u		672	(35)
B_g		623.5	
B_u		604	(30)

* From Hass and Sutherland [11].

tions have higher values in the solid than in the solution. For the ν_2 and ν_4 vibrations, on the other hand, the removal of degeneracy and interactions within the unit cell have produced a distribution of frequencies around the unperturbed modes of the free ion. Since the ν_1 and ν_3 modes arise primarily from the stretching of the S—O bonds, it may be concluded that the S—O force constant is slightly higher when the sulfate ions are in the crystal than when isolated. The degeneracy is almost completely removed in the ν_2, ν_3, and ν_4 modes. The Raman frequencies corresponding to ν_2 are split by 79 cm^{-1}. Rousset and Lochet have suggested that such a large splitting may indicate that the SO_4 ion is no longer tetrahedral in the gypsum crystal but is distorted to a lower symmetry class. This may indeed be the case since two of the sulfate oxygens and the water molecules are most probably hydrogen bonded in the crystal.

The three nondegenerate fundamental vibration frequencies of the water molecule in the gaseous state are $\nu_1 = 3652\,\text{cm}^{-1}$, $\nu_2 = 1595\,\text{cm}^{-1}$, and $\nu_3 = 3756\,\text{cm}^{-1}$ [222]. In the crystal, each of them is split into four modes, two of which are infrared active (*ungerade* representations), and the other two, Raman active (*gerade*). Near each of the above frequencies, one may therefore expect to find two absorption bands. One of these should be polarized parallel to the C_2 axis (A_u), with the other perpendicular to this axis (B_u). All but two of the H_2O modes in gypsum have been assigned by Hass and Sutherland. These are presented in Table XXV.

It may be noted that the hydroxyl stretching modes (ν_1 and ν_3) in the crystal have considerably lower frequencies than the corresponding free molecular modes. On the other hand, the ν_2 bending modes have higher values, which is indicative of appreciable hydrogen bonding between the H_2O molecules. The splitting of each H_2O mode in the crystal into four is due solely to the correlation field effect. Here the neighboring H_2O molecules are coupled together through hydrogen bonding to the same oxygen atom of a sulfate ion. This is in contrast to the sulfate ions themselves, which have no direct coupling. Thus the splitting of the water modes in general is considerably larger than the splitting of the corresponding modes of the sulfate ions.

Table XXV
Water Fundamentals*

Vapor		Gypsum	K
$3657.1\nu_1$	A_g	3404.5	
	A_u	3430	(8)
	B_g	3402.5	
	B_u	3410	(42)
$1595\nu_2$	A_g	–	
	A_u	1685	(6.5)
	B_g	–	
	B_u	1623	(12)
$3755.8\nu_3$	A_g	3496.5	
	A_u	3537	(60)
	B_g	3498	
	B_u	3490	(2)

* From Hass and Sutherland [11].

Three reflection bands at $450 \text{cm}^{-1}(B_u)$, $580 \text{cm}^{-1}(A_u)$, and $1205 \text{cm}^{-1}(B_u)$ and a number of absorption bands observed between 2100 and 2250 cm^{-1} are yet to be assigned. Hass and Sutherland assigned the first two of these bands to rotatory lattice modes of the water molecules, in analogy with those found for ice. The rest of the bands are assigned as combination bands on the assumption that the factor-group selection rules are obeyed, and these assignments may be regarded as tentative.

F.   The Infrared Spectrum of Crystalline Cyclopropane

The choice of cyclopropane as an example serves two purposes. First, it is a molecular crystal with small and fairly symmetrical molecules, which makes it possible to gain a clear understanding of the influence of the crystal-line field on the molecular modes. Second, the crystal structure of solid cyclopropane is not known, which makes it interesting to determine if such information can be inferred from the study of its infrared spectrum.

The cyclopropane molecule C_3H_6 belongs to the D_{3h} point group. The 21 molecular modes of vibration are distributed among the following irreducible representations: $3A_1' + 1A_1'' + 1A_2' + 2A_2'' + 4E' + 3E''$. Barring degeneracies, there are thus 14 distinct fundamental frequencies. In the infrared spectrum only the A_2'' and E' modes are permitted. In the Raman spectrum the allowed modes belong to the A_1', E', and E'' symmetry species. The modes belonging to the A_1'' and A_2' are inactive in both the spectra. The vibrational spectrum of the cyclopropane molecule has been studied by Lord and his co-workers [223] and by Mathai, Shepherd, and Welsh [224] and a satisfactory assignment of the molecular modes has been presented (see Table XXVI).

The infrared spectra of single crystals of C_3H_6 have been investigated by Brecher, Krikorian, Blanc, and Halford [225]. They have grown and examined two specimens with different orientations. In the absence of any X-ray crystallographic data, the relative orientations of the two samples were determined by comparing the polarized and unpolarized spectra of the two single crystals, the spectrum of a polycrystal, and also their extinctions by visible light. It was established that the two samples had (001) and $(0kl)$ cross sections, respectively. Spectra were recorded at $-160 °C$ and $-190 °C$.

Brecher et al. noted that the spectra in the condensed phases were characterized by the following features when compared with the vapor phase spectrum: (1) The frequency of the absorption bands shifted very little ($< 1\%$) from phase to phase, which indicated a weak static field effect, as expected for van der Waals binding; (2) the number of bands observed increased as one went from vapor to liquid to solid; and (3) the PQR structure of the vapor bands was not resolved in the liquid state, which produced

Table XXVI
The Fundamental Vibrations of Cyclopropane*

Mode	Species	Motion	Vapor, cm⁻¹		Liquid, cm⁻¹		Crystal, cm⁻¹
			Raman	IR	Raman	IR	IR
ν_1	A_1'	C-H stretch	3038		3027		
ν_2	A_1'	CH₂ deformation	1454	f	1453	1453	1454
ν_3	A_1'	Ring breathing	1188	f	1188	1191	1194
ν_4	A_1''	CH₂ twist	f(1133)	f	1131	1129	1133
ν_5	A_2'	CH₂ wag	f	f			1078
ν_6	A_2''	C-H stretch	f	3101		3081	3073
ν_7	A_2''	CH₂ rock		852			855
ν_8	E'	C-H stretch	3020	3025	3009	3013	3004
ν_9	E'	CH₂ deformation	1442	1442	1434	1432	1434
ν_{10}	E'	Ring deformation		1028	1023	1026	1027
ν_{11}	E'	CH₂ wag	866	866	866	865	865
ν_{12}	E''	C-H stretch	3082	f	3075		3073
ν_{13}	E''	CH₂ twist	1188	f	1178	1191	1200
ν_{14}	E''	CH₂ wag	739	f	741	741	749

* From Brecher et al. [225]. Bands forbidden in the vapor are denoted by the letter f.

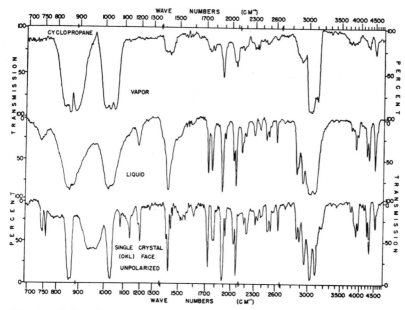

Fig. 33.   Infrared spectra of cyclopropane vapor (at 25 °C), liquid (at −125 °C), and crystal (at −190 °C), recorded with equivalent absorbing paths. (Reprinted by permission of C. Brecher et al. [225].)

rather broad absorption bands, whereas the bands originating from the solid were quite sharp. Figure 33 gives such a comparison.

Since the static field shift was limited, the assignment of the internal modes of the crystal was relatively simple. However, a number of modes that are inactive in the infrared spectrum of the free molecule became active in the solid state. Some of the modes also showed correlation field splitting, which was of the same magnitude as that due to the static field at the site. A large number of bands in addition to the fundamentals were recorded for the single-crystal samples. Most of these were assigned by Brecher et al. as combinations of molecular modes. Satisfactory numerical agreement was obtained in most cases by using the factor group (**k** = 0) fundamentals, which is not surprising in view of the limited static field effect. The observed and expected (from the polarizations of the combining frequencies) dichroism of the combination bands agreed in most cases.

A number of bands that were not assigned as either internal fundamentals or their combinations were assigned to the combination of internal with external modes. However, since no far-infrared spectrum of the solid was recorded, the frequencies of the combining lattice modes were deduced from the combination modes. The six external degrees of freedom of a cyclopropane molecule belong to the following irreducible representations of the D_{3h} point group: $1A'_2 + 1A''_2 + 1E' + 1E''$. These degrees of freedom of the isolated molecule give rise to the translatory and rotatory lattice modes in the crystal. However, since the crystal structure of the solid is not known, the exact number and nature of the lattice modes could not be enumerated with certainty. Brecher *et al.* were able to locate four primary lattice modes from the combination tones and the combining molecular mode frequencies. These are: 55 cm^{-1} (E' translation), 70 cm^{-1} (A'_2 libration), 78 cm^{-1} (A''_2 translation), and 92 cm^{-1} (E'' libration). These values were ascertained from both sum and difference bands arising from combination with one or more internal modes. The internal–external combination band assignment was also consistent with the following intensity considerations: The difference (hot) bands originate from higher lattice vibrational levels and hence are weaker and much more temperature-dependent than the sum bands. The complete assignment by Brecher *et al.* is given in Table XXVII.

By a judicious use of infrared spectral data, Brecher *et al.* have been successful in deducing a reasonable structure for cyclopropane crystals. This is not known from any other type of measurement. It was done by systematically testing the subgroups of the molecular point group (which is known) for selection rules which are in accord with the spectral data on the crystal. The acceptable subgroups were further examined within possible factor groups to predict the observed polarization properties of the absorption bands. It was found that only two site groups C_2 and $C_s(\sigma v)$ explain the appearance in the crystal spectra of the bands that are forbidden by the molecular selection rules. Now, by considering the dichroism of the absorption bands, the first of the two possibilities may be eliminated, which leaves $C_s(\sigma v)$ as the unique site group. The occurrence of a set of three crystal fundamentals corresponding to a single molecular fundamental (e.g., v_{14}), such that each component appears exclusively along one of the three orthogonal and nonequivalent axes, suggests that the cyclopropane crystal belongs to the orthorhombic system with only three possible isomorphous factor groups: C_{2v}, D_2, and D_{2h}. Of these, only C_{2v} and D_{2h} are compatible with the $C_s(\sigma v)$ site symmetry. There are 22 (10 primitive) space groups corresponding to the factor group C_{2v} and 28 (16 primitive) space groups isomorphous with the factor group D_{2h}. A few of these are eliminated on the ground that they do not contain reflection planes as sites. This still leaves 30 possible space groups to which the cyclopropane crystal may belong.

To supplement the spectroscopic data, Brecher *et al.* now assume that, in the crystal, the cyclopropane molecules are packed as closely as their geometry allows. This is usually the case with most hydrocarbons. It is also noted that crystalline cyclopropane is a trifle heavier than the liquid. Since

Table XXVII
The Infrared Absorption Bands of Cyclopropane*

Vapor, cm⁻¹	Liquid, cm⁻¹	Crystal, cm⁻¹	Polar	Assignment
	741 m, bd	742 m	I	$\nu_{14}(E'')$
		747 m	II	
		756 m	III $\ggg$ II $>$ I	$\nu_{11} - \nu_{(Rx,\,Ry)}(A_1'' + A_2'' + E'')$
		776† vvw, bd	I $>$ II $\approx$ III	$\nu_{11} - \nu_{Rz}(E')$
		798† w, bd	I $\approx$ II $\approx$ III	
841 vs P 852 pd Q' 866 vs Q 895 vs R	865 vs, bd	855 vs	III $\gtrsim$ II	$\nu_7(A_2'')$
		865 vs	I $\approx$ II $\approx$ III	$\nu_{11}(E')$
		937† s, bd	I $\approx$ II $\approx$ III	$\nu_{11} + \nu_{Rz}(E')$
		941‡ s, bd		
		960† s, bd	I $\gtrsim$ II	$\nu_{11} + \nu_{(Rz,\,Ry)}(A_1'' + A_2'' + E'')$
		965‡ s, bd		
1009 vs P 1028 vs Q 1051 vs R	1026 vs, bd	1027 vs	I $\approx$ II $\approx$ III	$\nu_{10}(E')$
		1078 m	I $\approx$ II $\approx$ III	$\nu_5(A_2')$
		$\approx$ 1096§ vw, bd	I $\approx$ II $\approx$ III	$\nu_{10} + \nu_{Rz}(E')$
		$\approx$ 1124§ vw, bd	I $\approx$ II $\approx$ III	$\nu_{10} + \nu_{(Rx,\,Ry)}(A_1'' + A_2'' + E'')$
	$\approx$ 1129 vw, sh	1133 m	I	$\nu_4(A_1')$

* From Brecher et al. [225]. Translatory and rotatory lattice modes are denoted ν_T and ν_R, respectively. Molecular directions with which the motions are associated are included in the subscript. Symbols: vs, very strong; s, strong; m, medium; w, weak; vw, very weak; vvw, very very weak; bd, broad; pd, poorly defined; sh, shoulder.

† Appears only in the spectrum taken at −160°C.

‡ Appears only in the spectrum taken at −190°C.

§ Appears at both temperatures, but is too poorly defined for precise measurement.

Table XXVII (Continued)

Vapor, cm⁻¹	Liquid, cm⁻¹	Crystal, cm⁻¹	Polar	Assignment
	1191 m, bd	$\left\{\begin{array}{l}1189\text{ m}\\1200\text{ m}\end{array}\right.$	$\left.\begin{array}{l}\text{II}\\\text{I}>\text{III}\end{array}\right\}$	$\nu_{13}(E'')$ and $\nu_3(A_1')$
		$\approx 1270\text{-}75$ vvw, pd		$\nu_3+\nu_{Tz}$
$\left.\begin{array}{l}1422\text{ m, sat }P\\1442\text{ m, }Q\\1465\text{ m, sat }R\end{array}\right\}$	1432 s	$\left\{\begin{array}{l}1424\text{ m}\\1434\text{ s}\end{array}\right.$	$\left.\begin{array}{l}\text{II}\\\text{III}>\text{I}\end{array}\right\}$	$\nu_9(E')$
	1453 w, sh	$\left\{\begin{array}{l}1449\text{ m}\\1459\text{ w}\end{array}\right.$	$\left.\begin{array}{l}\text{II}\\\text{III}\end{array}\right\}$	$\nu_2(A_1')$
	1504 m, sh, bd	1496 vw sat	$\text{I}\approx\text{II}\approx\text{III}$	$2\nu_{14}(E')$
		$\left.\begin{array}{l}\approx 1509^{\dagger}\text{ vvw, pd}\\\approx 1513^{\ddagger}\text{ vvw, pd}\end{array}\right\}$	$\text{I}\approx\text{II}\approx\text{III}$	$\nu_2+\nu(Tx,Ty)(E')$
		1518 m	$\text{II}>\text{III}$	$2\nu_{14}(A_1')$
		$\left.\begin{array}{l}1532^{\dagger}\text{ w, sat}\\1538^{\ddagger}\text{ w}\end{array}\right\}$	$\text{II}\approx\text{III}$	$\nu_2+\nu_{Tz}(A_2'')$
		1600 w	$\text{I}>\text{II}\approx\text{III}$	$\nu_7+\nu_{14}(E')$
		1714 m, sh	$\text{II}>\text{III}$	$2\nu_7(A_1')$
1740 w		1724 s	$\text{I}\approx\text{II}\gtrsim\text{III}$	$2\nu_{11}(A_1'+E')$
1776 w	1764 m	$\left\{\begin{array}{l}1774\text{ w}\\1786\text{ m}\end{array}\right.$	$\text{II}\approx\text{III}\gg\text{I}$	$\nu_{10}+\nu_{14}(A_1''+A_2''+E'')$
		1812 vvw	$\text{II}\approx\text{III}>\text{I}$	$\nu_5+\nu_{14}(E'')$
1881 m	1882 s	$\left\{\begin{array}{l}1883\text{ s}\\1894\text{ s, sh}\end{array}\right.$	$\left.\begin{array}{l}\text{I}\approx\text{II}\approx\text{III}\\\text{I}\end{array}\right\}$	$\nu_{10}+\nu_{11}(A_1'+A_2'+E')$
	1918 w	1923 m	$\text{I}>\text{II}\gtrsim\text{III}$	$\nu_5+\nu_{11}(E')$
	2040 m	$\left\{\begin{array}{l}2035\text{ m, sh}\\2049\text{ m}\end{array}\right.$	$\text{I}\approx\text{II}\approx\text{III}$ $\text{II}>\text{III}>\text{K}$	$\nu_7+\nu_{13}(E')$ $\nu_{11}+\nu_{13}(A_1''+A_2''+E'')$

Table XXVII (Continued)

Vapor, cm^{-1}	Liquid, cm^{-1}	Crystal, cm^{-1}	Polar	Assignment
2058 m, sh P 2083 m Q 2102 m, sh R	2082 s	2083 s 2091 m, sh	II $\approx$ III I	$\nu_5 + \nu_{10}(E')$
		2126 vw	II $>$ III	$2\nu'_5(A_1')$
2183 w	2169 m	2172 w	II $>$ III	$\nu_9 + \nu_{14}(A_1'' + A_2'' + E'')$
	2209 w, sat	2196 m 2206 m 2217 w sh	II $>$ III I II $\gtrsim$ III	$\nu_3 + \nu_{10}(E')$
2314 w	2304 w	2294 w 2311 w 2319 w	I $\approx$ II $\approx$ III II $\approx$ III I	$\nu_{10} + \nu_{13}(A_1'' + A_2'' + E')$ $\nu_9 + \nu_{11}(A_1' + A_1' + E')$ $\nu_2 + \nu_{11}(E')$ or $\nu_4 + \nu_{13}(E')$
2372 vw		2370 vw	I $\approx$ II $\approx$ III	$2\nu_{13}(A_1' + E')$
2456 vw	2450 w	2444 m 2456 m	I $>$ II $>$ III I $>$ II $>$ III	$\nu_9 + \nu_{10}(A_1' + A_2' + E')$ $\nu_2 + \nu_{10}(E')$
	2486 m	2487 m	II $>$ I $\approx$ III	$\nu_5 + \nu_9(E')$
	2518 w	2523 w 2540	I $\approx$ III II	$\nu_{10} + 2\nu_{14}(A_1' + A_2' + 2E')$
2612 w	2610 m	2608 m 2726 vw	I $\approx$ II $\approx$ III I $\approx$ II $\approx$ III	$\nu_3 + \nu_9(E')$ $\nu_{10} + 2\nu_{11}(A_1' + A_2' + 2E')$
2862 w, sh	2859 s	2855 m	II $>$ I $\approx$ III	$2\nu_9(A_1' + E')$

Table XXVII (Continued)

Vapor, cm^{-1}	Liquid, cm^{-1}	Crystal, cm^{-1}	Polar	Assignment
2875 w, sh	2870 s	2877 m	II > I ≈ III	$\nu_2 + \nu_9(E')$
2921 m, sat	2928 s	2934 s	I ≈ II ≈ III	$\nu_9 + 2\nu_{14}(A_1' + A_2' + 2E')$
3003 vs, pd P 3025 vs Q 3048 vs, pd R	3013 vs	3004 vs	I ≈ II ≈ III	$\nu_8(E')$ and $\nu_1(A_1')$
3101 s, sat Q 3120 s, sat R	3081 vs	3073 vs	II ≈ III > I	$\nu_6(A_2'')$ and $\nu_{12}(E')$
	3178 vw, sh	3168 m	I ≈ III > II	$\nu_2 + 2\nu_{11}(A_1' + E')$
3805 vs	3814 w	3824 m	I ≈ II ≈ III	$\nu_6 + \nu_{14}(E')$ or $\nu_{12} + \nu_{14}(A_1' + A_2' + E')$
3892 vw	3874 m	3863 m	I > II ≈ III	$\nu_1 + \nu_{11}(E')$ or $\nu_8 + \nu_{11}(A_1' + A_2' + E')$
	3929 w	3918 m	I > II ≈ III	$\nu_7 + \nu_{12}(E')$
4206 w	4201 m	4194 m	II ≈ III > I	$\nu_3 + \nu_8(E')$ and $\nu_8 + \nu_{13}(A_1'' + A_2'' + E')$
4281 w	4276 m	4264 s	I ≈ II ≈ III	$\nu_6 + \nu_{13}(E')$ or $\nu_{12} + \nu_{13}(A_1' + A_2' + E')$
4522 w	4517 s	4510 m	II ≈ III ≫ I	$\nu_9 + \nu_{12}(A_1' + A_2' + E')$
		4560 m, sh	II ≈ III	$\nu_2 + \nu_6(A_2'')$ and $\nu_{12} + 2\nu_{14}(A_1'' + A_2'' + 2E')$
4621 vw		4614 w, sh	II ≈ III	$\nu_6 + 2\nu_{14}(A_1'' + A_2'' + 2E')$

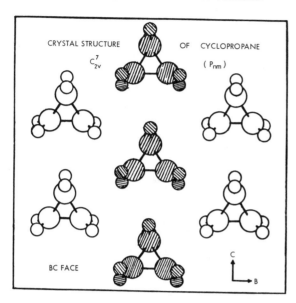

CRYSTAL STRUCTURE   OF   CYCLOPROPANE
C_{2v}^7 (P_{nm})

BC FACE

Fig. 34. Proposed crystal structure of cyclopropane, belonging to space group $C_{2v}^7(P_{nm})$. The shaded molecules are not in the same plane as the unshaded ones and are inclined oppositely. (Reprinted by permission of C. Brecher et al. [225].)

the structure of a single molecule is completely known, all the possible space groups are examined against the above criteria. Brecher et al. concluce that $C_{2v}^7 (Pnm)$ is the most likely space group.

There are two symmetry elements in the C_s site group, whereas the C_{2v} factor group consists of four elements. Since the number of spatially nonequivalent molecules per primitive unit cell is given by the ratio of the order of the factor group to the order of the site group, there must be two molecules per primitive (Bravais) unit cell. Furthermore, the C_{2v}^7 space group in an orthorhombic unit cell requires that one molecule be at the corner and the other at the body center. The crystal structure of the cyclopropane as proposed by Brecher et al. is shown in Fig. 34.

G. The Long-Wavelength Lattice Modes of Naphthalene and Anthracene [226]

1. Factor Group Analysis

Both naphthalene and anthracene crystallize in the monoclinic system, space group $P2_1/c-C_{2h}^5$ with two molecules in the unit cell [227]. The isolated molecules belong to the D_{2h} point group. The cleavage plane is the (ab) plane, which is the one normally developed. Each molecule in the crystal occupies a site of symmetry C_i. Schematic representation of the unit cell of anthracene and naphthalene is shown in Fig. 35. The distribution of the lattice fundamentals between the rotatory R' and translatory (T') types of modes along

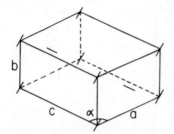

Fig. 35.   The unit cell of anthracene and naphthalene.

with their selection rules is presented in Table XXVIII. The six rotatory lattice modes are Raman active and infrared inactive. Of the three infrared-active translatory lattice modes, two belong to the A_u species with transition moment primarily parallel to the b axis of the crystal, while the remaining one is of the B_u type and should have a transition moment in the (ac) plane, i.e., perpendicular to the b axis.

2.   *Librational Modes*

In the first approximation, one may assume that a molecule is free of interactions with neighboring molecules, but its rotation is restricted by a force field which represents the combined effect of all other groups. If a simple harmonic-potential function in the librational angle θ is assumed, the wave equation will be given by

$$ H\psi = \tfrac{1}{2} f \theta^2 \ \psi - \frac{\hbar^2}{2\mu} \left[\frac{1}{r^2 \sin \theta} \frac{\partial}{\partial \theta} \left(\sin \theta \frac{\partial \psi}{\partial \theta} \right) + \frac{1}{r^2 \sin^2 \theta} \frac{\partial^2 \psi}{\partial \phi^2} \right] \quad (14.137) $$

Since interactions between internal and librational modes are neglected, r may be considered constant. By expanding the trigonometric functions of θ and neglecting the terms of the order θ^3 and higher, the wave equation can be simplified to essentially that of a two-dimensional harmonic oscillator.

Table XXVIII
Distribution of Lattice Modes for Naphthalene and Anthracene

C_{2h}^5	T	T'	R'	Infrared	Raman
A_g	0	0	3	f	$p(R_z)$
A_u	1	2	0	$p(M_b)$	f
B_g	0	0	3	f	$p(R_x, R_y)$
B_u	2	1	0	$p(M_a, M_c)$	f

The energy levels are given by

$$
E = (n+1) \left[\hbar\omega + \frac{2B^2}{\hbar\omega} \left(\frac{m^2}{15} - \frac{1}{60} \right) - \frac{2B^4}{\hbar^3 \omega^3} \left(\frac{m^2}{15} - \frac{1}{60} \right)^2 \right.
$$

$$
\left. + \frac{4B^6}{\hbar^5 \omega^5} \left(\frac{m^2}{15} - \frac{1}{60} \right)^3 + \dots \right] + \frac{B}{3} (m^2 - 1) \tag{14.138}
$$

where n is the librational quantum number; $B = \hbar^2/2I$, the librational fundamental $\omega = \sqrt{f/I}$; and I, the moment of inertia. The quantum numbers n and m are restricted to values

$$
n = 0, 1, 2, 3, \dots
$$
$$
|m| = n, (n-2), (n-4), \dots \tag{14.139}
$$

For a heavy molecule such as naphthalene or anthracene and for the presently available spectroscopic resolution, only the first term of equation (14.138) need be considered. Since a molecule of naphthalene (or anthracene) is capable of executing distinct and independent hindered rotations around the three molecular axes, there should be three librational modes (in the first approximation) given by

$$
\omega \text{ (long axis)} = \sqrt{\frac{f_x}{I_x}} \tag{14.140}
$$

$$
\omega \text{ (short axis)} = \sqrt{\frac{f_y}{I_y}} \tag{14.141}
$$

and

$$
\omega \text{ (perpendicular to molecular plane)} = \sqrt{\frac{f_z}{I_z}} \tag{14.142}
$$

where I_x, I_y, and I_z respectively represent the moments of inertia for rotations around the long and short axes of the molecule and an axis normal to the molecular plane.

However, from Table XXVIII, it is evident that there are six Raman-active rotatory lattice modes in naphthalene or anthracene instead of three. This is because each unit cell contains two molecules. The three modes belonging to the A_g species differ from those belonging to the B_g species only so far as the phase difference between the two molecules is concerned. Since the intermolecular forces are not appreciable, it would appear that the frequencies of the two types of motion may not differ very much. By assuming a force constant f', where $f' \ll f$, for the interaction between two molecules within a unit cell, the equations (14.140), (14.141), and (14.142) will now be modified to

$$\omega \text{ (long axis)} \begin{cases} A_g = \sqrt{\dfrac{f_x - f_x'}{I_x}} \\[2ex] B_g = \sqrt{\dfrac{f_x + f_x'}{I_x}} \end{cases} \qquad (14.140\text{a})$$

$$\omega \text{ (short axis)} \begin{cases} A_g = \sqrt{\dfrac{f_y - f_y'}{I_y}} \\[2ex] B_g = \sqrt{\dfrac{f_y + f_y'}{I_y}} \end{cases} \qquad (14.141\text{a})$$

and

$$\omega \ (\perp) \begin{cases} A_g = \sqrt{\dfrac{f_z - f_z'}{I_z}} \\[2ex] B_g = \sqrt{\dfrac{f_z + f_z'}{I_z}} \end{cases} \qquad (14.142\text{a})$$

Since there are only six observable frequencies and six unknown force constants, an unequivocal assignment of the Raman-active librational modes is not possible unless recourse is taken to crystals of isotopically substituted molecules, for which it may be assumed to a very good approximation that the force constants remain unchanged, whereas the moments of inertia change.

With the use of Kastler and Rousset's data [228] on naphthalene and our unpublished data [226] on deuterated naphthalene $C_{10}D_8$, single-crystal anthracene, and deuterated anthracene $C_{14}D_{10}$, the assignment of the librational lattice fundamentals are presented in Table XXIX.

3. Translational Modes

The far-infrared spectrum of naphthalene is shown in Fig. 36 for two different thicknesses of randomly oriented samples. Three low-frequency bands at 102, 66, and 49 cm^{-1} may be regarded as the three translatory lattice modes. Spectra of a single crystal of naphthalene measured by Hadni [229] with the use of polarized radiation are shown in Fig. 37. The band at 99 cm^{-1} is polarized parallel to the b axis and should be the logical choice for an A_u mode. The band at 66 cm^{-1}, on the other hand, is polarized perpendicular to the b axis and hence should be assigned to the B_u mode. The remaining band at 49 cm^{-1} observed by us should then be assigned to the other A_u mode. Harada and Shimanouchi [230] have observed this band at 53 cm^{-1}. Hadni, on the other hand, observes at least four very weak low-

Table XXIX
Assignment of Librational Modes (Comparison of Calculated
and Experimental Values for the Deuterated Compounds)

		Anthracene		
		d_0 (obs)	d_{10} (calc)	d_{10} (obs)
Around long axis	A_g	122	114	112
	B_g			
Around short axis	A_g	63	60	
	B_g	68	65	63
Around normal to the plane	A_g	38	36	
	B_g	44	42	

		Naphthalene		
		d_0 (obs)	d_8 (calc)	d_8 (obs)
Around long axis	A_g	109	99	98
	B_g	127	116	112
Around short axis	A_g	74	68	68
	B_g	76	70	
Around normal to the plane	A_g	46	43	42
	B_g	54	50	

frequency bands in this region. Some of these may be spurious or impurity bands. The assignment presented above is in agreement with Harada and Shimanouchi's independent conclusion with the use of an unpolarized spectrum.

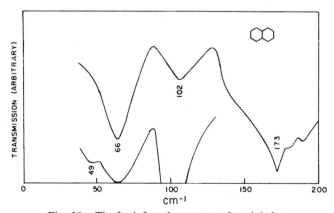

Fig. 36.   The far-infrared spectrum of naphthalene.

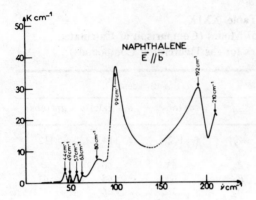

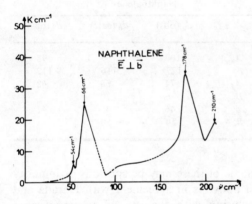

Fig. 37. The effect of polarization on the spectrum of monocrystalline naphthalene. (Reprinted from A. Hadni [229].)

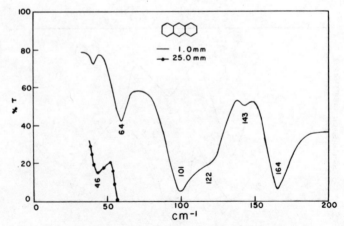

Fig. 38. The far-infrared spectrum of anthracene.

The far-infrared spectra of monocrystalline anthracene of two different thicknesses are shown in Fig. 38. One may note four low-frequency bands at 122, 101, 64, and 46 cm^{-1}. Of these, only three are to be assigned to translational lattice modes. The polarized spectrum is shown in Fig. 39. The band at 121 cm^{-1} is seen to be polarized parallel to the b axis, while the band at 65 cm^{-1} seems to be polarized perpendicular to the b axis. Thus, one will be tempted to assign the band at 121 cm^{-1} to the A_u species and the 65 cm^{-1} band to B_u species. However, there remains one puzzle. The strong band at 101 cm^{-1} ($\|b$) [105 cm^{-1}($\perp b$)] seems to be present in both polarizations. Since this band is also observed [229] in a solution of anthracene in benzene, it must be ascribed to molecular (internal) origin. The low-frequency band at 46 cm^{-1} is assigned to the remaining A_u mode.

The assignments of the translatory lattice modes of naphthalene and anthracene are presented in Table XXX.

In order to estimate the frequencies of the translatory lattice modes, we use the Lennard-Jones potential between a pair of appropriate mass points representing the molecule.

$$\phi_{ij} = 4\varepsilon\left[\left(\frac{\sigma}{r_{ij}}\right)^{12} - \left(\frac{\sigma}{r_{ij}}\right)^{6}\right] \tag{14.143}$$

where σ and ε are constants to be determined, and r_{ij} is the distance between

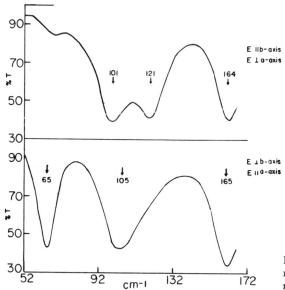

Fig. 39. The effect of polarization on the spectrum of monocrystalline anthracene.

Table XXX
Measured and Calculated Values of the Translatory Lattice Vibrations

	Measured, cm^{-1}	Calculated,* cm^{-1}
Naphthalene		
$\nu_1(A_u)$	99	88
$\nu_2(B_u)$	66	65
$\nu_3(A_u)$	49	36
Anthracene		
$\nu_1(A_u)$	121	89
$\nu_2(B_u)$	65	63
$\nu_3(A_u)$	46	29

* From equation (14.147).

the ith and the jth points. The total energy at any one point due to all the others is

$$\phi = \sum_j \phi_{ij} \quad (j \neq i) \tag{14.144}$$

By neglecting surface effects and kinetic energy of the atoms the total cohesive energy of the crystal is given by $U = N_0 \phi$ (N_0 instead of $2N_0$ occurs in order to avoid counting each pair twice). Thus,

$$U = 4 N_0 \varepsilon \left[A \left(\frac{\sigma}{r_0} \right)^{12} - B \left(\frac{\sigma}{r_0} \right)^6 \right] \tag{14.145}$$

where r_0 is the equilibrium nearest-neighbor distance. Lennard-Jones and Ingham [231] have performed the lattice sums to obtain A and B for cubic crystals. Mennerich, Mitra, and Singh [232] have performed these lattice sums for monoclinic crystals. In particular, they have obtained potential-energy parameters for naphthalene, anthracene, durene, dibenzyl, chrysene, coronene, ovalene, and biphenyl.

For the three lattice modes, the corresponding nearest-neighbor force constants are obtained approximately (ignoring anisotropy) as

$$f_{1,2,3} = \tfrac{1}{3} \left[\Phi''(r) + \frac{2}{r} \Phi'(r) \right]_{r=r_1, r_2, r_3} \tag{14.146}$$

In terms of the heat of sublimation, one obtains for the frequency

$$v_i = \frac{23.8}{r_i} \left(\frac{U_0}{M}\right)^{\frac{1}{2}} \tag{14.147}$$

where M is the molecular weight, and U_0 the heat of sublimation in calories per mole. Heat of sublimation for anthracene and naphthalene have been listed by Bondi [233]. They are U_0 (anthracene) = 22.8 $\pm$ 1.1 kcal/mole and U_0 (naphthalene) = 16 $\pm$ 0.8 kcal/mole. For the calculation of the three translatory lattice modes, we use the following values of r as an approximation:

$$r_1 = \frac{a}{2} \qquad r_2 = \frac{b}{2} \qquad \text{and} \qquad r_3 = c \sin \alpha$$

where α is the monoclinic angle. The calculated values are compared with observed values of the translatory lattice mode frequencies in Table XXX. In view of the approximations involved, i.e., negligence of steric effects and not using the symmetry properties of the unit cell, the agreement should be judged fair except perhaps for the low-frequency mode. In the other hand, the assignment of the low-frequency A_u mode by no means could be regarded as certain.

IX. REFERENCES

1. S. S. Mitra, *Solid State Phys.* **13**:1 (1962).
2. W. Veddar and D. F. Hornig, *Advan. Spectry.* **2**:189 (1961).
3. P. Mathieu, *J. Phys. Radium* **16**:219 (1955).
4. M. Born and T. von Kármán, *Z. Physik* **13**:297 (1912).
5. S. Bhagavantam and T. Venkatarayudu, *Proc. Indian Acad. Sci. A* **9**:224 (1939); S. Bhagavantam, *Proc. Indian Acad. Sci. A* **13**:543 (1941).
6. J. E. Rosenthal and G. M. Murphy, *Rev. Mod. Phys.* **8**:317 (1936).
7. R. S. Halford, *J. Chem. Phys.* **14**:8 (1946).
8. D. F. Hornig, *J. Chem. Phys.* **16**:1063 (1948).
9. H. Winston and R. S. Halford, *J. Chem. Phys.* **17**:607 (1949).
10. S. S. Mitra, *Z. Krist.* **116**:149 (1961).
11. M. Hass, Dissertation, University of Michigan, 1955; M. Hass and G. B. B. M. Sutherland, *Proc. Roy. Soc. (London) A* **236**:427 (1956).
12. E. J. Ambrose, A. Elliott, and R. B. Temple, *Proc. Roy. Soc. (London) A* **206**:192 (1951).
13. G. C. Pimentel and A. L. McClellan, *J. Chem. Phys.* **20**:270 (1952).
14. C. L. Angell, *Progress in Infrared Spectroscopy*, Vol. **2**, Plenum Press (New York), 1964, pp. 1–45.
15. D. F. Horning, *J. Chem. Phys.* **16**:1063 (1948).
16. W. West and R. T. Edwards, *J. Chem. Phys.* **5**:14 (1937).
17. H. Bauer and M. Magat, *J. Phys. Radium* **9**:319 (1938).
18. L. J. Bellamy, *Spectrochim. Acta* **14**:193 (1959).
19. A. D. E. Pullin, *Spectrochim Acta* **13**:125 (1958); *Proc. Roy. Soc. (London) A* **255**:39 (1960).

20. A. D. Buckingham, *Proc. Roy. Soc. (London) A* **248**:169 (1958) and *A* **255**:32 (1960).
21. R. M. Hexter, *J. Chem. Phys.* **25**:1286 (1956).
22. D. A. Dows, *J. Chem. Phys.* **29**:484 (1958).
23. D. A. Dows, *J. Chem. Phys.* **32**:1342 (1963).
24. K. Huang, *Proc. Roy. Soc. (London) A* **208**:352 (1951); M. Born and K. Huang, *Dynamical Theory of Crystal Lattices*, Oxford, University Press, (New York), 1954, p. 82.
25. M. Born and K. Huang, *Dynamical Theory of Crystal Lattices*, Oxford University Press (New York), 1954, p. 396.
26. R. H. Lyddane, R. G. Sachs, and E. Teller, *Phys. Rev.* **59**:673 (1941); T. N. Casselman, S. S. Mitra, and H. N. Spector, *J. Phys. Chem. Solids* **26**:529, (1965).
27. C. Haas and J. A. A. Ketelaar, *Phys. Rev.* **103**:564 (1956).
28. W. J. Turner and W. E. Reese, *Phys. Rev.* **127**:126 (1962).
29. T. H. Havelock, *Proc. Roy. Soc. (London) A* **105**:488 (1924); the idea of centro-frequency developed by Plendl [*Phys. Rev.* **119**:1598 (1960)] is sometimes useful in characterizing the lattice vibration of a solid by a single frequency.
30. W. G. Spitzer, D. Kleinman, and D. Walsh, *Phys. Rev.* **113**:127 (1959).
31. W. G. Spitzer and D. A. Kleinman, *Phys. Rev.* **121**:1324 (1961); D. A. Kleinman and W. G. Spitzer *Phys. Rev.* **125**:16 (1962).
32. A. S. Barker, Jr., *Phys. Rev.* **132**:1474 (1963).
33. W. Cochran, *Z. Krist.* **112**:465 (1959).
34. D. R. Renneke and D. W. Lynch, *Phys. Rev.* **138 A**: 530 (1965).
35. I. F. Chang and S. S. Mitra, *Phys. Status Solidi*, in press.
36. A. D. B. Woods, W. Cochran, and B. N. Brockhouse, *Phys. Rev.* **119**:980 (1960); A. D. B. Woods, B. N. Brockhouse, R. A. Cowley, and W. Cochran, *Phys. Rev.* **131**:1025 (1963); R. A. Cowley, W. Cochran, B. N. Brockhouse, and A. D. B. Woods, *Phys. Rev.* **131**:1030 (1963); C. Smart, G. R. Wilkinson, A. M. Karo, and J. R. Hardy "Two Phonon Infrared Absorption in NaCl Structure Ionic Crystals,. presented at the *International Conference on Lattice Dynamics, Copenhagen, Denmark, August 5–9, 1963*, Pergamon Press (New York), 1964.
37. D. W. Berreman, *Phys. Rev.* **130**:2193 (1963).
38. U. Fano, *Phys. Rev.* **103**:1202 (1956); J. J. Hopfield, *Phys. Rev.* **112**:1555 (1958).
39. H. B. Rosenstock, *Phys. Rev.* **121**:416 (1961).
40. T. H. K. Barron, *Phys. Rev.* **123**:1995 (1961).
41. J. Hardy, *Phil. Mag.* **7**:315 (1962).
42. T. N. Casselman, S. S. Mitra, and H. N. Spector, *J. Phys. Chem. Solids* **26**:529 (1965).
43. C. H. Henry and J. J. Hopfield, *Phys. Rev. Letters* **15**:964 (1965).
44. D. A. Kleinman and W. G. Spitzer, *Phys. Rev.* **118**:110 (1960); D. A. Kleinman, *Phys. Rev.* **118**:118 (1960).
45. R. Loudon, *Proc. Phys. Soc. (London)* **82**:393 (1963).
46. S. P. S. Porto, B. Tell, and T. C. Damen, *Phys. Rev. Letters* **16**:450 (1966).
47. A. A. Maradudin and R. F. Wallis, *Phys. Rev.* **125**:4 (1962).
48. V. V. Mitskevich, *Fiz. Tverd. Tela* **4**:3035 (1962). [English transl. *Soviet Phys.-Solid State* **4**:2224 (1963).]
49. J. Neuberger and R. D. Hatcher, *J. Chem. Phys.* **34**:5 (1961).
50. R. Wehner, *Phys. Status Solidi* **15**:725 (1966).
51. J. R. Jasperse, A. Kahan, J. N. Plendl, and S. S. Mitra, *Phys. Rev.* **146**:526 (1966).
52. R. Kubo, *J. Phys. Soc. Japan* **12**:570 (1957).
53. I. Šimon, *J. Opt. Soc. Am.* **41**:336 (1951).
54. T. S. Robinson, *Proc. Phys. Soc. (London)* **B 65**:910 (1952); T. S. Robinson and W. C. Price, *Proc. Phys. Soc. (London)* **B 66**:969 (1953).
55. R. de L. Kronig, *J. Opt. Soc. Am.* **12**:547 (1926) and *Phys. Rev.* **30**:521 (1929);

H. A. Kramers, *Atti Congr. Intern. Fis.*, *Como* **2**:545 (1927).
56. P. N. Schatz, S. Maeda, and K. Kozima, *J. Chem. Phys.* **38**:2658 (1963).
57. H. J. Bowlden, J. K. Wilmshurst, *J. Opt. Soc. Am.* **53**:1073 (1963).
58. J. Fahrenfort, *Spectrochim. Acta* **17**:698 (1961).
59. M. O. McMahon, *J. Opt. Soc. Am.* **40**:376 (1960).
60. C. Haas, Dissertation, Amsterdam, 1956; J. A. A. Ketelaar and C. Haas, *Physica* **22**:1283 (1956).
61. D. L. Stierwalt and R. F. Potter, *Proceedings of the International Conference on Physics of Semiconductors, Exeter*, **1962**, The Institute of Physics and Physical Society (London) 1962, p. 513.
62. H. Smith, *Phil. Trans. Roy. Soc. London* A **241**:105 (1948).
63. T. M. K. Nedungadi, *Proc. Indian Acad. Sci.* A **10**:197 (1939).
64. S. Bhagavantam, *Proc. Indian Acad. Sci.* A **11**:62 (1940).
65. B. D. Saksena, *Proc. Indian Acad. Sci.* A **11**:229 (1940).
66. V. Chandrasekharan, *Z. Physik* **175**:63 (1963).
67. R. Loudon, *Advan. Phys.* **13**:423 (1964).
68. S. P. S. Porto, J. A. Giordmaine, and T. C. Damen, *Phys. Rev.* **147**:608 (1966); T. C. Damen, S. P. S. Porto, and B. Tell, *Phys. Rev.* **142**:570 (1966); J. F. Scott and S. P. S. Porto, *Phys. Rev.* **142**:903 (1967).
69. S. S. Mitra, C. Postmus, and J. R. Ferraro, *Phys. Rev. Letters* **18**:455 (1967); C. Postmus, J. R. Ferraro, and S. S. Mitra, to be published.
70. M. Born and K. Huang, *Dynamical Theory of Crystal Lattices*, Oxford University Press (New York), 1954, Chap. 2.
71. E. W. Kellermann, *Phil. Trans. Roy. Soc. London* **238**:513 (1940).
72. S. S. Mitra and S. K. Joshi, *Physica* **26**:284 (1960).
73. S. S. Mitra, *Phys. Status Solidi* **9**:519 (1965).
74. A. Bienenstock and G. Burley, *J. Phys. Chem. Solids* **24**:1271 (1963).
75. C. Postmus, J. R. Ferraro, and S. S. Mitra, to be published.
76. M. Born and K. Huang, *Dynamical Theory of Crystal Lattices*, Oxford University Press (New York), 1954, p. 106.
77. A. A. Maradudin, *Phys. Stat. Solidi* **2**:1493 (1963).
78. A. A. Maradudin and A. E. Fein, *Phys. Rev.* **128**: 2589 (1962).
79. R. A. Cowley, *Advan. Phys.* **12**:421 (1963); D. H. Martin, *Advan. Phys.* **14**:39 (1965).
80. I. P. Ipatova, A. A. Maradudin, and R. F. Wallis, *Fiz. Tverd. Tela.* **8**:1064 (1966) and *Phys. Rev.* **155**:882 (1967).
81. G. O. Jones, D. H. Martin, P. A. Mawer, and C. H. Perry, *Proc. Roy. Soc. (London)* A **261**:10 (1961).
82. D. E. Schuele and C. S. Smith, *J. Phys. Chem. Solids* **25**:801 (1964).
83. R. A. Cowley, *J. Phys. (Paris)* **26**:659 (1965); G. Dolling and R. A. Cowley, *Proc. Phys. Soc. (London)* **88**:463 (1966).
84. B. Szigeti, *Proc. Roy. Soc. (London)* A **252**:217 (1959); *Proc. Roy. Soc. (London)* A **258**:377 (1960).
85. G. Heilmann, *Z. Physik* **152**:368 (1958).
86. M. Gottlieb, *J. Opt. Soc. Am.* **50**:343 (1960).
87. B. Piriou, *Compt. Rend.* **259**:1052 (1964).
88. B. Piriou, *Compt. Rend.* **260**:841 (1965).
89. M. Blackman, *Phil. Trans. Roy. Soc. London* A **236**:103 (1936).
90. M. Hass, *Phys. Rev.* **117**:6 (1960).
91. J. R. Hardy, *Phil. Mag.* **7**:315 (1962); A. M. Karo and J. R. Hardy, *Phys. Rev.* **129**:2024 (1963).
92. E. U. Condon, *Phys. Rev.* **41**:759 (1932).
93. M. F. Crawford and I. R. Dagg, *Phys. Rev.* **91**:1569 (1953).

94. E. Anastassakis, S. Iwasa, and E. Burstein, *Phys. Rev. Letters* **17**:1051 (1966).
95. H. Hartman, *Phys. Rev.* **147**:663 (1966).
96. R. Loudon, *Proc. Roy. Soc. (London)* A **275**:218 (1963).
97. P. A. Fleury and J. M. Worlock, *Phys. Rev. Letters* **18**:665 (1967).
98. R. C. Miller and W. G. Spitzer, *Phys. Rev.* **129**:94 (1963).
99. W. Cochran, *Advan. Phys.* **9**:387 (1960).
100. R. F. Schaufele, M. J. Weber, and B. D. Silverman, *Phys. Rev. Letters* **25A**:47 (1967).
101. J. M. Worlock and P. A. Fleury, *Phys. Rev. Letters* **19**:1176 (1967).
102. M. Born and M. Blackman, *Z. Physik* **82**:551 (1933); M. Blackman, *Z. Physik* **86**:421 (1933) and *Phil. Trans. Roy. Soc. London* A **236**:102 (1936); R. B. Barnes, R. R. Brattain, and F. Seitz, *Phys. Rev.* **48**:582 (1935); D. A. Kleinman, *Phys. Rev.* **118**:118 (1960).
103. M. Lax and E. Burstein, *Phys. Rev.* **97**:39 (1955).
104. J. L. Birman, *Phys. Rev.* **127**:1093 (1962) and **131**:1489 (1963).
105. S. S. Mitra, *Phys. Letters* **11**:119 (1964).
106. F. A. Johnson and R. Loudon, *Proc. Roy. Soc. (London)* A **231**:274 (1964).
107. E. Burstein, F. A. Johnson, and R. Loudon, *Phys. Rev.* **139**:A 1239 (1965).
108. S. Ganesan and E. Burstain, *J. Phys. (Paris)* **26**:645 (1965).
109. M. A. Nusimovici and J. L. Birman, *Phys. Rev.* **156**:925 (1967).
110. J. Ziomek and S. S. Mitra, *Solid State Phys.*, to be published.
111. J. M. Ziman, *Electrons and Phonons*, Clarendon Press (Oxford), 1960, Chap. 3.
112. S. S. Mitra, *J. Chem. Phys.* **39**:3031 (1963).
113. S. S. Mitra, *Z. Krist.* **116**:149 (1961).
114. G. S. Pawley, *Phys. Status Solidi* **20**:347 (1967).
115. L. Vegard, *Nature* **124**: 267 (1929) and **125**:14 (1930).
116. A. L. Smith, W. E. Keller, and H. L. Johnston, *Phys. Rev.* **79**:728 (1950).
117. G. E. Ewing and G. C. Pimentel, *J. Chem. Phys.* **35**:925 (1961); G.E. Ewing, *J. Chem. Phys.* **37**:2250 (1962).
118. D. A. Dows, *Spectrochim. Acta* **13**:308 (1959).
119. A. P. Gush, W. F. J. Hare, E. J. Allin, and H. L. Welsh, *Can. J. Phys.* **38**:176 (1960).
120. S. S. Mitra, *Solid State Phys.* **13**:47–53 (1962), and Dissertation, University of Michigan (Ann Arbor, Mich., 1957.
121. S. S. Mitra, *Solid State Phys.* **13**:66–78 (1962); K. A. Wickersheim, *J. Chem. Phys.* **31**:863 (1959); R. A. Buchanan, E. L. Kinsey, and H. H. Caspers, *J. Chem. Phys.* **36**:2665 (1962); R. A. Buchanan and H. H. Caspers, *J. Chem. Phys.* **38**:1025 (1963).
122. S. S. Mitra, in: *Crystallography and Crystal Perfection*, G. N. Ramachandran, ed., Academic Press (London), 1963, pp. 347–357.
123. F. A. Johnson and W. Cochran, *Proceedings of the International Conference on Physics of Semiconductors, Exeter, 1962*, The Institute of Physics and Physical Society (London), 1962, p. 498.
124. R. M. Hexter and D. A. Dows, *J. Chem. Phys.* **25**:504 (1956).
125. R. A. Buchanan, *J. Chem. Phys.* **31**:870 (1959).
126. R. G. Snyder, J. Kumamoto, and J. A. Ibers, *J. Chem. Phys.* **33**:1171 (1960).
127. M. Balkanski and W. Nazarewicz, *J. Phys. Chem. Solids* **23**:573 (1962); P. G. Dawber and R. J. Elliott, *Proc. Phys. Roy. Soc. (London)* **81**:453 (1963); B. Szigeti, *J. Phys. Chem. Solids* **24**:225 (1963); A. Hadni, J. Claudel, D. Chanal, P. Strimer, and P. Vergnat, *Phys. Rev.* **163**:836 (1967).
128. B. Szigeti, *Trans. Faraday Soc.* **45**:155 (1949).
129. R. Brout, *Phys. Rev.* **113**:43 (1959); S. S. Mitra and R. Marshall, *J. Chem. Phys.* **41**:3158 (1964).
130. W. Cochran, S. J. Fray, F. A. Johnson, J. E. Quarrington, and N. Williams, *J. Appl. Phys.* **32**:2102 (1961).

131. R. H. Parmenter, *Phys. Rev.* **100**:573 (1955).
132. M. Hass and B. W. Henvis, *J. Phys. Chem. Solids* **23**:1099 (1962).
133. S. Iwasa, I. Balslev, and E. Burstein, *Bull. Am. Phys. Soc.* **9**:237 (1964).
134. J. L. T. Waugh and G. Dolling, *Phys. Rev.* **132**:2410 (1963).
135. J. C. Phillips, *Phys. Rev.* **113**:147 (1959).
136. M. Czerny, *Z. Physik.* **65**:600 (1940).
137. R. B. Barnes and M. Czerny, *Z. Physik* **72**:447 (1931).
138. R. B. Barnes, *Z. Physik* **75**:723 (1932).
139. H. W. Hohls, *Ann. Physik* **29**:433 (1937).
140. M. Klier, *Z. Physik* **150**:49 (1958).
141. G. Heilmann, *Z. Physik* **152**:368 (1958).
142. F. Abeles and J. Mathieu, *Ann. Phys. (Paris)* **3**:5 (1958).
143. D. Frohlich, *Z. Physik* **169**:114 (1962).
144. M. Hass, *J. Phys. Chem. Solids* **24**:1159 (1963).
145. C. M. Randall, R. M. Fuller, and D. J. Montgomery, *Solid State Commun.* **2**:273 (1964).
146. W. Kaiser, W. G. Spitzer, R. H. Kaiser, and L. E. Howarth, *Phys. Rev.* **127**:1950 (1962); R. S. Krishnan and P. S. Narayanan, *Indian J. Pure and Appl. Phys.* **1**:196 (1963).
147. M. Hass, *Phys. Rev.* **117**:1497 (1960).
148. S. S. Mitra, *Solid State Phys.* **13**:57–59 (1962).
149. A. I. Stekhanov and M. B. Eliashberg, *Opt. Spectry. (USSR) (English Transl.)* **10**:174, (1961).
150. A. I. Stekhanov and M. B. Eliashberg, *Soviet Phys.–Solid State (English Transl.)* **5**:2185 (1964).
151. A. S. Filler and E. Burstein, *Bull. Am. Phys. Soc.* **5** (II):198 (1960).
152. H. W. Hohls, *Ann. Physik* **29**:433 (1937).
153. G. L. Bottger and A. L. Geddes, *J. Chem. Phys.* **46**:3000 (1967).
154. P. J. Gielisse, J. N. Plendl, L. C. Mansur, R. Marshall, S. S. Mitra, R. Mykolajewycz, and A. Smakula, *J. Appl. Phys.* **36**:2446 (1965).
155. S. S. Mitra and I. F. Chang, unpublished.
156. J. E. Hanlon and A. W. Lawson, *Phys. Rev.* **113**:472 (1959).
157. M. Born and K. Huang, *Dynamical Theory of Crystal Lattices*, Oxford University Press (New York), 1954, p. 106.
158. A. Mitsuishi, Paper presented at United States–Japan Cooperative Seminar on Far Infrared Spectroscopy, Columbus, Ohio, Sept. 15–17, 1965, *Nippon Buturi Gakkai Kaisi* (in Japanese) **19**:562 (1964).
159. R. Geick, C. H. Perry, and S. S. Mitra, *J. Appl. Phys.* **37**:1994 (1966).
160. M. Balkanski, M. Nusimovici, and R. Le Toullec, *J. Phys. Radium* **25**:305 (1964); M. Balkanski, J. M. Besson, and R. Le Toullec, *Proc. Intern. Conf. Semicond. Phys. Paris*, 1964, p. 1091.
161. S. Narita, H. Harada, and K. Nagasaka, *J. Phys. Soc. Japan* **22**:1176 (1967).
162. S. S. Mitra, *J. Phys. Soc. Japan* **21** (*Suppl.*):61 (1966).
163. C. Haas and J. P. Mathieu, *J. Phys. Radium* **15**:492 (1954).
164. E. C. Heltemes and H. L. Swinney, *J. Appl. Phys.* **38**:2386 (1967).
165. H. Poulet and J. P. Mathieu, *Ann. Phys. (Paris)* **9**:549 (1964).
166. B. Tell, T. C. Damen, and S. P. S. Porto, *Phys. Rev.* **144**:771 (1966).
167. R. Marshall and S. S. Mitra, *Phys. Rev.* **134**:A 1019 (1964).
168. S. S. Mitra and J. I. Bryant, *Bull. Am. Phys. Soc.* **10**:333 (1965).
169. T. C. Damen, S. P. S. Porto, and B. Tell, *Phys. Rev.* **142**:570 (1966).
170. S. S. Mitra and R. Marshall, *Proc. Intern Conf. Semicond. Phys., Paris*, 1964, p. 1085.
171. T. Deutsch, *Proc. Intern. Conf. Semicond. Phys., Exeter*, 1964, p. 505.

172. M. Aven, D. T. F. Marple, and B. Segall, *J. Appl. Phys.* **32**:1261 (1961).
173. S. S. Mitra, *Phys. Rev.* **132**:986 (1963).
174. R. E. Nahory and H. Y. Fan, *Phys. Rev.* **156**:825 (1967).
175. O. M. Stafsudd, F. A. Haak, and K. Radisavljevic, *J. Opt. Soc. Am.* **57**:1475 (1967).
176. G. L. Bottger and A. L. Geddes, *J. Chem. Phys.* **47**:4858 (1967). Also see A. Mooradian and G. B. Wright, *Bull. Am. Phys. Soc.* **13**:450 (1968) for the Raman spectrum of CdTe.
177. G. A. Slack. F. S. Ham, and R. M. Chrenko, *Phys. Rev.* **152**:376 (1966).
178. M. Hass and B. W. Henvis, *J. Phys. Chem. Solids* **23**:1099 (1962).
179. D. A. Kleinman and W. G. Spitzer, *Phys. Rev.* **118**:110 (1960); D. A. Kleinman, *Phys. Rev.* **118**:118 (1960).
180. P. J. Gielisse, S. S. Mitra, J. N. Plendl, R. D. Griffis, L. C. Mansur, R. Marshall, and E. A. Pascoe, *Phys. Rev.* **155**:1039 (1967).
181. A. T. Collins, E. C. Lightowlers, and P. J. Dean, *Phys. Rev.* **158**: 833 (1967). Also see I. Akasaki and M. Hashimoto, *Solid State Commun.* **5**: 851 (1967).
182. A. Mooradian and G. B. Wright, *Solid State Commun.* **4**:431 (1966).
182a. S. Z. Beer, J. F. Jackovitz, D. W. Feldman, and J. H. Parker, *Phys. Letters* **26A**: 331 (1968).
183. O. Brafman, S. S. Mitra, and P. J. Gielisse, *Solid State Commun.*, in press.
184. W. G. Spitzer, in *Semiconductors and Semimetals*, R. K. Willardson and A. C. Beer, eds., Vol. 3, Academic Press (New York), 1967, pp. 17-69.
185. D. L. Stierwalt and R. F. Potter, in *Semiconductors and Semimetals*, R. K. Willardson and A. C. Beer, eds., Vol. 3, Academic Press (New York), 1967, pp. 71-90.
186. S. S. Mitra, *Solid State Phys.* **13**:59–62 (1962).
187. B. N. Brockhouse and P. K. Iyengar, *Phys. Rev.* **111**:747 (1958).
188. B. N. Brockhouse, *Phys. Rev. Letters* **2**:256 (1959).
189. J. H. Parker, D. W. Feldman, and M. Ashkin, *Phys. Rev.* **155**:712 (1967).
190. J. R. Hardy and S. D. Smith, *Phil. Mag.* **6**:1163 (1961).
191. F. A. Johnson, *Proc. Phys. Soc. (London)* **73**:265 (1959).
192. W. G. Spitzer, D. A. Kleinman, and C. J. Frosch, *Phys. Rev.* **113**:133 (1959).
193. L. Patrick and W. J. Choyke, *Phys. Rev.* **123**:813 (1961).
194. J. L. Warren, R. G. Wenzel, and J. L. Yarnell, in *Symposium on the Inelastic Scattering of Neutrons*, Intern. Atomic Energy Agency (Bombay), 1964.
195. R. W. Keyes, *J. Chem. Phys.* **37**:72 (1962).
196. S. S. Mitra and R. Marshall, *J. Chem. Phys.* **41**:3158 (1964).
197. A. A. Maradudin, E. W. Montroll, and G. H. Weiss, *Theory of Lattice Dynamics in the Harmonic Approximation*, Academic Press (New York), 1963, p. 115; H. B. Rosenstock, *Phys. Rev.* **129**:1959 (1963).
198. F. Kruger, O. Reinkober, and E. Koch-Holm, *Ann. Physik* **85**:110 (1928).
199. P. J. Gielisse, J. N. Plendl, L. C. Mansur, R. Marshall, S. S. Mitra, R. Mykolajewycz, and A. Smakula, *J. Appl. Phys.* **36**:2446 (1965).
200. A. Mitsuishi, paper presented at *United States–Japan Cooperative Seminar on Far Infrared Spectroscopy* (Columbus, Ohio), Sept. 15–17, 1965.
201. R. K. Chang, B. Lacina, and P. S. Pershan, *Phys. Rev. Letters* **17**:755 (1966).
202. H. W. Verleur and A. S. Barker, *Solid State Commun.* **5**:695 (1967).
203. R. F. Potter and D. L. Stierwalt, *Proc. Intern. Conf. Semicond. Phys. Paris*, 1964, p. 1111.
204. G. Lucovsky, E. Lind, and E. A. Davis, *Proceedings of the International Conference on the Physics of II–VI Semiconductors*, W. A. Benjamin (New York), 1968.
205. O. Brafman, I. F. Chang, and S. S. Mitra, unpublished.
206. F. Oswald, *Z. Naturforsch* **14A**:374 (1959).
207. D. W. Feldman, M. Ashkin, and J. H. Parker, *Phys. Rev. Letters* **17**:1209 (1966).

208. Y. S. Chen, W. Shockley, and G. L. Pearson, *Phys. Rev.* **151**:648 (1966).
209. H. W. Verleur and A. S. Barker, *Phys. Rev.* **149**:715 (1966).
210. M. Balkanski, R. Beserman, and J. M. Besson, *Solid State Commun.* **4**:201 (1966).
211. J. Parrish, C. H. Perry, O. Brafman, I. F. Chang, and S. S. Mitra, *Proceedings of the International Conference on the Physics of II–VI Semiconductors*, W. A. Benjamin (New York), 1967.
212. H. W. Verleur and A. S. Barker, *Phys. Rev.* **155**:750 (1967).
213. O. Brafman, I. F. Chang, G. Lengyel, and S. S. Mitra, *Phys. Rev. Letters* **19**:1120 (1967).
214. G. Lucovsky, M. Brodsky, and E. Burstein, *Proceedings of the International Conference on Localized Excitations in Solids*, Plenum Press (New York), 1968.
215. L. Genzel, this volume, pp. 453–487.
216. I. F. Chang and S. S. Mitra, *Phys. Rev.*, August 15, 1968.
217. R. Braunstein, H. R. Moore, and F. Herman, *Phys. Rev.* **109**:695 (1958).
218. R. S. Krishnan, *Proc. Indian Acad. Sci.* A **22**:274 (1945).
219. A. Rousset and R. Lochet, *J. Phys. Radium* **6**:57 (1945).
220. A. I. Stekhanov, *Dokl. Acad. Nauk SSSR* **92**:281 (1953).
221. K. W. F. Kohlrausch, "Ramanspektren", in *Hand- und Jahrbuch der Chemischen Physik*, Vol. 9, Akademische Verlags Gessellschaft (Leipzig), 1943, p. 399.
222. G. Herzberg, *Infrared and Raman Spectra of Polyatomic Molecules*, D. Van Nostrand, Co., (New York), 1945, p. 281.
223. A. W. Baker and R. C. Lord, *J. Chem. Phys.* **23**:1636 (1955); H. H. Gunthard, R. C. Lord and T. K. McCubbin, *J. Chem. Phys.* **25**:768 (1956).
224. P. M. Mathai, G. G. Shepherd, and H. L. Welsh, *Can. J. Phys.* **34**:1448 (1956).
225. C. Brecher, E. Krikorian, J. Blanc, and R. S. Halford, *J. Chem. Phys.* **35**:1097 (1961).
226. S. S. Mitra, unpublished.
227. V. C. Sinclair, J. M. Robertson, and A. Mathieson, *Acta. Cryst.* **3**:251 (1950).
228. A. Kastler and A. Rousset, *J. Phys. Radium*, **2**:49 (1941).
229. A. Hadni, unpublished data.
230. I. Harada and T. Shimanouchi, *J. Chem. Phys.* **44**:2016 (1966).
231. J. E. Lennard-Jones and A. E. Ingham, *Proc. Roy. Soc. (London)* **107**:636 (1925).
232. D. Mennerich, S. S. Mitra, and R. S. Singh, unpublished.
233. A. Bondi, *J. Chem. Eng. Data* **8**:371 (1963).

CHAPTER 15

Impurity-Induced Lattice Absorption

L. Genzel

Physikalisches Institut
Freiburg, Germany

I. INTRODUCTION

In recent years much work has been done in studying the behavior of impurity particles in crystals with respect to their influence on lattice dynamics and optical absorption. Both are changed considerably by introducing defects in crystals mainly because they destroy the translational symmetry. Hence the normal modes of vibration are modified from their usual plane wave form. Although in a perfect crystal the conservation law for all wave vectors involved in any process of phonons and photons governs its relatively simple behavior, this wave vector conservation breaks down in crystals with impurities just due to the lack of periodic symmetry. With respect to electromagnetic absorption, this means that nearly all lattice modes might become optically active as long as these modes have a nonvanishing dipole moment over the periodicity interval. This dipole moment can have its origin simply in a charge of the impurity different from that of the perfect lattice on the same lattice site, but it can also be due to the changed eigenvector especially near the defect, The latter case applies, of course, only in ionic crystals. If the defect concentration is small enough to avoid interactions between defects, then the absorption is proportional to their concentration. Besides that, the density of lattice modes will then be nearly the same as in the host crystal, and the defect-induced one-phonon absorption will be approximately proportional to this phonon density. Generally, many characteristic features of the perfect crystal will remain for low concentrations of defects, such as the possible existence of an optically active dispersion oscillator (reststrahlen oscillator), or the spectral and thermal behavior of multiphonon absorption due to anharmonicity or nonlinear dipole moment. All this is true only because the lattice modes are not really harmonic and therefore have a finite lifetime which is of the order of 10^{-10} to 10^{-11} sec, at least for the modes of high density. This short lifetime reduces the dynamical influence of a defect to a sphere of about 10 to 100 lattice constants in diameter, so that some volume parts of the defective crystal are still "perfect" and can show, therefore, the normal behavior of the crystal. Thus, for the study of the additional absorption due to impurities in ionic crystals, one is limited to those spectral regions where the absorption of the host crystal is low or can be frozen out, i.e., for frequencies well above the highest lattice frequency and also on the low-frequency side of the dispersion oscillator. In case of homopolar crystals, such as Ge and Si, nearly the whole spectral range is available for the impurity-induced absorption. Since the dominating process here is that of the first order, the impurity absorption is mainly a "one-phonon absorption" occurring therefore in the middle and far infrared. So long as this absorption falls within the continuous region of the acoustic or optical band modes, one speaks about band mode absorption, which always exists in crystals with defects. Occasionally, however, a small number of lattice modes in a crystal with special defects might be outside of the continuum and form localized modes, which are well separated in their frequency from the band modes. The more their eigenvector is localized around the defect, the more their frequency is different from the band frequencies. These local modes can cause sharp absorptions, similar to a line of an isolated molecule, as long as they are optically active. One speaks, in this case, of local mode absorption. It can occur either in the gap between the acoustic and the optical lattice bands or above the latter. Due to its sharpness, one observes associated higher-order defect-induced processes in the form of phonon side bands.

Theoretical understanding of band mode and local mode absorption is based mainly on the lattice dynamics of a crystal with defects. Only in a very few and simple cases a three-dimensional calculation of this problem exists. Most of the theoretical work has been done for one-dimensional models, and they are therefore only of qualitative value.

II. LATTICE DYNAMICS OF CRYSTALS WITH IMPURITIES

This section gives only a brief survey of the methods used to obtain the eigenvalues and eigenvectors of the defect crystal. For a more correct and elaborate treatment, see the literature [1-21].

It is assumed that the eigenvalues λ^0 and eigenvectors $\mathbf{u}^0 = \boldsymbol{\eta}^0 e^{i\omega t}$ of the perfect lattice are known, that is, we have the solution of

$$\mathbf{L}\boldsymbol{\eta}^0 \equiv \mathbf{F}\boldsymbol{\eta}^0 - \lambda^0 \boldsymbol{\eta}^0 = 0$$

where $\boldsymbol{\eta}^0$ is the column matrix of the displacements and $\mathbf{F}$ is the force constant matrix. The eigenvalues λ° will result from $\text{Det} |\mathbf{F} - \lambda^\circ \mathbf{E}| = 0$, where $\mathbf{E}$ is the unit matrix. Since the perfect crystal has translational symmetry, the eigenvector η° can be represented by plane wave solutions

$$\eta_k^{\,0}(\mathbf{q},j) \sim \exp(-i\mathbf{q}\mathbf{r}_k)$$

where k labels the atoms, the cell, and the coordinate direction. The eigenvalues $\lambda_j^{\,0}$ of the branches j are dependent on the wave vector $\mathbf{q}$, the number of which equals the number of degrees of freedom of the particles per unit cell of the crystal. In this way the dispersion curves $\omega_j^{\,2}(\mathbf{q}) \sim \lambda_j^{\,0}(\mathbf{q})$, the eigenvectors $\eta_k^{\,0}(\mathbf{q},j)$ at each point of the lattice, and also the density of phonon states $D(\omega)$ of the perfect lattice are known.

The crystal shall now contain a well-defined impurity particle, which is put at the origin by replacing one of the host lattice particles. The spatial extension of the impurity depends mainly on the change of force constant parameters around it, but it can be assumed that this change is rather limited around the impurity. Under these assumptions, the equations of motion become

$$\mathbf{L}\eta \equiv \mathbf{F}\eta - \lambda\eta = \mathbf{S}\eta$$

where $\mathbf{F}$ is again the force constant matrix of the perfect lattice, but $\mathbf{S}$ is a matrix of low rank according to the spatial extension of the defect, measuring the change of mass and force constants around it.

Let us consider, as an example, a diatomic linear chain having an impurity at the origin, subject to nearest neighbor forces (see Fig. 1). In this simple case, the matrices can be easily written by using the abbreviations $\Delta m = m_0 - m_I$ and $\Delta f = f - f_I$:

$$
\begin{bmatrix}
\ddots & & & & & \\
(2f - m_0\omega^2) & (-f) & & & \bigcirc & \\
(-f) & (2f - m_1\omega^2) & (-f) & & & \\
& (-f) & (2f - m_0\omega^2) & (-f) & & \\
& & (-f) & (2f - m_1\omega^2) & (-f) & \\
\bigcirc & & & (-f) & (2f - m_0\omega^2) & \\
& & & & & \ddots
\end{bmatrix}
\begin{bmatrix}
\vdots \\
\eta_{-2} \\
\eta_{-1} \\
\eta_0 \\
\eta_1 \\
\eta_2 \\
\vdots
\end{bmatrix}
= \mathbf{L}\eta
$$

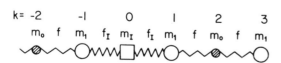

k= -2   -I   0   I   2   3

m_0 f m_1 f_I m_I f_I m_1 f m_0 f m_1

Fig. 1.   Diatomic linear chain model with nearest neighbor force constants and an impurity with mass defect and force constant defect.

and

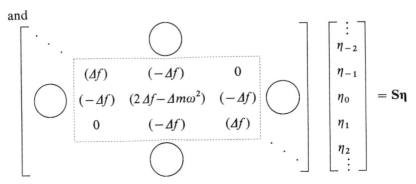

$$= S\eta$$

The term η_0 here denotes the displacement of the impurity itself, which will be later denoted by η_I. The limited extension of the perturbation matrix S is further reduced in the isotopic case where $\Delta f = 0$, and therefore S has only the element $(-\Delta m \cdot \omega^2)$ at the center. Due to the important fact that S is limited, the Green's function method can be used to solve the problem [1]. With the "Green's function matrix" G, which is defined by $LG = E$, the solutions can be formally expressed by

$$\eta = GS\eta = (F - \lambda E)^{-1} S\eta$$

or

$$\eta_k = \sum_{k'} g_{kk'} \sum_{k''} S_{k'k''} \eta_{k''}$$

where the elements $g_{kk'}$ of G are now functions of ω. The eigenvalues, that is, the eigenfrequencies, are obtainable from $\mathrm{Det} |GS - E| = 0$.

Before being able to use this rather formal way of solution, one needs an explicit representation of the elements $g_{kk'}$ of the Green's function matrix. This can be achieved in the following way: First multiply $F\eta - \lambda\eta = S\eta$ by $\eta^0 = \eta^0(\mathbf{q}, j)$ of the perfect lattice

$$(\eta^0, F\eta) - \lambda(\eta^0, \eta) = (\eta^0, S\eta)$$

Now since F is hermitian, transform the first term into

$$(\eta^0, F\eta) = (F\eta^0, \eta) = \lambda^0 (\eta^0, \eta)$$

getting the scalar product from above

$$(\eta^0, \eta) = \frac{(\eta^0, S\eta)}{\lambda^0 - \lambda}$$

Next, develop the η's with respect to the η^0's:

$$\eta = \sum_{\mathbf{q}, j} \alpha(\mathbf{q}, j) \eta^0(\mathbf{q}, j)$$

By multiplying this with $\eta^{0'} \equiv \eta^0(q', j')$ and using the orthonormality of the η^0's, it follows that

$$(\eta^{0'}, \eta) = \alpha(q', j') = \frac{(\eta^{0'}, S\eta)}{\lambda^{0'} - \lambda}$$

Therefore,

$$\eta = \sum_{q,j} \frac{(\eta^0, S\eta)}{\lambda^0 - \lambda} \eta^0$$

or

$$\eta_k = \sum_{q,j} \sum_{k'} \frac{\eta_{k'}^{0*}(S\eta)_{k'}}{\lambda^0 - \lambda} \eta_k^0 = \sum_{k'} \left(\sum_{q,j} \frac{\eta_{k'}^{0*} \eta_k^0}{\lambda^0 - \lambda} \right) (S\eta)_{k'} = \sum_{k'} \left(\sum_{q,j} \frac{\eta_{k'}^{0*} \eta_k^0}{\lambda^0 - \lambda} \right) \sum_{k''} S_{k'k''} \eta_{k''}$$

This can be compared to our formal solution with the Green's function

$$\eta_k = \sum_{k'} g_{kk'} \sum_{k''} S_{k'k''} \eta_{k''}$$

and it gives the representation of Green's function matrix

$$g_{kk'} = \sum_{q,j} \frac{\eta_{k'}^{0*}(q, j) \eta_k^0(q, j)}{\lambda^0(q, j) - \lambda}$$

The essential point of this method for solving the perturbed lattice is the limited extension of the impurity matrix S. As a result, one has to solve now the equations of motions only in the perturbed region. If this is done, the calculation of the rest of the lattice is simple arithmetic. This can be illustrated again with our diatomic chain having an impurity with Δm and Δf defects. S has here the elements

$$S_{0,0} = 2\Delta f - \Delta m \omega^2$$

$$S_{1,1} = S_{-1,-1} = \Delta f$$

$$S_{-1,0} = S_{1,0} = S_{0,-1} = S_{0,1} = -\Delta f$$

while the other elements are zero. After rearrangement, this gives the component η_k:

$$\eta_k = \eta_{-1} \Delta f(g_{k,-1} - g_{k,0}) + \eta_1 \Delta f(g_{k,1} - g_{k,0})$$
$$+ \eta_0 [(2\Delta f - \Delta m \omega^2) g_{k,0} - \Delta f(g_{k,1} + g_{k,-1})]$$

By setting the subscript k equal to -1, 0, and $+1$, three equations for η_{-1}, η_0, and η_1 evolve. However, the impurity is a center of inversion in our problem, in addition we have the relation $\eta_{-1} = \pm\eta_1$. This divides the solutions into two groups, namely, the odd solutions with the plus sign and the even solutions with the minus sign. Since the $g_{kk'}$'s have the same symmetry property, two equations for η_0 and $\eta_{\pm 1}$ result. After obtaining these, all the other η_k's are obtained by recursion. This implies, of course, knowledge of the eigenvectors λ of the perturbed lattice which occur in the $g_{kk'}$'s. These are determined by Det $|GS - E| = 0$, or by the determinant of the coefficients of our two homogeneous equations for η_0 and $\eta_{\pm 1}$.

In this way, one obtains after some tedious calculations odd solutions for our chain in the following form (see Genzel, Renk, and Weber [16] and note that only the odd solutions have a dipole moment and give rise to optical absorption):

Eigenvalue equation

$$\text{tg } N\phi = \pm \frac{\rho(\omega)\, F_1(\omega)}{\sigma(\omega)\, F_0(\omega)}$$

$$\begin{cases} + \text{sign for } \phi > 0 \text{ in acoustic branch} \\ \phi < 0 \text{ in optical branch} \\[1em] - \text{sign for } \phi > 0 \text{ in optical branch} \\ \phi < 0 \text{ in acoustic branch} \end{cases}$$

Eigenvector

$$\eta_k = A_0 \cos(|k|\, \phi + \psi) \qquad \begin{array}{l} A_0 \text{ for } k = \text{even and} \neq 0 \\ A_1 \text{ for } k = \text{odd} \end{array}$$

$$\eta_I \equiv \eta_0 = A_I \cos \psi$$

where N is the number of unit cells in the periodicity interval and

$$F_0 = \left(\frac{m_1}{\mu} (1 - \Omega^2) \left| 1 - \frac{m_0}{\mu} \Omega^2 \right| \right)^{\frac{1}{2}}$$

$$F_1 = \left(\frac{m_0}{\mu} \Omega^2 \left| 1 - \frac{m_1}{\mu} \Omega^2 \right| \right)^{\frac{1}{2}}$$

with

$$\Omega^2 = \frac{\omega^2}{\omega_R^2} \qquad \omega_R^2 = \frac{2f}{\mu} \qquad \mu = \frac{m_0 m_1}{m_0 + m_1}$$

where ω_R is the frequency of the optical branch at $\mathbf{q} = 0$, $\sqrt{\mu/m_0}\, \omega_R$ and $\sqrt{\mu/m_1}\, \omega_R$ are the frequencies of the two branches at the zone boundaries, and where, in the eigenvalue equation above,

$$\sigma = 1 - \frac{\gamma}{\varkappa}\Omega^2 \qquad \rho = \varepsilon - \frac{\gamma}{\varkappa}\Omega^2$$

with

$$\varepsilon = \frac{\Delta m}{m_0} \qquad \gamma = \frac{\Delta f}{f} \qquad \varkappa = \frac{\mu}{m_0} \cdot \frac{1-\gamma}{1-\varepsilon}$$

In these equations, ϕ runs from $-\pi \cdots +\pi$ in the first Brillouin zone and gives the phase difference of neighboring particles. It is determined by the relation

$$\cos^2 \phi = \left(1 - \frac{m_0}{\mu} \Omega^2\right)\left(1 - \frac{m_1}{\mu} \Omega^2\right)$$

In the perfect lattice, ϕ equals q^a (a is the nearest neighbor particle distance and $q = 2\pi/\lambda$ is the wave vector). This then is the familiar dispersion relation $\Omega = \Omega(\mathbf{q})$, which gives the acoustic and optical branches.

The phase shift ψ is dependent on frequency and contains in a complicated way the impurity parameters. By means of the cyclic boundary conditions, ψ is related to ϕ by

$$\psi = s\pi - N\phi \qquad s = \text{integers}$$

and

$$tg\,\psi = \mp \frac{\rho F_1}{\sigma F_0}$$

as ψ goes to zero in the perfect lattice. The A's are related as shown below:

$$\frac{A_I}{A_0} = \frac{1}{\sigma} \qquad \frac{A_0}{A_1} = \frac{\left[\left(1 - \frac{m_0}{\mu} \Omega^2\right)\left(1 - \frac{m_1}{\mu} \Omega^2\right)\right]^{\frac{1}{2}}}{\left(1 - \frac{m_0}{\mu} \Omega^2\right)}$$

This type of solution, written in the above form, was first discussed by Szigeti [7]. Figure 2 indicates how the eigenvector might look around the impurity. To illustrate the eigenvalue equation, Fig. 3 presents the way in which this transcendental equation can be solved and the direction the points $\phi = s\pi/N$ may shift due to the defect.

It is of special interest for the case of the impurity-induced optical absorption to know the ratio of η_I to the displacement η_0 at the same lattice site in the perfect lattice. Since η_0 for the odd modes is just equal to A_0, it follows that

$$\left|\frac{\eta_I}{A_0}\right| = \left[\sigma^2 + \rho^2 \left(\frac{F_1}{F_0}\right)^2\right]^{-\frac{1}{2}}$$

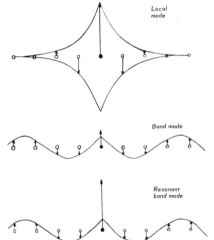

Local
mode

Band mode

Resonant
band mode

Fig. 2. Schematic representation of the eigenvector in a linear chain around the impurity for three cases : local mode, band mode, and resonant band mode.

Figure 4 shows this relation as a function of the reduced frequency Ω for a NaCl chain influenced by impurity parameters $\varepsilon = \Delta m/m_0$ and $\gamma = \Delta f/f$. One finds a "resonant behavior" in cases where the defect force constant f_I

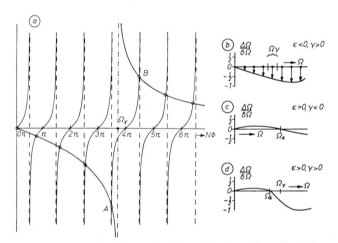

Fig. 3. (a) Determination of the eigenvalues in a diatomic chain model with an impurity having mass defect $\varepsilon = \Delta m/m_0$ and force constant defect $\gamma = \Delta f/f$. The ordinate is $tgN\phi$ resp. $\rho F_1/\sigma F_0$. At frequency $\Omega_\gamma = (\varkappa/\gamma)^{1/2}$ a resonant mode might occur starting a new gap of width $2\delta\Omega$. After Renk [35]. (b) Relative frequency shift $\Delta\Omega/\delta\Omega$ of eigenvalues; $\Delta\Omega/\delta\Omega$ in a resonant mode case near Ω_γ, where $\sigma = 0$ and $\psi = \pm \pi/2$. (c) Relative frequency shift $\Delta\Omega/\delta\Omega$ of eigenvalues; $\Delta\Omega/\delta\Omega$ near $\Omega_\varepsilon = (\varepsilon\varkappa/\gamma)^{1/2}$, where $\rho = 0$ and $\psi = 0$. (d) Relative frequency shift $\Delta\Omega/\delta\Omega$ of eigenvalues; $\Delta\Omega/\delta\Omega$ in general.

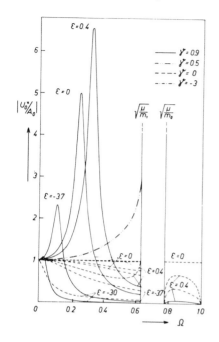

Fig. 4. Reduced amplitude $|\eta_I/A_0|$ of the impurity *versus* $\Omega = \omega/\omega_R$ for a NaCl chain under different cases of defect parameters $\varepsilon = \Delta m/m_0$ and $\gamma = \Delta f/f$. After Genzel, Renk, and Weber [16].

is weak, the "resonant frequency" shifting to lower frequencies with heavier impurity masses. These are called the "resonant band modes" or "pseudo-localized modes." The latter name results from the fact that here the relative difference of η_I and A_0 is localized spatially around the defect. For comparison with normal band modes, the eigenvector of a resonant band mode around the defect is shown schematically in Fig. 2. Brout and Visscher [4] were the first to point out that these resonant modes can occur, and Dawber and Elliott [10] were the first to make a three-dimensional calculation of these modes for a monatomic Debye crystal. Their result for the corresponding ratio is (see Fig. 5)

$$\left|\frac{\eta_I}{A_0}\right| \sim \left\{\left[1+\varepsilon\omega^2\int\frac{D(\omega')\,d\omega'}{\omega'^2-\omega^2}\right]^2+\varepsilon^2\left(\frac{\pi}{2}\right)^2\omega^2\,D^2(\omega)\right\}^{-\frac{1}{2}}$$

showing that "resonant frequency" of the mode occurs where

$$1 = -\varepsilon\omega\int\frac{D(\omega')\,d\omega'}{\omega'^2-\omega^2}$$

This is just the frequency where real localized modes (compare with the following section) would occur, if there were no band of continuous modes.

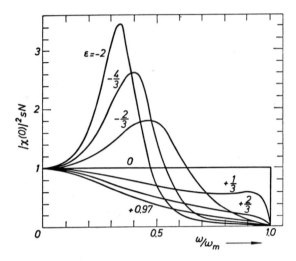

Fig. 5. Squared amplitude of band mode at the isotopic impurity site in a three-dimensional monatomic Debye crystal. Parameter is $\varepsilon = \Delta m/m$. After Dawber and Elliott [10].

The Dawber–Elliott equation shows, moreover, the important result that the half-width of the resonant mode is determined by the density of states $D(\omega)$, which for low impurity concentrations is just the same as $D(\omega)$ of the perfect lattice. The more resonant these modes are the smaller $D(\omega)$ is at the frequency where the maximum occurs. Such resonant modes therefore are predominantly found at the low-frequency part of the acoustic branches, that is, for heavy impurity masses and weak impurity force constants. Moreover, they can also occur at higher frequencies where $D(\omega)$ is accidentally small, as Fig. 6 shows for a case in silicon [29].

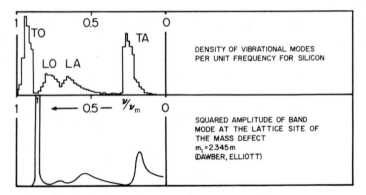

Fig. 6. Occurrence of a resonant band mode in silicon at a frequency where the density of vibrational modes is small. After Angress, Goodwin, and Smith [29].

It might be of interest to reformulate our equation for $|\eta_I/A_0|$ for the diatomic chain into the Dawber–Elliott form. This can be done with the help of the normalized density of states for that model, namely,

$$D(\Omega) = \frac{1}{\pi} \frac{m_0 + m_1}{\mu} \frac{\Omega(1 - 2\,\Omega^2)}{F_0 \, F_1}$$

One gets in this way

$$\left| \frac{\eta_I}{A_0} \right| = \left[\sigma^2 + \rho^2 \pi^2 \Omega^2 D^2(\Omega) \left(\frac{m_0}{m_0 + m_1} \frac{1 - m_1/\mu\,\Omega^2}{1 - 2\,\Omega^2} \right)^2 \right]^{-\frac{1}{2}}$$

This reduces in the isotopic case to

$$\left| \frac{\eta_I}{A_0} \right| = \left[1 + \varepsilon^2 \left(\frac{\pi}{2} \right)^2 \Omega^2 D^2(\Omega) \right]^{-\frac{1}{2}}$$

for $\gamma = 0$ and $m_0 = m$. The latter equation shows that no resonant mode can occur for the one-dimensional lattice in the isotopic case, in contrast to the three-dimensional lattice.

The one-dimensional lattice, however, may have resonant modes for an impurity with a force constant defect (as the former equation shows). It occurs in the case of $1 - \gamma \ll 1$ and for not too heavy impurities. The resonant frequency is then

$$\omega_{RM} \simeq \left(\frac{2f_I}{m_I} \right)^{\frac{1}{2}}$$

which is just that frequency where the weakly bound impurity would vibrate and the host lattice would be at rest.

The half-width of the resonant mode is determined by the density of phonon states, analogous to the Dawber–Elliott equation. The phase ψ becomes $\pi/2$ at the resonant frequency. Since ψ is the phase difference between the oscillation of the perturbed and the unperturbed lattice at the impurity site, one can consider the resonant mode as a forced oscillation of the impurity oscillator with the host lattice acting as the driving force.

Recently, explicit three-dimensional calculations of frequency and half-width of very low-lying resonant band modes have been made [18]. These consider the A_{2u} mode (Fig. 7) of an impurity with an ε defect and γ defect, on the lattice site of a monatomic simple cubic lattice, having only nearest neighbor force constants. The result for ω_{RM} and $\Delta\omega/\omega_{RM}$ in the

Fig. 7. A_{2u} vibration of an impurity with mass and nearest neighbor force constant f_I in a simple cubic lattice of mass m. After Sievers [20a].

harmonic approximation is

$$\left(\frac{\omega_{RM}}{\omega_D}\right)^2 = \frac{1-\gamma}{2\gamma+\varepsilon\gamma-3\varepsilon}$$

$$\frac{\Delta\omega}{\omega_{RM}} = \frac{3\pi}{2}|\varepsilon|\left(\frac{\omega_{RM}}{\omega_D}\right)^3$$

where ω_D is the Debye frequency. This shows the drastic decrease of the relative half width with decreasing values of ω_{RM}, a result which has been first pointed out by Brout and Visscher [4].

Thus, only band modes have been considered. Occasionally, however, a small number of lattice modes might emerge from the band continuum forming localized modes. This depends on the impurity parameters ε and γ. These localized modes may occur above the top of the bands or in the acoustic–optical gap, if such a gap exists.

For the case of the three–dimensional monatomic Debye crystal with one particle per cubic unit cell, Wallis and Maradudin [3], Maradudin *et al.* [6], and Dawber and Elliott [10] have shown that the eigenvalue equation for the position of ω_{loc} takes the form

$$1 = -\varepsilon\omega^2 \int_0^{\omega_{max}} \frac{D(\omega')\,d\omega'}{\omega'^2 - \omega^2}$$

Dawber and Elliott have further shown that the squared eigenvector of the impurity is given by

$$|\eta_I(\omega)|^2 = \left\{\varepsilon m\left[\varepsilon\omega^4\int \frac{D(\omega')\,d\omega'}{(\omega'^2-\omega^2)^2} - 1\right]\right\}^{-1} \qquad \varepsilon > 0$$

Figure 8 shows the quantities ω_{loc} and $|\eta_I|^2$ as functions of ε [10].

Recent calculations made by Jaswal [19] on NaI (a crystal with an acoustic–optical gap) give the positions of ω_{loc} in the gap for isotopic impurities (Fig. 9).

A large number of calculations have been done for localized modes with one-dimensional models (cf. [12]). We want to repeat the results of such a

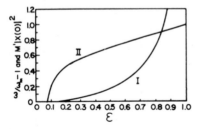

Fig. 8. Monatomic cubic Debye crystal with an isotopic defect. Curve I, position of local mode frequency *versus* $\varepsilon = \Delta m/m$. Curve II, squared amplitude of local mode at the impurity site *versus* ε. After Dawber and Elliott [10].

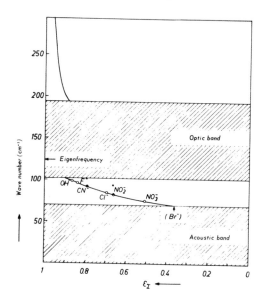

Fig. 9. Position of local mode frequencies *versus* $\varepsilon = \Delta m/m_I$ for an isotopic impurity in a three-dimensional NaI crystal. The impurity replaces an iodine atom. Calculation after Jaswal [19]. The gap mode curve shows the position for various isotopic substituents (open circles). Solid circles indicate experimental values [49]. After Renk [35].

calculation [35] for our diatomic chain model (Fig. 1). The eigenvector for localized modes is already contained in our band mode solutions for η_k and η_I. Namely, one has to set $\psi = 0$ and to replace ϕ by a complex phase $\tilde{\phi} = \phi \pm i\delta$, where $\phi = 0$, $\pi/2$ and where δ is positive. $\phi = 0$ gives the local modes above the highest lattice frequency, while $\phi = \pi/2$ determines the gap modes. This comes about by setting the complex phase $\tilde{\phi}$ into the phase frequency relation

$$\cos^2 \phi = \left(1 - \frac{m_0}{\mu} \Omega^2\right) \left(1 - \frac{m_1}{\mu} \Omega^2\right)$$

which yields for the left side

$$(\cos^2 \phi \cosh^2 \delta - \sin^2 \phi \sinh^2 \delta) \pm i\, 2 \sin \phi \cos \phi \sinh \delta \cosh \delta$$

Since the right side is real, one has either

$$\phi = 0$$

$$\cosh^2 \delta = \left(1 - \frac{m_0}{\mu} \Omega^2\right) \left(1 - \frac{m_1}{\mu} \Omega^2\right) > 0$$

$$\Omega_{\text{loc}} > 1$$

or

$$\varphi = \frac{\pi}{2}$$

$$\sinh^2 \delta = -\left(1 - \frac{m_0}{\mu} \Omega^2\right)\left(1 - \frac{m_1}{\mu} \Omega^2\right) > 0$$

$$\Omega_{\text{loc}} \text{ in gap between } \left(\frac{\mu}{m_0}\right)^{\frac{1}{2}} \text{ and } \left(\frac{\mu}{m_1}\right)^{\frac{1}{2}}$$

With this, the eigenvector becomes for $\Omega_{\text{loc}} > 1$

$$\eta_I = A_I \qquad \eta_k = A_0\, e^{-|k|\delta} \qquad k \text{ even} \neq 0$$

$$\eta_k = A_1\, e^{-|k|\delta} \qquad k \text{ odd}$$

and for Ω_{loc} in the gap

$$\eta_I = A_I \qquad \eta_k = (-1)^{k/2} A_0\, e^{-|k|\delta} \qquad k \text{ even} \neq 0$$

$$\eta_k = (-1)^{(|k|-1)/2} A_1\, e^{-|k|\delta} \quad k \text{ odd}$$

The ratio of the A's is

$$\frac{A_I}{A_1} = \mp\, \frac{e^{-\delta}}{1 - \dfrac{1}{\varkappa}\, \Omega^2}$$

$$\frac{A_1}{A_0} = \mp\, \left(\left|\frac{1 - \dfrac{m_0}{\mu} \Omega^2}{1 - \dfrac{m_1}{\mu} \Omega^2}\right|\right)^{\frac{1}{2}}$$

where the minus sign is for $\Omega_{\text{loc}} > 1$ and the plus sign is for Ω_{loc} in the gap. The important result is the exponential drop of amplitude by going from the impurity into the host lattice. The spatial localization is greater the more Ω_{loc} differs from the band frequencies. "Localization" $\exp(\delta)$ and the ratio A_1/A_0 as a function of frequency are plotted in Fig. 10. The eigenvector around the impurity is shown in Fig. 11.

The eigenvalue equation for the position of Ω_{loc} can be written in the form

$$\Omega^2(\Omega^2 - \Omega_0^2)(\Omega^2 - \Omega_s^2)^2 = (\Omega^2 - \Omega_1^2)(\Omega^2 - 1)(\Omega^2 - \Omega_k^2)^2$$

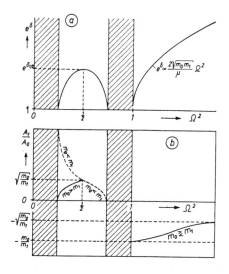

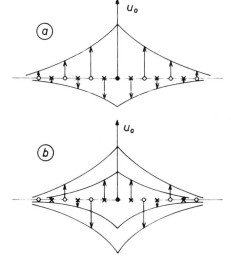

Fig. 10. Localization exp (δ) and amplitude ratio A_1/A_0 *versus* the reduced frequency squared for local modes in a diatomic chain. After Renk [35].

with the abbreviations

$$\Omega_0^2 = \frac{\mu}{m_0} \qquad \Omega_1^2 = \frac{\mu}{m_1} \qquad \Omega_s^2 = \frac{2f_I}{\mu_I} \cdot \frac{\mu}{2f} \qquad \Omega_k^2 = \frac{2f_I}{m_I} \cdot \frac{\mu}{2f}$$

$$\mu = \frac{m_0 m_1}{m_0 + m_1} \qquad \mu_I = \frac{m_I m_1}{m_I + m_1}$$

Fig. 11. Eigenvector of infrared-active local modes near the impurity in a diatomic chain. (a) Local mode above the lattice bands. (b) Local mode in the acoustic–optical gap. After Renk [35].

In addition, the requirement of the conservation of momentum for the oscillation of the local mode leads to the relation

$$
e^{-\delta} = \left(\left| \frac{1 - \dfrac{\Omega^2}{\Omega_1^2}}{1 - \dfrac{\Omega^2}{\Omega_0^2}} \right| \right)^{\frac{1}{2}} \left[1 + \frac{1+\varepsilon}{\dfrac{1}{\varkappa}\Omega^2 - 1} \left(\frac{\Omega^2 |\Omega^2 - 1|}{\Omega_0^2 \Omega_1^2} \cdot \left| \frac{1 - \dfrac{\Omega^2}{\Omega_0^2}}{1 - \dfrac{\Omega^2}{\Omega_1^2}} \right| \right)^{\frac{1}{2}} \right]
$$

This last equation, together with the above-given eigenvalue equation, can give only two local modes or less, in accordance with the number of degrees of freedom.

In some limiting cases, one can write down the solutions analytically. The easiest one is the monatomic chain ($m_0 = m_1$) with an isotopic defect ($\gamma = 0$), which gives the famous result of Mazur, Montroll, and Potts [2]:

$$
\Omega_{loc}^2 = \frac{1}{1-\varepsilon^2} \qquad e^\delta = \frac{1+\varepsilon}{1-\varepsilon} \qquad \varepsilon > 0
$$

Another simple case is the diatomic chain with a very light impurity ($m_I \ll m_0$) for a very strong defect force constant ($f_I \gg f$). In this case, one has $\Omega_{loc} \gg 1$ and

$$
\Omega_{loc}^2 \simeq \tfrac{1}{2}(\Omega_k^2 + \Omega_s^2) = \frac{2f_I}{m_I} \cdot \frac{\mu}{2f} \cdot \frac{m_I + 2m_1}{2m_1}
$$

For other cases, compare the literature [12, 20b, 35, 54]. Figures 12a and 12b show calculations of Ω_{loc} as functions of the defect parameters $\varepsilon = \Delta m/m_0$ and $\gamma = \Delta f/f$ for the case of a NaI chain [35]. Comparison with Jaswal's three-dimensional calculation of NaI (Fig. 9) shows, of course, great differences. In particular, the curve above the band modes is too flat and the curve for the gap mode does not reach the optical band for $\varepsilon = 1$.

III. OPTICAL LATTICE ABSORPTION INDUCED BY IMPURITIES

This section deals with the additional optical absorption in a crystal due to impurities. Other than for exceptional cases, one considers only "one-phonon absorption" in harmonic approximation. Since the lattice frequencies range up to about 10^{13} cps, the effects of interest fall into the middle and far infrared region of the electromagnetic spectrum.

It is assumed that one has a well-defined impurity, statistically distributed in the lattice. The concentration N_I should be small enough to avoid interaction between the impurities, so that the lattice dynamics described in

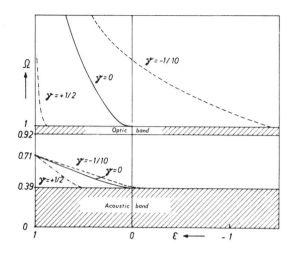

Fig. 12a. Frequencies of infrared-active local modes in a one-dimensional NaI lattice *versus* $\varepsilon = \Delta m/m_I$. $\gamma = \Delta f/f$ is the parameter. The iodine atom is replaced. After Renk [35].

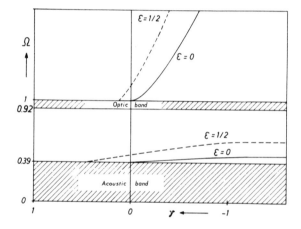

Fig. 12b. Frequencies of infrared-active local modes in a one-dimensional NaI lattice *versus* $\gamma = \Delta f/f$. $\varepsilon = \Delta m/m_I$ is the parameter. The iodine atom is replaced. After Renk [35].

section II can be applied. The absorption can then be written in the form [7, 10, 20a]

$$\omega \varepsilon''(\omega) = \frac{2\pi^2}{3} N_I \left(\frac{\varepsilon'+2}{3}\right)^2 [M_{LM}'^2(\omega) S(\omega) + M_{BM}'^2(\omega) D(\omega)]$$

$$M'(\omega) = \frac{dM(\omega)}{dQ_\omega} \qquad Q_\omega = \text{normal coordinate}$$

where M_{LM} is the dipole moment which is associated at ω with local modes and M_{BM} is that for band modes. The shape function $S(\omega)$ accounts for anharmonicity or impurity interaction causing a line width of the local mode which is generally also temperature-dependent. $D(\omega)$ is the density of phonon states for the host lattice (low concentration N_I) but it could also include anharmonicity broadening in the case of sharp resonant band modes. The factor $[(\varepsilon'+2)/3]^2$ is the local field correction, with the dielectric constant ε' valued at the frequency ω. Additionally, there is a sum rule [10] for the absorption coefficient $\alpha_I = \omega\varepsilon''/c\sqrt{\varepsilon'}$, namely,*

$$\int_0^\infty \alpha_I(\omega)\,d\omega = \frac{2\pi^2 e^2}{\sqrt{\varepsilon'}\,c^2\,m_I}\left(\frac{\varepsilon'+2}{3}\right)^2 N_I f$$

where m_I is the effective impurity mass and f is the oscillator strength equal to 1 for an allowed electric dipole transition. This sum rule shows that an existing local mode absorption grows at the expense of the ever-present band mode absorption.

The dipole moment $M(\omega) = \Sigma e_k u_k(\omega)$ can be calculated if the charges e_k of all particles and especially of the impurity are known and if the lattice dynamics have been carried through to get the perturbed eigenvector $\mathbf{u}$.

The expression $M'(\omega)$ in the equation for $\omega\varepsilon''$ can then be calculated if the normal coordinates Q_ω are known for the mode ω. For low concentration N_I, the Q_ω's of the unperturbed lattice can be used. In the diatomic chain case, these are [7, 16]

$$Q_\Omega = A_0\left[\frac{N}{2}(m_0+m_1)\,\frac{1-2\Omega^2}{1-\dfrac{m_1}{\mu}\Omega^2}\right]^{\frac{1}{2}}$$

The dipole moment of the mode ω could formally be written as

$$M(\omega) = e_{\text{eff}}(\omega)\cdot\eta_I(\omega)$$

so that

$$M'^2(\omega) \equiv \left[\frac{dM(\omega)}{dQ_\omega}\right]^2 = e_{\text{eff}}^2\left(\frac{\eta_I}{A_0}\right)^2\frac{2}{N}\frac{1}{m_0+m_1}\frac{1-\dfrac{m_1}{\mu}\Omega^2}{1-2\Omega^2}$$

This formulation reintroduces the amplitude ratio η_I/A_0 and expresses M so that the host lattice is uncharged and only the impurity with charge e_{eff} and amplitude η_I determines the dipole moment.

For the diatomic chain of section II one gets [16], for example,

$$e_{\text{eff}}(\Omega) = \Delta e + e\,\frac{\Omega^2}{1-\Omega^2}\left(\frac{\gamma}{\varkappa}-\varepsilon\right)$$

---

* This formula holds only for covalent crystals having not an optically active dispersion oscillator causing reststrahl absorption.

where Ω is again equal to ω/ω_R and where the impurity should have an uncompensated charge defect Δe in addition to e. This expression shows that covalent crystals, such as Ge and Si, can show a defect-induced dipole absorption only if the impurity has a charge defect which can be neutralized nearby or farther away (cf. Szigeti [7]). The equation shows furthermore that ionic crystals can have impurity-induced absorption with defects having the same charge as the replaced host particle. In this case, the additional absorption is caused solely by the changed eigenvector around the impurity. Calculations of this sort have been made by several authors [3, 6, 9, 16, 17, 20b, 54]. Figure 13 shows an example which has been calculated with the diatomic chain model of section II for band mode absorption. The density of states $D(\omega)$ will normally influence to a considerable extent the overall shape of the band mode absorption. For extremely sharp resonant band modes, however, the relative impurity amplitude η_I/A_0 (see Fig. 4) may already show the essential features of the absorption. This is demonstrated in Fig. 14 for the system KBr–Li, where the measurements of Sievers and Takeno [31] together with a theoretical curve using our chain model are shown. It may happen, however, that the density of states $D(\omega)$ represents directly the optical absorption. This can accidentally happen* as in solidified

---

* Already the very first observations of the impurity-induced band-mode absorption by Kaiser and Bond [22] and by Smith and Hardy [23] on diamond with natural impurities showed clearly the connection with the density of phonon states.

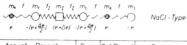

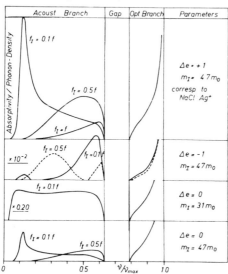

Fig. 13.   Band mode absorption of a NaCl chain for various combinations of defect mass m_I and defect force constant f_I. Values of $(dM/dQ_\omega)^2$ are plotted on the ordinate. After Genzel, Renk, and Weber [16].

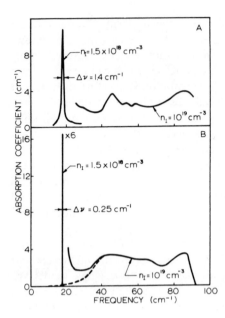

Fig. 14. (a) Absorption of KBr–⁶Li. After Sievers and Takeno [31]. (b) Calculation of the absorption for KBr–⁶Li with the linear chain model by fitting the data of part (a) with a force constant of $f_I = 0.005 f$ and using the density of KBr for the broad one-phonon absorption in the acoustic region (dashed curve). After Genzel, Renk, and Weber [16].

argon with xenon impurities, as measured by Jones and Woodfine [30] (Fig. 15). However, this can also be used purposely to get information for $D(\omega)$ of the host lattice. For this case, one must apparently smear out by many kinds of defects all the characteristic features which are associated with a well-defined defect. A good methodical way to achieve this seems to be the neutron irradiation of crystals as shown by Angress, Smith, and Renk [24] in the case of silicon (Fig. 16). Other information can be obtained by band-mode absorption in cases where the host crystal has an acoustic–

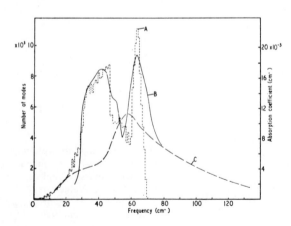

Fig. 15. Curve A, calculated phonon spectrum of pure argon. Curve B, impurity-induced absorption in Ar + 0.5% Xe at 55 °K. Curve C, impurity-induced absorption in Ar + 1% Kr at 80 °K. After Jones and Woodfine [30].

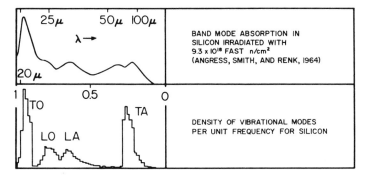

Fig. 16. Band mode absorption in silicon irradiated with fast neutrons. After Angress, Smith, and Renk [24]. For comparison, the density of vibrational modes in silicon is included.

optical gap. Here, one expects an abrupt break-off of the band absorption at the band edges, even for very different kinds of impurities. Figure 17 shows such an example for KI, where the acoustic band edge could be determined to higher precision than that possible from inelastic neutron scattering.

Since most of the defect-induced absorption measurements have been done on alkali halides, our considerations will be confined to representing schematically some of the pertinent data in Figs. 18 and 19, together with the density of phonon states of these crystals. Furthermore, interest will now be concentrated on the resonant band mode absorption on the one hand and the local mode absorption on the other.

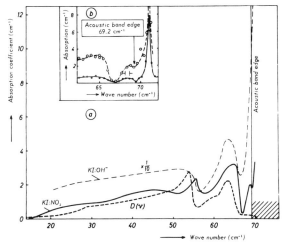

Fig. 17. (a) Acoustic band mode absorption in KI (host crystal absorption subtracted), compared with the density of phonon states in KI. (b) Location of the acoustic band edge of KI at 69.2 cm^{-1} by the impurity absorption of KI–NO$_2^-$. Solid curve, 0.25×10^{19} NO$_2^-$ per cm^3. Dashed curve, 1×10^{19} NO$_2^-$ per cm^3. ⊣⊢ Resolution.

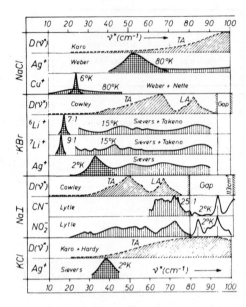

Fig. 18. Impurity-induced absorption in NaCl, NaI, KCl, and KBr.

Table I gives a summary of the data of the known resonant modes in alkali halides.

The force constant ratio is determined by the equations of Mitani and Takeno [18] to give the best fit for the measured position of ω_{RM}.

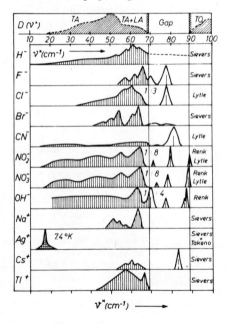

Fig. 19. Impurity-induced absorption in KI. Due to newer experimental results, the gap extends to about 100 cm^{-1} instead of 90 cm^{-1}.

Table I
Resonant Modes in Alkali Halides

Lattice–impurity combination	ν_{RM} (cm^{-1})	$\dfrac{\nu_{RN}}{\Delta\nu}$	$\dfrac{m_I}{m_0}$	$\dfrac{f_I}{f}$	Ionic radius ratio r_I/r_0
NaCl–Li$^+$	44.0	60	0.304	—	0.632
	59.0				
NaCl–Cu$^+$	23.7	39	2.76	0.04	1.0
NaCl–Ag$^+$	52.5	5	4.7	0.6	1.33
NaBr–Ag$^+$	48.0	—	4.7	1.1	1.33
NaI–Ag$^+$	36.7	—	4.7	1.1	1.33
KCl–Li$^+$	42	—	0.179	—	0.450
KCl–Ag$^+$	38.8	7	2.77	0.30	0.947
KBr–^{6}Li$^+$	17.9	18	0.154	0.0052	0.450
KBr–^{7}Li$^+$	16.3	20	0.179	0.0051	0.450
KBr–Ag$^+$	33.5	7.5	2.77	0.36	0.947
KI–Ag$^+$	17.4	19	2.77	0.172	0.947
RbCl–Ag$^+$	21.4	—	1.26	—	0.852
	26.4				
	36.1				

The ionic radius ratio r_I/r_0 of the impurity and host particle shows in some cases a relation to the force constant ratio f_I/f. Trends such are that, if r_I/r_0 is small, there is some possibility that f_I/f is small also and *vice versa*. Exceptions such as NaCl–Cu$^+$ are probably due to the polarizability of the impurity. This could be taken into account in the theory by means of a shell model with a weak spring constant between the nucleus and shell of the impurity ion.

Table I shows moreover the impossibility of describing, in most cases, the resonant behavior with an isotopic defect, that is, with $f_I = f$. Furthermore, one observes that normally the relative half-width of the resonant mode is smaller, the lower ν_{RM}. This is what one expects from a decrease of the density $D(\omega)$ in the lower acoustic branch. Recently, Sievers *et al.* [37] found an interesting example of this sort in the system MnF$_2$–Eu^{2+} with a resonant band mode at 15.95 cm^{-1} and a relative half-width of the order of 10^{-2}. This system might already be observable with the Mössbauer effect, a possibility which was first pointed out by Brout and Visscher [4, 5]. The system KBr–Li, also measured by Sievers *et al.*, is especially interesting due to the effects which are involved with two isotopic substitutions of ^{6}Li and ^{7}Li. Since no force constant difference should be expected in these two cases, one can predict rather easily the isotopic shift of ν_{RM} as well as the change in the relative half-width (compare this with experiment [31, 16] and also with Fig. 14). Such theoretical predictions of the half-width are significant, of course, only if the harmonic approximation used in section II is still valid. Anharmonicity will give each mode within the resonant mode region a finite width, which is even present at absolute zero and might change with temperature. In particular, the narrow resonant band modes can be influenced considerably in their life times by an anharmonic decay in other lattice modes. A realistic microscopic theory of such effects does not exist at present.*

On the other hand, there is some experimental evidence for the influence of temperature on resonant modes. In the harmonic approximation, there should be no influence because it is a one-phonon absorption process. Indeed, in the system NaCl–Ag$^+$ measured by Weber [26], no temperature influence could be observed in the mode at 52.5 cm^{-1}. This is not too surprising if one views this mode as a rather broad resonant mode. Very drastic temperature influences have been found, however, in the systems KBr–Li$^+$ [31], KI–Ag$^+$ [31], and NaCl–Cu$^+$ [36], which all have narrow resonant band modes. Here, the last system behaves quite differently from the other two in that the integrated absorption strength is rather constant. The half-width, however, decreases rapidly with temperature as shown in Fig. 20. This behavior is similar to that observed for U-centers and can be interpreted by a flat anharmonic potential and a two-phonon decay. The former two systems, however, show on the contrary a strong temperature dependence of the line strength (see Figs. 21, 22, and 23). In all three systems, one observes a small shift of the resonant mode frequency to lower values.

Attempts have been made to describe the temperature dependence of frequency and line width by a phenomenological theory similar to that used for the real local mode absorption, particularly for U-centers [20a, 31, 52].

* Recent calculations on the anharmonic broadening of resonant modes by K. H. Timmesfeld and H. Bilz (to appear in the *Proceedings of the International Conference on Localized Excitations in Solids*, Sept. 1967, Irvine, California) describe the case of NaCl:Cu$^+$ (Fig. 20) quite well.

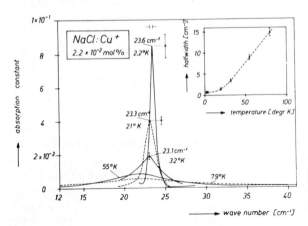

Fig. 20. The resonant band mode absorption of NaCl–Cu$^+$ and its temperature behavior. After Weber and Nette [36].

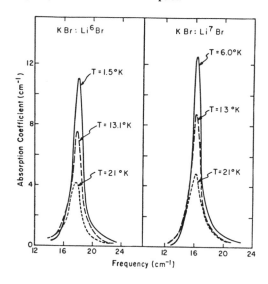

Fig. 21. Temperature dependence of the absorption coefficient in the neighborhood of the resonant mode for [6]Li and and [7]Li in KBr. After Sievers [20a].

These theories are similar to the treatment of Silsbee and Fitchen [53], considering the local mode absorption and the sharp resonant mode absorption as an analog to the Mössbauer effect. This seems to be rather plausible

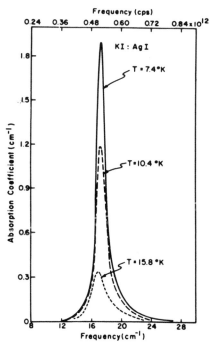

Fig. 22. Temperature dependence of the absorption coefficient in the neighborhood of the resonant mode for KI–Ag[+]. After Sievers [20a].

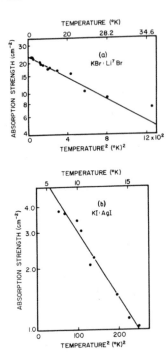

Fig. 23.   Temperature dependence of the integral line strength for the resonant modes of (a) KBr–Li[+] and (b) KI–Ag[+]. After Sievers [20a].

for the zero–phonon lines in vibronic spectra and also for the local mode absorption with its side bands (see below). For resonant band modes, however, it is of rather doubtful significance.

The main assumption in this model is that one can use the adiabatic condition for the local mode. This means that mode transition occurs much faster than the phonon transition of the host crystal, which is surely not true for the low-lying resonant band modes. Furthermore one cannot separate the normal coordinates of band modes and local modes.

Other minor assumptions, however, of the Silsbee model seem to be very useful for a successful treatment of the temperature dependence of resonant band modes. The Raman scattering processes of phonons on the weakly coupled impurity influence the potential of the impurity due to the vibrations of the nearest neighbor ions. For low-lying resonant modes, a good approximation is to use the Debye spectrum of acoustic phonons.

Some of the low-temperature behavior of the resonant modes can be understood qualitatively by the form of the potential in which the impurity vibrates. It can be either a flat anharmonic potential, where the impurity particle stays on a lattice site even at very low temperatures, or it can be a many-valley potential, where the impurity shifts at low temperatures out of the lattice site and creates tunnel-split states. Such behavior has been proved by static dielectric measurements as well as by thermal conductivity measurements at low temperatures. The coupling of the resonant mode to lattice distortions of different symmetries can be successfully studied by the application of uniaxial stress along different crystallographic directions. This is due mainly to a stress-induced frequency shift of the resonant mode, which is moreover dependent on the polarization of the radiation [20a].

Our discussion of the impurity-induced band mode absorption shall be concluded with mention of an experimental observation that has not yet been explained. This is the case of the absorption spectrum of NaCl with impurities, such as Ag^+, Mg^{2+}, or Ca^{3+} when the reststrahlen eigenfrequency ω_R is approached from below (Fig. 24). Careful measurements of the absorption show a strong increase of the absorption coefficient near ω_R. This behavior can partly be explained by the density of the phonon states, but also it is not clear whether it has also to do with the theoretical prediction that defect-induced absorption in ionic crystals should diverge approaching the reststrahlen frequency ω_R (see Fig. 13).

We turn now to the local mode absorption. Some of the localized gap modes in alkali halides have already been shown (Figs. 18 and 19). Table II gives the data for these localized gap modes insofar as they are known. Most of these modes have been found in KI. This compound is not too hygroscopic, can be easily grown and handled, and has a well-known broad acoustic–optical gap ranging from 69.2 to about 97.0 cm^{-1}.

The frequencies of the gap modes in KI normally cannot be explained by isotopic substitution, as shown in Figs. 25 and 26. The data require

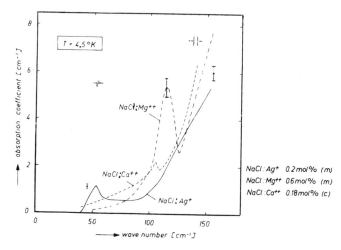

Fig. 24. The band mode absorption of NaCl–Ag$^+$, NaCl–Mg^{2+}, and NaCl–Ca^{2+} approaching the reststrahlen frequency of NaCl at 161 cm^{-1}. After Weber and Siebert, to appear in Z. Physik (1968).

Table II
Localized Gap Modes in Alkali Halides
(After Sievers [20a], Lytle [49], and Renk [35])

Lattice–impurity combination	ν_{loc} (cm^{-1})	Oscillator strength f	$\dfrac{m_I}{m_0}$	$\dfrac{f_I}{f}$	Ionic radius ratio r_I/r_0
KI–F$^-$	70 78		0.149		
KI–Cl$^-$	77.0	0.17* 0.31†	0.280	0.38	0.838
KI–Br$^-$	73.8	0.04* 0.17†	0.630	0.81	0.903
KI–Cs$^+$	83.5	4†	3.40	1.5	1.27
KI–CN$^-$	81.2	0.07	0.204	0.312	0.547 0.884
KI–NO$_2^-$	71.2 79.5	0.015* 0.026* 11†	0.362	0.40 0.56	0.585 0.682
KI–NO$_3^-$	73.0 79.4 88.0		0.51		
KI–OH$^-$	69.7 77 86.2 86.9 89		0.13	0.7	0.724
KI–OD$^-$	69.9‡ 77.0 82.5 86.2 88.8 94.3		0.14		0.724
NaI–CN$^-$	93.5 100.5	0.0078* 0.14†	0.205		0.547 0.884
NaI–NO$_2^-$	83.6 93.3	0.0017* 0.0025* 0.024†	0.362		0.585 0.682

* Strength of line absorption.
† Strength of line plus band absorption.
‡ Data by R. G. J. Grisar et al., Phys. Status Solidi 23: 613 (1967).

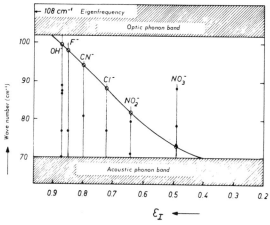

Fig. 25. Positions of the localized gap modes in KI for various substituents replacing the iodine atom. Open circles, for isotopic substitution. Solid circles, experimental values [32–35]. Solid line, the calculation for NaI [19] for isotopic substitution has been carried over to the KI case. After Renk [35].

a rather more detailed model, similar to that of Mitani and Takeno [18] that takes into account the change of coupling due to nearest neighbors of the impurity. However, even this model fails to satisfy the experimental results quantitatively. In the above-mentioned figures, the isotopic model calculations of Jaswal [19] on NaI have been simply applied to KI, since the band edges as well as the dispersion curves for KI are very similar to those of NaI. Molecular impurities, in contrast to monatomic impurities, show several optically active local modes. These more complex defects lift the threefold degeneracy of a monatomic defect at a lattice site of cubic symmetry due to their anisotropic coupling to the lattice. It is possible, furthermore, for these molecular defects to have librational transitions and also a tunnel splitting of the states due to a many-valley librational potential. We refer to the literature [20a, 35] for these open questions.

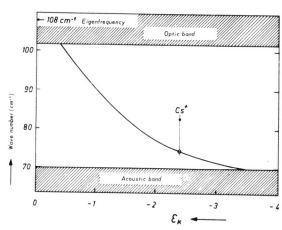

Fig. 26. Position of the localized gap mode of KI–Cs$^+$. Open circles, for isotopic substitution. Solid circles, experimental values [33]. Solid line, the calculation for NaI [19] for isotopic substitution has been carried over to the KI case. After Renk [35].

Much attention has been paid to the sharp local mode transitions above the lattice bands since Schäfer [40] first observed the local mode absorption of a U-center [39]. Figure 27 shows the U-center in KCl and its dependence on temperature and isotopic substitution, as measured by Mitsuishi and Yoshinaga [41]. Table III gives known data on U-centers in alkali halides. In addition local mode absorption above the lattice bands has been found for other impurities, although it is mainly restricted to covalent crystals such as silicon [29, 56, 58]. Ionic crystals have too strong a fundamental multiphonon absorption at the frequencies where the particles heavier than H, D, or Li could have local vibrations.

The line width of the U-center absorption is strongly temperature-dependent as shown in Fig. 27 and Table III. When the spectral resolution of the spectrometers can be made sufficiently high, one finds at low temperature a rest half-width (see for instance [46]) which seems to be determined by the lifetime of the excited state of the local mode due to a multiphonon decay. Whether these are two-phonon, * three-phonon, or still higher multiphonon decays depends on the frequency separation of ω_{loc} from the highest lattice frequency, which is approximately the frequency of the LO branch at wave vector zero (cf. Table III). At increasing temperature, the line becomes broadened from thermal averaging of the vibrational quantum numbers of the phonons involved. Thus, for example, for a three-phonon summation decay

$$\Delta v \sim (1+\bar{v}_1)(1+\bar{v}_2)(1+\bar{v}_3) - \bar{v}_1\bar{v}_2\bar{v}_3 \simeq 3\bar{v}^2\left(\frac{\omega_{loc}}{3}\right) + 3\bar{v}\left(\frac{\omega_{loc}}{3}\right) + 1 \sim T^2$$

$$\text{for } T > \Theta$$

---

* Bilz et al. [59] showed that the two-phonon decay is forbidden for the H⁻ centers, but allowed for the D⁻ centers.

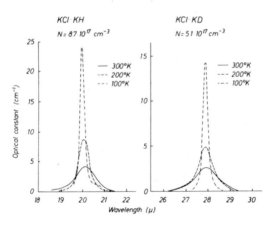

Fig. 27. U-center absorption of KCl–H⁻ and KCl–D⁻ and its temperature dependence. After Mitsuishi and Yoshinaga [41].

Table III
U-Centers in Alkali Halides

U-center, crystal	ν_{loc} (cm^{-1})		$\Delta\nu$ (cm^{-1})	$\dfrac{\nu_{loc}}{\nu_{LO}}$	Reference
LiF–H$^-$	1027	(20)*	4	1.5	[47]
	1015	(295)	19		
NaF–H$^-$	859.5	(20)	0.6	2.1	[47]
	846.7	(295)	12		
NaF–D$^-$	615.0	(20)	1.3	1.5	[47]
	607.5	(295)	10		
NaCl–H$^-$	565	(90)	5.6	2.1	[46]
NaCl–D$^-$	408	(90)	4.0	1.5	[46]
NaBr–H$^-$	498	(90)	17	2.4	[46]
NaBr–D$^-$	361	(90)	10	1.7	[46]
KCl–H$^-$	502	(90)	2.3	2.4	[41,46]
	496.5	(300)	26.2		
KCl–D$^-$	360	(90)	2.3	1.7	[41,46]
	357.5	(300)	30.8		
KBr–H$^-$	446	(90)	6.0	2.7	[41,46]
	447	(300	—		
KBr–D$^-$	318	(12)	—	1.9	Timusk and Klein (private
	319.7	(100)	—		communication)
KI–H$^-$	446	(90)	6.0	3.4	[46]
RbCl–H$^-$	476	(90)	4.8	2.5	[46]
RbCl–D$^-$	339	(90)	3	1.8	[46]
RbBr–H$^-$	425	(90)	8	3.3	[46]
CsBr–H$^-$	363	(80)	10		†

* The number in parentheses is the temperature of measurement in degrees Kelvin.
† S. S. Mitra and Y. Brada, *Phys. Rev.* **145**: 626 (1966).

where

$$\bar{v}_i = \frac{1}{e^{\dfrac{\hbar\omega_i}{kT}} - 1}$$

In addition, broadening is due to the beginning of multiphonon difference processes.* The integrated line strength is also temperature-dependent, i.e.,

* U-centers in KCl and NaCl show roughly a T^2 increase of the half-width [46], while the integral line strength remains constant.

it decreases with increasing temperature. Here, the optical analog to the Mössbauer effect [31, 52, 53] mentioned earlier gives an adequate treatment of the problem. The sharp U-center line corresponds to the zero-phonon transition, the strength of which is determined by the Debye–Waller factor. Transitions of the local mode, however, which are accompanied by phonon transitions occur as a continuous side-band absorption of the U-center. Figure 28 gives beautiful examples of this effect. The processes where one phonon is destroyed appear on the low-frequency side of the U-center line, and the processes where a phonon is created together with the excitation of the local mode appear on the high-frequency side. At low temperatures, the low-frequency absorption must freeze out primarily since no phonons are excited. The high-frequency absorption, however, remains even at absolute zero showing all the details of the host lattice density of phonon states, such as for instance the acoustic–optical gap (these are marked in Fig. 28). The coupling causing the side bands is mainly due to the anharmonicity, as shown by Bilz et al. [59]. This means that the optically active local mode of the U-center acts as a dispersion oscillator (TO oscillator at wavevector zero) in ionic crystals. The local mode gets excited virtually in the electromagnetic field by means of dipole interaction and decays within the uncertainty time into three modes from anharmonicity, two of which are band phonons and the third is the local mode itself.

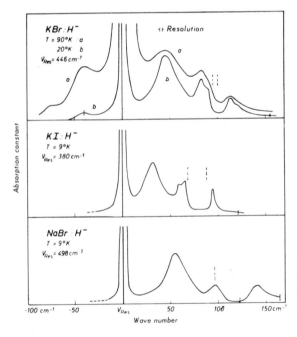

Fig. 28. One-phonon side bands at the U-centers of KBr–H⁻, KI–H⁻, and NaBr– H⁻. After Fritz, Gross, and Bäuerle [46].

IV. CONCLUSION

After about a ten-year period of research on impurity-induced absorption, it is clear that this is a rather general field in solid state physics. In principle, not only point defects will cause impurity absorption, but also any defect that destroys the translational symmetry of the crystal—the surface, dislocations, electronically excited lattice particles, etc. Moreover, some phase transitions of solids (such as order–disorder transitions in ferroelectrics or the transition to the liquid state) destroy the translational symmetry and will exhibit impurity-induced absorption. Indeed, the statement can be made therefore that no liquid can be free of absorption in the region of its "phonon transitions".

Experimental evidence of this defect-induced absorption in liquids already exists [61, 62], although no theoretical approach has yet been able to give us a deeper understanding. Those liquids whose elementary translational excitations have already been measured by inelastic neutron scattering are particularly interesting in this connection. Figure 29 shows the case of liquid bromine where both phenomena, far infrared absorption and neutron scattering, have been measured. There seems to be little doubt that both methods observe the same thing, namely, the density of short-living oscillatory

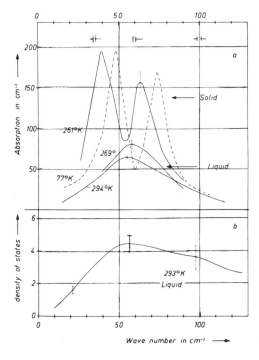

Fig. 29. (a) Far infrared absorption of liquid and polycrystalline bromine. After Wagner [62]. (b) Frequency distribution of the density of states of liquid bromine as determined by inelastic neutron scattering. After Coote and Haywood, *Inelastic Scattering of Neutrons in Solids and Liquids, Vol. I,* International Atomic Energy Agency (Vienna), 1963.

excitations. The main deviation of the neutron scattering spectrum in Fig. 29 from the infrared spectrum at higher frequencies is probably due to two-phonon excitations by the neutrons.

V. REFERENCES

A. Theory of Vibrations and Optical Absorption in Lattices with Impurities

1. I. M. Lifshitz, *Nuovo Cimento Suppl.* **3**:716 (1956).
2. P. Mazur, E. W. Montroll, and R. B. Potts, *J. Wash. Acad. Sci.* **46**:2 (1956).
3. R. F. Wallis and A. A. Maradudin, *Progr. Theoret. Phys. (Kyoto)* **24**:1055(1960).
4. R. Brout and W. M. Visscher, *Phys. Rev. Letters* **9**:54 (1962).
5. W. M. Visscher, *Phys. Rev.* **129**:28 (1963).
6. A. A. Maradudin, E. W. Montroll, and H. G. Weiss, in: F. Seitz and D. Turnbull (eds.), *Solid State Phys. Suppl.*, Vol. 3, Academic Press (New York), 1963.
7. B. Szigeti, *J. Phys. Chem. Solids* **24**:225 (1963).
8. S. Takeno, S. Kashiwamura, and E. Teramoto, *Progr. Theoret. Phys. (Kyoto) Suppl.* **23**:124 (1962).
9. S. Takeno, *Progr. Theoret. Phys. (Kyoto)* **29**:191 (1963); **38**: 995 (1967).
10. P. G. Dawber and R. J. Elliott, *Proc. Roy. Soc. (London)* A **273**:122 (1963); *Proc. Phys. Soc.* **81**:453 (1963).
11. B. Lengeler and W. Ludwig, *Z. Physik* **171**:273 (1963).
12. W. Ludwig, *Ergeb. Exakt. Naturw.* **35**:1 (1964).
13. R. Loudon, *Proc. Phys. Soc.* **84**:379 (1964).
14. G. S. Zavt and E. E. Tyutkson, *Soviet Phys. Solid State (English Transl.)* **11**:2561 (1965).
15. H. Dettmann and W. Ludwig, *Phys. Status Solidi* **10**:689 (1965).
16. L. Genzel, K. F. Renk, and R. Weber, *Phys. Status Solidi* **12**:639 (1965).
17. M. Wagner and W. E. Bron, *Phys. Rev.* **139**:A223 (1965).
18. Y. Mitani and S. Takeno, *Progr. Theoret. Phys. (Kyoto)* **33**:77 (1965).
19. S. S. Jaswal, *Phys. Rev.* **137**:A302 (1965).
20a. A. J. Sievers, *NATO Advanced Study on Elementary Excitations and their Interactions*, Cortina d'Ampezzo, July 1966.
20b. G. F. Nardelli, *NATO Advanced Study on Elementary Excitations and their Interactions*, Cortina d'Ampezzo, July 1966.
20c. M. Balkanski, *NATO Advanced Study on Elementary Excitations and their Interactions*, Cortina d'Ampezzo, July 1966.
21. L. Genzel, *Festkörper-Probleme*, VI, O. Madelung (ed.), Vieweg u. Sohn (Braunschweig), 1967.

B. Band Mode Absorption

22. W. Kaiser and W. L. Bond, *Phys. Rev.* **115**:857 (1959).
23. S. D. Smith and J. R. Hardy, *Phil. Mag.* **5**:1311 (1960).
24. J. F. Angress, S. D. Smith, and K. F. Renk, *Proc. Intern. Conf. Latt. Dyn.*, Copenhagen, 1963, p. 467.
25. A. J. Sievers, *Phys. Rev. Letters* **13**:310 (1964).
26. R. Weber, *Phys. Letters* **12**:311 (1964).
27. M. Balkanski and M. Nusimovici, *Phys. Status Solidi* **5**:635 (1964).
28. C. D. Lytle, thesis, Cornell University, Ithaca, New York, 1965.
29. J. F. Angress, A. R. Goodwin, and S. D. Smith, *Proc. Roy. Soc. (London)* A **287**:64 (1965).

30. G. O. Jones and J. M. Woodfine, *Proc. Phys. Soc.* **86**:101 (1965).
31. A. J. Sievers and S. Takeno, *Phys. Rev.* **140**:A1030 (1965); *Phys. Rev. Letters* **15**:1020 (1965).
32. K. F. Renk, *Phys. Letters* **14**:281 (1965).
33. A. J. Sievers, *Low Temp. Phys.* L T 9 (Part B): 1170 (1965).
34. K. F. Renk, *Phys. Letters* **20**:137 (1966).
35. K. F. Renk, thesis, University of Freiburg, Germany, 1966; *Z. Physik* **201**: 445 (1967).
36. R. Weber and P. Nette, *Phys. Letters* **20**:493 (1966).
37. A. J. Sievers, R. W. Alexander, Jr., and S. Takeno, *Solid State Comm.* **4**:483 (1966).
38. I. G. Nolt and A. J. Sievers, *Phys. Rev. Letters* **16**:1103 (1966).

C. Local Mode Absorption and U-Centers

39. R. Hilsch and R. W. Pohl, *Nachr. Akad. Wiss. Goettingen, II. Math. Physik Kl. Neue, Folge* **2**:139 (1936). *Trans. Faraday Soc.* **34**:883 (1938).
40. G. Schäfer, *J. Phys. Chem. Solids* **12**:233 (1960).
41. A. Mitsuishi and H. Yoshinaga, *Progr. Theoret. Phys. Suppl. (Kyoto)* **23**:241 (1962).
42. B. Fritz, *J. Phys. Chem. Solids* **23**:375 (1962).
43. W. Hayes, G. D. Jones, R. J. Elliott, and E. F. McDonald, *Proc. Intern. Conf. Latt. Dyn., Copenhagen,* 1963, p. 475.
44. D. N. Mirlin and I. I. Reshina, *Soviet. Phys. Solid State (English Transl.)* **6**:2454 (1964).
45. W. Hayes, *Phys. Rev. Letters* **13**:275 (1964); *Phys. Rev.* **138**:A1227 (1965).
46. B. Fritz, U. Gross, and D. Bäuerle, *Phys. Status Solidi* **11**:231 (1965).
47. H. Dotsch, W. Gebhardt, and Ch. Martius, *Solid State Comm.* **3**:297 (1965).
48. K. F. Renk, *Phys. Letters* **14**:281 (1965).
49. C. D. Lytle, thesis, Cornell University, Ithaca, New York, 1965.
50. A. J. Sievers and C. D. Lytle, *Phys. Letters* **14**:271 (1965).
51. S. Takeno and A. J. Sievers, *Phys. Rev. Letters* **15**:1020 (1965).
52. S. S. Mitra and R. S. Singh, *Phys. Rev. Letters* **16**:694 (1966).
53. R. H. Silsbee and D. B. Fitchen, *Rev. Mod. Phys.* **36**:432 (1964).
54. A. J. Sievers, A. A. Maradudin, and S. S. Jaswal, *Phys. Rev.* **138**:A272 (1965).
55. S. S. Mitra and Y. Brada, *Phys. Letters* **17**:19 (1965).
56. A. R. Goodwin and S. D. Smith, *Phys. Letters* **17**:203 (1965).
57. K. F. Renk, *Phys. Letters* **20**:137 (1966).
58. M. Balkanski and W. Nazarewicz, *J. Phys. Chem. Solids* **27**:671 (1966).
59. H. Bilz, B. Fritz, and D. Strauch, *J. Phys. Radium* (1966). Issue Conf. on Spectroscopy, Montpellier.
60. A. Mitsuishi, A. Manabe, and H. Yoshinaga, *Semicond. Conf. Kyoto,* 1966.

D. "Lattice Absorption" in Liquids

61. G. E. Ewing and S. Trajmar, *J. Chem. Phys.* **42**:4038 (1965).
62. V. Wagner, *Phys. Letters* **22**:58 (1966).

CHAPTER 16

Pseudo–Brewster Angle Technique for Determining Optical Constants

R. F. Potter

Infrared Div.
Corona Laboratories
Corona, California

I. INTRODUCTION

The optical properties of solids have proven to be extremely useful in determining the physical characteristics of crystalline solids. Although a great deal of information has been gained from normal reflectance measurements on opaque samples, including the application of the Kramers–Kronig dispersion relations to such data, much additional information can be gained from the direct determination of the real and imaginary parts of the dielectric constant. There have been several techniques used in the past, based on similar approaches, for measuring reflectances as a function of the angle of incidence. However, they all have one common drawback in that they require graphical solutions of the expressions containing related coefficients. This requirement increases the difficult of acquiring sufficient data to perform optical constant "spectroscopy" with the required resolution.

In the first part of this chapter, we shall brieflly review Maxwell's equations for an isotropic medium and the corresponding development of the Fresnel relations from those expressions. Next, the ratio of the reflected intensities for parallel and perpendicular linearly polarized radiation will be shown to have a unique solution at the pseudo-Brewster angle, which can be related to the desired optical constants. The experimental arrangement for measuring the required parameters will then be shown, along with some typical data. This will be followed by a discussion of the correction necessary when a thin film is present, and, finally, a very brief review will be given of some recent measurements for germanium in the spectral region from 0.5 to 2.5 eV.

II. FRESNEL RELATIONS AND THE OPTICAL CONSTANTS*

We shall consider in this section a homogeneous, optically isotropic medium only. For a dimension large compared with that of the lattice constant but very much smaller than the wavelength of the incident radiation, the following macroscopic equations apply:

$$\nabla \times \mathbf{E} = -\mu\mu_0 \frac{\partial \mathbf{H}}{\partial t}$$

$$\nabla \times \mathbf{H} = \sigma\mathbf{E} + \varepsilon\varepsilon_0 \frac{\partial \mathbf{E}}{\partial t} \tag{16.1}$$

$$\nabla \cdot \mathbf{E} = 0 \text{ (no permanent charge density)}$$

$$\nabla \cdot \mathbf{H} = 0$$

Application of suitable operators to these equations leads to

$$\nabla \times \nabla \times \mathbf{E} = -\mu\mu_0 \left(\sigma \frac{\partial \mathbf{E}}{\partial t} + \varepsilon\varepsilon_0 \frac{\partial^2 \mathbf{E}}{\partial t^2} \right)$$

and

$$\nabla \times \nabla \times \mathbf{E} = \nabla(\nabla \cdot \mathbf{E}) - \nabla^2 \mathbf{E}$$

and finally to the wave equation for the electric field vector $\mathbf{E}$

$$\nabla^2 \mathbf{E} - \sigma\mu\mu_0 \frac{\partial \mathbf{E}}{\partial t} - \mu\mu_0 \varepsilon\varepsilon_0 \frac{\partial^2 \mathbf{E}}{\partial t^2} = 0 \tag{16.2}$$

A similar expression for the magnetic field vector $\mathbf{H}$ can also be written. A solution for the wave equation is given by

$$\mathbf{E}_z = \mathbf{E}_0 \exp \left[i\omega \left(t - \frac{z}{v} \right) \right] \tag{16.3}$$

which describes a wave traveling in the z direction with a complex "velocity"

$$\frac{1}{v^2} = \mu\varepsilon\mu_0 \varepsilon_0 - i\sigma\mu_0 \frac{\mu}{\omega} \tag{16.4}$$

---

* More detailed discussions are found in texts on optics and optical studies, e.g., T. S. Moss, *Optical Properties of Semiconductors*, Butterworths Scientific Publications, Ltd. (London), 1959.

The square of the refractive index in the medium is given by

$$N_2{}^2 = \frac{c^2}{v^2}$$

In free space the parameters have the following values:

$$N_0 = 1 \qquad \varepsilon = 1 \qquad \mu = 1 \qquad \sigma = 0$$

Because the wave velocity in the medium can be a complex quantity [equation (16.4)], we write the complex refractive index as

$$N_2{}^2 = \mu\varepsilon - \frac{i\sigma\mu}{\omega\varepsilon_0} = \varepsilon' = \varepsilon_1 + i\varepsilon_2$$

where

$$N_2 = n - ik$$

$$\mu = 1$$

$$\varepsilon_1 = n^2 - k^2$$

$$2nk = \frac{\sigma}{\omega\varepsilon_0} = -\varepsilon_2$$

When the plane wave is incident at the interface at an oblique angle and the X–Z plane is the plane of incidence (Fig. 1), equation (16.3) becomes, respectively, for the three cases of incident wave, reflected wave, and transmitted wave

$$E = E_0 \exp\left[i\omega\left(t - \frac{x \sin \phi_0 + z \cos \phi_0}{v_0}\right)\right]$$

$$E' = E_0 \exp\left[i\omega\left(t - \frac{x \sin \phi_0 - z \cos \phi_0}{v_0}\right)\right] \qquad (16.3a)$$

$$E'' = E_0 \exp\left[i\omega\left(t - \frac{x \sin \phi_2 + z \cos \phi_2}{v_2}\right)\right]$$

where v_2 is the "velocity" in the refracting medium, i.e., $N_2 = c/v_2$. By resolving the components at the boundary where the tangential components of the electric field are continuous and setting equal all those components which vary with the same function of x, we find

$$N_0 \sin \phi_0 = N_2 \sin \phi_2$$

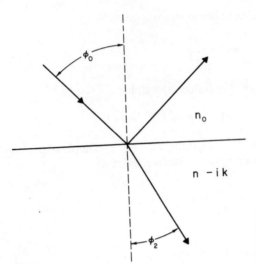

Fig. 1. The figure is in the plane of incidence and represents the interface between conducting and nonconducting dielectrics.

which is Snell's law. When we apply the boundary condition that the magnetic field is continuous through the boundary $z = 0$, we get for the reflected electric field amplitude coefficients $E'_{//}$ (parallel) and $E'_\perp$ (perpendicular) the following relations:

$$E'_{//} = E_{0//} \frac{N_0 \cos \phi_2 - N_2 \cos \phi_0}{N_0 \cos \phi_2 + N_2 \cos \phi_2} = E_{0//} \frac{\tan (\phi_0 - \phi_2)}{\tan (\phi_0 + \phi_2)}$$

$$E'_\perp = E_{0\perp} \frac{N_0 \cos \phi_0 - N_2 \cos \phi_2}{N_0 \cos \phi_0 + N_2 \cos \phi_2} = E_{0\perp} \frac{\sin (\phi_2 - \phi_2)}{\sin (\phi_2 + \phi_0)}$$

(16.5)

These are the Fresnel coefficients that are the basis for the so-called Drude expressions, which are used in ellipsometry.

We are interested, however, in the intensities of those beams which are linearly polarized parallel to the plane of incidence ($R_{//}$) and those which have electric vector **E** polarized perpendicular to the plane of incidence ($R_\perp$). From the Fresnel relations above [equation (16.5)], we can show that

$$R_{//} = R_\perp \left| \frac{\cos(\phi_2 + \phi_0)}{\cos(\phi_2 - \phi_0)} \right|^2$$

(16.6)

where, in general, ϕ is a complex angle. The fact that a complex refractive angle is used is a statement that the refracted wave is not a homogeneous one, i.e., that the planes of constant amplitude are not coplanar with those of constant phase. Thus, we write Snell's law in a complex form as

$$\frac{N_2 \cos \phi_2}{N_0} = \frac{N_2}{N_0} \left(1 - \frac{N_0{}^2}{N_2{}^2} \sin^2 \phi_0 \right)^{\frac{1}{2}} = \alpha + i\beta \tag{16.7}$$

Using this form of Snell's law, we write

$$\mathcal{R} = \frac{R_{//}}{R_\perp} = \frac{(\alpha - X)^2 + \beta^2}{(\alpha + X)^2 + \beta^2} \tag{16.8}$$

where $X = \sin \phi \tan \phi$ and the quantities α and β are related to the optical constants in the following manner:

$$\alpha^2 - \beta^2 = \frac{n^2 - k^2}{N_0{}^2} - \sin^2 \varphi_0$$

$$\alpha^2 \beta^2 = \frac{n^2 k^2}{N_0{}^4} \tag{16.9}$$

In the remainder of the chapter, we shall consider that the incident medium is free space, set $N_0 = 1$, and call $\phi_0 = \phi$ and $B_2 = n - ik$.

III. ANALYTICAL EVALUATION OF THE OPTICAL CONSTANTS*

Although the expression giving the ratio $\mathcal{R}$ is well-known and has been used in several experimental determinations for the optical constants, it was only recently shown that a knowledge, of the pseudo-Brewster angle ϕ_B (or $X_B{}^2$) and $\mathcal{R}_B$ can be used to determine explicitly the optical constants of an isotropic medium. The subscript B refers to the fact that these quantities have been determined at the so-called pseudo-Brewster angle, i.e., ϕ_B of the second type, where the ratio $\mathcal{R}$ is at its minimum. This minimum occurs when the following expression is satisfied:

$$(\alpha_2 - X^2) \frac{X d\alpha}{d\phi} - (\alpha^2 - X^2) \frac{\alpha dX}{d\phi} - \beta^2 \left(\frac{X d\alpha}{d\phi} + \frac{\alpha dX}{d\phi} \right) + 2\alpha\beta \frac{X d\beta}{d\phi} = 0 \tag{16.10}$$

This condition can also be expressed as

$$X_0{}^2 = X_B{}^2 \left(1 - 4 \frac{\beta_0{}^2 \gamma_B}{X_B{}^2} \right) \tag{16.11}$$

* The material contained in this section originally appeared in *J. Opt. Soc. Am.* **54**: 904 (1964).

where $X_0^2 = \alpha_0^2 + \beta_0^2$ at ϕ_B, and $\gamma_B = K/(1-K)$. Here

$$K = \frac{d\alpha}{\alpha_0}\bigg|_B \left(\frac{dX}{X}\right)^{-1}\bigg|_B = -\frac{\sin^2 \phi_B}{(2+\tan^2 \phi_B)\left(1-4\dfrac{\beta_0^2 \gamma_B}{X_B^2}\right)X_B^2}$$

$$= -[(1-4q^2 \gamma_B)f]^{-1} \qquad (16.12)$$

We also define the following parameters:

$$f = (2+\tan^2 \phi_B)\tan^2 \phi_B$$

$$q^2 = \frac{\beta_0^2}{X_B^2}$$

$$p^2 = \frac{\alpha_0^2}{X_B^2}$$

$$C^2 = \frac{(1-\mathscr{R}_B)^2}{(1+\mathscr{R}_B)^2}$$

$$D^2 = \frac{4\mathscr{R}_B}{(1+\mathscr{R}_B)^2}$$

The minimum ratio $\mathscr{R}_B$ at ϕ_B is written in these terms as

$$\mathscr{R}_B = \frac{X_B(1-2q^2 \gamma_B)-\alpha_0}{X_B(1-2q^2 \gamma_B)+\alpha_0} \qquad (16.13)$$

Algebraic manipulation yields the following expressions for the quantities p^2 and q^2:

$$p^2 = (1-2q^2 \gamma_B)^2 C^2$$

$$q^2 = 1-4q^2 \gamma_B-(1-2q^2 \gamma_B)^2 C^2 = D^2(1-4q^2 \gamma_B)-4q^4 \gamma_B^2 C^2 \qquad (16.14)$$

The quantity $q^2 \gamma_B$ is required in order to determine p^2, q^2, and, in turn, α_0^2 and β_0^2 and the associated optical constants. The quantity γ_B is evaluated at the pseudo-Brewster angle and can be written as

$$q^2 \gamma = -\frac{q^2}{1+(1-4q^2 \gamma_B)f} \qquad (16.15)$$

We define the quantity $S = q^2 \gamma_B$, and equations (16.14) and (16.15) give us two simultaneous equations for q^2 in terms of the quantity S. The quantity S, in turn, is given by the expression

$$S = \frac{1+4S_0-(1-8S_0)^{\frac{1}{2}}}{8(1+S_0)} \tag{15.16}$$

where $S_0 = -D^2/(1+f)$.

The quantity S is generally small, and in many instances a close approximation is

$$S \cong S_0 \left(1+4S_0^2\frac{1+5S_0}{1+S_0}\right) = S_0(1+\eta_2) \tag{16.17}$$

$$\eta_2 \leqslant 4S_0^2(1+4S_0) \qquad S_0 \leqslant \tfrac{1}{8}$$

Figure 2 shows that, for values of ϕ_B greater than 60°, $S \cong S_0$ with an error of less than 1%. In many practical cases this condition prevails for angles of ϕ_B greater than 40°. When the expressions for p^2 and q^2 are written in terms of S, the following equations result:

$$q^2 = D^2(1-4S_0)(1-\eta_3) \qquad \eta_3 \leqslant \frac{4C^2D^2}{(1+f)^2(1-4S_0)}$$

$$p^2 = C^2(1-2S_0)^2(1+\eta_4) \qquad |\eta_4| \leqslant |16S_0^3(1+2S_0)| \tag{16.18}$$

The limits for η_3 and η_4 are also shown in Fig. 2. Indeed, one can write the quantity p^2 and q^2 in terms of the constant S_0 with an error of less than 1% for values of ϕ_B greater than 65°. The same figure illustrates that in all instances one can ignore S_0 itself for all values of ϕ_B greater than 75°

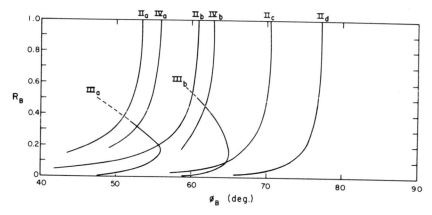

Fig. 2. Curves for which certain approximation limits are valid. In regions to the right of labeled curves, II_a and II_b give limits for $\eta_2 \leqslant 0.03$ and 0.01, respectively; III_a and III_b give limits for η_3 of 0.01 and 0.0001, respectively; IV_a and IV_b show the same limits for η_4; II_c gives an upper limit for $|4S_0| < 0.05$ and II_d gives an upper limit for $|4S_0| < 0.01$.

with a resulting error in the dielectric constant of less than 1%. Such precision is generally considered good in experiments of this type. The data of Fig. 2, especially the curves II_c and II_d, provide additional important information inasmuch as it indicates those angles of ϕ_B about which the quantities α and β are essentially constant.

Equivalent expressions for S and the polarization have also been given by Damany [1].

IV. EXPERIMENTAL DETERMINATIONS
OF THE QUANTITIES $\mathscr{R}$ AND X_B*

It has been demonstrated that a knowledge of the pseudo-Brewster angle ϕ_B or the quantity X_B (which is equal to $\sin \phi_B \tan \phi_B$) and the value of the ratio $\mathscr{R}_B$ at that pseudo-Brewster angle is sufficient to give an explicit analytical determination of the optical constants. The next problem is the development of the experimental techniques required to obtain those quantities.

Representative of previous techniques have been those of Tousey [2], Šimon [3], and Avery [4]. All are similar in concept; the reflectance is measured at two or more predetermined angles of incidence that include the pseudo-Brewster angle ϕ_B, and graphical solutions of the expressions for pre-determined angles of incidence are developed in terms of the real and imaginary components of the optical dielectric constant. Tousey used measurements of the reflectance of unpolarized incident light, and Šimon extended the procedure to include reflectance of polarized incident radiation. Avery pointed out the advantages of measuring the ratio of the polarized reflectances of the beams that were initially polarized parallel $(R_{//})$ and perpendicular $(R_\perp)$ to the plane of incidence. This method permits a relative measurement at each angular setting and minimizes the effect of sample size on the limiting aperture, the effect of any movement of the image on the detector as a function of angle of incidence, and the effects from long-term fluctuations in source, detector, and monochromator characteristics.

More recently, Lindquist and Ewald [5] described a geometric construction technique for determining the optical constants. They also make use of the relative measurements by determining the ratio of the two reflectances at two or more angles of incidence, which include the pseudo-Brewster angle ϕ_B. Their technique has a distinct advantage over the others in that they are free to choose those angles that are appropriate to the material being studied.

---

* The material contained here originally appeared in *Appl. Opt.* **4**:52 (1965).

A. Determination of $\mathscr{R}$

The quantity to be measured is the ratio $R_{//}/R_\perp = \mathscr{R}$ as a function of the angle of incidence, including the pseudo-Brewster angle ϕ_B, which, for most vacuum–sample interfaces, is between 45° and 90°. The reflectometer achieves this by rotating the sample about a fixed axis and at the same time keeping the source, polarizer, and detector in fixed positions. In Fig. 3, the sample M_S is set at an angle of incidence ϕ with respect to the collimated source beam. The reflected beam strikes the lower portion of the mirror M_1 at an angle of 45° in the plane of incidence. The image center now strikes the mirror M_2. Since M_2 is mounted perpendicular to M_1, the light is again incident on mirror M_3 at ϕ but is displaced vertically by the distance h. The beam is again reflected from M_3 at ϕ, after which it is collected and placed on the detector D. The mirror system $M_1 M_2$ is on a rotating arm that is geared to the axis A in a 2-to-1 ratio. The arm B moves through an angular distance at twice the angular rate that the sample is rotated about the axis A. Thus, with a 45° angle of incidence, arm B is perpendicular to the incident

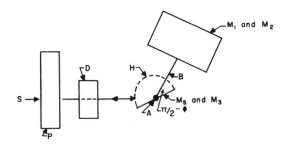

(a) Top View

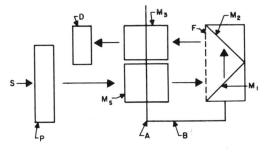

(b) Side View

Fig. 3. Schematic diagram of reflectometer showing beam from S passing polarizer P, reflecting from M_S, M_1, M_2, and M_3, and being collected at the detector D. (a) Top view (plane of incidence); (b) side elevation.

beam; with a 90° angle of incidence, arm B is parallel to the beam. The present apparatus permits an almost unlimited selection of detectors for distinct spectral regions and sensitivity requirements.

If metallic mirrors are used at M_1 and M_2, some polarization will be introduced at certain portions of the spectral region at wavelengths short of 5 μ, depending upon the metal used. This is in addition to whatever polarization has been introduced in the beam by the monochromator. Although the angular incidence on the sample changes, the polarization at each wavelength is constant because the angle of incidence on the mirrors M_1 and M_2 is always 45°. Since one must calibrate and determine the amount of polarization from the monochromator, the effect of mirrors M_1 and M_2 can be determined at the same time. Even this requirement can be removed if one uses a 90° roof prism that is made of a material with a refractive index such that the critical angle for total internal reflection is less than 45° in the spectral region of interest. When such a prism is used, its hypotenuse face is mounted so that it is always perpendicular to the reflected beam, that is, the angle of incidence on the prism face is zero. No polarization is introduced in the beam because, when the beam strikes the back surface at a 45° angle of incidence, both polarizations, which are totally internally reflected from the surface M_1 to the surface M_2, exit at zero angle of incidence from the hypotenuse face of the prism.

The mirror M_3 presents a somewhat different problem. In the spectral regions, where metallic mirrors are expected to show a great deal of dispersion effect, a transparent hemicylinder prism H has been used, with the flat specular surface of the hemicylinder serving as M_3. The collimated beam strikes the cylindrical surface at an angle of incidence very near to zero. Again, the refractive index of the hemicylinder is such that the rays which strike the surface M_3 are at angles of incidence greater than the critical angle, and both polarized beams are totally internally reflected, exiting from the hemicylinder at either zero or very small angles of incidence. In this manner, any polarization effects that might be introduced by a flat metallic mirror at M_3 are either eliminated or minimized to such a degree that they can be considered negligible. In certain cases where the minimum of the ratio $R_{//}/R_{\perp}$ is not very pronounced, the mirror M_3 can be replaced with a second sample of the material to be studied. For data reduction, the samples are considered to be identical. To date, prism materials of quartz and KRS 5 have been used for hemicylinders and roof prisms.

The gear system that controls the motion of arm B relative to axis A does not have stringent requirements and can take several mechanical forms. The principal criterion is that the surface F always be normal to the reflected beam. However, the gears and vernier readouts necessary for measuring the amount of rotation about the axis A can have a wide range, depending upon the accuracy required.

Any polarizer with sufficient aperture size can be placed at P. Sheet Polaroid, stacked AgCl plates, and Glan–Thompson prisms have been used. A distinct advantage accrues with this reflectometer because only a single polarizer is required.

The entire assembly—consisting of the axis A with sample S, the arm B, and the mirrors M_1, M_2, and M_3—has been mounted in a cryogenic apparatus with a single fixed window through which the light enters and exits at zero angle of incidence, permitting the sample to be cooled to temperatures near $77°K$. The apparatus is shown in Fig. 4. For accuracy, the present apparatus has a dial setting which can be read to a least count of three minutes of arc.

In Fig. 5 we see four examples of the type of data that one can obtain of $\mathscr{R}$ as a function of the angle of incidence. The materials range from a transparent insulator (SiO_2), through partially absorbing semiconductors (GaSb and Ge), to an opaque evaporated metal layer (Al). The minimum value of $\mathscr{R}_B$ at the pseudo-Brewster angle ϕ_B must be corrected for any polarization that is present in the incident beam or introduced at M_1, M_2, or M_3, or combinations thereof. As discussed above, polarization from the three mirrors can be minimized to become negligible. With data such as those given in Fig. 5, an interpretation can be made by using the technique

Fig. 4. The reflectometer. (a) Reflectometer portion including M_S, M_1, M_2, M_3, the axis A, and a modified arm B. The dial at the top gives the angle ϕ. Also shown are the cryogenic dewar mount and a window–detector arrangement with a PbS photoconductor. (b) Reflectometer mounted in the cryogenic dewar.

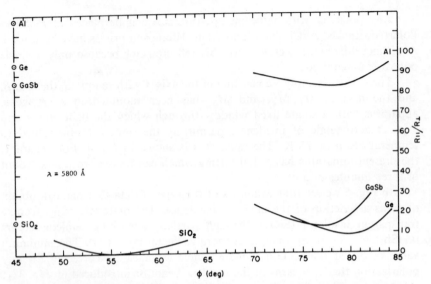

Fig. 5.   Typical curves of the ratio $\mathscr{R} = R_{//}/R_{\perp}$ for several materials. The curves must be corrected for beam polarization by multiplying by 0.67. The values for Al shown are for $\mathscr{R}^2$.

outlined by Avery, or variations of it, or by geometric construction such as that described by Lindquist and Ewald. However, such data also permit the determination of the optical constants from a straightforward analysis of the reflectance expressions, based on the Fresnel equations given earlier in equation (16.13) and repeated here

$$\mathscr{R}_B = \frac{X_B(1-2S)-\alpha_0}{X_B(1-2S)+\alpha_0} \tag{16.13}$$

Two techniques for extracting the experimental values required from data such as those of Fig. 5 will now be given. The first leads to an accurate measure of ϕ_B; the second gives the quantity $X_B{}^2$.

B.   Determination of ϕ_B

The ratio $\mathscr{R}$ [see equation (16.8)] can be expanded about ϕ_B, or the point where $\mathscr{R}' = 0$, by means of a Taylor's expansion

$$\mathscr{R}(\phi_k) = \mathscr{R}(\phi_B+\varDelta\phi_k) = \mathscr{R}_B + \frac{(\varDelta\phi_k)^2}{2}\,\mathscr{R}_B'' + \frac{(\varDelta\phi_k)^3}{6}\,\mathscr{R}_B''' + \dots$$

$$\mathscr{R}(\phi_j) = \mathscr{R}(\phi_B-\varDelta\phi_j) = \mathscr{R}_B + \frac{(\varDelta\phi_j)^2}{2}\,\mathscr{R}_B'' - \frac{(\varDelta\phi_j)^3}{6}\,\mathscr{R}_B''' + \dots \tag{16.19}$$

Also, we find

$$\phi_k - \phi_j = \Delta\phi_k + \Delta\phi_j = \Delta\phi_k \left(1 + \frac{\Delta\phi_j}{\Delta\phi_k}\right) = \Delta\phi_j \left(1 + \frac{\Delta\phi_k}{\Delta\phi_j}\right)$$

By setting $\mathcal{R}(\phi_j) = \mathcal{R}(\phi_k)$, the following expressions can be written:

$$\left(\frac{\Delta\phi_k}{\Delta\phi_j}\right)^2 \left(1 + \frac{\Delta\phi_k + \Delta\phi_j}{3} \frac{\mathcal{R}_B'''}{\mathcal{R}_B''} + \frac{\Delta\phi_k{}^2 - \Delta\phi_j{}^2}{12} \frac{\mathcal{R}_B''''}{\mathcal{R}_B''} + \ldots\right) = 1$$

$$\frac{\Delta\phi_j}{\Delta\phi_k} \cong 1 + \frac{\Delta\phi_k + \Delta\phi_j}{6} \frac{\mathcal{R}_B'''}{\mathcal{R}_B''}$$

$$\frac{\Delta\phi_k}{\Delta\phi_j} \cong 1 - \frac{\Delta\phi_k + \Delta\phi_j}{6} \frac{\mathcal{R}_B'''}{\mathcal{R}_B''} \qquad (16.20)$$

$$\frac{\Delta\phi_k}{\Delta\phi_j} + \frac{\Delta\phi_j}{\Delta\phi_k} \cong 2$$

and from equations (16.19) and (16.20)

$$(\Delta\phi_k{}^2 - \Delta\phi_j{}^2)\,\mathcal{R}_B'' + (\Delta\phi_k{}^3 + \Delta\phi_j{}^3)\,\frac{\mathcal{R}_B'''}{3} \cong 0 \qquad (16.21)$$

The substitution of

$$(\Delta\phi_k{}^2 - \Delta\phi_j{}^2)\,(\Delta\phi_k - \Delta\phi_j) + \Delta\phi_k\,\Delta\phi_j\,(\Delta\phi_k + \Delta\phi_j)$$

in equation (16.21) for $\Delta\phi_k{}^3 + \Delta\phi_j{}^3$ results in an equation quadratic in $\Delta\phi_k - \Delta\phi_j$:

$$\Delta\phi_k - \Delta\phi_j \cong -\tfrac{1}{3} \Delta\phi_k\,\Delta\phi_j\,\frac{\mathcal{R}_B'''}{\mathcal{R}_B''}$$

$$= \frac{(\Delta\phi_k + \Delta\phi_j)^2}{12} \frac{\mathcal{R}_B'''}{\mathcal{R}_B''} \cong -\frac{(\phi_k - \phi_j)^2}{12} \frac{\mathcal{R}_B'''}{\mathcal{R}_B''} \qquad (16.22)$$

Equations (16.19) and (16.20) have also been used. The mean value of ϕ_j and ϕ_k is given by

$$\bar{\phi} = \frac{\phi_j + \phi_k}{2} = \frac{\phi_B + \Delta\phi_k + \phi_B - \Delta\phi_j}{2}$$

$$= \phi_B \left(1 + \frac{\Delta\phi_k - \Delta\phi_j}{2\phi_B}\right) \qquad (16.23)$$

$$\cong \phi_B \left[1 - \frac{(\phi_k - \phi_j)^2}{24\phi_B} \frac{\mathcal{R}_B'''}{\mathcal{R}_B''}\right]$$

The ratio $\mathscr{R}_B'''/\mathscr{R}_B''$ in the situation considered here is of the order of 3 tan ϕ_B/rad. As ϕ_B has a value of approximately 1 rad, the mean value of ϕ_j and ϕ_k is equal to ϕ_B with an error of less than 1% for a difference between ϕ_j and ϕ_k of as much as 10°.

The symmetrical, or nearly symmetrical, character of the curves (as shown in Fig. 5) has been discussed previously. Johnson [6], for instance, analyzed such data very close to ϕ_B; he considered the shape of the curve to be a parabola. This is equivalent to setting $\mathscr{R}'''$ and higher terms to zero in equation (16.19). However, the assumption that the function $\mathscr{R}$ is parabolic becomes less valid at lower values of $\Delta\phi$ than the assumption that the ratio $\mathscr{R}$ is symmetrical about ϕ_B, that is, when the correction term of equation (16.23) is ignored.

With data of the type shown in Fig. 5, both parameters ϕ_B and $\mathscr{R}_B$ (which are required to establish the optical constants ε_1 and ε_2) can be determined. Because the signal-to-noise ratio $\mathscr{R}_B$ is too low for accurate angle determination, this capability is especially important when the minimum is very broad, as well as when it is necessary to determine ϕ_j and ϕ_k at large values of $\Delta\phi$.

As an example, data for a GaSb sample at a wavelength of 5800 Å is used. The quantity $\bar{\phi} = (\phi_j + \phi_k)/2$ is plotted as a function of $(\phi_j - \phi_k)^2$ in Fig. 6. The value of $\phi_B = 76.88°$ is given at the intercept of the extrapolated line at $(\phi_j - \phi_k)^2 = 0$.

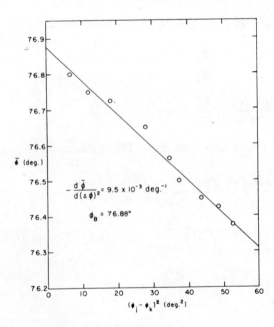

Fig. 6. A plot of ϕ versus $(\phi_j - \phi_k)^2$ for GaSb data of Fig. 5. The line was fitted by the least squares method. The intercept $\phi = 76.88°$ is a very precise measure of ϕ_B.

C. Determination of $X_B{}^2 = \sin^2 \phi_B \tan^2 \phi_B$

If the values of α and β are expanded about their values α_0 and β_0 at the pseudo-Brewster angle ϕ_B, we find

$$\alpha_j = \alpha_0\left(1 - \frac{\Delta_j}{\alpha_0}\right) \qquad \alpha_k = \alpha_0\left(1 + \frac{\Delta_k}{\alpha_0}\right)$$

$$\beta_j = \beta_0\left(1 + \frac{\Delta_j}{\alpha_0}\right) \qquad \beta_k = \beta_0\left(1 - \frac{\Delta_k}{\alpha_0}\right) \tag{16.24}$$

where the subscripts j and k refer to measurements and values at ϕ_j and ϕ_k, respectively, with ϕ_B at an intermediate value. Inserting these quantities in equation (16.8), setting $\mathscr{R}(\phi_j)$ equal to $\mathscr{R}(\phi_k)$, and clearing fractions, we get

$$(X_j X_k - \alpha_j \alpha_k)(\alpha_j X_k - \alpha_k X_j) = \beta_j{}^2 \alpha_k X_k - \beta_k{}^2 \alpha_j X_j$$

$$= \beta_j \beta_k \left(\frac{\beta_j \alpha_k}{\beta_k \alpha_j}\alpha_j X_k - \frac{\beta_k \alpha_j}{\beta_j \alpha_k}\alpha_k X_j\right) \tag{16.25}$$

The right-hand side of this expression can be rewritten as

$$\beta_j \beta_k(\alpha_j X_k - \alpha_k X_j) + 2\beta_j\beta_k(\alpha_j X_k + \alpha_k X_j)\frac{\Delta_j + \Delta_k}{\alpha_0} \tag{16.26}$$

Now $\alpha_j\alpha_k = \alpha_0{}^2[1 - (\Delta_j - \Delta_k)/\alpha_0]$, and there is a similar expression for $\beta_j\beta_k$. At angles close to the pseudo-Brewster angle ϕ_B, $\alpha_j\alpha_k$ is very closely approximated by $\alpha_0{}^2$, and $\beta_0{}^2$ is a very close approximation to $\beta_j\beta_k$; that is to say, the difference between Δ_j and Δ_k divided by α_0 is very much less than 1. It is of the order $[(\phi_k - \phi_j)^2/4][(\cos^2 \phi_B - \sin^2 \phi_B)/X_n{}^2]$. If we define the product $X_j X_k = X_n{}^2$ and the sum $\alpha_0{}^2 + \beta_0{}^2 = X_0{}^2$, equation (16.25) can be written as

$$X_0{}^2 - X_n{}^2 = \frac{-2\beta_0{}^2(X_k + X_j)}{\alpha_0}\frac{\Delta\alpha}{X_k - X_j}\frac{1 - \dfrac{\Delta_j X_k - \Delta_k X_j}{(X_k + X_j)\alpha_0}}{1 - \dfrac{\Delta_j X_k + \Delta_k X_j}{(X_k - X_j)\alpha_0}} = 4\beta_0{}^2\gamma_n \tag{16.27}$$

where

$$\gamma_n = \frac{X_k + X_j}{2\alpha_0}\frac{\Delta\alpha}{X_k - X_j}\frac{1 - \dfrac{\Delta_j X_k - \Delta_k X_j}{(X_k + X_j)\alpha_0}}{1 - \dfrac{\Delta_j X_k + \Delta_k X_j}{(X_k - X_j)\alpha_0}}$$

Equation (16.11) is written as $X_0{}^2 - X_B{}^2 = -4\beta_0{}^2 \gamma_B$, where

$$\gamma_B = \frac{\dfrac{d\alpha}{\alpha_0} \dfrac{X}{dX}\Big|_B}{1 - \dfrac{d\alpha}{\alpha_0} \dfrac{X}{dX}\Big|_B} \qquad (16.28)$$

Thus, $X_n{}^2$ is equivalent to $X_B{}^2$ in value, as γ_B is equivalent to γ_n. The limit will be determined by the precision of measurements being made and the values of ϕ_B about which the measurements are being made. In most instances γ_B is equal to γ_n, which means that $X_B{}^2$ can be determined with a high degree of precision by the data, as is shown in Fig. 6. By determining X_j and X_k at angles where the reflectance ratio is at a suitable signal-to-noise ratio and by taking a series of such measurements, the precision is improved to a considerable extent.

Table I lists the data used for Fig. 6 along with the corresponding $X_n{}^2$ values, the average of which gives $X_B{}^2 = 17.32$, whereas for $\phi_B = 76.88°$,

Table I
Values of ϕ_j and ϕ_k for GaSb at $\lambda = 5800$ Å

ϕ_k (deg)	ϕ_j (deg)	$X_n{}^2$
78.1	75.5	17.38
78.5	75.0	17.36
78.85	74.60	17.42
79.30	74.00	17.43
79.50	73.62	17.32
79.65	73.45	17.38
79.75	73.15	17.20
79.90	72.95	17.23
80.00	72.75	17.18

$$X_n{}^2 = 17.32 = X_B{}^2$$

Table II
Optical Constants for GaSb Sample at $\lambda = 5800$ Å

ϕ_B (deg)	$X_B{}^2$	ε_1	ε_2	n	k
76.88	17.40	7.63	16.1	3.56	2.26

$X_B{}^2 = 17.46$. Although the mean 17.40 was used in determining the constants, the total difference of less than 1% corresponds to an angular discrepancy of less than 2 minutes of arc in ϕ_B for the two methods. Table II gives the optical constants.

V. CORRECTING FOR THE PRESENCE OF A SINGLE-LAYER FILM

We have considered the highly idealized case of a plane electromagnetic wave incident on a perfectly specular sample surface immersed in a non-dispersive dielectric. In Fig. 7 another idealized picture is shown in which the specular surface has a uniform film having a third refractive index n_1. This idealization is somewhat closer to the real surface which has a thin layer placed there by chemisorption, adsorption, etc. Using this model, we can derive a suitable correction to be applied to the reflection measurements of the pseudo-Brewster angle technique. This should result in a set of values for the optical constants approaching those for the substrate material in the absence of any such film.

We have already briefly outlined the boundary value solutions from Maxwell's equations for a plane electromagnetic wave incident upon a dielectric surface. As shown earlier, these are usually given in terms of the Fresnel coefficients r_1 and r_2, which give the amplitude of reflectivities at the interfaces 0–1 and 1–2, respectively. The coefficients, in general, are dependent on the angles ϕ_0, ϕ_1, ϕ_2 and on the state of polarization of the incident electromagnetic wave. In addition, there are phase relations between the coefficients which result from the polarized character of the incident wave.

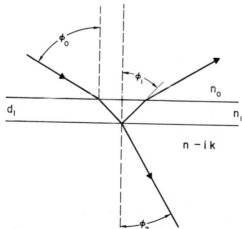

Fig. 7. The figure is in the plane of incidence and represents a conducting dielectric with a parallel-sided layer of thickness d_1 and a refractive index $n_1 \neq n_0$.

While these phase differences must be considered in detail for ellipsometric measurements, they will not be treated explicitly in the present instance.

Two states of linear polarization will be considered: (1) **E** vector parallel to the plane of incidence

$$r_{1_{//}} = r_{//} = \frac{n_0 \cos \phi_1 - n_1 \cos \phi_0}{n_0 \cos \phi_1 + n_1 \cos \phi_0}$$

$$r_{2_{//}} = g + ih = \frac{n_1 \cos \phi_2 - n_2 \cos \phi_1}{n_1 \cos \phi_2 + n_2 \cos \phi_1}$$

(16.29)

and (2) **E** vector perpendicular to the plane of incidence

$$r_{1_{\perp}} = r_{\perp} = \frac{n_0 \cos \phi_0 - n_1 \cos \phi_1}{n_0 \cos \phi_0 + n_1 \cos \phi_1}$$

$$r_{2_{\perp}} = s + it = \frac{n_1 \cos \phi_1 - n_2 \cos \phi_2}{n_1 \cos \phi_1 + n_2 \cos \phi_2}$$

(16.30)

In general, g, h, s, and t are all functions of ϕ_0. The reflected amplitude from the composite transparent film–substrates shown in Fig. 7 is given by the following:

$$r_{12} = \frac{r_1 + r_2 e^{-i\delta}}{1 + r_1 r_2 e^{-i\delta}}$$

(16.31)

where δ represents a phase factor

$$\delta = \frac{2 \pi n_1 d_1 \cos \phi_1}{\lambda_0}$$

(16.32)

and in the case considered here n_1 is a real quantity. The reflected intensity (reflectance) for the parallel polarization is given by

$$R_{f_{//}} = r_{12} r_{12}^* = \frac{r_{//}^2 + g^2 + h^2 + 2 r_{//}(g \cos 2\delta + h \sin 2\delta)}{1 + r_{//}^2 (g^2 + h^2) + 2 r_{//}(g \cos 2\delta + h \sin 2\delta)}$$

$$= \frac{(r_{//} + g)^2 + h^2 - 4 r_{//} \sin \delta (g \sin \delta - h \cos \delta)}{(1 + r_{//} g)^2 + r_{//}^2 h^2 - 4 r_{//} \sin \delta (g \sin \delta - h \cos \delta)}$$

(16.33)

where $R_{f_{//}}$ is the reflectance for the entire film–substrate combination. The expression for $R_{f_{\perp}}$ is obtained by substituting $r_{\perp}$, s, and t for $r_{//}$, g, and h, respectively. The reflectances for the film–free surfaces are given by

$$R_{0_{//}} = \frac{(r_{//}+g)^2 + h^2}{(1+r_{//}g)^2 + h^2 r_{//}^2}$$

$$R_{0_{\perp}} = \frac{(r_{\perp}+s)^2 + t^2}{(1+r_{\perp}s)^2 + t^2 r_{\perp}^2}$$

(16.34)

while the quantities that contain the effects of the presence of a film are given by

$$C_{//} = \frac{4 r_{//} \sin \delta (g \sin \delta - h \cos \delta)}{(1+r_{//}g)^2 + (r_{//}h)^2}$$

$$C_{\perp} = \frac{4 r_{\perp} \sin \delta (s \sin \delta - t \cos \delta)}{(1+r_{\perp}s)^2 + (r_{\perp}t)^2}$$

(16.35)

The polarized reflectance for the film-coated surface, then, is given by

$$R_f = \frac{R_0 - C}{1 - C}$$

$$R_0 = R_f + C(1 - R_f)$$

(16.36)

Up to this point no approximations have been made in the solution for the thin film–substrate model. These formulas provide good indications concerning the corrections that one must apply for optical measurements of the reflectance when a film surface is present: (1) As $R_f \to 1$ the required correction becomes smaller, i.e., $R_{f_{\perp}}$ at large angles of incidence and R_f for highly reflecting metals; and (2) the correction becomes 0 as $C \to 0$, in the following four instances:

$2nd_1 \cos \phi_1 = m\lambda_0$	$m = 1, 1, 2, \ldots$
$r_{//} = 0$ at $\phi_0 = \tan^{-1}(n_1/n_0)$	$C_{//} = 0$
$\tan \delta = (h/g)$	$C_{//} = 0$
$\tan \delta = (t/s)$	$C_{\perp} = 0$

In order to apply these corrections for the case of the pseudo-Brewster angle technique, we follow the procedure described below. For a film-covered sample the ratio $\mathscr{R}_0$ for the film–free surface is given by

$$\mathscr{R}_0 = \frac{R_{0_{//}}}{R_{0_{\perp}}} = [\mathscr{R}_f + C_{//}(R_{f_{\perp}}^{-1} - \mathscr{R}_f)] [1 + C_{\perp}(R_{f_{\perp}}^{-1} - 1)]^{-1} \quad (16.37)$$

As discussed previously, we have a nonhomogeneous wave; i.e., n_2 and ϕ_2 are both complex quantities, because the substrate is a highly dispersive medium. Thus, in terms of previously defined quantities, we can write

$$n_2 \cos \phi_2 = \alpha + i\beta \qquad n_2 = n - ik \qquad \cos \phi_2 = A + iB$$

$$A = \frac{\alpha n - \beta k}{n^2 + k^2} \qquad B = \frac{\beta n + \alpha k}{n^2 + k^2}$$

By applying Snell's law, $n_2 \sin \phi_2 = n_0 \sin \phi_0$, we get

$$
\begin{aligned}
2\alpha^2 &= n^2 - k^2 - n_0^2 \sin^2 \phi_0 + [(n^2 + k^2)^2 \\
&\qquad + n_0^4 \sin^4 \phi - 2(n^2 - k^2) n_0^2 \sin^2 \phi_0]^{\frac{1}{2}}
\end{aligned}
$$

$$
\begin{aligned}
2\beta^2 &= -(n^2 - k^2 - n_0^2 \sin^2 \phi_0) + [(n^2 + k^2)^2 \\
&\qquad + n_0^4 \sin^4 \phi - 2(n^2 - k^2) n_0^2 \sin^2 \phi_0]^{\frac{1}{2}}
\end{aligned}
$$

$$\alpha\beta = -nk$$

(16.38)

We must now express g, h, s, and t in terms of α, β, n, and k, as well as ϕ_1 and n_1.

$$r_{2//} = \frac{n_1 \cos \phi_2 - n_2 \cos \phi_1}{n_1 \cos \phi_2 + n_2 \cos \phi_1} = g + ih$$

$$g = \frac{\dfrac{\alpha^2 + \beta^2}{n^2 + k^2} - (n^2 + k^2) \dfrac{\cos^2 \phi_1}{n_1^2}}{D_{//}}$$

$$h = \frac{\dfrac{\cos \phi_1}{n_1} \dfrac{(2\alpha^2 - n^2 + k^2)\, 2nk}{(n^2 + k^2)\, \alpha}}{D_{//}}$$

$$D_{//} = \frac{\alpha^2 + \beta^2}{n^2 + k^2} + \frac{(n^2 + k^2) \cos^2 \phi_1}{n_1^2} + \frac{\cos \phi_1}{\alpha n_1} \frac{2[\alpha^2(n^2 - k^2) + 2n^2 k^2]}{n^2 + k^2}$$

$$r_{2\perp} = \frac{n_1 \cos \phi_1 - n_2 \cos \phi_2}{n_1 \cos \phi_1 + n_2 \cos \phi_2} = s + it$$

$$s = \frac{n_1^2 \cos^2 \phi_1 - (\alpha^2 + \beta^2)}{D_\perp}$$

$$t = \frac{2\,nk\,n_1\cos\phi_1}{\alpha D_\perp}$$

$$D_\perp = \alpha^2 + \beta^2 + n_1{}^2\cos^2\phi_1 + 2\alpha n_1\cos\phi_1$$

$$r_{//} = \frac{n_0\cos\phi_1 - n_1\cos\phi_0}{n_0\cos\phi_1 + n_1\cos\phi_0} = \frac{(n_1{}^2 - n_0{}^2\sin^2\phi_0)^{\frac{1}{2}} - n_1{}^2\cos\phi_0}{(n_1{}^2 - n_0{}^2\sin^2\phi_0)^{\frac{1}{2}} + n_1{}^2\cos\phi_0}$$

$$r_\perp = \frac{n_0\cos\phi_0 - n_1\cos\phi_1}{n_0\cos\phi_0 + n_1\cos\phi_1}$$

Using Snell's law again, we find

$$n_1\cos\phi_1 = (n_1{}^2 - n_0{}^2\sin^2\phi_0)^{\frac{1}{2}}$$

Thus, all the required quantities can be expressed in terms of n, k, n_1, and ϕ_0, the angle of incidence, and no approximations have been made.

In order to gain a suitable expression for the correction for a thin film layer on the surface of a sample, we expand $\sin\delta$ and $\cos\delta$ to first-order terms, i.e., δ^2 is very much less than 2; $\sin\delta \to \delta$; $\cos\delta \to 1$; and the quantity $C_{//}$ becomes

$$C_{//} \cong \frac{4\,r_{//}\,\delta(\delta g - h)}{(1 + r_{//}g)^2 + (r_{//}h)^2} \tag{16.39}$$

with a similar expression for $C_\perp$.

In many cases $r_{//}$ is the only term in $C_{//}$ which has a strong dependence on ϕ_0, being negative for $\phi_0 < \tan^{-1}n_1$ and positive for $\phi_0 > \tan^{-1}n_1$.

For example, the useful range of angles for a germanium substrate is between 74° and 83°. Assuming an oxide film with a refractive index of 2.0., we find that the quantity $n_1{}^2\cos^2\phi_1$ varies less than 1.5% over this range from the mean value of 3.042. The quantities g and h are affected less than 0.33% by variation in ϕ_0 in this range, whereas both $r_{//}$ and $r_\perp$ are strongly affected because of the terms involving $\cos\phi_0$. It also turns out for a high index substrate that the correction for $R_{f\perp}$ is quite small (of the order of 1% or less) for $\phi_0 \geq 60°$. Hence, a very good approximation for $\mathscr{R}_0(\phi_0)$ is given by

$$\mathscr{R}_0(\phi_0) \cong \mathscr{R}_f(\phi_0) + C_{//}(\phi_0)\,[R_{f\perp}^{-1}(\phi_0) - \mathscr{R}_f(\phi_0)] \tag{16.40}$$

where $\mathscr{R}_f$ is a measured quantity and $R_{f\perp}$ is calculated from the experimental quantities X_{nf2} and $\mathscr{R}_{nf}$. Thus the correction term $C_{//}$ requires the knowledge of film thickness, film refractive index, and the specimen's optical constants.

The first two quantities are assumed to be known from independent measurements, and all that remains is to obtain reliable estimates of the quantities g and h.

When the optical constants of the substrate are unknown, a set of values for g and h is reached in the following manner: A set of optical constants is determined by setting $C_{//} = 0$, and the quantities g and h are calculated and used for a correction term. This, of course, provides an overcorrection, and the new set of optical constants is used for determining a second set of g and h to provide an undercorrection for the values of the ratio $\mathcal{R}$ as a function of ϕ_0. This series of approximations consisting of undercorrections, overcorrections, undercorrections, etc., is continued until a consistent set of optical constants for g and h is achieved. The process is shortened if one has a reasonable first estimate of the quantities g and h to begin with. In practice, the correction can be programmed on a computer, the interval and degree of approximation being determined by the storage and speed of the calculating system.

Figure 8 shows the quantities g and h and the quantity $\delta(g\delta - h)$ obtained from a series of measurements on germanium for a 50 Å oxide layer having an index of 2.0. Corrections in the quantities X_n^2 and $\mathcal{R}_B$ were found to be quite large—as much as 40% in the case of X_n^2 in the spectral region covered.

Figure 9 shows the results of the iteration process for a germanium sample having a 50 Å film of refractive index 2. The zero curve is as measured, and the numbered curves show the subsequent iterations which stopped at number 6. The curves for cases 4 and 5 are the same as 6 on this scale. The final average values for g and h are indicated also. These data represent germanium at 2.50 eV.

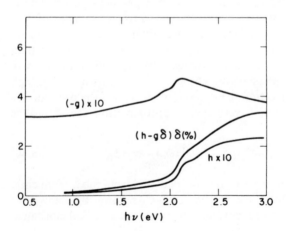

Fig. 8. Spectral behavior of quantities $-g$, h, and $(h - g\delta)\delta$ for germanium at 300 °K. Here $r_{12_{//}} = g + ih$ for parallel polarized light.

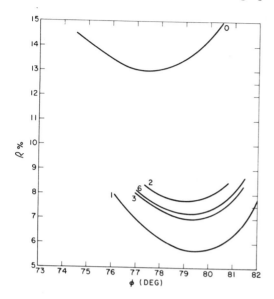

Fig. 9. The uppermost curve gives $\mathscr{R}$ as a function of the angle of incidence; the numbered curves demonstrate the iterative process described in the text. (Curves 4 and 5 are equivalent to 6 on this scale.) The correction factors for $-g$ and h are 0.40 and 0.21, respectively. These data correspond to those of germanium at 2.5 eV.

VI. OPTICAL CONSTANTS FOR GERMANIUM AT 0.5–2.5 eV

Measurements have been made on an electropolished sample of intrinsic germanium in the spectral region 0.5 to 2.5 eV [7]. The room-temperature spectra for X_{nf}^2 and $\mathscr{R}_{nf}$ are shown in Fig. 10. The subscript f indicates that no correction has been made for the presence of a film. Even in this "raw data" form, the power of this spectroscopy can be seen. Distinct doublet features are seen at 0.80 and 1.10 eV, at 1.74 and 1.94 eV, and at 2.10 and 2.30 eV.

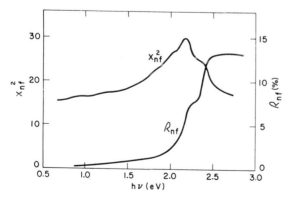

Fig. 10. Details of the experimentally determined quantity X_{nf}^2 in the spectral region 0.70–3.00 eV. Each feature is preceded by a positive slope.

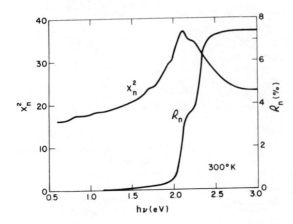

Fig. 11. The corrected quantities X_n^2 and $\mathcal{R}_n$ at 300 °K.

When the correction for a 50-Å-thick oxide layer with a refractive index of 2.0 is applied, the spectra appear as in Fig. 11. These data are sufficient to determine the optical constants for germanium when the formulas described earlier in equations (16.11) and (16.13) are applied.

The quantities ε_1 and ε_2 are shown in Fig. 12. There is a marked degree of correspondence between X_n^2 and ε_1. Although some similarity exists, the quantities $\mathcal{R}_n$ and ε_2 do not have so close a correspondence. The open circles indicate calculated values of ε_2 based on absorption measurements by Hobden [8].

The real and imaginary parts of the refractive index $(n - ik)$ were also calculated, and the room-temperature values are shown in Fig. 13. Room-temperature values given by Archer [9] are also shown. The discrepancy in n over most of the comparable spectral range is less than 2%, whereas that for k is a maximum of 14% at the 2.4 eV peak with good agreement for $h\nu \leqslant 2.25$ eV.

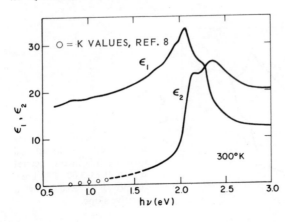

Fig. 12. The optical dielectric constants for electropolished germanium at 300 °K, based on data of Fig. 11.

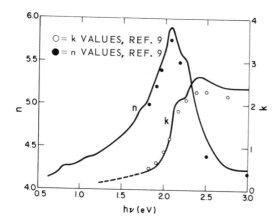

Fig. 13. The refractive index constants for germanium at $300\,°K$, based on data of Fig. 11.

VII. SUMMARY

The pseudo-Brewster angle technique for measuring the optical constants of an opaque sample with a specular surface has been described. A determination of the angle ϕ_B at which the ratio $\mathscr{R} = R_{//}/R_\perp$ is minimum and the minimum ratio is sufficient to characterize the solid. Indeed, these two quantities are optical constants. An experimental procedure has also been described, as well as a method for correcting for the presence of an oxide layer (and other films if present). This technique has been applied to germanium and some results have been shown. It is noteworthy that the quantity $X_n^2 = \sin^2 \phi_B \tan^2 \phi_B$ has a very close resemblance to the real part of the dielectric constant ε_1.

VIII. REFERENCES

1. H. Damany, *J. Opt. Soc. Am.* **55**:1558 (1965).
2. R. Tousey, *J. Opt. Soc. Am.* **29**:235 (1939).
3. I. Šimon, *J. Opt. Soc. Am.* **41**:336 (1951).
4. D. G. Avery, *Proc. Phys. Soc. (London)* B **65**:425 (1952).
5. R. E. Lindquist and A. W. Ewald, *J. Opt. Soc. Am.* **53**:247 (1963).
6. J. R. Johnson, *University of Pennsylvania, Tech. Rept. No. 15 N6-ONR* 24914, August 1958 (unpublished).
7. R. F. Potter, *Phys. Rev.* **150**:562 (1966).
8. M. V. Hobden, *J. Phys. Chem. Solids* **23**:821 (1962).
9. R. J. Archer, *Phys. Rev.* **110**:354 (1958).

CHAPTER 17

Introductory Notes to Electron Spin Resonance Absorption Spectroscopy *

S. A. Marshall

Argonne National Laboratory
Argonne, Illinois

I. INTRODUCTION

Electron spin resonance (ESR) absorption spectroscopy is that branch of radio-frequency spectroscopy which deals with the lowest-lying terms of systems having some degree of electronic angular momentum. This momentum may originate from the intrinsic spin, orbital motion, or both, of electrons. In either case, terms possessing angular momentum must necessarily be degenerate [1], and it is the removal of this degeneracy that forms the basis for ESR spectroscopy. Examples of systems having this ground-term degeneracy are: (1) most free atoms; (2) free-radical molecules and molecule-ions; (3) ions of the three transition series as well as those of the lanthanide and actinide series, the last two being characterized by partially filled inner electronic shells; (4) various species trapped in solids of which the alkali halide F-centers are examples; and finally, (5) conduction electrons.

Aside from its intrinsic interest, magnetic-resonance absorption spectroscopy has many applications, some of which relate to the theory of solids. These applications vary in their scopes of interest and extend, for example, from identification and concentration determinations of paramagnetic species to the areas of many-bodied systems exhibiting cooperative properties. With reference to the optical properties of solids, ESR spectroscopy and temperature-dependent magnetic-susceptibility measurements represent two very useful supplementary tools in the analysis of spectra due to magnetically active crystal–ion species.

With such applications in mind, a series of topics which relate to the spectroscopy at radio frequencies associated with atomic and molecular

* Based upon work performed under the auspices of the U. S. Atomic Energy Commission.

systems in their ground electronic terms is developed and discussed. These begin with two examples of paramagnetic systems in isolation, that is, systems having angular-momentum degeneracy and subject to no external perturbations other than a magnetic field and a transition-inducing radio-frequency field. Next, the effects of crystalline electric fields such as might be encountered by an atom or ion imbedded in a solid are discussed. The notion of an irreducible tensor operator is briefly mentioned to introduce operator equivalents. Finally, various interactions that might occur in an orbitally nondegenerate electronic ground term are reduced to their electron spin angular momentum operator equivalent forms, and a spin Hamiltonian is substituted for the energy operator. A few examples of ions in crystalline electric fields are worked out.

II. ISOLATED PARAMAGNETIC SYSTEMS

Simple systems in isolation, that is, those of a single component experiencing either no external interactions or perhaps a few specified interactions, represent useful ideals from which may be developed complex systems which more closely approach those that are normally encountered in the laboratory. With this in mind, it seems useful to consider the primative system of an isolated magnetic dipole of moment μ immersed in a homogeneous magnetic field of strength H. The energy of the dipole is given by

$$\mathscr{E} = -\mu \cdot H \tag{17.1}$$

and if μ were a classical moment, its energy could in a continuous fashion assume all values from $-\mu H$ to μH. An electron in an atom which is otherwise isolated might be taken as an example. The atomic magnetism may arise from three sources all of which are related to angular momentum. The first two and most intense sources of magnetism relate to the angular momentum of the electron. Of these, one arises as a consequence of the orbital motion of the electron just as a closed loop of electrical current generates a magnetic field whose characteristics depend upon the strength of the current and geometry of the loop. The second arises from the intrinsic or spin angular momentum of the electron whose interpretation requires a relativistic quantum-mechanical treatment. If the convention is agreed upon that to each unit of orbital angular momentum there is generated one unit of magnetic moment, it may be demonstrated that to each unit of spin angular momentum there will be generated very nearly two (2.00229) units of magnetic moment. If the unit of angular momentum is set as $h/2\pi$, where h is Planck's constant, the corresponding unit of magnetic moment, the Bohr magneton, is

$$\beta = \frac{eh}{4\pi mc} = 0.92734 \times 10^{-20} \text{ ergs/G} \qquad (17.2)$$

where e and m are the electronic charge and mass, respectively, and c is the velocity of light. The third source of atomic magnetism is that which arises from the spin angular momentum of the nucleus. By analogy with the electronic spin magnetism, a unit of nuclear magnetism, the nuclear magneton, may be obtained from equation (17.2) by replacing the electronic mass with the nuclear mass unit, the latter being of the order of eighteen hundred times greater.

If L, S, and I are defined as the electronic orbital, electronic spin, and nuclear angular momentum operators, respectively, an energy operator expression may be given as a specialized quantum-mechanical generalization of equation (17.1)

$$\mathscr{H} = g\beta H \cdot J + \gamma \beta'' H \cdot I + AI \cdot J \qquad (17.3)$$

where $J = L + S$ is the total electronic angular momentum operator, γ is the nuclear gyromagnetic ratio, β'' is the nuclear magneton, A is the hyperfine structure parameter, and g, the Landé factor, is given by [2]

$$g = 1 + \frac{J(J+1) + S(S+1) - L(L+1)}{2J(J+1)} \qquad (17.4)$$

A relatively simple and yet not at all trivial example that may be introduced is that of the isolated atom of hydrogen. As the ground term of this atom has zero orbital angular momentum, the spin may replace the total angular momentum in equation (17.3) to provide an energy operator whose dynamical variables are the components of the electron and nuclear spin vector operators. Using these variables, equation (17.3) may be rewritten as

$$\mathscr{H} = g\beta H S_z + \gamma \beta'' H I_z + A I_z S_z + \tfrac{1}{2} A(I_+ S_- + I_- S_+) \qquad (17.5)$$

where the z axis is taken to be along the direction of the magnetic field. The matrix representation of this operator evaluated from within the four-dimensional union of the two spin spaces corresponding to $S = \tfrac{1}{2}$ and $I = \tfrac{1}{2}$ is given by Table I. Generally, eigenvalues to such an operator would be obtained by approximation methods; however, in this case exact eigenvalue expressions are a relatively simple matter to develop, and, in addition, an exact analysis of this problem has been used to provide a means of testing the theory of quantum electrodynamics. The four eigenvalues to the matrix of

Table I
Representation of the Atomic Hydrogen ($S = \frac{1}{2}, I = \frac{1}{2}$)
Ground-Term Energy Operator*

	$\lvert \frac{1}{2}, \frac{1}{2} \rangle$	$\lvert \frac{1}{2}, -\frac{1}{2} \rangle$	$\lvert -\frac{1}{2}, \frac{1}{2} \rangle$	$\lvert -\frac{1}{2}, -\frac{1}{2} \rangle$
$\langle \frac{1}{2}, \frac{1}{2} \rvert$	$\frac{1}{2} g\beta H(1+\zeta)$ $+\frac{1}{4}A$			
$\langle \frac{1}{2}, -\frac{1}{2} \rvert$		$\frac{1}{2} g\beta H(1-\zeta)$ $-\frac{1}{4}A$	$\frac{1}{2}A$	
$\langle -\frac{1}{2}, \frac{1}{2} \rvert$		$\frac{1}{2}A$	$-\frac{1}{2} g\beta H(1-\zeta)$ $-\frac{1}{4}A$	
$\langle -\frac{1}{2}, -\frac{1}{2} \rvert$				$-\frac{1}{2} g\beta H(1+\zeta)$ $+\frac{1}{4}A$

* $\zeta = \gamma\beta^n/g\beta$.

Table I are given by [3]

$$\mathscr{E}_1 = \pm\frac{1}{2} g\beta H(1+\zeta)+\frac{1}{4}A$$

$$\mathscr{E}_2 = \pm\frac{1}{2} g\beta H(1-\zeta)\left[1 + \frac{A^2}{g^2 \beta^2 H^2(1-\zeta)^2}\right]^{\frac{1}{2}} - \frac{1}{4}A \qquad (17.6)$$

For finite magnetic field strengths there will be three transition possibilities, two for the operators $S_\pm$ and one for the operator S_z whose expectation values are given by

$$S_\pm \lvert S, m \rangle = [(S \mp m)(S \pm m+1)]^{\frac{1}{2}} \lvert S, m\pm 1 \rangle$$

$$S_z \lvert S, m \rangle = m \lvert S, m \rangle \qquad (17.7)$$

so that, in principle, the parameters of the spectrum (g, ζ, and A) may be determined without approximation. Of these three spectral parameters, the experimentally determined value of g is of special interest because of its deviation Δg from the value of 2. In a series of experiments performed by Kusch et al., by Beringer and Heald, and by Franken and Liebes [4–8], this deviation was determined to be

$$\Delta g = 2(0.001165 \pm 0.000011) \tag{17.8}$$

which agrees to within four parts in one thousand of the deviation predicted by the theory of quantum electrodynamics [9,10]. The value of A, for an electron in an s state is given by the expression [11]

$$A = \frac{8\pi}{3} g\beta\gamma\beta'' |\psi(r = 0)|^2 \tag{17.9}$$

which is also of interest because it provides a measure of the probability of the electron being at the nucleus. In the present case of the ground term of atomic hydrogen, the value of this spectral parameter has been very precisely measured to be $A = 1420.40577$ Mcps. Eigenvalues for this system are given in Fig. 1 as functions of the magnetic field strength H, and, in addition, positions of the three transitions discussed earlier are indicated for transition-inducing radiation of 3-cm wavelength.

Another interesting example may be mentioned of an isolated system which is in a sense one degree more complex. This is the oxygen atom which has a ^3P ground term, that is to say, $L = 1$ and $S = 1$. For these values of L and S, J may take on the three values 2, 1, and 0, and, since atomic oxygen is a system for which the forces are no longer central, the energy spectrum will not be independent of angular momentum [12]. Of these three ^3P low-lying terms, one for each value of J, the lowest is that associated with $J = 2$ and it has five magnetic sublevels, the first excited term corresponds

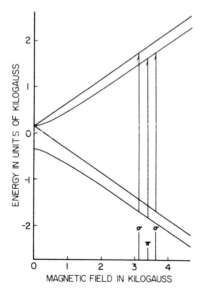

Fig. 1.   Zeeman field-dependent energy-level diagram for the atom hydrogen $(S = \frac{1}{2}, I = \frac{1}{2})$ system.

to $J = 1$ which has three magnetic sublevels, and the highest is the $J = 0$ term which is nonmagnetic (no degeneracy in electronic angular momentum). One would thus expect to observe six transitions, four due to the $^3P_2$ term and two due to the $^3P_1$ term with the former being the more intense because the $J = 2$ and $J = 1$ terms are separated by an energy of $158.5 \, \text{cm}^{-1}$ with both sets of transitions being characterized by a Landé factor of $g = 1.5$ [13, 14]. This is precisely what is observed and, furthermore, because of what has been referred to as an incipient Paschen–Bach effect [14], all six components are resolved and may be observed in magnetic fields of sufficient homogeneity and at pressures ($\sim 0.05 \, \text{mm Hg}$) low enough to reduce collision broadening [15].

III. CRYSTALLINE ELECTRIC FIELDS

A. Orbital Angular Momentum

In an ionic solid, the constituent ions may to first order be considered as motionless point charges located at sites of minimum electric potential. If the solid is crystalline, these sites will be distributed throughout the space of the solid in some regular and periodic fashion so that each point, whether it be an ion site or an arbitrary field point, will possess overall as well as local symmetries which may be determined from the crystal structure details. Thus, each space point in this idealized distribution of point charges will be subject to electric fields of specific symmetry, and furthermore the ion sites will be distinguished by electric fields of relatively high symmetry (lower than or at most as high as that of the overall structure) as well as by specifiable coordination. At each charge-free point of this structure, the electric field potential function $\mathscr{V}$ will necessarily satisfy Laplace's equation $\nabla^2 \mathscr{V} = 0$. This is a consequence of the assumptions of point charge immobility and of total ionicity. Arguments based upon these idealizations will therefore be of limited utility since totally ionic as well as completely rigid structures are not realizable. Such idealizations are, however, quite useful not only as models for supplying first-order approximations but also as devices for elaborating upon various interaction processes. In addition, an important consequence of the primitive character of these models is their flexibility, thus making them valuable as foundations for more complicated models.

To begin, it might be well to consider an idealized crystalline structure constituted of, for example, atomic ions for which one ion, that located at the origin, has been deleted thereby removing the coulomb singularity at the origin. The potential function at any field point (r, θ, ϕ) within a charge-free region bounded by a sphere whose radius is equal to the magnitude of the position vector of the ion nearest the origin is given by

$$\mathscr{V}(r, \theta, \phi) = \sum_i e_i/|r - r_i| \tag{17.10}$$

This expression, which is real, may be simplified a great deal by taking advantage of the fact that each product function $r_n Y_n^m(\theta, \phi)$ is a solution to Laplace's equation so that equation (17.10) may be rewritten as

$$\mathscr{V}(r, \theta, \phi) = \sum_i \sum_{nm} (e_i/r_i)(r/r_i)^n Y_n^m(\theta_i, \phi_i) Y_n^m(\theta, \phi) \tag{17.11}$$

where r_i, θ_i, ϕ_i and e_i are the coordinates and electric charge of the ith crystal site, the index i being summed over all sites. Expressing the potential in this manner has two advantages; the first is that it provides a ready device for evaluating the coefficients of the spherical harmonics provided structure data are available. The second advantage relates to the manner in which the potential is expressed into groups of polynomials of common order. This makes it possible to identify the various polynomial groups with the multipole nature of the potential as well as deducing certain relations which may exist between the coefficients of the various spherical harmonics as may be required by reality and symmetry. For example, the first two terms in the n summation of equation (17.11) are of particular interest. The first, due to $n = 0$, is just the sum of (e_i/r_i) taken over all the lattice sites and which may, for spectroscopic purposes, be substracted out of the potential since it has no dependence on dynamical variables. The second group of terms due to $n = 1$ may be shown to have vanishing coefficients because the potential gradient must vanish at each lattice site as a precondition for equilibrium. In addition, it may be pointed out that, if the site in question happens to be a center of symmetry, all the spherical harmonics of odd order will have vanishing coefficients as required by inversion symmetry

$$\mathscr{V}(r, \theta, \phi) = \mathscr{V}(r, \pi - \theta, \pi + \phi)$$

Suppose that an ion (and to avoid undue complications let this be an atomic ion) whose various terms are described by the quantum numbers (S, L, J, M) is brought into a crystal and located at a normal lattice site. The energy spectrum of this ion will, in varying degrees, be altered as a consequence of the electric field generated by the ions of the other crystal. Depending upon its strength and symmetry, this field, whose potential is of the form given by equation (17.11), will in principle produce some splitting of the $(2L+1)(2S+1)$-fold degenerate terms. This removal of degeneracy, also referred to as quenching of angular momentum, was first treated in a fundamental manner by Bethe, who took account of crystalline point symmetry through the application of the theory of groups [16]. Aside from symmetry considerations, crystalline electric fields may be classified as being weak, intermediate, or strong, the classification referring to a comparison of the effects of the crystalline field relative to the Russell–Saunders and fine-structure coupling interactions. Examples of weak crystalline electric fields are given by the rare-earth series ions incorporated in various hosts. For these ions, the active electrons are of the incomplete inner $4f$ shell and consequently are rather well shielded from the influence of external fields by the completed shell of outer electrons. In similar hosts, ions of the incomplete $3d$, $4d$, and $5d$ series provide examples of intermediate and strong crystalline electric fields. In these examples, the active electrons are in the outermost shell permitting interaction with the external fields to be much more effective than spin-orbit interaction and in some cases to compete with Russell–Saunders coupling. It thus appears that the strength classification of a crystalline electric field must take account of the nature of the interacting ions [17]. As a consequence, the nomenclature *crystal–ion*, due to Bethe, will occasionally be used to indicate the composite nature of the crystal and ion.

As an example, trivalent titanium located at a crystalline site having cubic (octahedral) symmetry may be considered as a crystal–ion system. Under complete isolation the titanium ion will, according to Hund's rule, have a ^2D ground term which has twofold spin degeneracy and fivefold orbital degeneracy. Neglecting spin-orbit interaction, the five degenerate orbital term vectors may be associated with the familiar set of second-degree polynomial functions $Y_2^{\pm 2}(\theta, \phi)$, $Y_2^{\pm 1}(\theta, \phi)$, and $Y_2^{0}(\theta, \phi)$. Application of the crystalline electric field to this ion will produce some removal of orbital degeneracy as well as the generation of a new set of eigenvectors whose union will span the original five-dimensional function space. To demonstrate this, consider first the potential function whose form may, from symmetry considerations, be deduced to be

$$\mathscr{V}_{\text{cubic}} = A_4 \left\{ Y_4^{0}(\theta, \phi) + \sqrt{\frac{5}{14}} \left[Y_4^{4}(\theta, \phi) + Y_4^{-4}(\theta, \phi) \right] \right\} \quad (17.12)$$

where the potential component associated with $n = 0$ has been subtracted out and all components associated with $n > 4$ have been ignored because of their vanishing matrix elements. This last statement may be seen to be a consequence of the rule $<\lambda, \mu|\ Y_n^{m}\ |\lambda, \mu'> = 0$ unless $n \leqslant 2\lambda$ and $m = \mu - \mu'$ [12]. The eigenvalues, eigenvectors, and symmetry-transformation designations to this potential operator evaluated over the spherical harmonics of $L = 2$ are given in Table II where the symmetry designations Γ_i are those of Bethe for the cubic group [16]. It will be noted that the total energy of this system, obtained by adding together each eigenvalue multiplied by its degeneracy, vanishes. This is quite general since the only potential component with the property of shifting the total energy of an ion is the

Table II
Eigenvalues and Eigenvectors for the Cubic (Octahedral) Crystalline Electric Field Potential Evaluated from Within the Manifold $L = 2$

Eigenvalues	Eigenvectors	
	Tetragonal	Trigonal
$\Gamma_3 : \frac{3}{5}\Delta$	Y_2^0	$1/\sqrt{3}(Y_2^{-2} - \sqrt{2}\,Y_2^1)$
	$1/\sqrt{2}(Y_2^2 + Y_2^{-2})$	$1/\sqrt{3}(Y_2^2 + \sqrt{2}\,Y_2^{-1})$
$\Gamma_5 : -\frac{2}{5}\Delta$	Y_2^1	$1/\sqrt{3}(\sqrt{2}\,Y_2^{-2} + Y_2^1)$
	Y_2^{-1}	$1/\sqrt{3}(\sqrt{2}\,Y_2^2 - Y_2^{-1})$
	$1/\sqrt{2}(Y_2^2 - Y_2^{-2})$	Y_2^0

monopole term ($n = 0$). To demonstrate this, it will be recalled that the total energy of a system is just the trace of the matrix representing the energy operator and that this sum is invariant under a principal-axis transformation. One such transformation is that which brings the eigenvectors of the crystalline electric field into the eigenvectors of the operator L_z. Upon having performed this transformation, the trace of each component of the potential function may be evaluated. However, in the L_z representation, the only components of the potential to have diagonal matrix elements are the Y_n^0, so that the trace of $\mathscr{V}$ may be given as the sum of its n components which in turn are given by

$$T_r \mathscr{V}_n = \sum_{\mu=-\lambda}^{\lambda} A_n^0 \langle \lambda, \mu | Y_n^0 | \lambda, \mu \rangle$$

$$= A_n^0 \left[\sum_{\mu=-\lambda}^{\lambda} (\psi_\lambda^\mu)^* \psi_\lambda^\mu \right] \int Y_n^0 \, d\tau \tag{17.13}$$

where the summation within the brackets may be shown to be a constant quantity [18, 19] and where the integration is taken over the unit sphere. The vanishing of the trace of $\mathscr{V}_n$ for $n > 0$ is now seen to be a consequence of the orthogonality condition which exists between the Y_n^0 functions [20]. An alternate proof may be obtained by utilizing the relation between the matrix elements of the spherical harmonics Y_n^m and the vector coupling coefficients [21]

$$\langle \lambda, \mu | Y_n^m | \lambda', \mu' \rangle = (\lambda \mu \lambda' \mu' | \lambda \lambda' nm) \tag{17.14}$$

as well as the orthogonality condition which exists between vector coupling coefficients of different n

$$\sum_\mu (\lambda \mu \lambda' \mu' | \lambda \lambda' nm)(\lambda \mu \lambda' \mu' | \lambda \lambda' n' m) = \delta_{n,n'} \tag{17.15}$$

where, to complete the proof, n' is set equal to zero.

Proceeding with this example, let the symmetry of the crystalline electric field be lowered from cubic to, say, trigonal by producing a distortion of the cube body diagonal. How the term degeneracies and eigenvectors are altered may be seen by performing the lowering of symmetry in two steps. The first is to rotate the crystal so that its body diagonal is brought into the z axis. This rotation will transform the potential expression exhibiting fourfold rotational symmetry into one exhibiting threefold rotational symmetry. New eigenvectors may then be developed thus leaving the second and final step to be the lowering of symmetry by distinguishing the z axis through a compression or elongation distortion.

Let $r (\alpha, \beta, \gamma)$ be a general three-dimensional rotation transformation with α, β, and γ being the three Euler angles. Under such a transformation, a given spherical harmonic Y_n^m will go over into the function $P_r Y_n^m$ which will in general be a linear combination of spherical harmonics of similar n

$$P_r Y_n^m = \sum_{m'=-n}^{n} D^{(n)} (\alpha, \beta, \gamma)_{m',m} Y_n^{m'} \tag{17.16}$$

where the elements of the matrix representing $D^{(n)} (\alpha, \beta, \gamma)$ are given by

$$D^{(n)} (\alpha, \beta, \gamma)_{m',m} = \sum_x (-1)^x \frac{[(n+m)!\,(n-m)!\,(n+m')!\,(n-m')!]^{\frac{1}{2}}}{(n-m'-x)!\,(n+m-x)!\,x!\,(x+m'-m)!}$$

$$\times e^{i(m'\alpha + m\gamma)} (\cos \tfrac{1}{2} \beta)^{2n+m-m'-2x} (\sin \tfrac{1}{2} \beta)^{2x+m'-m} \tag{17.17}$$

By using the matrix elements given by this expression for that rotation which brings the crystal body diagonal into the z-axis, another crystalline electric field potential expression may be developed which is given by

$$\mathcal{V}_{\text{cubic}} = A_4' \{ Y_4^0 (\theta, \phi) + \sqrt{\tfrac{10}{7}} [Y_4^3(\theta, \phi) + Y_4^{-3}(\theta, \phi)] \} \tag{17.18}$$

whose eigenvectors are given under the trigonal column of Table II. The functions in this column are of a form that requires no further manipulations to provide correct eigenvectors to a crystalline electric field having trigonal symmetry. To obtain eigenvectors from this column of functions, those which exhibit similar z-coordinate dependence should be grouped together as should those which exhibit an x–y equivalence. These eigenvectors are given in Table III as are those for the noncubic, tetragonally symmetric, crystalline electric field. Eigenvalues for both forms of distorted octahedra may be obtained by introducing an axial distortion into either of the potentials given by equations (17.12) and (17.18). These distortions have the effect of generating some Y_2^0 and altering the coefficient of Y_4^0 in the potential expression. For the tetragonal potential eigenvectors, both Y_2^0 and Y_4^0 are diagonal operators so that their perturbation to the cubic eigenvalues although independent of one another may be determined. For the trigonal potential, the situation is somewhat more complicated since the distortion component of the potential produces a slight (ζ is assumed to be small compared to Δ) mixing of the two sets of Γ_3 eigenvectors.

If it is assumed that the signs of Δ and ζ are positive, the ground term for the trivalent titanium ion located in a trigonally distorted octahedral crystalline electric field is seen to be $^2\Gamma_1$, which is a term of twofold degeneracy. This degeneracy is identified with spin angular momentum so that the liberty is taken in assigning to this term an effective spin of $S' = \frac{1}{2}$. The nomenclature *effective spin* is used as a reminder that the residual degeneracy of the $^2\Gamma_1$ term, for that matter all paramagnetic crystal–ion terms, is not a consequence of spin angular momentum alone since a crystalline electric field of finite strength cannot with total success quench orbital angular

Table III

Eigenvalues and Eigenvectors for the Tetragonally and Trigonally Distorted Crystalline Electric Field Potential Evaluated from Within the Manifold $L = 2$*

Tetragonal		Trigonal	
Γ_3 : $\frac{3}{5}\Delta+2\eta$	$1/\sqrt{2}(Y_2^2+Y_2^{-2})$	Γ_3 : $\frac{3}{5}\Delta$	$1/\sqrt{3}(Y_2^{-2}-\sqrt{2}Y_2^1)$
			$1/\sqrt{3}(Y_2^2+\sqrt{3}Y_2^{-1})$
Γ_1 : $\frac{3}{5}\Delta-2n$	Y_2^0		
Γ_4 : $-\frac{2}{5}\Delta+2\eta$	$1/\sqrt{2}(Y_2^2-Y_2^{-2})$	Γ_3 : $-\frac{2}{5}\Delta+\zeta$	$1/\sqrt{3}(\sqrt{2}Y_2^{-2}+Y_2^1)$
			$1/\sqrt{3}(\sqrt{2}Y_2^2-Y_2^{-1})$
Γ_5 : $-\frac{2}{5}\Delta-\eta$	Y_2^1		
	Y_2^{-1}	Γ_1 : $-\frac{2}{5}\Delta-2\zeta$	Y_2^0

* Both η and ζ are assumed to be small compared to Δ.

momentum. To demonstrate this, spin-orbit interaction, which has so far been neglected, may be brought into the problem. This interaction has the form

$$\mathscr{V}_{so} = \lambda L_z S_z + \frac{\lambda}{2}(L_+ S_- + L_- S_+) \tag{17.19}$$

and may be seen to connect terms of composition $< M_L, M_S|$ with those of composition $< M_L, M_S|$ and $< M_L \pm 1, M_S \mp 1|$. Thus, the $^2\Gamma_1$ ground term mixes with the nearby $^2\Gamma_3$ term so that a more nearly correct set of ground-term vectors would be

$$\langle \Gamma_1, \tfrac{1}{2}| = \langle 2, \tfrac{1}{2}; 0, \tfrac{1}{2}| + k \langle 2, \tfrac{1}{2}; 1, -\tfrac{1}{2}|$$

$$\langle \Gamma_1, -\tfrac{1}{2}| = \langle 2, \tfrac{1}{2}; 0, -\tfrac{1}{2}| + k \langle 2, \tfrac{1}{2}; -1, \tfrac{1}{2}| \tag{17.20}$$

where k is of the order of $\lambda/6\zeta$ and the nomenclature $<L, S; M_L, M_S|$ is used on the right-hand side of equation (17.20).

In addition to acting as a mechanism which connects and thereby mixes terms, spin-orbit interaction has the property of lifting orbital angular momentum degeneracy. To demonstrate this, it would be useful to reconsider the tetragonal crystalline electric field. From Tables II and III it may be seen that the cubic term Γ_5, which is threefold degenerate in orbital angular momentum, goes over to the orbitally nondegenerate Γ_4 and the twofold degenerate Γ_5 tetragonal terms, the latter lying lowest for a positive distortion parameter η. If it is assumed that the strength of the spin-orbit interaction is of the same order as that of the tetragonal distortion, then an interaction matrix may be constructed for the threefold orbitally degenerate Γ_5 cubic term using the spin vector convention $\alpha = <\tfrac{1}{2}, \tfrac{1}{2}|$, $\beta = <\tfrac{1}{2}, -\tfrac{1}{2}|$ as well as the short-hand $\phi = Y_2^1$, $\phi_{-1} = Y_2^{-1}$, and $\phi_0 = 1/\sqrt{2}(Y_2^2 - Y_2^{-2})$. This combined interaction matrix is given in Table IV, as are its eigenvalues. It will be seen that the effect of the combined interaction is to lift partially the six-fold degeneracy of the Γ_5 cubic term into three twofold degenerate terms (see Fig. 2) where the residual twofold degeneracy should not be specifically identified with either the spin or orbital angular momentum of the ion but rather with a mixture of the two. It is interesting to note that although the expectation values of the operator $(L_z + S_z)$ are $\tfrac{1}{2}$ and $-\tfrac{1}{2}$ for the two term vectors $\phi_1 \beta$ and $\phi_{-1} \alpha$, respectively, the expectation values of the operator $(L_z + 2S_z)$ vanish, suggesting that, to within the approximation $\Delta \gg \eta$ and λ, the twofold degenerate term at $\mathscr{E} = -\tfrac{2}{5}\Delta - \eta - \tfrac{1}{2}\lambda$ is non-magnetic.

Divalent copper has a $3d^9$ configuration and by Hund's rule its ground term is designated ^2D, which is the same designation as that for trivalent titanium whose configuration is $3d^1$. It will be noted that since there are ten $3d$ electronic terms, those of divalent copper and trivalent titanium are

Table IV
Matrix of the Combined Spin-Orbit and Tetragonal Distortion Interaction for the Cubic Γ_5 ($L = 2$, $S = \frac{1}{2}$) Term *

	$\phi_1\beta$	$\phi_{-1}\alpha$	$\phi_1\alpha$	$\phi_0\beta$	$\phi_{-1}\beta$	$\phi_0\alpha$
$\phi_1\beta$	$-\eta-\frac{1}{2}\lambda$					
$\phi_{-1}\alpha$		$-\eta-\frac{1}{2}\lambda$				
$\phi_1\alpha$			$-\eta+\frac{1}{2}\lambda$	$\sqrt{2}\lambda$		
$\phi_0\beta$			$\sqrt{2}\lambda$	2η		
$\phi_{-1}\beta$					$-\eta+\frac{1}{2}\lambda$	$-\sqrt{2}\lambda$
$\phi_0\alpha$					$-\sqrt{2}\lambda$	2η

* $\mathscr{E}_1 = -\eta - \frac{1}{2}\lambda$ and $\mathscr{E}_{2,3} = \frac{1}{2}\eta + \frac{1}{4}\lambda \pm \frac{1}{2}[(3\eta - \frac{1}{2}\lambda)^2 + 8\lambda^2]^{\frac{1}{2}}$.

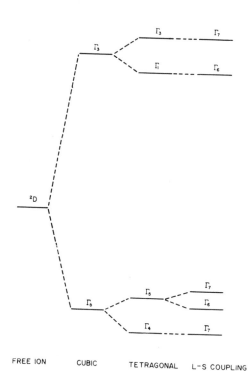

Fig. 2. Schematic representation of the trivalent titanium ion ground-term splittings in an octahedral crystalline electric field.

conjugate images of one another as a consequence of the fact that, subject to the Pauli exclusion principle, they may be coupled together in only one manner, that is, to yield a ^1S term. It will be recognized that this is a constant term or one which is independent of the spherical coordinates θ and ϕ. Consequently, the $3d^9$ ground term may be thought of as being the effective nullifier of the $3d^1$ ground term. However, a $3d^1$ hole, an electron of opposite charge and spin orientation, has exactly the same nullifying capacity. Thus, it appears that, except for the θ and ϕ independent component of the potential energy (the monopole component), the matrix elements of $e\mathcal{V}$ evaluated over the $3d^1$ ground-term manifold are equal and opposite to the elements evaluated over the $3d^9$ ground-term manifold. Similar arguments may be used to demonstrate the conjugate character of the $3d^n$ and $3d^{10-n}$ ion ground terms as well as conjugate relations for ions of the other paramagnetic series.

Up to this point, a rather elementary crystal–ion system and its conjugate have been considered, namely, an ion having a single active electron in a crystalline environment of relatively high symmetry. To proceed to more complex crystal–ion systems, the application of physical principles and mathematical techniques which are basically the same as those already demonstrated or their generalizations and extensions is required.

For example, the analysis of multielectronic terms, such as are to be found for the $3d$ ions, influenced by crystalline electric fields of cubic or lower symmetry requires application of the Pauli exclusion principle, use of Hund's rule for ground-term designation, and application of vector coupling of angular momenta. Thus, in building up multielectron term vectors from single-electron vectors, the step-by-step process of coupling angular momenta two at a time may be used successively to arrive at terms composed of a given number of electrons with the coupling scheme at each step of the way being given by

$$\Psi_n^m = \sum_\mu (\lambda\mu\,\lambda'\,m-\mu\,|\,\lambda\lambda'\,nm)\,\psi_\lambda^\mu\,\psi_{\lambda'}^{m-\mu} \tag{17.21}$$

subject to the condition that each term vector be antisymmetric and where the possible values of n are limited to the interval $|\lambda-\lambda'| \leqslant n \leqslant (\lambda+\lambda')$.

As an illustration of the procedure in building up an angular momentum configuration, a $3d^3$ ion, such as trivalent chromium or divalent vanadium, might be considered. The first step in the process would be to build up from two single-electron d vectors all those composite vectors which could couple with a single-electron d vector to form a ^4F Hund's rule ground term. Using the coupling scheme given in equation (17.21), the three d electrons will form product vectors given by

$$\phi_n^m = (\lambda\nu\,2\,m-\nu\,|\,\lambda\,2\,nm)(2\,\mu\,2\,\nu-\mu\,|\,22\,\lambda\nu)\,\psi_2^\mu\,\psi_2^{\nu-\mu}\,\psi_2^{m-\nu} \tag{17.22}$$

there being a total of one thousand such products when electron spin is taken into account. However, not all product vectors $\phi_n{}^m$ may appear in the expression for $\Psi_n{}^m$ because of the exclusion principle as well as the restrictions placed on the indices of the vector-coupling coefficients that neither μ, $(v-\mu)$, nor $(m-v)$ be greater than 2 in magnitude. For the case being considered, the ground term of the $3d^3$ ion is described by the $n = 3$ vectors each having maximum spin multiplicity. These vectors form a seven-dimensional orbital manifold from which, for example, the $m = 3$ vector may be selected to illustrate these restrictions. Since μ, $(v-\mu)$, and $(m-v)$ must be no greater than 2 in magnitude, the specifying of m restricts the values of v to 4, 3, 2, and 1 which in turn restricts the values of μ to 2, 1, 0, and -1. With use of these restrictions, a list of indices may be constructed which illustrates the possible choices of product vectors for $m = 3$ (see Table V). Of these ten entries, the Pauli principle requires that a, e, g, and j be dropped thereby leaving six entries which will be recognized as being equivalent to one another because of the indistinguishability of electrons. If account is taken of all restrictions, the $m = 3$ component of the ^4F term is given by

$$\Psi_3{}^3 = \tfrac{1}{6} [\psi_2{}^0(1)\,\psi_2{}^1(2)\,\psi_2{}^2(3) - \psi_2{}^1(1)\,\psi_2{}^0(2)\,\psi_2{}^2(3)$$

$$+ \psi_2{}^2(1)\,\psi_2{}^0(2)\,\psi_2{}^1(3) - \psi_2{}^0(1)\,\psi_2{}^2(2)\,\psi_2{}^1(3) \qquad (17.23)$$

$$+ \psi_2{}^1(1)\,\psi_2{}^2(2)\,\psi_2{}^0(3) - \psi_2{}^2(1)\,\psi_2{}^1(2)\,\psi_2{}^0(3)]\,\alpha(1)\,\alpha(2)\,\alpha(3)$$

which is antisymmetric to electron interchange.

In a cubic crystalline electric field such as given by the expressions in equations (17.12) and (17.18), the sevenfold orbital degeneracy of the ^4F term will be lifted into three terms, one being an orbital singlet Γ_2 and two being

Table V
List of Indices for the Coupling of Three Nonequivalent d Electrons to Form the Product Vector $\Psi_3{}^3$

v	$(m-v)$	$(v-\mu)$	μ	
4	-1	2	2	(a)
3	0	1	2	(b)
3	0	2	1	(c)
2	1	0	2	(d)
2	1	1	1	(e)
2	1	2	0	(f)
1	2	-1	2	(g)
1	2	0	1	(h)
1	2	1	0	(i)
1	2	2	-1	(j)

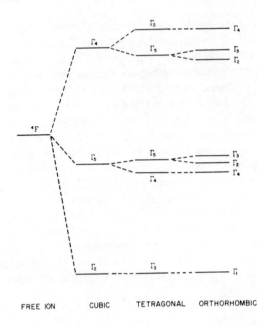

Fig. 3. Schematic representation of the trivalent chromium
ion ground-term splittings in an
octahedral crystalline electric
field.

orbital triplets Γ_4 and Γ_5. As the symmetry is lowered, for example, by
causing distortions along specific crystallographic directions, the lifting of
degeneracy continues until all terms are reduced to orbital singlets in fields
of orthorhombic or lower symmetry (see Fig. 3). The details of this degeneracy lifting in crystalline electric fields of cubic and lower symmetries may

Table VI
Character System for the Octahedral Double Group

Representations	Classes*							
	E	R	$6C_2$	$6C_3$	$6C_3'$	$12C_4$	$8C_5$	$8C_5'$
Γ_1	1	i	1	1	1	1	1	1
Γ_2	1	1	1	-1	-1	-1	1	1
Γ_3	2	2	2	0	0	0	-1	-1
Γ_4	3	3	-1	1	1	-1	0	0
Γ_5	3	3	-1	-1	-1	1	0	0
Γ_6	2	-2	0	$\sqrt{2}$	$-\sqrt{2}$	0	1	-1
Γ_7	2	-2	0	$-\sqrt{2}$	2	0	1	-1
Γ_8	4	-4	0	0	0	0	-1	1

* See Bethe [16].

be deduced from the character tables of the irreducible representations of groups of various symmetries which have been given in detail by Bethe and some of which are listed in Tables VI–XII.

Table VII
Character System for the Tetragonal Double Group

Representations	Classes*						
	E	R	$2C_2$	$2C_3$	$2C_3'$	$4C_4$	$4C_5$
Γ_1	1	1	1	1	1	1	1
Γ_2	1	1	1	1	1	-1	-1
Γ_3	1	1	1	-1	-1	1	-1
Γ_4	1	1	1	-1	-1	-1	1
Γ_5	2	2	-2	0	0	0	0
Γ_6	2	-2	0	$\sqrt{2}$	$-\sqrt{2}$	0	0
Γ_7	2	-2	0	$-\sqrt{2}$	$\sqrt{2}$	0	0

* See Bethe [16].

Table VIII
Character System for the Trigonal Double Group

Representations	Classes*					
	E	R	$2C_2$	$2C_2'$	$3C_3$	$3C_3'$
Γ_1	1	1	1	1	1	1
Γ_2	1	1	1	1	-1	-1
Γ_3	2	2	-1	-1	0	0
Γ_4	1	-1	1	1	i	$-i$
Γ_5	1	-1	-1	1	$-i$	i
Γ_6	2	-2	1	-1	0	0

* See Bethe [16].

Table IX
Character System for the Orthorhombic Double Group

Representations	Classes*				
	E	R	$2C_2$	$2C_3$	$2C_4$
Γ_1	1	1	1	1	1
Γ_2	1	1	-1	1	-1
Γ_3	1	1	-1	-1	1
Γ_4	1	1	1	-1	-1
Γ_5	2	-2	0	0	0

* See Bethe [16].

Table X
Reduction of the Octahedral Irreducible Representations*

Octahedral		Tetragonal	Trigonal	Orthorhombic
Γ_1	$\rightarrow$	Γ_1	Γ_1	Γ_1
Γ_2	$\rightarrow$	Γ_4	Γ_2	Γ_1
Γ_3	$\rightarrow$	$\Gamma_1+\Gamma_4$	Γ_3	$2\Gamma_1$
Γ_4	$\rightarrow$	$\Gamma_2+\Gamma_5$	$\Gamma_2+\Gamma_3$	$\Gamma_2+\Gamma_3+\Gamma_4$
Γ_5	$\rightarrow$	$\Gamma_3+\Gamma_5$	$\Gamma_1+\Gamma_3$	$\Gamma_2+\Gamma_3+\Gamma_4$
Γ_6	$\rightarrow$	Γ_6	Γ_6	Γ_5
Γ_7	$\rightarrow$	Γ_7	Γ_6	Γ_5
Γ_8	$\rightarrow$	$\Gamma_6+\Gamma_7$	$\Gamma_4+\Gamma_5+\Gamma_6$	$2\Gamma_5$

* For a more complete list of reductions, see E. B. Wilson, J. C. Decius, and P. C. Cross, *Molecular Vibrations*, McGraw-Hill Book Co. (New York), 1955.

Table XI
Resolution of the Three-Dimensional Rotation Representations into the Irreducible Representations of the Octahedral Double Group

j	Resolution of $D^{(j)}$	Number of terms
0	Γ_1	1
1/2	Γ_6	1
1	Γ_4	1
3/2	Γ_8	1
2	$\Gamma_2+\Gamma_5$	2
5/2	$\Gamma_7+\Gamma_8$	2
3	$\Gamma_2+\Gamma_4+\Gamma_5$	3
7/2	$\Gamma_6+\Gamma_7+\Gamma_8$	3
4	$\Gamma_1+\Gamma_3+\Gamma_4+\Gamma_5$	4
9/2	$\Gamma_6+2\Gamma_8$	3
5	$\Gamma_3+2\Gamma_4+\Gamma_5$	4
11/2	$\Gamma_6+\Gamma_7+2\Gamma_8$	4
6	$\Gamma_1+\Gamma_2+\Gamma_3+\Gamma_4+2\Gamma_5$	6

Table XII
Resolution of the Direct-Product Representations of the Regular Octahedron

$$\Gamma_1 \times \Gamma_i = \Gamma_i$$

$$\Gamma_2 \times \Gamma_2 = \Gamma_1$$
$$\Gamma_2 \times \Gamma_3 = \Gamma_3$$
$$\Gamma_2 \times \Gamma_{4,5} = \Gamma_{5,4}$$
$$\Gamma_2 \times \Gamma_{6,7} = \Gamma_{7,6}$$
$$\Gamma_2 \times \Gamma_8 = \Gamma_8$$

$$\Gamma_3 \times \Gamma_3 = \Gamma_1+\Gamma_2+\Gamma_3$$
$$\Gamma_3 \times \Gamma_{4,5} = \Gamma_4+\Gamma_5$$
$$\Gamma_3 \times \Gamma_{6,7} = \Gamma_8$$
$$\Gamma_3 \times \Gamma_8 = \Gamma_6+\Gamma_7+\Gamma_8$$

$$\Gamma_4 \times \Gamma_{4,5} = \Gamma_{1,2}+\Gamma_3+\Gamma_4+\Gamma_5$$
$$\Gamma_4 \times \Gamma_{6,7} = \Gamma_{6,7}+\Gamma_8$$
$$\Gamma_4 \times \Gamma_8 = \Gamma_6+\Gamma_7+\Gamma_8$$

$$\Gamma_5 \times \Gamma_5 = \Gamma_1+\Gamma_3+\Gamma_4+\Gamma_5$$
$$\Gamma_5 \times \Gamma_{6,7} = \Gamma_{7,6}+\Gamma_8$$
$$\Gamma_5 \times \Gamma_8 = \Gamma_6+\Gamma_7+2\Gamma_8$$

$$\Gamma_6 \times \Gamma_6 = \Gamma_1+\Gamma_4$$
$$\Gamma_6 \times \Gamma_7 = \Gamma_2+\Gamma_5$$
$$\Gamma_6 \times \Gamma_8 = \Gamma_3+\Gamma_4+\Gamma_5$$

$$\Gamma_7 \times \Gamma_7 = \Gamma_1+\Gamma_4$$
$$\Gamma_7 \times \Gamma_8 = \Gamma_3+\Gamma_4+\Gamma_5$$

$$\Gamma_8 \times \Gamma_8 = \Gamma_1+\Gamma_2+\Gamma_3+2\Gamma_4+2\Gamma_5$$

B. Spin Angular Momentum

So far, electron spin has been used in the treatment of the crystalline electric field as a device to impose the restrictions necessary to generate the antisymmetric term vectors required by the Pauli principle. The quenching of orbital angular momentum may be visualized as being the result of electrostatic interactions with the variously designated charge distributions arising from within a manifold of $n =$ constant vectors. Thus, in a cubic crystalline electric field having sixfold coordination with the nearest fixed charges located at $(\pm a, \pm a, \pm a)$, the $n = 2$ charge distributions belonging to Γ_3 will be more energetic than those belonging to Γ_5 because the former vectors represent charge distributions which have, on the average, higher densities at the fixed charges than do the latter vectors. On the other hand, unlike the orbital angular momentum which has, so to speak, an associated gyrating electrical charge which directly samples its environment through electrostatic interactions, no such direct interaction exists for the spin angular momentum, excluding of course the relatively weak magnetic dipole–dipole interactions between the electron spin and the spins of nuclei associated with the neighboring ions. There are, however, interactions which may, in principle, contribute to the lifting of spin angular momentum degeneracy in an indirect manner. These indirect processes may be visualized as being magnetic interactions between the spin angular momentum of the electron and an internal magnetic field arising from a nonspherical electron charge distribution. That is, as an electron gyrates through its orbits of constant energy, it generates a distribution of charge lobes which reflect the symmetry of the local environment and which may have associated with them internal magnetic fields capable of interacting with the spin angular momentum thereby lifting some degeneracy. These internal fields, being locked into the crystal, do not directly contribute to magnetic susceptibility. Furthermore, since a crystalline electric field cannot distinguish between clockwise or counterclockwise gyration of electronic charge, these internal fields will be expressed by multipole expansions which have no magnetic field component and consequently will not be capable of removing north–south degeneracy. Thus, for example, a crystal–ion system having a single active electron will exhibit at the very least a twofold degeneracy due to the inability of the crystalline electric field to distinguish energetically between the two possible orientations of the electron spin.

It might be worthwhile to work out the details of at least one specific example to demonstrate the manner in which these indirect interactions may lift degeneracy attributed to electron spin angular momentum. For this purpose, suppose that an ion of the $3d^n$ series, for example, is subject to the action of a crystalline electric field and that the resultant ground term of the crystal–ion is orbitally nondegenerate. A nondegenerate orbital term must be expressible by a vector which is reducibly real* and, consequently, to first order the expectation value of L_z must vanish [1]. To show this, consider a given orbital term which is nondegenerate, that is, to a given eigenvalue of the energy operator there corresponds one and only one eigenvector. That this eigenvector must be reducibly real may be seen by the fact that, if it were not, its complex conjugate would be a linearly independent eigenvector of the energy operator, the latter being assumed real. Thus the vanishing of the expectation value of the operator L_z is guaranteed by the reality of the eigenvector and by the definition $L_z = -i\hbar(\partial/\partial\phi)$. Consequently, the expectation values of the spin-orbit operator $\lambda L \cdot S$ taken over nondegenerate orbital terms also vanish in first order. However, to second order these expectation values may be nonvanishing since this operator does not in general commute with the crystalline electric field. The second-order expectation value of the spin-orbit operator taken over a nondegenerate crystal–ion ground term is given by

$$\mathscr{E}_{so}^{(2)} = -\lambda^2 \sum_n \frac{|\langle 0| L\cdot S |n\rangle|^2}{\varepsilon_n - \varepsilon_0} \tag{17.24}$$

where $|0\rangle$ and $|n\rangle$ designate the ground and excited orbital term vectors which issue from an $L = $ constant manifold as eigenvectors of the crystalline electric field operators. It will be seen that the spin-orbit interaction has the effect of mixing into the nondegenerate crystal–ion ground term some of the characteristics of excited terms. That is, to second order, the ground-term eigenvector is given by

$$\psi_0^{(2)} = |0\rangle - \lambda \sum_n \frac{\langle 0| L\cdot S |n\rangle}{\varepsilon_n - \varepsilon_0} |n\rangle \tag{17.25}$$

Returning to equation (17.24), we can see that the spin-orbit interaction energy may be put into the form

$$\mathscr{E}_{so}^{(2)} = -\lambda^2 \Lambda_{ij} S_i S_j \tag{17.26}$$

where the symmetric tensor $\tilde{\Lambda}$ has components given by

$$\Lambda_{ij} = \sum_n \frac{\langle 0| L_i |n\rangle \langle n| L_j |0\rangle}{\varepsilon_n - \varepsilon_0} \tag{17.27}$$

The spin-orbit interaction energy expression given by equation (17.26) is in effect one for which the orbital variables have been integrated out leaving a symmetric second-order spin polynomial. In a similar manner, the eigenvalues of the Zeeman operator $\beta H \cdot (L + 2S)$ may to second order be given by

$$\mathscr{E}_z^{(2)} = -2\beta\lambda\Lambda_{ij} H_i S_j \tag{17.28}$$

---

* A vector is defined to be reducibly real if it belongs to the same ray of vectors as its complex conjugate, that is, if at most only its amplitude is complex [22].

so that, to this order, an energy operator for the ground orbital term, being $(2S+1)$-fold degenerate, may be expressed as

$$\mathcal{H} = g\beta(\delta_{ij} - \lambda\Lambda_{ij})\,H_i S_j - \lambda^2 \Lambda_{ij}\,S_i S_j \qquad (17.29a)$$

or

$$\mathcal{H} = \beta S \cdot \tilde{g} \cdot H - S \cdot \tilde{\Lambda} \cdot S \qquad (17.29b)$$

where $\tilde{g}$ and $\tilde{\Lambda}$ are tensor quantities which may provide information about the crystalline environment. The expressions given in equations (17.29a) and (17.29b) are electron spin polynomial operators which operate over a space of dimension $(2S+1)$ so that the ground term of the crystal–ion may be characterized by an effective spin. However, this is not to be thought of as a pure electronic spin angular momentum since the ground term has mixed into it some excited orbital term characteristics [23-26].

When the indices i and j of the orbital angular momentum operators correspond to symmetry directions of the crystalline electric field it may be demonstrated that the tensor Λ_{ij} is diagonal and furthermore that the diagonal elements reflect the symmetry of the field. This may be done by operating on Λ_{ij} with an appropriate element of symmetry first to demonstrate that $\Lambda_{ij} = 0(i \neq j)$ and then to discuss the relationships between Λ_{ii}, Λ_{jj}, and Λ_{kk}. A useful example which may be given is that of a crystalline electric field having orthorhombic symmetry. In this example the appropriate elements of symmetry are chosen to be the three mutually orthogonal axes each having twofold rotation symmetry. Applications of such a rotation about the ith axis, for example, will bring the crystal–ion into itself. However, while L_i goes into itself, L_j will go into $-L_j$ so that Λ_{ij} will transform into its negative and hence $\Lambda_{ij} = 0(i \neq j)$ since the crystal–ion and its properties must remain invariant under such a transformation. Consequently, the energy operator becomes

$$\mathcal{H} = \beta(g_{zz}H_z S_z + g_{xx}H_x S_x + g_{yy}H_y S_y) - \lambda^2(\Lambda_{zz}S_z^2 + \Lambda_{xx}S_x^2 + \Lambda_{yy}S_y^2) \ (17.30)$$

and the term quadratic in the spin variables may be put into the form $D[S_z^2 - \tfrac{1}{3}S(S+1)] + E[S_x - S_y^2]$ where $D = -\lambda^2[\Lambda_{zz} - \tfrac{1}{2}(\Lambda_{xx} + \Lambda_{yy})]$ and $E = -\tfrac{1}{2}\lambda^2(\Lambda_{xx} - \Lambda_{yy})$. In the case of axial symmetry, for example tetragonal or trigonal, an expression for the energy operator may be obtained by simply permitting an x–y degeneracy in equation (17.30) which then reduces to

$$\mathcal{H} = g_{\parallel}\beta H S_z \cos\theta + \tfrac{1}{2}g_{\perp}\beta H \sin\theta(e^{i\phi}S_- + e^{-i\phi}S_+)$$
$$+ D\,[S_z^2 - \tfrac{1}{3}S(S+1)] \qquad (17.31)$$

since E vanishes. Finally for cubic symmetry, all the components of the $\tilde{g}$ tensor are equal to one another as are the components of the $\tilde{\Lambda}$ tensor so that D vanishes.

The perturbation energy given by equation (17.26) has been carried out to second order in the spin-orbit interaction and is consequently a polynomial of second degree in the spin variables. Higher-order perturbations in this interaction as well as perturbations involving a series of distinct interactions which connect the ground and excited terms may be brought in to provide energy-operator expressions having spin polynomials of higher degree. Each such perturbation development will involve quantities which reflect the symmetry of the local environment so that the resulting spin polynomials of various degree will transform in the same manner as the crystalline electric field [27-30].

Angular momentum polynomial operators of degree n having m-fold rotational symmetry may be developed from the corresponding spherical harmonics $Y_n^m(\theta, \phi)$ through a procedure based upon a corollary to the Wigner–Eckart theorem. This theorem states that if T_n is an irreducible tensor operator of order n, the matrix elements of its components are given by

$$\langle \lambda, \mu | T_n^m | \lambda', \mu' \rangle = \langle \lambda \| T_n \| \lambda' \rangle \, (\lambda \mu \lambda' \mu' | \lambda \lambda' \, nm) \qquad (17.32)$$

where the double bar or reduced matrix element multiplying the vector coupling coefficient is independent of the projection quantum numbers μ and μ' [21, 31]. This corollary to the theorem states that, if two irreducible tensor operators U and V exhibit similar transformation properties, then their matrix elements will be proportional to one another

$$\langle \lambda, \mu | U \, \lambda', \mu' \rangle = \alpha \, \langle \lambda, \mu | V | \lambda', \mu' \rangle \qquad (17.33)$$

and that the proportionality parameter will be independent of the projection quantum numbers μ and μ'. Thus, as a consequence of the fact that the spherical harmonics of order one as well as the angular momentum operators J_x, J_y, and J_z form the components of irreducible tensor operators of order one, their corresponding matrix elements will be proportional. From this it follows that an angular momentum polynomial operator of order n having m-fold rotational symmetry may be formed from the spherical harmonic $Y_n^m(\theta, \phi)$ by making the substitution $x \to J_x$, $y \to J_y$, and $z \to J_z$ while taking account of the noncommutative nature of the angular momentum operators [32, 33]. Thus, for example, the angular momentum operator equivalents of the spherical harmonics of order two are given by the relations

$$2z^2 - x^2 - y^2 = \beta(2J_z^2 - J_x^2 - J_y^2)$$

$$x^2 - y^2 = \beta(J_x^2 - J_y^2)$$

$$xy = \tfrac{1}{2}\beta(J_x J_y + J_y J_x) \qquad (17.34)$$

$$yz = \tfrac{1}{2}\beta(J_y J_z + J_z J_y)$$

$$zx = \tfrac{1}{2}\beta(J_z J_x + J_x J_z)$$

With the use of these techniques, a complete set of angular momentum operator equivalents to the spherical harmonics may be developed for each value of the index n.

To show how these equivalents may be developed, use is made of the properties of irreducible tensor operators. Thus, for example, an operator of degree n having $(2n+1)$ components $O_n{}^m$ transforms according as the corresponding spherical harmonic and satisfies the following commutation relations with components of the angular momentum operator [12, 34]:

$$[(J_x \pm iJ_y), O_n{}^m] = [(n \mp m)(n \pm m + 1)]^{\frac{1}{2}} O_n{}^{m \pm 1} \tag{17.35}$$

$$[J_z, O_n{}^m] = m\, O_n{}^m$$

Using these relations, one may obtain all components of a given irreducible tensor operator, provided one component is given, for example, that corresponding to $m = 0$. From the three polynomials given in Table XIII, all the operator equivalents of order up to six that relate to electron spin polynomial representations of crystalline electric fields may be obtained by repeated application of equation (17.35).

Consideration will now be given to the consequences of time reversal $t \to -t$ on the various terms of the crystal–ion [35, 36]. When applied to the time-dependent Schroedinger equation, this operation sends

$$i\hbar \frac{\partial}{\partial t} \psi_a = \mathscr{H} \psi_a \quad \text{into} \quad -i\hbar \frac{\partial}{\partial t} \psi_{-a} = \mathscr{H} \psi_{-a} \tag{17.36}$$

where ψ_{-a} is the time reversal of ψ_a (the Hamiltonian operator is itself

Table XIII
The Spin Polynomial Operator Equivalents Corresponding to $n = 2$, 4, and 6 with $m = 0$*

$$O_2{}^0 = \tfrac{1}{2}[3S_z{}^2 - S(S+1)]$$

$$O_4{}^0 = \tfrac{1}{8}[35S_z{}^4 - 30S(S+1)S_z{}^2 + 25S_z{}^2 - 6S(S+1) + 3S^2(S+1)^2]$$

$$O_6{}^0 = \tfrac{1}{16}[231S_z{}^6 - 315S(S+1)S_z{}^4 + 735S_z{}^4 + 105S^2(S+1)^2 S_z{}^2$$

$$- 525S(S+1)S_z{}^2 + 294S_z{}^2 - 5S^3(S+1)^3$$

$$+ 40S^2(S+1)^2 - 60S(S+1)]$$

* For a complete list of spin polynomial operator equivalents of order one to six, see S. A. Al'tshuler and B. M. Kozyrev, *Electron Paramagnetic Resonance*, C. P. Poole, Jr. (ed.), Academic Press (New York and London), 1964.

assumed to be time-reversal invariant). Taking the complex conjugate of the Schrödinger equation one obtains

$$-i\hbar \frac{\partial}{\partial t} \psi_a^* = \mathcal{H}^* \psi_a^* \qquad (17.37)$$

and if P is a unitary operator having the property of bringing $\mathcal{H}^*$ into $\mathcal{H}$ through a similarity transformation then

$$-i\hbar \frac{\partial}{\partial t} (P\psi_a^*) = \mathcal{H}(P\psi_a^*) \qquad (17.38)$$

On comparing this with the time-reversed Schrödinger equation [see equation (17.36)], the relation $\psi_{-a} = PK\psi_a$ is obtained where K is the complex conjugation operator. Thus the time-reversal operator is given by $T = PK$. Now consider a system involving spin angular momentum; the operator P must be given as the product of two operators, one which transforms space coordinates and another which transforms spin coordinates, $P = P_r P_s$. Since angular momentum operators must alter sign on time reversal, there results

$$P_s S^* = -SP_s \qquad (17.39)$$

If the three Pauli spin matrices

$$S_x = \begin{bmatrix} 0 & 1 \\ 1 & 0 \end{bmatrix} \qquad S_y = \begin{bmatrix} 0 & -i \\ i & 0 \end{bmatrix} \qquad S_z = \begin{bmatrix} 1 & 0 \\ 0 & -1 \end{bmatrix} \qquad (17.40)$$

are considered, it will be recognized that

$$P_s = iS_y \qquad (17.41)$$

Thus the time-reversal operator for a system of n electrons is given by

$$T_n = P_r (i)^n S_y^1 \cdot S_y^2 \dots S_y^n K \qquad (17.42)$$

and by inspection it may be seen that $T_n \cdot T_n = (-1)^n$. If the Hamiltonian is time-reversal invariant and if ψ is an eigenvector belonging to the eigenvalue $\mathcal{E}$, then $T_n \psi$ will also be an eigenvector belonging to the same eigenvalue. Suppose now that ψ and $T_n \psi$ are either the same, or vectors belonging to the same ray, then

$$T_n \psi = a\psi \qquad |a|^2 = 1 \qquad (17.43)$$

and an application of T_n yields

$$T_n^2 \psi = a^* a\psi \qquad (17.44)$$

Thus, the statement leading to equation (17.43) is valid only if n is an even integer. If n is odd, $T_n \psi$ will be an eigenvector belonging to $\mathscr{E}$ but will be linearly independent of ψ so that the term in question will at the very least be twofold degenerate. This is the statement of Kramers' theorem [35, 36].

It will be noted that Table XIII lists spin angular momentum operator polynomials of even order only. The reason for this is that if an operator is to represent a crystalline electric field, the latter being real, it must be invariant under the time-reversal operator. However, as may readily be demonstrated through manipulation of the Pauli matrices, a spin polynomial operator of odd order is not invariant under time reversal while one of even order is. Thus, a crystalline electric field may be represented by spin polynomial operators of even order only.

IV. NUCLEAR HYPERFINE STRUCTURE

As for a classical distribution of charges, a nucleus may have its electromagnetic properties described by a series expansion of multipole operators. From the requirement that nuclear energy terms be of either even or odd parity, it may be shown that the first nonvanishing component in the multipole expansion is a second-rank operator which contains the nuclear magnetic dipole and electric quadrupole moment operators (for the same reasons given in the case of the atom, the nuclear monopole component is dropped) [37]. Furthermore, from the definitive parity requirement for nuclear energy terms it may be shown that the next nonvanishing nuclear multipole component will be an operator which contains the nuclear magnetic octopole moment and the nuclear electric hexadecapole moment. On taking into account approximate nuclear dimensions, it is not difficult to argue that the components of these last two moments of the nucleus will be small compared to the magnetic dipole and electric quadrupole moments. Thus, considering that for most cases hyperfine interactions arising from the nuclear magnetic dipole and electric quadrupole moments are small compared to fine-structure interactions, the interactions due to the sixteenth- and higher-order multipoles may be ignored. The nuclear magnetic dipole and electric quadrupole moment operators, while operating within a given manifold of nuclear spin angular momentum, may be replaced by operator equivalents of order one and two, respectively (corollary to the Wigner–Eckart theorem). As a consequence, the components of the nuclear magnetic dipole moment operator may be set proportional to the components of the nuclear spin angular momentum. The constant of proportionality is defined to be the magnitude of the dipole moment. The nuclear electric quadrupole moment operator may be given by expressions analogous to those in equation (17.32) where the J_i are replaced by I_i.

Since the electron is a particle of spin one-half, its multipole moments of order two or greater, if they in fact exist, would be spectroscopically unobservable because two angular momenta of $S = \frac{1}{2}$ can couple to yield angular momentum vectors of S no greater than one. This restriction, however, does not exclude the magnetic dipole moment operator of the electron from being an observable and, as a consequence, hyperfine interaction is allowed to take place through the magnetic dipole moments of the electron and nucleus. The form of the electron–nucleus interaction is given by

$$I \cdot \tilde{A} \cdot S = A_{zz} I_z S_z + A_{xx} I_x S_x + A_{yy} I_y S_y \qquad (17.45)$$

where $\tilde{A}$ is the hyperfine tensor. If z' is defined to be the direction of the Zeeman field, this interaction may to first order be approximated by

$$I \cdot \tilde{A} \cdot S = A I_{z'} S_{z'} \qquad (17.46)$$

with the scalar A defined by [38]

$$A^2 = \frac{1}{g^2} (g_{zz}^2 A_{zz}^2 \cos^2 \theta + g_{xx}^2 A_{xx}^2 \sin^2 \theta \cos^2 \phi + g_{yy}^2 A_{yy}^2 \sin^2 \theta \sin^2 \phi) \qquad (17.47)$$

and

$$g^2 = g_{zz}^2 \cos^2 \theta + g_{xx}^2 \sin^2 \theta \cos^2 \phi + g_{yy}^2 \sin^2 \theta \sin^2 \phi$$

The interaction of the nuclear magnetism with the spin of the electron arises from two distinct sources. One is the isotropic contact or Fermi interaction and the other is an action at a distance. The isotropic contact interaction is given by

$$\mathcal{H} = \frac{8}{3} \pi (g \beta \gamma \beta'') |\psi(r = 0)|^2 (I \cdot S) \qquad (17.48)$$

where $\psi(r = 0)$ is the electronic wave-function amplitude evaluated at the nucleus. This interaction may be deduced from the uncertainty principle and interpreted as an electronic *Zitterbewegung* occurring in the neighborhood of the nucleus [11]. The other interaction is a dipole–dipole action at a distance between the electron and nuclear spin angular momenta. The form of this interaction is given by

$$\mathcal{H}' = (g \beta \gamma \beta'') \left[\frac{I \cdot S}{|r|^3} - 3 \frac{(I \cdot r)(r \cdot S)}{|r|^5} \right] \qquad (17.49)$$

where r is the distance vector between the nucleus and the electron. In general this interaction will be anisotropic because of the second term whose numerator may be looked upon as a second-rank tensor product. This interaction can be put into operator-equivalent form by properly substituting orbital angular momentum operators for the space coordinates. On having done this, it may be demonstrated that the trace of the matrix of this interaction, whose elements are taken from within an orbital angular momentum manifold of $L = $ constant, vanishes. To do this, the expression in brackets of equation (17.49) is expanded and letting [ ] represent the bracketed expression of this equation there results [39, 40]

$$
\begin{aligned}
\left[\quad\right] = & \; I_z S_z + I_x S_x + I_y S_y \\
& - 3(S_z \cos\theta + S_x \sin\theta \cos\phi + S_y \sin\theta \sin\phi) \\
& \times (I_z \cos\theta + I_x \sin\theta \cos\phi + I_y \sin\theta \sin\phi)
\end{aligned} \tag{17.50}
$$

Upon expansion and substitution by orbital angular momentum operators for appropriate space coordinates, the bracket becomes

$$
\begin{aligned}
\left[\quad\right] = & \; \zeta \{ [3L_z^2 - L(L+1)] I_z S_z + [3L_x^2 - L(L+1)] I_x S_x \\
& + [3L_y^2 - L(L+1)] I_y S_y + \tfrac{3}{2}(L_z L_x + L_x L_z)(I_z S_x + I_x S_z) \\
& + \tfrac{3}{2}(L_y L_z + L_z L_y)(I_y S_z + I_z S_y) + \tfrac{3}{2}(L_x L_y + L_y L_x)(I_x S_y + I_y S_x) \}
\end{aligned} \tag{17.51}
$$

which may also be given by the more concise expression

$$
\left[\quad\right] = \tfrac{3}{2}\zeta\,[(L\cdot S)(L\cdot I) + (L\cdot I)(L\cdot S) - L^2(I\cdot S)] \tag{17.52}
$$

The trace of the dipole–dipole hyperfine interaction matrix may be obtained from equation (17.51)

$$
\text{Trace } \mathcal{H}' = 3(g\beta\gamma\beta'')\,\zeta\,[L_z^2 + L_x^2 + L_y^2 - L(L+1)] \tag{17.53}
$$

which vanishes within a manifold of $L = $ constant of orbital angular momentum. Since the trace is an invariant under a principal-axis transformation, the vanishing trace of the dipole–dipole interaction will be maintained even for that representation in which the total hyperfine interaction matrix is in diagonal form. Thus, if $\tilde{K}$ is used to denote the isotropic contact or Fermi interaction, the hyperfine tensor will have the form

$$
\tilde{A} = \tilde{K} + \tilde{T} \tag{17.54}
$$

where T is traceless, so that the trace of $\tilde{A}$ is determined solely by the isotropic contact interaction.

There are two remaining components to the hyperfine-structure interaction, these being the nuclear electric quadrupole interaction and the nuclear Zeeman interaction, neither one of which involves the spin of the electron. Of these, the latter is simply the interaction of the nuclear magnetic moment with the externally applied magnetic field. Assuming the nucleus to be well isolated from its surroundings, the nuclear Zeeman interaction will have the form $\gamma \beta^n H \cdot I$, where $\gamma \beta^n$ is assumed to be a scalar quantity. The electric quadrupole moment of the nucleus will be sensitive to the gradient component of the surrounding electric field which is a second-order tensor field and originates from the nonspherical electron distribution within which the nucleus is imbedded. The form of the nuclear electric quadrupole interaction is $I \cdot \tilde{Q} \cdot I$ which is similar in form to the crystalline electric field interaction given in equations (17.29a) and (17.29b).

Collecting all the terms taken into consideration, a spin Hamiltonian or energy operator for an orbitally nondegenerate electronic ground term may be constructed in terms of electron and nucleus spin angular momentum operator equivalents. This operator is given by an expression of the form

$$\mathscr{H} = \beta H \cdot \tilde{g} \cdot S + I \cdot \tilde{A} \cdot S + \gamma \beta^n H \cdot I + I \cdot \tilde{Q} \cdot I + \sum_n \sum_m b_n{}^m S_n{}^m \quad (17.55)$$

where S is a fictitious electron spin angular momentum operator introduced into the theory as a convenience to account for the degeneracy in the ground term not attributable to orbital angular momentum.

V. APPLICATION OF THE SPIN HAMILTONIAN

Rather than attempting to produce a survey of ESR absorption spectra of free radicals and ions of the various paramagnetic series, three examples of spectra in the solid will be sketched out in the hope that they will provide a notion of how the spin Hamiltonian is applied to interpreting spectra. To do more would not appear to be fruitful in view of the extensive literature on this subject [39-44]. The first of these examples will be a free radical having effective spin $S = \frac{1}{2}$ and exhibiting nuclear hyperfine structure. Next, divalent nickel located in a crystalline environment having octahedral symmetry will be given as an example of a $3d^n$ ion exhibiting loss of some of its spin degeneracy through the action of a crystalline electric field. Finally, an S-state ion, trivalent gadolinium located in a crystalline environment having orthorhombic symmetry, will be considered to demonstrate problems associated with interpreting spectra of high spin ions under the influence of fields having low symmetry.

In a sense, an example of the $S = \frac{1}{2}$ case has already been given in the section on isolated systems. However, the study of gaseous atomic species at low pressure may be classified as a field of study in itself by virtue of the extraordinary potential for precision spectroscopic measurements. In comparison to investigations carried out on the interactions occurring within a solid, those for the gas appear to be at a somewhat higher level

Table XIV

Matrix Representation of the Energy Operator Given by Equation (17.55) : The Total Matrix Can be Reduced to the Direct Sum of Two 4×4 Matrices, One for Each of the Alternating Signs *

	$\lvert \pm\tfrac{1}{2}, \tfrac{3}{2}\rangle$	$\lvert \mp\tfrac{1}{2}, \tfrac{1}{2}\rangle$	$\lvert \pm\tfrac{1}{2}, -\tfrac{1}{2}\rangle$	$\lvert \mp\tfrac{1}{2}, -\tfrac{3}{2}\rangle$
	$\pm\alpha \pm \tfrac{3}{4} A_{zz} + \tfrac{3}{2}\delta + eqQ$	$(\sqrt{3}/4)(A_{zz} \mp A_{yy})$		
	$(\sqrt{3}/4)(A_{zz} \mp A_{yy})$	$\mp\alpha \mp \tfrac{1}{4} A_{zz} + \tfrac{1}{2}\delta - eqQ$	$\tfrac{1}{2}(A_{xx} \pm A_{yy})$	
		$\tfrac{1}{2}(A_{xx} \pm A_{yy})$	$\pm\alpha \mp \tfrac{1}{4} A_{zz} - \tfrac{1}{2}\delta - eqQ$	$(\sqrt{3}/4)(A_{xx} \mp A_{yy})$
			$(\sqrt{3}/4)(A_{xx} \mp A_{yy})$	$\mp\alpha \pm \tfrac{3}{4} A_{zz} - \tfrac{3}{2}\delta + eqQ$

* For the present case, $A_{zz} = A_{\parallel}$ and $A_{xx} = A_{yy} = A_{\perp}$.

of sophistication and generally relate to studies having to do with the testing of fundamental physical theories. Examples are: establishing a test to the theory of quantum electrodynamics, studying the details of excited-atom energy levels which are of great importance to certain cosmological problems, and perhaps precision spectroscopic examinations of atomic ground terms with a view to obtaining information relating to the local isotropy of space.

The first example is that of an ion or atom characterized by $S = \frac{1}{2}$ and $I = \frac{3}{2}$ imbedded in a crystalline host having axial symmetry. When the magnetic field is directed along the crystal symmetry axis (for the sake of convenience, it is assumed that the spectrum exhibits symmetry about this axis) the spin Hamiltonian operator takes the form

$$\mathscr{H} = g_{\parallel}\beta H S_z + A_{\parallel} I_z S_z + \tfrac{1}{2} A_{\perp}[I_+ S_- + I_- S_+)$$

$$+ \gamma\beta'' H I_z + eqQ[I_z^2 - \tfrac{1}{3}I(I+1)] \tag{17.56}$$

where eqQ is the nuclear electric quadrupole coupling coefficient defined as the product of Q, the strength of the nuclear electric quadrupole moment, and eq, the strength of the gradient of the electric field as seen by the nucleus. The matrix of this operator evaluated from within the double manifold $S = \frac{1}{2}$ and $I = \frac{3}{2}$ is of dimension 8×8 and may readily be shown to be the direct sum of two 1×1 matrices and three 2×2 matrices (see Table XIV). The eight eigenvalues to this matrix are given by

$$\mathscr{E}(\pm\tfrac{1}{2}, \pm\tfrac{3}{2}) = \pm\alpha + \tfrac{3}{4}A_{\parallel} \pm \tfrac{3}{2}\delta + eqQ$$

$$\mathscr{E}(\pm\tfrac{1}{2}, \mp\tfrac{1}{2} = -\tfrac{1}{4}A_{\parallel} - eqQ \pm [(\alpha - \tfrac{1}{2}\delta)^2 + A_{\perp}^2]^{\frac{1}{2}}$$

$$\mathscr{E}(\pm\tfrac{1}{2}, \pm\tfrac{1}{2}) = -\tfrac{1}{4}A_{\parallel} \pm \delta \pm \tfrac{1}{2}[(2\alpha \pm A_{\parallel} - \delta \mp 2eqQ)^2 + 3A_{\perp}^2]^{\frac{1}{2}} \tag{17.57}$$

$$\mathscr{E}(\pm\tfrac{1}{2}, \mp\tfrac{3}{2}) = -\tfrac{1}{4}A_{\parallel} \mp \delta \pm \tfrac{1}{2}[(2\alpha \mp A_{\parallel} - \delta \pm 2eqQ)^2 + 3A_{\perp}^2]^{\frac{1}{2}}$$

where $\alpha = \frac{1}{2}g_{\parallel}\beta H$, $\delta = \gamma\beta'' H$, and the labeling of terms corresponds to the strong magnetic field limit system as shown in Fig. 4, where the three mixing parameters are given by

$$\sin\xi = \tfrac{1}{2}\sqrt{3}\, A_{\perp}/[\mathscr{E}(\tfrac{1}{2}, \tfrac{1}{2}) - \mathscr{E}(-\tfrac{1}{2}, \tfrac{3}{2})]$$

$$\sin\eta = A_{\perp}/[\mathscr{E}(\tfrac{1}{2}, -\tfrac{1}{2}) - \mathscr{E}(-\tfrac{1}{2}, \tfrac{1}{2})] \tag{17.58}$$

$$\sin\zeta = \tfrac{1}{2}\sqrt{3}A_{\perp}/[\mathscr{E}(\tfrac{1}{2}, -\tfrac{3}{2}) - \mathscr{E}(-\tfrac{1}{2}, -\tfrac{1}{2})]$$

The four transitions shown by vertical arrows on the left-hand side of Fig. 4 obey the selection rules $\Delta M = \pm 1$, $\Delta m = 0$, which apply to the case for which H is perpendicular to H_{rf}. Using equation (17.57) and taking appropriate energy differences, the expressions relating Zeeman field-strength

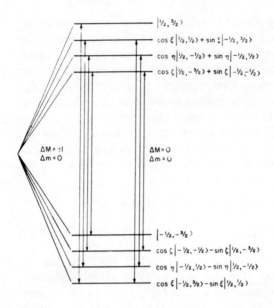

Fig. 4. Representation of the $S = \frac{1}{2}$, $I = \frac{3}{2}$ Zeeman terms with the π and σ transitions indicated.

positions to spin Hamiltonian parameters at constant microwave frequency are given by

$$hv = \alpha_1 + A_{\parallel} + \tfrac{1}{2}\delta_1 + eqQ$$
$$\qquad + \tfrac{1}{2}[(2\alpha_1 + A_{\parallel} - \delta_1 - 2eqQ)^2 + 3A_{\perp}^{2}]^{\frac{1}{2}}$$
$$hv = eqQ + \delta_2 + \tfrac{1}{2}[(2\alpha_2 + A_{\parallel} - \delta_2 - 2eqQ)^2 + 3A_{\perp}^{2}]^{\frac{1}{2}}$$
$$\qquad + [(\alpha_2 - \tfrac{1}{2}\delta_2)^2 + A_{\perp}^{2}]^{\frac{1}{2}} \qquad\qquad (17.59)$$
$$hv = -eqQ + \delta_3 + [(\alpha_3 - \tfrac{1}{2}\delta_3)^2 + A_{\perp}^{2}]^{\frac{1}{2}}$$
$$\qquad + \tfrac{1}{2}[(2\alpha_3 - A_{\parallel} - \delta_3 + 2eqQ)^2 + 3A_{\perp}]^{\frac{1}{2}}$$
$$hv = \alpha_4 - A_{\parallel} + \tfrac{1}{2}\delta_4 - eqQ$$
$$\qquad + \tfrac{1}{2}[(2\alpha_4 - A_{\parallel} - \delta_4 + 2eqQ)^2 + 3A_{\perp}^{2}]^{\frac{1}{2}}$$

where the numerical subscripts on α and δ refer to the transitions corresponding to the four values of the nuclear spin quantum number m. These equations are such that the terms in eqQ and γ do not appear in first order because of the selection rule $\Delta m = 0$. It thus seems of advantage to make use of transitions in which the nuclear quantum number m effectively changes by ± 1 giving equations which contain eqQ and δ in first order. For the case of H parallel to H_{rf}, the $\Delta M = \pm 1$, $\Delta m = 0$ transitions go to zero intensity giving way to those corresponding to the selection rules $\Delta M = 0$, $\Delta m = 0$. Here, the intensities depend upon the degree of spin states mixing caused by the off-diagonal operator $\frac{1}{2}A_{\perp}(I_{-}S_{-}+I_{-}S_{-})$ of the spin Hamiltonian. With use of the normalized eigenvectors given in Fig. 4, the probabilities for these transitions are calculated to be proportional to $\sin^2\omega \cos^2\omega$ ($\omega = \xi, \eta, \zeta$). The expression obtained from equations (17.57) which apply to these transitions are given by

$$hv = [(2\alpha_1 + A_{\|} - \delta_1 - 2\,eqQ)^2 + 3\,A_{\perp}{}^2]^{\frac{1}{2}}$$

$$hv = [(2\alpha_2 - \delta_2)^2 + 4\,A_{\perp}{}^2]^{\frac{1}{2}} \qquad (17.60)$$

$$hv = [(2\alpha_3 - A_{\|} - \delta_3 + 2\,eqQ)^2 + 3\,A_{\perp}{}^2]^{\frac{1}{2}}$$

Generally, the more dominant spectral parameters $g_{\|}$, $g_{\perp}$, $A_{\|}$, and $A_{\perp}$ may be obtained from the normally permitted transitions corresponding to the selection rules $\Delta M = \pm 1$ and $\Delta m = 0$ and whose magnetic field strength positions are given by equations (17.59). This leaves eqQ and $\gamma\beta^n$ which are parameters describing the electric field gradient at the nucleus interacting with the free-radical electron and the magnetic moment of the nucleus. The latter is especially important in assisting in the determination of the nuclear species as is its spin which is obtained from the multiplicity of hyperfine-structure components. The $\Delta M = \pm 1$, $\Delta m = 0$ transitions may be seen to be rather ineffective in determining these parameters for in first order they do not appear in the corresponding field-strength expressions. However, if the normally forbidden transitions corresponding to the selection rules $\Delta M = 0$, $\Delta m = 0$ are used as given in equations (17.60), both eqQ and $\gamma\beta^n$ may be evaluated to within a precision commensurate with that of determining field-strength positions, provided of course that $A_{\|}$ and $A_{\perp}$ are large enough to supply these transitions with measurable strengths through mixing of spin vectors.

The next system to be considered is divalent nickel which is a $3d^8$ ion and by Hund's rule has a ^3F ground term which is threefold degenerate in spin angular momentum and sevenfold degenerate in orbital angular momentum. Under the action of an octahedral electric field, as shown in Fig. 3, this orbital degeneracy will be lifted leaving a ground term having spin-only degeneracy. Since naturally occurring nickel does not have a significant abundance of nonzero spin isotopes, its spin Hamiltonian will be of the form

$$\mathscr{H} = \beta H \cdot \tilde{g} \cdot S + D[S_z{}^2 - \tfrac{1}{3}S(S+1)] + E(S_x{}^2 - S_y{}^2) \qquad (17.61)$$

In cubic symmetry, the tensor $\tilde{g}$ is proportional to the identity matrix while the two spectral parameters D and E vanish. The eigenvalues to the energy operator are then linear functions of the magnetic field being given by $\mathscr{E}(M) = g\beta HM$. For the $S = 1$ manifold, two equally intense transitions will be induced by the operators $S_\pm = (S_x \pm iS_y)$ corresponding to $M = -1 \rightarrow 0$ and $M = 0 \rightarrow 1$ which occur at the same value of the magnetic field.

For axial symmetry, for example, tetragonal or trigonal, the tensor $\tilde{g}$ will have two components $g_\parallel = g_{zz}$ and $g_\perp = g_{xx} = g_{yy}$ while the spectral parameter E will vanish. In this case, the eigenvalues to the energy operator will be given by the roots to

$$\mathscr{E}^3 - \left(\frac{D^2}{3} + g^2\beta^2H^2\right)\mathscr{E} + \frac{D}{3}\left[\frac{2}{g}D^2 + g^2\beta^2H^2\left(1 - 3\frac{g_\parallel^2}{g^2}\cos^2\theta\right)\right] = 0 \quad (17.62)$$

where $g^2 = g_\parallel^2\cos^2\theta + g_\perp^2\sin^2\theta$. At $\theta = 0$ the roots to this equation are given by $\varepsilon_{1,3} = \frac{1}{3}D \pm g_\parallel\beta H$ and $\mathscr{E}_2 = -\frac{2}{3}D$. For this orientation the state functions are the unmixed spin vectors corresponding to the electron-spin eigenvalues $M = 0, \pm 1$. At $\theta = \pi/2$, the roots to equation (17.62) are given by $\mathscr{E}_{1,3} = -\frac{1}{6}D \pm (D^2/4 + g_\perp^2\beta^2H^2)^{\frac{1}{2}}$, and $\mathscr{E}_2 = \frac{1}{3}D$. In this case, the state functions are mixed because of the off-diagonal operators which are generated in the rotated spin Hamiltonian. By measuring resonance-absorption field-strength positions at $\theta = 0$ and $\theta = \pi/2$, the three spectral parameters D, $g_\parallel$, and $g_\perp$ may be determined. In addition, the sign of D may be determined by ascertaining which of the two terms lies lowest in zero magnetic field. This in turn may be accomplished by observing the relative intensities of the spectral components at temperatures low enough to cause an observable population preference for the lowest lying spin term. Thus at the $\theta = 0$ orientation, for example, if two transitions are observed and identified as $M = 0 \leftrightarrow \pm 1$, then at sufficiently low temperature ($kT < D$) one of these transitions will be more intense. Since the transition moments have identical expectation values, the intensity difference will be a consequence of preferential term population.

In orthorhombic symmetry the spectral parameter E will no longer vanish so that the spin operator $(S_x^2 - S_y^2)$ which may be rewritten as $\frac{1}{2}(S_+^2 + S_-^2)$ will mix spin vectors. At $\theta = 0$ the eigenvalues to the spin Hamiltonian operator will be given by $\mathscr{E}_{1,3} = \frac{1}{3}D \pm (g_\parallel^2\beta^2H^2 + E^2)^{\frac{1}{2}}$ and $\mathscr{E}_2 = -\frac{2}{3}D$, which are insensitive to the sign of E. It will be noted that, since $\mathscr{E}_1$ and $\mathscr{E}_3$ are conjugate eigenvalues, the parameter E has no effect upon the center of gravity of the eigenvalue spectrum. These eigenvalue expressions show that even in the absence of an applied magnetic field the orthorhombic crystalline electric field removes all the spin degeneracy of the divalent nickel ground orbital term (see Table X for the orthorhombic resolution of the $D^{(1)}$ representation of the three-dimensional rotation group). It may at first appear that Kramers' theorem has been violated, thereby suggesting an error in determining the influence of an orthorhombic crystalline field on the ground orbital term of the divalent nickel ion.

This is not the case, however, since the theorem on time-reversal symmetry distinguishes between systems having even or odd numbers of electrons.

It might be asked how, for the case of orthorhombic symmetry, a decision is reached in labeling the three spectroscopically distinct and mutually orthogonal axes. A convention which seems quite simple and reasonable is to choose the z axis to be along that crystal direction for which the diagonal spin polynomial operators of the spin Hamiltonian have their largest coefficients or, to put it in another way, to choose the z axis to be along that direction for which the spectrum has its greatest magnetic field strength extension. This convention also reduces the coefficients of the off-diagonal spin polynomial operators to their smallest values. Having decided upon a z axis designation, the spectra along each of the three orthogonal coordinates can be studied so that a determination of the three components to the g tensor and the parameters D and E may be obtained. For example, along any of the three coordinates not more than two transitions may be observed unless of course D and E are sufficiently large compared to the transition energy to produce heavy spin-vector mixing. In any event, for an anisotropic spectrum, there will always be more than a sufficient number of transitions to determine the spectral parameters.

As the crystal is rotated relative to the magnetic field, the spin polynomial operators $[S_z^2 - \frac{1}{3}S(S+1)]$ and $[S_x^2 - S_y^2]$ transform into their partner polynomials according as the matrix given by equation (17.16). For the rotation $r(0, \pi/2, 0)$ which sends z into x, these two spin polynomials transform into

$$[S_z^2 - \tfrac{1}{3}S(S+1)] \to -\tfrac{1}{2}[S_z^2 - \tfrac{1}{3}S(S+1)] + \tfrac{1}{2}(S_x^2 - S_y^2)$$
$$(S_x^2 - S_y^2) \to \tfrac{3}{2}[S_z^2 - \tfrac{1}{3}S(S+1)] + \tfrac{1}{2}(S_x^2 - S_y^2)$$

$$(17.63)$$

which is equivalent to replacing D by $\frac{1}{2}(-D+3E)$ and E by $\frac{1}{2}(D+E)$ so that, in principle, the sign of E relative to that of D may be determined.

An interesting example of a crystal–ion system whose interpretation is facilitated by application of the transformation given by equation (17.16) is that of the trivalent gadolinium ion located in a crystalline electric field of orthorhombic symmetry. This ion has a $4f^7$ configuration and by Hund's rule its ground term is designated ^8S. Under the influence of a crystalline electric field of orthorhombic symmetry, this eightfold spin degeneracy will be lifted to yield four twofold degenerate Γ_5 terms (see Tables IX–XI). The spin Hamiltonian for this crystal–ion system is given by

$$\mathcal{H} = \beta H \cdot \tilde{g} \cdot S + b_2^0 S_2^0 + b_2^2 S_2^2$$
$$+ b_4^0 S_4^0 + b_4^2 S_4^2 + b_4^4 S_4^4$$
$$+ b_6^0 S_6^0 + b_6^2 S_6^2 + b_6^4 S_6^4 + b_6^6 S_6^6 \quad (17.64)$$

Table XV
Partial Matrix Representation Entries of the Rotation Operations $r\,(0,\,\pi/2,\,0)$ and $r\,(\pi/2,\,\pi/2,\,\pi/2)$

$D^{(2)}(0,\,\pi/2,\,0)_{m',\,m}$

m' \ m	2	1	0
2	$\frac{1}{4}$		
1	$-\frac{1}{2}$	$-\frac{1}{2}$	
0	$\frac{\sqrt{6}}{4}$	0	$-\frac{1}{2}$

$D^{(2)}(\pi/2,\,\pi/2,\,\pi/2)_{m',\,m}$

m' \ m	2	1	0
2	$\frac{1}{4}$		
1	$\frac{i}{2}$	$\frac{1}{2}$	
0	$-\frac{\sqrt{6}}{4}$	0	$-\frac{1}{2}$

$D^{(4)}(0,\,\pi/2,\,0)_{m',\,m}$

m' \ m	4	3	2	1	0
4	$\frac{1}{16}$				
3	$-\frac{\sqrt{2}}{8}$	$-\frac{3}{8}$			
2	$\frac{\sqrt{7}}{8}$	$-\frac{\sqrt{14}}{8}$	$\frac{1}{4}$		
1	$-\frac{\sqrt{14}}{8}$	$-\frac{\sqrt{7}}{8}$	$\frac{\sqrt{2}}{8}$	$\frac{3}{8}$	
0	$\frac{\sqrt{70}}{16}$	0	$-\frac{\sqrt{10}}{8}$	0	$\frac{3}{8}$

$D^{(4)}(\pi/2,\,\pi/2,\,\pi/2)_{m',\,m}$

m' \ m	4	3	2	1	0
4	$\frac{1}{16}$				
3	$\frac{i\sqrt{2}}{8}$	$\frac{3}{8}$			
2	$-\frac{\sqrt{7}}{8}$	$-\frac{i\sqrt{14}}{8}$	$\frac{1}{4}$		
1	$-\frac{i\sqrt{14}}{8}$	$-\frac{\sqrt{7}}{8}$	$-\frac{i\sqrt{2}}{8}$	$-\frac{3}{8}$	
0	$\frac{\sqrt{70}}{16}$	0	$\frac{\sqrt{10}}{8}$	0	$\frac{3}{8}$

Matrix Elements Related to $D^{(n)}(\alpha, \beta, \gamma)_{m',\,m}$

(1) $D^{(n)}(\alpha, \beta, \gamma)_{m',\,m} = \exp{(im'\alpha)}\, d^{(n)}(\beta)_{m',\,m}\, \exp{(im\gamma)}$

and

$d^{(n)}(\beta)_{m',\,m} = \langle\, n,\, m' \,|\, \exp{[(i\beta/h)J_y]} \,|\, n,\, m \,\rangle$

(2) From these definitions, there follow the relations

(a) $D^{(n)}(\alpha, \beta, \gamma)_{-m',\,-m} = (-1)^{m'-m}\, D^{(n)}(-\alpha, \beta, -\gamma)_{m',\,m}$

(b) $D^{(n)}(\alpha, \beta, \gamma)_{-m',\,m} = (-1)^{n-m}\, D^{(n)}(-\alpha, \pi-\beta, \gamma)_{m',\,m}$

(c) $D^{(n)}(\alpha, \beta, \gamma)_{m',\,-m} = (-1)^{n+m'}\, D^{(n)}(\alpha, \pi-\beta, -\gamma)_{m',\,m}$

(d) $D^{(n)}(\alpha, \beta, \gamma)_{m,\,m'} = (-1)^{m'-m}\, D^{(n)}(\gamma, \beta, \alpha)_{m',\,m}$

and

(e) $d^{(n)}(-\beta)_{m',\,m} = (-1)^{m'-m}\, d^{(n)}(\beta)_{m',\,m}$

Table XV. Continued

$D^{(6)}(0, \pi/2, 0)_{m', m}$

$m' \backslash m$	6	5	4	3	2	1	0
6	$\frac{1}{64}$						
5	$-\frac{\sqrt{3}}{32}$	$-\frac{5}{32}$					
4	$\frac{\sqrt{66}}{64}$	$\frac{\sqrt{22}}{16}$	$\frac{13}{32}$				
3	$-\frac{\sqrt{55}}{32}$	$-\frac{\sqrt{165}}{32}$	$-\frac{\sqrt{30}}{16}$	$-\frac{1}{32}$			
2	$\frac{3\sqrt{55}}{64}$	$\frac{\sqrt{165}}{32}$	$\frac{\sqrt{30}}{64}$	$-\frac{9}{32}$	$-\frac{17}{64}$		
1	$-\frac{\sqrt{198}}{32}$	$-\frac{\sqrt{66}}{32}$	$\frac{\sqrt{3}}{8}$	$\frac{3\sqrt{10}}{32}$	$-\frac{\sqrt{10}}{32}$	$-\frac{5}{16}$	
0	$\frac{\sqrt{231}}{32}$	0	$-\frac{\sqrt{126}}{32}$	0	$\frac{\sqrt{105}}{32}$	0	$-\frac{5}{16}$

$D^{(6)}(\pi/2, \pi/2, \pi/2)_{m', m}$

$m' \backslash m$	6	5	4	3	2	1	0
6	$\frac{1}{64}$						
5	$\frac{i\sqrt{3}}{32}$	$\frac{5}{32}$					
4	$-\frac{\sqrt{66}}{64}$	$\frac{i\sqrt{22}}{16}$	$\frac{13}{32}$				
3	$-\frac{i\sqrt{55}}{32}$	$-\frac{\sqrt{165}}{32}$	$\frac{i\sqrt{30}}{16}$	$\frac{1}{32}$			
2	$\frac{3\sqrt{55}}{64}$	$-\frac{i\sqrt{165}}{32}$	$-\frac{\sqrt{30}}{64}$	$-\frac{i9}{32}$	$\frac{17}{64}$		
1	$\frac{i\sqrt{198}}{32}$	$\frac{\sqrt{66}}{32}$	$\frac{i\sqrt{3}}{8}$	$\frac{3\sqrt{10}}{32}$	$\frac{i\sqrt{10}}{32}$	$\frac{5}{16}$	
0	$-\frac{\sqrt{231}}{32}$	0	$-\frac{\sqrt{126}}{32}$	0	$-\frac{\sqrt{105}}{32}$	0	$-\frac{5}{16}$

which is an energy operator having twelve spectral parameters. For an $S = \frac{7}{2}$ system, there will be seven $\Delta M = \pm 1$ normally allowed transitions unless some turn out to be unobservable because they occur either in the unreal region of negative magnetic fields or in magnetic field regions where excessive vertical broadening occurs.* Thus, unless very heavy mixing of spin vectors takes place as a consequence of large $b_n{}^m$ values, in which case spectral assignments would be of commensurate difficulty, spectral observations at more than one crystallographic orientation would be required to determine the magnitudes and signs of all the crystalline electric field parameters. To obtain all twelve parameters, spectral observations would be required along the x, y, and z coordinates. In Table XV partial matrix representation entries for the three-dimensional rotation operations $D^{(n)}(0, \pi/2, 0)$ and $D^{(n)}(\pi/2, \pi/2, \pi/2)$ are listed for $n = 2, 4,$ and 6.

As a first step to solving this spectrum, approximate eigenvalues to the spin Hamiltonian are obtained for the magnetic field oriented along the z axis. These approximate eigenvalues are given by

$$\begin{aligned}
\mathscr{E} = &\pm \tfrac{7}{2} g\beta H + 7\, b_2^0 + 7\, b_4^0 + b_6^0 \\
&\pm \tfrac{5}{2} g\beta H + b_2^0 - 13\, b_4^0 - 5\, b_6^0 \\
&\pm \tfrac{3}{2} g\beta H - 3\, b_2^0 - 3\, b_4^0 + 9\, b_6^0 \\
&\pm \tfrac{1}{2} g\beta H - 5\, b_2^0 + 9\, b_4^0 - 5\, b_6^0
\end{aligned} \qquad (17.65)$$

Spectral identification can be assisted by making use of relative intensities to the seven $\Delta M = \pm 1$ transitions for $S = \frac{7}{2}$ which stand in the ratios $7:12:15:16:15:12:7$. After approximate determinations of g_{zz} and $b_n{}^0$ are made, the spin Hamiltonian may be transformed to either the x or y coordinate, thereby mixing into each parameter $b_n{}^0$ some of the $b_n{}^m$ as prescribed by the entries in Table XV. The procedure may be repeated with refinements added in by taking account of off-diagonal contributions to the eigenvalues which may be done either by applying perturbation theory, provided the perturbation strengths are sufficiently small be ensure convergence requirements, or by making use of machine calculations.

VI. REFERENCES

1. J. H. Van Vleck, *The Theory of Electric and Magnetic Susceptibilities*, Oxford University Press (London and New York), 1932.
2. G. Herzberg, *Atomic Spectra and Atomic Structure*, Dover Publications (New York), 1944.
3. G. Breit and I. I. Rabi, *Phys. Rev.* **38**:2082 (1931).

---

* This broadening results from differences in slopes of the magnetic-field-dependent eigenvalues to the spin Hamiltonian.

4. A. G. Prodell and P. Kusch, *Phys. Rev.* **88**:184 (1952).
5. S. H. Koenig, A. G. Prodell, and P. Kusch, *Phys. Rev.* **88**:191 (1952).
6. P. Kusch, *Phys. Rev.* **100**:1188 (1955).
7. R. Beringer and M. A. Heald, *Phys. Rev.* **95**:1474 (1954).
8. P. Franken and S. Liebes, *Phys. Rev.* **104**:1197 (1956).
9. J. Schwinger, *Phys. Rev.* **73**:416 (1948).
10. J. Schwinger, *Phys. Rev.* **76**:790 (1949).
11. E. Fermi, *Z. Physik* **60**:320 (1930).
12. E. U. Condon and G. H. Shortley, *The Theory of Atomic Spectra*, Cambridge University Press (London and New York), 1953.
13. E. B. Rawson and R. Beringer, *Phys. Rev.* **88**:677 (1952).
14. A. Abragam and J. H. Van Vleck, *Phys. Rev.* **92**:1448 (1953).
15. S. Krongelb and M. W. P. Stransberg, *J. Chem. Phys.* **31**:1196 (1959).
16. H. A. Bethe, *Ann. Physik* **3**(5):133 (1929).
17. L. E. Orgel, *Introduction to Transition Metal Chemistry*, John Wiley and Sons, Inc. (New York and London), 1960.
18. T. M. MacRobert, *Spherical Harmonics and Elementary Treatise on Harmonic Functions with Applications*, second edition, Methuen and Co, Ltd. (London), 1947.
19. A. Unsöld, *Ann. Physik* **82**:355 (1927).
20. C. Kittel and J. M. Luttinger, *Phys. Rev.* **73**:162 (1948).
21. E. P. Wigner, *Group Theory and Its Application to the Quantum Mechanics of Atomic Spectra*, Academic Press (New York and London), 1959.
22. M. Hamermesh, *Group Theory and Its Application to Physical Problems*,, Addison-Wesley Publishing Company (Reading, Massachusetts), 1962.
23. B. Bleaney and K. W. H. Stevens, *Rept. Progr. Phys.* **16**:180 (1953).
24. K. D. Bowers and J. Owens, *Rept. Progr. Phys.* **18**:304 (1955).
25. M. H. L. Pryce, *Proc. Phys. Soc.* A **63**:25 (1950).
26. A. Abragam and M. H. L. Pryce, *Proc. Roy. Soc.* A **205**:135 (1951).
27. J. H. Van Vleck and W. G. Penney, *Phil. Mag.* **19**:961 (1934).
28. M. H. L. Pryce, *Phys. Rev.* **80**:1107 (1950).
29. H. Watanabe, *Progr. Theoret. Phys.* (*Kyoto*) **18**:405 (1957).
30. M. J. D. Powell, J. R. Gabriel, and D. P. Johnston, *Phys. Rev. Letters* **5**:145 (1960).
31. C. Eckart, *Rev. Mod. Phys.* **2**:305 (1930).
32. K. W. H. Stevens, *Proc. Phys. Soc.* A **65**:209 (1952).
33. B. R. Judd, *Proc. Roy. Soc.* A **227**:552 (1955).
34. G. Racah, *Phys. Rev.* **62**:438 (1942).
35. H. A. Kramers, *Proc. Amsterdam Acad. Sci.* **33**:959 (1930).
36. A. S. Davydov, *Quantum Mechanics*, Pergamon Press (London and New York), 1965.
37. M. A. Preston, *Physics of the Nucleus*, Addison-Wesley Publishing Company (Reading, Massachusetts), 1962.
38. C. Kikuchi, AFOSR TM 59-220 (1959).
39. D. J. E. Ingram, *Spectroscopy at Radio and Microwave Frequencies*, Butterworth Scientific Publications (London), 1955.
40. W. Low, *Paramagnetic Resonance in Solids*, Academic Press (New York and London), 1960.
41. G. E. Pake, *Paramagnetic Resonance*, W. A. Benjamin, Inc. (New York), 1962.
42. J. S. Griffith, *The Theory of Transition-Metal Ions*, Cambridge University Press (London and New York), 1961.
43. W. Low, "Paramagnetic Resonance," *Proceedings of the First International Conference, Jerusalem, Vols. I and II*, Academic Press (New York and London), 1963.
44. W. Low and E. L. Offenbacher, *Solid State Physics, Advances in Research and Application*, Vol. 17, F. Seitz and D. Turnbull (eds.), Academic Press (New York and London), 1965.

CHAPTER 18

Electronic Spectra of Molecular Crystals

Donald S. Mcclure *

Department of Chemistry
University of Chicago
Chicago, Illinois

I. ENERGY LEVELS OF MOLECULAR CRYSTALS—TIGHT BINDING THEORY

Before any experimental data on molecular crystals had become available, Frenkel [1-4] had formulated a theory of electronic energy levels of weakly interacting systems. He began with the tight binding approximation; the wave functions and energy levels of each molecule are unchanged from the free state; and, upon formation of the crystal, each state is shifted and split into a band of closely spaced levels. The electronic spectra of organic crystals, such as benzene and anthracene, show the features expected from Frenkel's theory, namely, the energy levels appear to be almost those of the free molecules, but band shifts and splittings are observed. Other examples of weakly interacting systems are the d-shell states of MnF_2 [5-7] or Cr_2O_3 [8] and f-shell states of $PrCl_3$, etc. "Molecules" such as MnO_4^- in $KMnO_4$ or CO_3^{2-} in $CaCO_3$ are weakly interacting in their lower electronic levels. In addition, molecules and molecular ions are also weakly interacting in their vibrational levels, and a Frenkel-type theory applies to these.

An interesting type of system which we expect to fit the Frenkel picture is the rare-gas solid [9]. The excited states of these solids overlap a great deal, however, and the best zero-order approximation seems to be the band model, although the localized model works well for ground state properties [10]. The spectra of rare gases are surprisingly similar to those of the alkali halides, but, after all, krypton, for example, is isoelectronic with RbBr.

Another type of system commonly encountered is an impurity in a host crystal. The impurity is subject only to "crystal field" of the surroundings and not to resonant energy exchange as is possible for an excited molecule

---

* Present address: Department of Chemistry, Princeton University, Princeton, N. J.

of a pure molecular crystal. At higher concentrations in these systems, pairs of molecules may become near neighbors, and resonant interchange of energy occurs with its attendant spectral features. Some molecules, such as biphenyl, consisting of two benzene rings joined by a C–C bond, also show spectral features caused by resonant energy exchange [11].

Figure 1 shows the absorption spectrum of crystalline naphthalene compared to a solid solution of naphthalene in durene [12, 13]. The durene is transparent throughout the first and part of the second transition of naphthalene and simply orients the naphthalene molecules in a particular way [14]. The principal feature of the spectrum of the crystal which differs from that of the solid solution is the large splitting of the first absorption band. One of the vibrational lines is split observably, but the others are not. All of these facts are now moderately well understood [15].

Since the elementary theory of excitons in molecular crystals is derived in many other places [15-20], only an outline will be given here, mainly for purposes of establishing notation. Furthermore, only the cases of one or two molecules per unit cell will be treated here. Then we will be ready for a direct comparison with naphthalene and other examples.

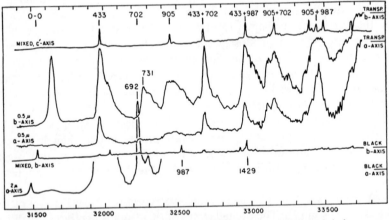

Fig. 1.   The absorption spectrum of naphthalene crystal and of naphthalene as an impurity in a durene crystal (labelled "mixed"). The mixed-crystal spectrum along the c' axis shows the vibrationally induced portion of the transition and is almost the same as the main peaks of the pure-crystal spectrum. The b-axis spectrum of the mixed crystal shows the direct electronic portion of the transition. The origin at 31,544 cm^{-1} of the mixed crystal falls almost exactly at the average of the two Davydov components of the origin of the pure crystal—31,476 cm^{-1} (a axis) and 31,642 cm^{-1} (b axis). The thickness of the crystal is given in microns, and the light was passing normal to the ab plane of the crystals used. The temperature employed was 20°K, and the spectra shown are microphotometer tracings of photographic plates [D. S. McClure and O. Schnepp, J. Chem. Phys. 23:1575 (1955)].

The ground state wave function is a simple product of single-molecule ground state wave functions:

$$\Phi^0 = \prod_i^N \psi_i^{\,0} \tag{18.1}$$

where i goes over the N molecules of crystal. One excited state can be written

$$\Phi_j^{\,g} = \psi_j^{\,g} \prod_{i \neq j} \psi_i^{\,0} \tag{18.2}$$

where a molecule at lattice point j has been raised to an excited state labelled g. Of course, this state is N-fold degenerate and should be combined with its partners in order to form wave functions which conform to the translational symmetry of the crystal. After applying the translation-projection operator, the correct functions are found to be Bloch waves wherein molecular excitation rather than electrons constitute the wavelike disturbance in the crystal:

$$\psi^g(K) = \frac{1}{\sqrt{N}} \sum_j^N e^{iK \cdot R_j} \psi_j^{\,g} \prod_{i \neq j} \psi_i^{\,0} \tag{18.3}$$

where R_j gives the position of molecule j and K is the wave vector. These wave functions are exact for noninteracting molecules where the Hamiltonian is

$$H = \sum_i H_i \tag{18.4}$$

but in the presence of an interaction $\sum_{i<j} V_{ij}$ the degeneracy in equation (18.3) is removed:

$$E^g(K) = \langle \psi^g(K) | \sum_i H_i + \sum_{i,j} V_{ij} | \psi^g(K) \rangle$$

$$= \Delta w^g + D + \sum_j J_{ij} \cos K \cdot R_j \tag{18.5}$$

where Δw^g is the molecular excitation energy from the ground state, D is the "crystal field" shift given by

$$D = \sum_i \langle \psi_j^{\,g}(1) \psi_i^{\,0}(2) | V_{ij} | \psi_j^{\,g}(1) \psi_i^{\,0}(2) \rangle$$

$$- \langle \psi_j^{\,0}(1) \psi_i^{\,0}(2) | V_{ij} | \psi_j^{\,0}(1) \psi_i^{\,0}(2) \rangle \tag{18.6}$$

and J is the excitation exchange energy given by

$$J_{ij} = \langle \psi_j^{\,g} \psi_i^{\,0} | V_{ij} | \psi_i^{\,g} \psi_j^{\,0} \rangle \tag{18.7}$$

Equation (18.4) gives in a simplified way the conclusions about the energy levels using the Frenkel scheme. The transition probability from the ground state is proportional to the square of the transition moment integral:

$$\langle \psi^0 | e \sum_i r_i | \psi^g(K) \rangle = e\sqrt{N} \langle \psi_0 | r_i | \psi_i^g \rangle \, \delta_{K0} \tag{18.8}$$

The only state which can be observed is the one for which $K \cong 0$. This means that only one sharp line should be seen in the spectrum instead of the entire band of levels given by equation (18.4).

This simple theory at least tells us that the spectrum should not be a complete smear, but otherwise it is too simple to cope with real data. It is easy to extend, however, to several molecules per unit cell. We simply add together functions such as equation (18.3) for the separate sublattices by using the projection operator for the space group representation of the function being constructed. For two molecules per cell as in naphthalene, C_{2h}^5, the two excitons for $K = 0$ are

$$\psi^{g\pm} = \frac{1}{\sqrt{N}} \sum_j^N \frac{1}{\sqrt{2}} (\psi_{j\alpha}^g \psi_{j\beta}^0 \pm \psi_{j\alpha}^0 \psi_{j\beta}^g) \prod_{i \neq j} \psi_{i\alpha}^0 \psi_{i\beta}^0 \tag{18.9}$$

where α and β designate the two sublattices, and the energy is

$$E^{g\pm} = \Delta w^g + D + \sum_{j=1}^N (J_{ij}^{\alpha\alpha} \pm J_{ij}^{\alpha\beta}) \tag{18.10}$$

This results in two energy levels separated by $2J^{\alpha\beta}$, the energy of interaction between the two sublattices. Davydov[16] was the first to obtain such a result, and the spectral splitting which equation (18.10) predicts is called Davydov splitting.

The molecules in a crystal are oriented and the transition moment M can therefore be calculated along the directions of vibration of the light wave travelling through the crystal. Note that the light is assumed to be only weakly coupled to the crystal, an assumption which is not always true. The molecules of the crystal are supposed to absorb light with its electric vector oriented in the same direction with respect to molecular axes as in the gas. On this basis, we can find the transition moment of the crystal for the $+$ and $-$ states of equation (18.9)

$$\mathbf{M}^{\pm} = \frac{1}{\sqrt{2}} (\mathbf{M}_\alpha \pm \mathbf{M}_\beta) \tag{18.11}$$

and the intensity of the transitions to the two states

$$f_g^{\pm} = 3.27 \times 10^{11} \, v_g \left(\frac{\mathbf{M}_\alpha^{\,2} \pm \mathbf{M}_\alpha \cdot \mathbf{M}_\beta}{2} \right) \tag{18.12}$$

in terms of the oscillator strength. Experimentally more important is the polarization ratio, the ratio of intensities measured along two crystal axes, particularly for the origin band of the spectrum

$$P = \left(\frac{\mathbf{M^+ \cdot a}}{\mathbf{M^- \cdot b}}\right)^2 \tag{18.13}$$

where $\mathbf{a}$ and $\mathbf{b}$ are the two crystal axes along which light is polarized in the crystal.

It is necessary to add the effects of the vibrational levels in order to achieve a theory good enough to interpret experiments with real molecular crystals. If the molecules were perfectly free, we would use the Born–Oppenheimer approximation and write

$$\psi_{\text{total}} = \psi_{\text{electronic}} \times \psi_{\text{vibrational}}$$
$$\equiv \psi^g \prod_m \chi_m^g \tag{18.14}$$

where $\chi_m{}^g$ is the vibrational wave function for the mth normal mode in the gth electronic state. If the rate of electronic energy interchange as expressed by $J_{ij}^{\alpha\beta}$ is slow compared to the vibrational periods, one should use the above product wave functions. This is the limit of weak coupling of vibrational motion to crystal excitation; the vibrational motion is unaffected by the intermolecular resonance. If equation (18.10) is derived with the product wave functions of equation (18.14), then

$$\Delta E_v{}^{g\pm} = \Delta w^g + D + e_v + \sum_{j=1}^N J_{ij}^{\alpha\alpha} \prod_m \langle \chi_m^{g\alpha} | \chi_m^{0\alpha} \rangle^2 \pm J_{ij}^{\alpha\beta} \prod_m \langle \chi_m^{g\alpha} | \chi_m^{0\alpha} \rangle \langle \chi_m^{g\beta} | \chi_m^{0\beta} \rangle \tag{18.15}$$

The vibrational overlap product $\prod_m \langle \chi_m{}^g | \chi_m{}^o \rangle_v^2$ appropriate to each vibrational energy level, e_v, of the upper state (ground state supposed to be unexcited) multiplies the Davydov splitting and reduces it to a value smaller than that in the absence of vibrational excitation. This product when summed over all vibrational levels of the upper state is unity

$$\sum_v \prod_m \langle \chi_m^g | \chi_m^0 \rangle_v{}^2 = 1 \tag{18.16}$$

so each member of the sum is a weighting factor between 0 and 1 giving the reduction of the Davydov splitting for its level. Furthermore, the spectral intensity of level v is proportional to the square of the vibrational overlap product, and therefore the splitting should be proportional to the intensity of the line.

Now, returning to the spectrum of crystalline naphthalene we can check off the observed features in comparison with the theory.

1. The mixed-crystal spectrum identifies the origin, and the pure-crystal origin is split into two bands 166 cm^{-1} apart. They are oppositely polarized, as equation (18.12) requires; one component is along the b axis in the ab plane, the other and weaker one is along the a direction. This must be the Davydov splitting.

2. One vibrational line, 702 cm^{-1}, is split into 692 and 731, measuring from the point halfway between the Davydov components of the origin. The 39 cm^{-1} splitting is reduced from 166 roughly by the reduction of intensity compared to the origin and so is in qualitative agreement with equation (18.15) containing the Franck–Condon overlap factor.

3. The strong vibrational band at 433 cm^{-1} and most of the other spectral bands do not split at all. This would appear to contradict equation (18.15), but actually there is no disagreement. Bands 433 and 905 are non-totally symmetric vibrations which distort the molecular framework in such a way that transitions are induced with polarization along the short molecular axis. The allowed component of the transition includes the origin band and the totally symmetric 702 band and is polarized in the long direction of the molecule. The overlap between the vibrational functions of different symmetry is identically zero, and equation (18.15) gives zero splitting for the non-totally symmetric modes.

Qualitatively, the spectrum is in agreement with the theory, and we can presume that it is a reasonable first approximation, worth taking more seriously. We now note some quantitative discrepancies.

1. The polarization ratio at the origin is not right. The projection of the electric moment vector corresponding to the long-axis molecular transition onto the a and b crystal axes gives $I_a/I_b = 4.2$. Actually the ratio is about $1/100$.

For the short-axis vibrationally induced transitions, the polarization ratio should be $I_a/I_b = 1/7.7$. This is just about what is observed in the 433 band.

2. The magnitude of the Davydov splitting is too large to be explained easily. Equation (18.15) gives a splitting

$$\Delta E = 2 \sum_{j}^{N} J_{ij}^{\alpha\beta} \prod_{m} \langle \chi_m{}^g | \chi_m{}^0 \rangle^2 \tag{18.17}$$

The excitation exchange integral J, a coulomb integral, can be expanded into a multipole series

$$V_{kl} = \frac{e^2}{R_{kl}^3} (x_k x_l + y_k y_l - 2 z_k z_l) + \text{higher terms} \tag{18.18}$$

Here only the dipole–dipole term is shown. The z direction connects centers of molecules k and l, and x and y are to be chosen in some convenient way, perpendicular to z. The matrix elements of V_{kl} are

$$\langle \psi^g | V_{kl} | \psi^g \rangle = \langle \psi_k{}^0 \psi_l{}^g | V_{kl} | \psi_k{}^g \psi_l{}^0 \rangle$$

$$= \frac{e^2}{R_{kl}^3} [\langle \psi_k{}^0 | x_k | \psi_k{}^g \rangle \langle \psi_l{}^g | x_l | \psi_l{}^0 \rangle + \ldots] \qquad (18.19)$$

$$= \frac{e^2 \mathbf{M}_{og}^2}{R_{kl}^3} (\cos \theta_k{}^x \cos \theta_l{}^x + \cos \theta_k{}^y \cos \theta_l{}^y - 2 \cos \theta_k{}^z \cos \theta_l{}^z)$$

The entire interaction energy $\sum_k V_{kl}$ can be evaluated by calculating the dipole sum [equation (18.19)] over the entire crystal. The interaction energy is proportional to the square of the transition moment $\mathbf{M}_{og}$ for the excitation moving in the crystal. Since the naphthalene transition is very weakly allowed ($f \cong 0.0005$), the splitting energy as calculated from equation (18.19) is very small (only a few cm^{-1}) while 166 cm^{-1} is observed.

Craig and Walmsley [15] calculated the next nonvanishing term in the series (18.18), the octupole–octupole term. For what appeared to be reasonable values of the octupole transition moment, a Davydov splitting in agreement with experiment was obtained. More recent work to be discussed later suggests another explanation [21].

The trouble with the polarization ratio was recognized in the early work on naphthalene to be related to the mixing of different molecular states by the crystal [12]. In the factor group C_{2h}, the molecular states at 3200 and 2900 Å have Davydov components in the same representation. They are therefore able to mix under the influence of the excitation exchange. We should therefore consider matrix elements such as those in equation (18.10), but between two different electronic states, say, f and g:

$$H_{fg}(0) = \langle \psi^g(0) | \sum_i H_i + \sum_{k<l} V_{kl} | \psi^f(0) \rangle$$

$$= (\Delta w + D) \delta_{fg} + \sum_{j=1}^{N} (J_{ij}^{\alpha\alpha, fg} \pm J_{ij}^{\alpha\beta, fg}) \qquad (18.20)$$

This permits setting up an energy matrix with matrix elements H_{fg} including all the molecular excited states and their crystal-induced interactions. The actual energy shifts due to these offdiagonal terms are not large for naphthalene, but the admixture of even a small component of a differently polarized transition can have a large effect on the polarization ratio. It is easy to explain the naphthalene data semiquantitatively in this way.

The first transition in naphthalene is therefore fairly well understood, but, because the molecular transition is weak, higher multipoles and other small effects have to be taken into account to explain the Davydov splitting. An entirely satisfactory explanation of this quantity has not yet been achieved.

II. DEPARTURES FROM THE TIGHT BINDING APPROXIMATION

The attempt to achieve a quantitative understanding of the spectra of molecular crystals leads us into a number of technical and fundamental problems. It was assumed that the sum in equations (18.10) or (18.17) could be performed, but, if V_{ij} is expressed as in equation (18.18) by a dipole–dipole interaction, the sum is only conditionally convergent. The number of molecules at a distance R increases as fast as the energy of interaction decreases. The sum then depends on crystal shape. It can be carried out in a unique manner only if the wave vector is not zero, or if electromagnetic retardation is taken into account. In these cases, the lattice sum is modulated by factors such as $\exp(ik \cdot R_i)$ which improve the rate of convergence. The basic problem in calculating such sums was treated by Heller and Marcus[22], while Silbey, Jortner, and Rice [23] have done a detailed study of it for the case of the anthracene crystal. The latter authors have largely dispensed with the multipole expansion and have carried out the interaction integrals over all pairs of molecules within a 50-Å sphere using molecular wave functions.

Table I
Davydov Splittings in Crystalline Anthracene*

	Theory (cm^{-1})	Experiment (cm^{-1})
P Band		
Dipole terms to 55 Å	+200	
Higher multipoles	−180	
Long-range dipole terms (from 55 Å to infinity)	+632	
Electron exchange	+30	
First-order splitting	622	
Second-order splitting (β state included)	350	≈400
β Band		
First-order splitting	33,800	
Second-order splitting	34,495	(15,000)

* Taken from R. Silbey, J. Jortner, and S. A. Rice, *J. Chem. Phys.* **42**:1515 (1965).

The interactions outside of this range were accounted for by using a continuous dielectric model of the crystal. The near contributions therefore included all multipoles, as well as exchange contributions. The latter are not very important for singlet–singlet transitions, but become dominant for the much smaller Davydov splittings of triplet states [24]. The far contributions were corrected for the finite wave vector of the light and for retardation. The latter was found to be not very important for the case of the anthracene crystal. Table I, taken from the paper by Silbey, Jortner, and Rice [23], shows the contributions of the various terms to the Davydov splitting of the first and second bands (P and β bands) of anthracene. The second-order effect due to crystal mixing of the two molecular states, P and β, is quite important here. The total splitting given is to be apportioned among the principal vibronic peaks of the spectrum; the experimental value given is the sum of all such splittings.

A fundamental problem with the theory presented so far is that it does not permit the crystal to absorb light. Neither an atom nor a crystal can absorb light unless there is a finite density of final states available. In the Frenkel model, the states of the crystal and of the radiation field are considered to be independent until the act of absorption occurs, and then only one crystal state is able to interact with light of a given wavelength. This can be seen by writing out the matrix element for a transition from the ground state

$$\mathbf{M}_{0 \to g} = \int \prod_i \psi_i \left(\sum_i e^{ikx} \frac{\partial}{\partial z_i} \right) \sum_i e^{i\mathbf{K} \cdot \mathbf{R}_i} \psi_i^g \prod_{j \neq i} \psi_j \, dv \qquad (18.21)$$

where the light is propagated in the direction x with electric vector in the direction z, and wavelength $\lambda = 2\pi/k$. If the only contributions to this matrix element are the ones coming from individual molecules, we can make the following expansion at each molecule i

$$e^{ikx} = e^{ikx_i} (1 + ik\eta + \cdots)$$

with $x = x_i + \eta$.

The wavelength of light is large compared to molecular dimensions, so the second and higher terms are small and can be ignored. The phase factor of the light wave is thus effectively a constant over the unit call. The molecular transition moment can be factored out and we have

$$\mathbf{M}_{0 \to g} = \left(\psi_j^0 \left| \frac{\partial}{\partial z_j} \right| \psi_j^g \right) \sum_i e^{ikx_i} e^{i\mathbf{K} \cdot \mathbf{R}_i} \qquad (18.22)$$

The sum is zero for a general K and can be nonzero only when the sum of the exponents vanishes. This occurs when $k = -K_x$ and $K_y = K_z = 0$, i.e., the absorbing molecules in the yz plane must be in phase, and the wave vector in the propagation direction must equal that of the light wave.

There is only one such state of the crystal and the density of final states is therefore infinitesimal. The energy should simply oscillate between the field and the crystal. This is what occurs for wavelengths which are far from the absorption band; within the absorption band, some dissipative process must occur which enables many final states to participate so that the photon can remain in the crystal.

Absorption by an atom occurs over a finite width in frequency determined by the varying phases between light and oscillator in classical theory, or by the uncertainty principle in quantum theory. By constructing Frenkel exciton states, we have made it impossible to have random amounts of recoil, or to have anything but the phase requirement $k = K$, where K is the wave vector of the crystal state and k is that of the light. In terms of the uncertainty principle, states have been constructed whose separation is of the order of

$$\frac{h(v_+ - v_-)}{N}$$

where $h(v_+ - v_-)$ is the Davydov splitting and N is the number of molecules in the crystal. The lifetime of such states is (from the uncertainty principle) $\tau \geqslant N/(v_+ - v_-)$ and for a macroscopic crystal this is of the order of 10^8 sec (~ 30 years). This is an absurdly large number, since the observed lifetimes of an excited crystal state are usually no longer than the lifetimes of the individual molecular states. On the other hand, one must remember that line narrowing corresponding to very long lifetimes is achieved in lasers by virtue of radiative coupling between atoms which are otherwise isolated.

If a molecular crystal were to behave as a laser, the absorption lines would be extremely narrow, whereas they are found to be quite broad in many cases. In naphthalene, the a-polarized Davydov component has a half-width of 10 cm^{-1}, and the b-polarized one is much greater. Quite narrow lines (about 1 cm^{-1}) are observed in the singlet–triplet absorption of the pyrazine crystal. Here the transition is quite weak and the molecules can be considered to be more isolated than in naphthalene. However, the line widths in these two cases are in about the same ratio as the Davydov splittings and therefore include the same large number of crystal states. The occurrence of these line widths can be explained at least qualitatively by crystal imperfections and lattice and molecular vibrations; these, therefore, act as the ultimate sinks for the incident radiation. For this reason, the line widths and shapes are of fundamental importance, although not much research has been devoted to them. Most of this work has been done by Maria and Zahlan [25, 26].

Even more far-reaching problems are encountered in trying to understand spectra of molecular crystals having strongly allowed molecular transitions. The Davydov splitting, if it can be recognized as such, may be larger than the Franck–Condon bandwidth of the molecular transition, and each Davydov component must have its own vibrations and vibrational force constants. In the strongest transitions in dense molecular solids, the phenomenon of metallic reflection is observed in a wavelength region near the molecular absorption band.

Anex and Simpson [27] have found criteria for the occurrence and nonoccurrence of metallic reflection from strongly absorbing crystals. Their semiclassical treatment of this phenomenon represents an entirely different approach to the spectra of molecular crystals; yet it is possible to relate it to the other approaches when proper account is taken of the interaction of radiation with the crystal. In these phenomena, it is not possible to describe the energy as being separately in either the photon field or the manifold of crystal levels. It is in the treatment of strongly absorbing crystals that the theory of molecular crystals resembles that of the absorption of infrared light by ionic crystals [28].

Anex and Simpson studied metallic reflection from crystals of certain dye molecules, such as

$$(CH_3)_2N^+ = CH–CH = CH–CH = CH–N(CH_3)_2ClO_4^-$$

or 1,5 bis-(dimethylamino)pentamethinium perchlorate, which will be referred to as BDP. The molecule has a single very strong transition ($f = 3.31$) at 24,500 cm^{-1}, polarized along the length of the carbon chain. Those faces of the crystal to which the molecular chain is most nearly parallel show metallic reflection from 22,000 to 32,000 cm^{-1}. A classical treatment of this phenomenon would begin with the dispersion formula

$$\frac{n^2 - 1}{n^2 + 2} = \frac{\omega_p^2/3}{\omega_0^2 - \omega^2 + i\gamma\omega} + G \tag{18.23}$$

where G takes care of higher states in an approximate way, $\omega_p^2 = 4\pi N f e^2/m$ is the "plasma frequency," ω_0 is the molecular absorption frequency, and γ is a damping constant. The value of ω_p depends on the density of molecules N and the number of electrons per molecule involved in the transition, f. The appearance of the plasma frequency in this problem means that we are considering the collective motion of all the electrons in the solid, rather than of a single molecule. Physically the plasma frequency arises as the result of the reaction of all the molecules to any displacement of electrons on one. Its value for BDP is $\omega_p^2 = 3.29 \times 10^{30}$ (rad/sec)^2. Anex and Simpson succeeded in fitting the observed reflectivity curve fairly well, using the Fresnel formula to calculate reflectivity, when the resonance frequency ω_0 is taken to be somewhat higher than the molecular value and when $\gamma = 0.36 \times 10^{15}$ rad/sec (see Fig. 2). Possibly the position of the resonance is shifted by the electrostatic field in the crystal. Most interesting is the fact

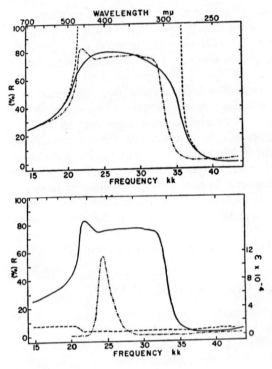

Fig. 2. Reflection spectrum of a crystal of an organic dye, 1,5 bis-(dimethylamino)pentamethinium perchlorate [B. G. Anex and W. T. Simpson, *Rev. Mod. Phys.* **32**:466 (1960)]. Top: - - - - - - - - Theoretical fit with $\omega_p{}^2 = 32.9 \times 10^{10}\ \mathrm{sec}^{-2}$ and $\gamma = 0.$ ———— Theoretical fit with same valve of ω_p and $\gamma = 0.361 \times 10^{15}\ \mathrm{sec}^{-1}.$ —·—·—·—. Experimental. Bottom: ———— Light polarized along one principal direction of a highly reflecting crystal face. - - - - - - - - Polarized at right angles to above axis on the same face. —·—·—·—. Absorption spectrum of the compound in water solution (isolated molecules).

that γ is equal to the Franck–Condon bandwidth of the molecular spectrum in solution; Anex and Simpson interpret this to mean that the damping mechanism is the trapping of the excitation on a single molecule (followed by other degradative processes leading to heat and fluorescence). A quantum interpretation of $1/\gamma$ is that it represents the mean lifetime of an electronic state in the crystal. All of the levels associated with the molecular transition in the crystal are strongly coupled. This is quite different from Frenkel's theory, presented earlier, where such interactions were left out. The classical theory presented here does not provide a description of the crystal states; instead it has used parameters of the crystal whose meaning must be interpreted by a more detailed theory, and it has concentrated on the behavior of the light.

In terms of the parameters of the semiclassical theory, metallic reflectivity is considered to occur whenever the width of the dispersion curve exceeds the normal width, given by γ, and it can be shown that this implies

$$\omega_p{}^2/2\omega_0 > \gamma \tag{18.24}$$

Insertion of the definition of ω_p gives

$$4\pi M^2/D^3 > \hbar\gamma \tag{18.25}$$

where **M** is the molecular electric transition moment and D is the volume per molecule. In other words, when the rate at which energy can pass from one molecule to another exceeds the rate at which it is degraded within the molecule, then metallic reflection can occur.

Another interesting aspect of the theory of molecular crystals leads to the relation between the Frenkel description and the band theory description of the crystal states. In the Frenkel type of exciton, the excitation is supposed to be confined to a single molecule; but it is quite conceivable that the excitation could take the form of an electron–hole pair occupying different molecules. Such pairs, if dissociated, would be able to carry an electric current, and in fact photoconductivity is observed in many organic crystals. Even the low-lying excited states may be partly hole–electron pair states. An admixture of such states in the naphthalene and anthracene lowest singlet and triplet has been shown to give important contributions to their Davydov splitting [29].

The theory of the transition between Frenkel and Wannier excitons is given by Knox [30].

Briefly, it is now necessary to include among the accessible states the case in which an electron and a hole are on different molecules and to define a wave vector for each. Let us call k the electron wave vector and $k - K$ the hole wave vector, where K is the total wave vector for the electron–hole state. Product wave functions of the hole and electron are written as products of Bloch functions $\Phi_e(k)\, \Phi_h(k - K)$. This product does not specify the separation between the hole and electron; a representation which does, called the exciton representation, is

$$\Phi(K, \beta) = \frac{1}{\sqrt{N}} \sum_k e^{-i\beta \cdot k}\, \Phi_e(k)\, \Phi_h(k - K) \qquad (18.26)$$

where β is the distance between hole and electron. Since β is only a parameter and not a quantum number, an eigenfunction of the Hamiltonian will be a linear combination of these functions corresponding to different values of β

$$\psi(K, E) \equiv \sum_\beta U_K(\beta)\, \Phi(K, \beta) \qquad (18.27)$$

where E is the energy eigenvalue and the $U_K(\beta)$'s are the "envelope" coefficients describing the correct combinations.

An equivalent way of writing equation (18.27) which emphasizes the molecular wave functions is

$$\Phi(K, R) = \frac{1}{\sqrt{N}} \sum_i e^{-iK \cdot R_i}\, \phi(R_i, R_i + \beta) \qquad (18.28)$$

These could be called charge transfer or ion-pair states. When β refers to a lattice translation between nearest neighbors, the wave function (18.28) could have appreciable overlap with neutral exciton functions. If the charge transfer state has an energy near enough to the neutral exciton state, then appreciable mixing can occur, with corresponding energy and optical polarization changes. In fact, the triplet Davydov splitting in naphthalene is calculated to be increased by 50% upon the inclusion of charge transfer states [24].

The Davydov splitting of the first band of naphthalene is not given correctly by the neutral exciton states [21]. The use of Pariser–Parr wave functions cancels out most of the coulomb interactions (to all multipole orders) because of the pairing property of the wave functions and the fact that the principal configurations in this molecular state give equal and opposite contributions. However, a calculation including ion-pair excitons can be made to yield the correct splitting and polarization ratio if the ion-pair state lies within about 500 cm^{-1}. The splitting is extremely sensitive to the position of the ion-pair state, and, since we cannot estimate the latter to better than about 0.5 eV, the position of the ion-pair state has to be treated as an adjustable parameter.

It would be interesting to locate the ion-pair states, but these calculations cannot yet be trusted to give this information. Physical methods other than ordinary absorption spectroscopy must be used to find them, as their transition probabilities from the ground state must be small, on account of the small overlap of molecular wave functions.

In this chapter, we have seen that the attempt to interpret spectra of molecular crystals quantitatively has led away from the initial approximation of nearly isolated molecules to considerations common to inorganic insulators and semiconductors. We can find examples which represent all degrees of the transition between weakly and strongly coupled systems.

III. REFERENCES

1. J. Frenkel, *Phys. Rev.* **37**:17, 1276 (1931).
2. J. Frenkel, *Physik. Z. Sowjetunion* **9**:158 (1936).
3. R. Peierls, *Ann. Physik* **13**:905 (1932).
4. F. Seitz, *Modern Theory of Solids*, McGraw-Hill Book Co. (New York), 1940, pp. 414-416.
5. J. W. Stout, *J. Chem. Phys.* **31**:709 (1959).
6. R. L. Greene, D. D. Sell, W. M. Yen, A. L. Schawlow, and R. M. White, *Phys. Rev. Letters* **15**:656 (1965).
7. P. G. Russell, D. S. McClure, and J. W. Stout, *Phys. Rev. Letters* **16**:176 (1966).
8. D. S. McClure, *J. Chem. Phys.* **38**:2289 (1963).
9. G. Baldini, *Phys. Rev.* **128**:1562 (1962).
10. J. C. Phillips, *Phys. Rev.* **136**:A1714 (1964).

11. D. S. McClure, *Can. J. Chem.* **36**:59 (1958).
12. D. S. McClure and O. Schnepp, *J. Chem. Phys.* **23**:1575 (1955).
13. D. P. Craig, L. E. Lyons, and J. R. Walsh, *Mol. Phys.* **4**:97 (1961).
14. D. S. McClure, *J. Chem. Phys.* **22**:1668 (1954).
15. D. P. Craig and S. H. Walmsley, *Mol. Phys.* **4**:113 (1961).
16. A. S. Davydov, *J. Exptl. Theoret. Phys. (U.S.S.R.)* **18**:210 (1948).
17. O. Schnepp, *Ann. Rev. Phys. Chem.* **14**:35 (1963).
18. D. S. McClure, *Solid State Phys.* **8**:1 (1959).
19. A. S. Davydov, *Theory of Molecular Excitons*, McGraw-Hill Book Co. (New York), 1962.
20. D. P. Craig and S. H. Walmsley, in: D. Fox, M. M. Labes, and A. Weissberger (eds.), *Physics and Chemistry of the Organic Solid State*, Vol. *I*, Interscience (New York), 1963.
21. R. Silbey, J. Jortner, M. T. Vala, Jr., and S. A. Rice, *J. Chem. Phys.* **42**:2948 (1965).
22. W. R. Heller and A. Marcus, *Phys. Rev.* **84**:809 (1951).
23. R. Silbey, J. Jortner, and S. A. Rice, *J. Chem. Phys.* **42**:1515 (1965).
24. J. Jortner, S. A. Rice, J. L. Katz, and Sang-Il Choi, *J. Chem. Phys.* **42**:54 (1965).
25. H. Maria and A. Zahlan, *J. Chem. Phys.* **38**:941 (1963).
26. H. J. Maria, *J. Chem. Phys.* **40**:551 (1964).
27. B. Anex and W. T. Simpson, *Rev. Mod. Phys.* **32**:466 (1960).
28. M. Born and K. Huang, *Dynamical Theory of Crystal Lattices*, Oxford (New York), 1954.
29. Sang Il Choi, J. Jortner, S. A. Rice, and R. Silbey, *J. Chem. Phys.* **41**:3294 (1964).
30. R. S. Knox, "The Theory of Excitons," Suppl. 5 in: F. Seitz and D. Turnbull (eds.), *Solid State Physics*, Academic Press (New York), 1963.

CHAPTER 19

Spectra of Ions in Crystals

D. L. Wood

Bell Telephone Laboratories, Incorporated
Murray Hill, New Jersey

I. INTRODUCTION

The absorption and emission spectra for ions in solids are usually explained today in terms of the crystal field theory, or ligand field theory, as it is sometimes called.

Historically crystal field theory began in the late nineteenth century with the elucidation of the structures of the highly colored amino-cobalt complexes where the newly formulated classical ideas of valence were seemingly not obeyed. Werner's ideas on stereochemistry solved this puzzle and now the concept of a coordination complex has been extended from molecules in solution to ions in crystals, where the ion and its neighbors form a complex. The behavior of the central ion with its ligands forming the united complex is the subject of this chapter.

The basic ideas of the ligand field theory in solids were first put forward by Becquerel [1] and by Bethe [2] in 1929, who thought of the central ion as being perturbed by the electrical field of the ligands. Through the symmetry properties of the complex, Bethe formulated the splitting of the energy levels of the central ion due to the presence of the ligands. At approximately the same time, Kramers [3] showed that the energy levels of molecules containing an odd number of electrons must remain at least twofold degenerate (the Kramers degeneracy) paralleling closely the requirements of the "double groups" introduced by Bethe [2].

The ligand field theory was applied with great success to chemical and physical problems by Van Vleck and co-workers [3a], and by 1940 a more quantitative expression of the theory had been developed, centered mainly on the magnetic properties of solids. With the advent of modern microwave and optical apparatus, many new detailed data were obtained, and the requirements of their interpretation led to the development of the theory which we follow here. It is convenient to date this phase of the historical picture from the work of Tanabe and Sugano [4] who in 1954 published two detailed studies on the transition elements. Other publications followed closely, and now several textbooks on the subject have been published [5-10].

We will begin with the Schrödinger equation for an ion in a crystal under the influence of the electric field of its neighbors. The Hamiltonian consists of several terms, the first four belonging to the free ion:

$$H = -\frac{\hbar}{2m} \sum_i \nabla_i^2 - \sum_i \frac{Ze^2}{r_i} + \tfrac{1}{2} \sum_{i \neq j} \frac{e^2}{r_{ij}} + \sum_i \xi_i(r)\, \mathbf{l}_i \cdot \mathbf{s}_i + V_c \qquad (19.1)$$

and

$$H\psi = E\psi \qquad (19.2)$$

is the equation determining the system. The first term corresponds to the kinetic energy of the ith electron in operator form, and the second to the coulomb potential energy of that electron in the field of the nucleus. The third term involves the interelectronic repulsion, and the fourth the coupling between the spin and orbital momenta of the electrons. Finally V_c is the crystal field potential which will be discussed in detail. First, however, we will consider briefly the energy levels of a free ion or atom as described by the solutions of Schrödinger's equation without the term V_c.

If we had just the first two terms, with no electrostatic interactions or spin-orbit coupling, the energy levels of the atom would depend only on a principal quantum number n, and all levels belonging to a given n would be degenerate. In the case of more than one electron, however, the electrostatic interaction term $\sum e^2/r_{ij}$ removes this degeneracy and the various energy levels for, say, $3d$ electrons ($n = 3$) are different, giving various terms characterized by the particular values of orbital angular momentum, S, P, D, F, etc. When spin-orbit coupling is included, the spin multiplet degeneracy is removed, and levels characterized by various values of $J = L+S$ lie at different energies. Thus a spin-orbit multiplet ^4F_J, for example, consists of $2S+1$ levels characterized by $J = L+S, L+S-1, ...,$ $L-S$ with an order characterized by Hund's rule [11] and an energy separation governed approximately by the Landé interval rule [12].

Hund's rule states that according to experience the lowest term of a configuration of equivalent electrons is that which has the highest value of S, and, if there are several of these, the one with the highest L lies lowest. Within the multiplet, the Landé rule states that the interval between successive levels is proportional to the larger of the two J values. For elements with less than half-filled shells, the terms with lower J values lie lower, while the reverse is true for more than half-filled shells.

Now the effect of the crystalline field which is represented by the remaining term V_c of our Hamiltonian will depend on the magnitude of the term V_c relative to the others in the equation, and for an evaluation of that we need results from physics. It turns out that cases are observed for all values of V_c relative to the other terms. One such case is that for which V_c is small compared to the other terms. Since $\sum \xi \mathbf{l} \cdot \mathbf{s}$ is the smallest free-ion term, $V_c < \sum \xi \mathbf{l} \cdot \mathbf{s}$ suffices for the criterion. This is the case of the trivalent $4f$ rare earth ions and some $5f$ elements mostly belonging to the transuranics. These ions are characterized by a preservation of the identity of the free-ion $^{2S+1}L_J$ multiplet levels with a smaller crystalline field splitting superimposed. This is one case which will be discussed in detail later.

A second group of ions commonly found is that for which

$$\sum \xi \mathbf{l} \cdot \mathbf{s} < V_c < \sum e^2/r_{ij}$$

Then the identity of the free-ion terms is lost, to be replaced by levels characterized by the crystalline field, but where the identity of the levels may be traced back to those of the free ion. The first series of transition elements falls into this classification, and the so-called "weak crystalline field" treatment is useful for these ions. We should note that the terms "weak field" and "strong field" must be used with care since a distinction is often made using these terms for $3d$ ions on the basis of the spin multiplicity of the ground state of the ion instead of the relative magnitude of the terms of the Hamiltonian as noted here. This will be referred to again later.

The third category of ions is that for which $V_c > \sum e^2/r_{ij}$. These are characterized by covalent bonding and the properties associated with that kind of chemical behavior. Such ions would properly be treated by a molecular orbital approach, and examples include the $4d$ and $5d$ transition elements. Little will be said here about this category, but there is much interesting work to be done with the spectra of these ions.

With this introduction then, we proceed to a discussion of the effect of V_c on the energy levels first of the $3d$ transition elements and second of the $4f$ rare earth ions.

II. LIGAND FIELD THEORY FOR $3d$ TRANSITION ELEMENTS

For the problem where $\sum \xi \mathbf{l} \cdot \mathbf{s} < V_c < \sum e^2/r_{ij}$ the perturbation theory permits an inquiry into the effect of V_c on the energy levels of an ion having electrostatic interaction but negligible spin-orbit coupling: Assume for the moment that this zero-order problem is solved. In fact the zero-order problem is not especially simple for many electron spectra, but it has been solved using as basis functions linear combinations of single-electron wave functions [13]. In practice it is found useful to assign electrostatic interaction parameters from experimental results for the free-ion spectra, but the values for ions in solids are not always the same. Some variation in the parameters occurs also in going from one crystal to another.

Because the zero-order solution of Schrödinger's equation is spherically symmetric, it is convenient to expand the crystalline field in terms of spherical harmonics using a generalized Fourier theorem

$$V_c = \sum_i \sum_n \sum_m A_n^m Y_n^m(\vartheta, \phi) R_{kn}(r_i) \tag{19.3}$$

where the potential for the ith electron is expressed as a sum over a radial part $R_{kn}(r_i)$, and an angular part $Y_n^m(\vartheta, \phi)$, and the A_n^m's are the generalized Fourier coefficients.

The requirements of symmetry and of orthogonality for the integrals forming the matrix elements in the perturbation problem severely limit the number of terms required to be considered in the expansion.

If it is assumed that the equation is separable into radial and angular parts, the radial part of the expansion of V_c may contribute to a large heat of solution or lattice energy, but it contributes a constant amount to all the energy levels and can be dropped where transitions between levels are being investigated. Thus,

$$V_c = \sum_i \sum_n \sum_m A_n^m Y_n^m(\vartheta, \phi) = \sum_{n, m} V_n^m \tag{19.4}$$

For the case of small crystal fields where n, S, L, J, and M are good quantum numbers, we may start with the zero-order solution wave functions characterized by these quantum numbers:

$$\psi_{\gamma M}^J \qquad \text{where} \qquad \gamma \to n, S, L$$

In matrix notation then, the perturbation problem is solved by

$$|\langle \psi_i | H_1 | \psi_j \rangle - E_{ii}| = 0 \tag{19.5}$$

where H_1 is in this case $V_c = \sum_{n,m} V_n^m$, and $\psi_i = \sum A_M^J \psi_{\gamma M}^J$ or perhaps a linear combination of single-electron wave functions belonging to the symmetry classes of the crystal field. The problem then reduces to the determination of the matrix elements of V_c and the solution of the secular equation.

The elements

$$\langle \psi_i | V_n^m | \psi_j \rangle \tag{19.6}$$

or

$$\langle \gamma J M | V_n^m | \gamma' J' M' \rangle$$

can be computed by several methods, including direct integration and the operator equivalent method [14] or from Wigner's coefficients [15]

$$\langle \gamma J M \,|V_n^m|\, \gamma' J' M' \rangle = (\alpha_n)_{\gamma J,\,\gamma' J'}\, C_{JnJ'}^{MmM'} \tag{19.7}$$

where $(\alpha_n)_{\gamma J,\,\gamma' J'}$ is independent of M, M' and m. It can also be shown (a useful summary is given by Fick and Joos [16]) that nonvanishing matrix elements can be obtained only if

$$M = M' + m$$

and

$$n < 2J$$

and

$$n < 2l$$

where l is the angular momentum of the electron involved. Finally it is also necessary that n be even. Thus,

$$n = 2, 4 \quad \text{for the } 3d^n \text{ iron series } l = 2$$
$$n = 2, 4, 6 \text{ for the } 4f^n \text{ rare earths } l = 3$$

Because of the symmetry of the ligand arrangement around the central ion site (site symmetry), there are even more restrictions on the values of n and m which give nonvanishing matrix elements. For example in the case of S_4 symmetry, only matrix elements for V_2^0, V_4^0, V_6^0, V_4^4, and V_6^4 are different from zero. For O_h, there are nonzero elements for V_4^0, V_6^0, V_4^4, and V_6^4, and certain restrictions pertain to the ratios of fourth-order to sixth-order terms as well. A useful tabulation of the values of n and m possible on the basis of symmetry alone has been given by Prather [17].

The most important result at the moment is that for one $3d$ electron in an octahedral cubic field of O_h symmetry, direct integration gives the energy of the two levels obtained [7] as

$$E_t = -4Dq$$
$$E_e = +6Dq \tag{19.8}$$

where Dq is a coefficient which measures the crystal field magnitude. The splitting of the ^2D level of the single $3d$ electron or the splitting of the $3d$ electron orbital in the O_h symmetry field of a perfect octahedron is thus $10Dq$. It can also be shown that, for a tetrahedral configuration of ligands,

$$E_t = 6Dq$$
$$E_e = -4Dq \tag{19.9}$$

and the same magnitude of splitting is observed, but with opposite sign. This can be understood qualitatively by looking at the $3d$ orbitals shown in Fig. 1. The twofold degenerate orbitals have maximum electron density along the coordinate axes where the six ligands occur, and thus the energy of the e species is raised above that of the threefold t species which has maximum density between the axes where no ligands are found. In the case of the tetrahedral coordination, the opposite is true since the ligands now occur in pairs in the xz, xy, and yz planes off the coordinate axes. Thus, the energy splitting for d orbitals would be that shown in Fig. 2, and in either case the energy difference is $10\,Dq$. If one electron is present in the system, it requires a photon of energy $10\,Dq$ to promote that electron from the lowest state to the next higher state. Similar results could be obtained by putting more electrons into the orbitals and asking then about the energy levels. This is useful in determining the number and symmetry species of levels expected, but for many electrons much better results are obtained if the matrices containing both the electrostatic and crystal field terms are constructed and then diagonalized directly. This case has been dealt with by Tanabe and Sugano [4] and the matrix elements for the $3d$ transition elements have been given in their papers. There are two electrostatic interaction parameters, B and C, neither of which have directly interpretable physical significance, although B is reduced in covalent compounds. In this scheme many spectra can be quite nicely explained using B, C, and Dq as experimentally determined parameters.

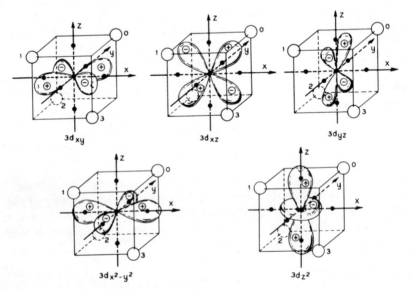

Fig. 1.   The $3d$ orbitals and tetrahedral or octahedral ligand configurations. The large open circles locate tetrahedral ligand positions, while the small solid circles locate the octahedral positions.

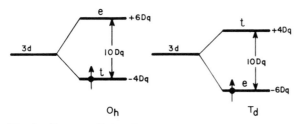

Fig. 2.   Energy splitting diagrams for 3 d orbitals in tetrahedral (T_d) and octahedral (O_h) coordination.

It is a simple step, in principle, to go from Tanabe and Sugano's
matrix elements to those required for the case where spin-orbit coupling
is large, too. The spin-orbit matrix elements for some of the transition elements have been given by Eisenstein [18], Tanabe and Kamimura [19],
and Griffith [20]. The large matrices obtained by combining spin-orbit,
electrostatic interaction, and crystal field terms have been diagonalized by
high-speed machine solution by Liehr and Ballhausen [21] and by Liehr [22].
Thus, for 3 d transition elements, a very sophisticated theory exists for the
interpretation of these spectra [20].

It should be emphasized that the results of group theory are essential
in the analysis of spectra since the wave functions describing the electronic
states in the crystal must be characterized by irreducible representations
of the point symmetry on which the ion is found. This is the so-called site
symmetry. The number and species of the possible crystal field wave functions
were tabulated for some point groups by Bethe [2] and for all the point
groups by Koster [24] and co-workers. It is merely necessary to know the
value of L, if spin-orbit effects are negligible, or of J, if the spin-orbit
coupling is large, and the number and species of levels possible are immediately known. Of course, the energy separation of these levels requires the
solution of the perturbation problem.

One more fundamental fact of crystal field spectroscopy concerns
intensities. The energy levels with which we are dealing arise within a single-
electron configuration, in the case of the transition elements, within the
3 d^n shell. For the rare earths, which we will take up later, the levels originate
within the 4 f^n shell. This means that the crystal field levels, being characterized by wave functions which are linear combinations of free-ion 3 d or 4 f
wave functions, all have the same parity. Because an electric dipole operator
is of odd parity, the expectation value of a dipole transition between two
levels of the same parity cannot have even parity. It therefore must be zero:

$$\int_{\substack{\text{even}}} \underset{\text{even}}{\psi_i} \; \underset{\text{odd}}{P} \; \underset{\text{even}}{\psi_j} \, dt = 0 \qquad (19.10)$$

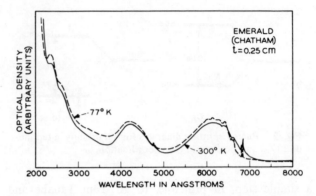

Fig. 3. Spectrum of emerald (beryl + Cr^{3+}) at 300 and 77 °K showing the lack of intensity change with temperature (from Wood *et al.* [26]).

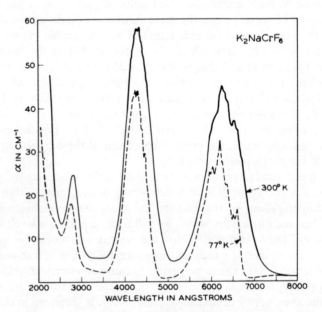

Fig. 4. Spectrum of K$_2$NaCrF$_6$ at 300 and 77 °K showing the decrease of intensity of the long-wavelength side of each band as the temperature is decreased (from Wood *et al.* [26]).

The fact that electric dipole transitions are observed means that in some way the Laporte rule [25] forbidding transitions between states having the same parity must be overcome.

There are two ways in which the electric dipole transition may occur and both act by introducing odd (or even) parity into the wave function of one or the other of the states involved. In the first case, for every ion of the type we are discussing there exist states of both parity, though often widely separated in energy, and in many point-group symmetries there are odd-parity crystal field terms which do not contribute to the energy matrix elements. These odd-parity crystal field terms may introduce enough parity mixture in one or both wave functions to make the transition occur. The intensities are very weak, however, being 10^2 to 10^7 times weaker than parity-allowed transitions. In the second case, electric dipole transitions occur between levels that are not purely electronic in nature, but are characterized by both electronic and vibrational wave functions. For this case, one level must have an appropriate number of quanta of the proper parity vibration in order for the transition to occur. When the electronic transition is accompanied by a minor rearrangement of the ligands, the strongest transitions involve that vibration of the complex which carries the ground state over into the excited state geometry and *vice versa*. These are called vibronic transitions and usually consist of progressions of bands having increasing separation from the locus of the pure electronic transition.

These two mechanisms for overcoming Laporte's rule can be distinguished in practice by the temperature dependence of absorption intensity. In the case where odd-parity crystal field terms mix in the high-energy odd-parity states, there is usually no temperature dependence, since a single state may result. In the case of vibronic transitions, the vibrational quantum usually has an appreciable value compared to kT at low temperature. Thus, the upward transition on the lower vibronic level can be damped out at low temperature, leaving that from the pure electronic ground state to the vibronic upper state. Examples of this effect are shown in Figs. 3 and 4 where the spectra of Cr^{3+} in beryl (emerald) and in K_2NaCrF_6, respectively, are given [26].

In emerald, where the site has low symmetry, the odd-parity crystal field terms mix high-lying odd-parity components into the wave functions, and the intensity does not change much with temperature. For the case of K_2NaCrF_6, however, there is a large change of intensity on the long-wavelength side with temperature, and vibronic transitions are important. In the latter compound, the Cr^{3+} ions are located on purely cubic sites where no odd-parity crystal field components exist.

Probably the best worked out example of a complicated crystal field spectrum is that of octahedral Cr^{3+} which has three $3d$ electrons: $3d^3$. We can note several principles governing the spectra of crystals by considering this example. Let us begin with the experimental analysis of the free-ion spectrum of Cr(IV), as the old atomic spectroscopy calls triply ionized chromium. Observe that in modern crystal field spectroscopy the notation Cr(III) is used. Some care is required to avoid confusion on this point when Cr(III) and Cr(IV) are used in the literature for the same ion.

The free-ion spectrum of Cr(IV) is known [27] and the results are summarized in Fig. 5. The observed levels are on the left, and the corresponding levels computed from crystal field theory with $Dq = 0$ are on the right. Obviously, the free ion can be considered as an ion with zero perturbation by the ligands. The theory used here is just that which has been described, namely, for an ion having electrostatic interaction, spin-orbit coupling, and crystal field. The limited agreement between the observed levels of Cr(IV) and the computed levels indicates the limitation of the basic theory for $3d$ electrons even if all were well with the crystal field part.

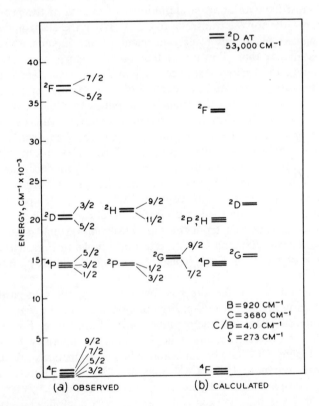

Fig. 5. Free ion spectrum of Cr^{3+} as calculated from crystal field theory with $Dq = 0$ and from experiment (from Wood et al. [26]).

Table I
Crystal Field Matrices for the Quartets of $3\,d^3$

$3\,d^3\;{}^4T_1({}^4P,\,{}^4F)$		
	$-2\,Dq-3\,B$	$6\,B$
	$6\,B$	$8\,Dq-12\,B$
${}^4A_2({}^4F)$	$-12\,Dq-15\,B$	
${}^4T_2({}^4F)$	$-2\,Dq-15\,B$	

Now in the crystal field of symmetry O_h, a level having $L=3$ will split into three components having symmetry species $A_2+T_1+T_2$. Thus, the $4\,F$ ground state is expected to split into three levels, and the energies are given by the matrix elements of Table I from the work of Tanabe and Sugano [4].

Thus, the separation ${}^4A_2\to{}^4T_2=10\,Dq$ gives the parameter Dq independent of B, and the value of B can be approximately determined from the separation of ${}^4T_2\to{}^4T_1\approx12\,B$. The splitting is plotted in Fig. 6 for this simple theory.

The value of the parameter C can be found approximately by locating one of the doublets at $9\,B+3\,C$ above the ground state. Using these para-

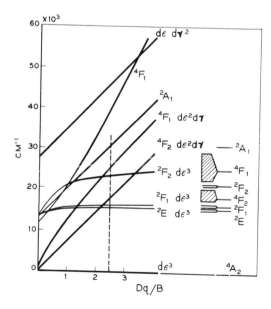

Fig. 6. Crystal field splitting of Cr^{3+} in octahedral coordination, without spin-orbit coupling. Levels changing rapidly with Dq give broad bands, while those changing slowly with Dq give sharp lines.

Table II
Crystal Field Parameters for Cr^{3+} in Various Host Lattices*

	Free ion	Yttrium gallium garnet	Yttrium aluminum garnet	Ruby	Emerald	K_2NaCrF_6	CrF_3	$CrCl_3$	$CrBr_3$
Dq (cm^{-1})	0	1650	1640	1630	1630	1610	1460	1370	1340
B (cm^{-1})	920	570	650	640	780	760	740	550	370
C (cm^{-1})	3680	3400	3250	3300	2960	3020	...	3400	3700
C/B	4.0	5.95	5.0	5.15	3.8	4.0	...	6.3	10.0
Cr–ligand distance (Å)	...	1.93	1.92	1.90	1.94	1.90	1.90	2.38	2.54

* From Wood et al. [26].

meters which may differ appreciably from the values in the free ion, a set of calculated energy levels can be made and compared with the experimental values. Three examples of octahedral Cr^{3+}, together with the fit using Liehr's theory [22], are shown in Figs. 7, 8, and 9.

It is quite evident that there is a variation of Dq, B, and C from one crystal to another, and many authors have made much of the interpretation of these variations. For Dq there is a spectrochemical series [8] which relates the expected crystal field magnitude to the ligand type. For simple anions, a list in order of increasing Dq value would be as follows:

$$I^- < Br^- < Cl^- < F^- < H_2O < \text{oxide}$$

A much more exhaustive treatment is given in the work of Jørgensen [8], but here we note that hydrated compounds show surprisingly large Dq values, testifying to the strength of the water dipole moment. A summary of values found for octahedral Cr^{3+} is given in Table II [26]. Another series

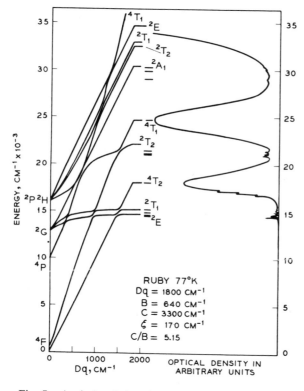

Fig. 7. Analysis of the absorption spectrum of Cr^{3+} in Al_2O_3 (ruby) from Wood *et al.* [26]. Calculated levels are at the left, the observed spectrum at the right.

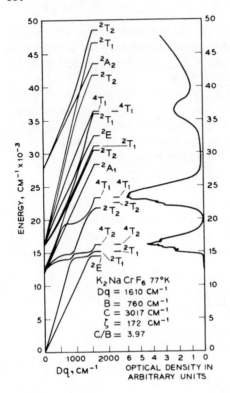

Fig. 8. Analysis of the absorption spectrum of K₂NaCrF₆ from Wood *et al.* [26]. Calculated levels are at the left, the observed spectrum at the right.

which depends on the value of B is called the nephelauxetic series [8], and it is measured by $\beta = B_{crystal}/B_{free\ ion}$.

Various ions can be ordered with regard to their ability to decrease B from the free-ion value, but this series is not as clear-cut as the spectrochemical series. In fact for the various oxides listed in Table II alone there is a large discrepancy in the degree of reduction of B. Deviations from these series are probably due to the effect of next nearest neighbors [28] and to the effect of varying degrees of covalency between central ion and ligand [8]. For example, in Table II a value of $Dq = 1460$ for CrF_3 is much lower than $Dq = 1610$ for K_2NaCrF_6, although both crystals contain CrF_6 polyhedra. The difference is that in CrF_3 next neighbors are Cr^{3+} ions, while in K_2NaCrF_6 the next neighbors are monovalent Na cations. The polarization of the fluorine ions is appreciable in the case of the higher charge. Another example [28] is that of Co^{2+} in MgO and in $CoWO_4$ where for MgO–Co $Dq = 920$, but for $CoWO_4$ $Dq = 690$. Both are oxides and there is in each an octahedral coordination around the Co^{2+} ion. In the case of $CoWO_4$ the highly charged W ion on the other side of the Co–O bound reduces the crystal field, all other factors being equal.

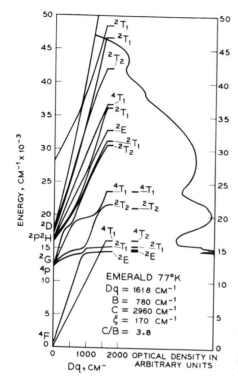

Fig. 9. Analysis of the absorption spectrum of Cr^{3+} in beryl (emerald) from Wood *et al.* [26]. Calculated levels are at the left, the observed spectrum at the right.

Another empirical rule for ions in solids is that, for an isoelectronic sequence, higher valence goes with higher values of Dq [6]. One such series is for $3d^3$ in V^{2+}, Cr^{3+}, and Mn^{4+} where typical values of Dq are given in Table III.

Another rule is that with a single ion higher valence goes with higher values of Dq, all other factors being equal. For example, octahedral cobalt might have typical values [29] of $Dq = 900$ for Co^{2+} but $Dq = 1600$ for

Table III
Values of Dq for an
Isoelectronic Sequence $3d^3$

Ion	Dq
V^{2+}	1250
Cr^{3+}	1600
Mn^{4+}	2200

Co^{3+}. Tetrahedral cobalt in a similar material has the values $Dq = 400$ for Co^{2+} and $Dq = 710$ for Co^{3+}. In practice these are qualitative rules since usually bond distances and next neighbors are different as well.

III. CRYSTAL FIELD SPLITTING FOR RARE EARTH IONS

Returning to the Hamiltonian for an ion in a crystalline field [equation (19.1)], we take up the case where $V_c \ll \sum \xi \mathbf{l} \cdot \mathbf{s}$ and the identity of the free-ion $^{2S+1}L_J$ manifolds are preserved. The reason often given for the weakness of V_c is that the screening effect of the $5s^2 5p^6$ electrons "outside" the $4f$ shell in which the transitions of interest arise reduces the effect of the ligands.

The first and immediate result of small values of V_c which has been experimentally known for a long time is that each of the $4f^n$ rare earths (which are the ones falling into this category) have practically the same spectrum superficially, no matter what the host, either in the solid or in solution. This means that once the free-ion spectrum is solved the broad features of the ion in a solid or liquid also will be understood. Only in the fine structure will there be important differences which we now wish to take up. A second result due to the weak influence of the crystalline field on the $4f$ electrons is that most of the rare earth spectra consist of sharp lines. A qualitative explanation of this fact involves the idea that if the energy does not vary much with crystal field, then the vibrational fluctuations in the field do not "smear out" the energy levels, and sharp lines result.

Again the reconciliation between experiment and theory derives from the assignment of solutions to equation (19.5)

$$|\langle \psi_i | V_c | \psi_j \rangle - E_{ii}| = 0$$

and the calculation of the proper matrix elements. In the case at hand, it is convenient to use $\psi_i = \psi_{LSJ}^M$ which are free-ion wave functions in a Russell–Saunders coupling scheme. This is a scheme characterized by coupling of $L = \Sigma l$ and $S = \Sigma s$ to give J and M. Again, the crystal field is conveniently expressed in terms of spherical harmonics, but many authors use the operator equivalent method [14] for matrix elements in which

$$V_c = \sum_{n, m} B_n^m \langle r^n \rangle \theta_n O_n^m \qquad (19.11)$$

where

$$\langle r^n \rangle \equiv \int [R(r)]^2 r^n r^2 \, dr$$

and where $B_n^m \langle r^n \rangle = A_n^m$ will be an experimentally determined set of parameters; $\theta_n = \alpha, \beta, \gamma$ depends on whether $n = 2, 4, 6$; and the O_n^m's are operator equivalents for the angular part of the spherical harmonics in which the electrostatic crystalline field is expanded. The operator equivalents are simply algebraic combinations of angular momentum operators whose matrix elements are the same as those for the spherical harmonics but whose values may be more easily calculated.

Others use the Wigner–Eckhart theorem [15] and calculate matrix elements from the Wigner coefficients which now are tabulated for the whole range required [30]. For those who choose the operator equivalent method, the tables of Hutchings [31] give all the matrix elements required for the O_n^m's and the secular equation can almost be completed with their aid. However, the θ_n's depend on the particular state of the particular ion for which the splitting is being computed. They depend on L, S, and J and can be computed for pure L, S, J states by reduction to one-electron wave functions for $4f^n$ states, and most of these have been tabulated. Unfortunately, however, the $4f^n$ rare earths are not characterized by pure L, S coupling, and the admixture of several L, S states in intermediate coupling must be explained. The changes in α, β, and γ due to intermediate coupling can be calculated easily since the new wave functions are linear combinations of the old pure L, S, J wave functions and the reduced matrix elements for these are known. However, it is then usually necessary to have an intermediate coupling calculation based on the positions of the centers of gravity of the various free-ion manifolds for the particular crystal, if good values for the so-called operator equivalent constants α, β, and γ are to be available. It is unfortunate that these "constants" may vary somewhat from one crystal to another, and it is useful to have a machine computation for the intermediate coupling as well as for the crystal field problem. In both problems, hand solutions are practical only in very special cases.

The number of crystal field parameters A_n^m is limited by the symmetry of the site of the ion in question and the l and J values involved, and one would expect that the parameters would be greatly overdetermined by the many energy differences observed in the typical rare earth spectrum. In fact it is not usually so simple. For example, in a machine solution with least-squares fitting, more than twenty minima were obtained with appreciably different A_n^m's. Some outside criteria must be brought in to reduce this number of sets of possible crystal field parameters. The many useful techniques available are illustrated by a commonly used stepwise procedure which requires data on many are earths in the same host lattice. The first idea involves the fact that low J value manifolds involve only low-order A_n^m's. Let us take an example from the case of $CaWO_4$–rare earths.

The rare earth site in $CaWO_4$ has S_4 point symmetry with eight oxygen ligands placed at the corners of a distorted cube. The character table [24] for S_4 and the selection rules for magnetic dipole radiation derived from data in the work of Koster *et al.* [24] are given in Tables IV and V. Table VI gives the number and species of levels for various values of J in S_4. Note that the single group has two complex conjugate states Γ_3 and Γ_4 which must be degenerate in an electrostatic field, but not if a magnetic field is applied. Thus, there is a linear Zeeman effect for integral J in $CaWO_4$ in contrast with many crystals for which the rare earth point group has no

Table IV
Character Table for S_4 Point Group

S_4	E	S_4^{-1}	C_2	S_4	
Γ_1	1	1	1	1	
Γ_2	1	-1	1	-1	z
$\Gamma_{3,4}$	1,1	$\pm i$	-1	$\pm i$	$\pm i(x-iy)$
$\Gamma_{5,6}$	1,1	$\begin{cases} \omega \\ -\omega^3 \end{cases}$	$\pm i$	$\begin{cases} -\omega^3 \\ \omega \end{cases}$	
$\Gamma_{7,8}$	1,1	$\begin{cases} -\omega \\ \omega^3 \end{cases}$	$\pm i$	$\begin{cases} \omega^3 \\ -\omega \end{cases}$	

Table V

S_4	Γ_1	Γ_2	$\Gamma_{3,4}$	$\Gamma_{5,6}$	$\Gamma_{7,8}$
Γ_1	σ		π		
Γ_2		σ	π		
$\Gamma_{3,4}$	π	π	σ		
$\Gamma_{5,6}$				π	$\pi\sigma$
$\Gamma_{7,8}$				$\pi\sigma$	π

Table VI
Number and Species of Levels for Various J Values in S_4

			$J =$		
0	1/2	1	3/2	2	4
Γ_1	$\Gamma_{5,6}$	$\Gamma_1+\Gamma_{3,4}$	$\Gamma_{5,6}+\Gamma_{7,8}$	$\Gamma_1+2\Gamma_2+\Gamma_{3,4}$	$3\Gamma_1+2\Gamma_2+2\Gamma_{3,4}$

complex conjugate states. Of course, there always is a linear Zeeman effect for an odd number of electrons and for these ions the crystal field states will belong exclusively to the representations of the double group. This is Bethe's equivalent of Kramers' theorem. Now for S_4 the only nonzero crystal field matrix elements [17] are those multiplied by A_2^0, A_4^0, A_6^0, A_4^4, and A_6^4. For manifolds with $J = 0, \frac{1}{2}$ there is no splitting, and for $J = 1, \frac{3}{2}$ only A_2^0 enters into the splitting. Therefore, by seeking the levels characterized by $J = 1$ or $J = \frac{3}{2}$, one can evaluate A_2^0 independently. It turns out that of all the rare earths europium ($^5D_1$ and $^7F_1$) and praseodymium ($^3P_1$) are most important for A_2^0, although neodymium, terbium, and erbium are also important. Then a first step in the estimation of the crystal field parameters is the following.

Consider the experimentally determined energy level diagram for europium shown in Fig. 10. Clearly the $\Gamma_{3,4}$ lies lower for both $^7F_1$ and $^5D_1$ because of the selection rules observed (see Table V). The matrix elements for $J = 1$ will be [31]

$$\langle M = \pm 1 \,|\alpha A_2^0\, O_2^0|\, M = \pm 1 \rangle = \alpha A_2^0 \tag{19.12}$$

$$\langle 0 \,|\alpha A_2^0\, O_2^0|\, 0 \rangle = -2\alpha A_2^0 \tag{19.13}$$

and

$$\begin{vmatrix} \alpha A_2^0 - \varepsilon & 0 & 0 \\ 0 & -2\alpha A_2^0 - \varepsilon & 0 \\ 0 & 0 & \alpha A_2^0 - \alpha \end{vmatrix} = 0 \tag{19.14}$$

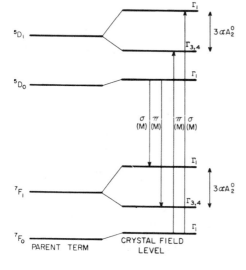

Fig. 10. Energy level diagram for Eu^{3+} in CaWO$_4$ with observed transitions indicated. For $\sigma(M)$ transitions, the oscillator strength is greater for the $E \perp S_4$ axis and weaker for the axial spectrum. For $\pi(M)$, the axial spectrum and that for the $E \| S_4$ axis are both stronger than for σ.

whose solutions are

$$E_1 = \alpha A_2{}^0 \text{ twice } \Gamma_{3,4}$$

$$E_2 = -2\alpha A_2{}^0 \text{ once } \Gamma_1 \tag{19.15}$$

and the energy level diagram would be just that which is observed. From this, the magnitude and sign of $\alpha A_2{}^0$ is determined. It might however, be different for other ions, so we check with praseodymium [32] and find it agrees both in sign and approximate magnitude with $A_2{}^0 = +250 \text{ cm}^{-1}$. It is a general principle that for a given host material the values of the crystal field parameters tend to remain constant or to vary slowly and monotonically with atomic number in the $4f^n$ rare earths. For example, some values for rare earth chlorides are given in Table VII.

Clearly it is desirable to have measurements of absorption and fluorescence as a function of temperature in polarized light supplemented by the Zeeman effect for several rare earths in the same lattice before attempting to determine the crystal field parameters.

So far we have indicated that a qualitative use may be made of the Zeeman effect in determining the degeneracy of various levels in a given symmetry. It is also useful to have values for the g factors of excited states as an additional help in determining the higher-order crystal field parameters, and there are two approaches to the computation of g factors.

In either approach, it is assumed that the crystal field parameters are known, but, in the first, the crystal field eigenvectors are constructed as linear combinations of the free-ion wave functions. Most computer programs have eigenvector subroutines as a bonus, and the eigenvectors are easily

Table VII
Crystal Field Parameters for Rare Earth Chlorides

Ion	$B_2{}^0$	$B_4{}^0$	$B_6{}^0$	$B_6{}^6$	Reference
Sm	81	−23	−44	426	*
Eu	89	−38	−51	495	†
Tb	92	−40	−·30	290	‡
Dy	91	−39	−23	258	*
Ho	122	−45	−28	280	§
Er	93	−35	−25	235	**

* J. D. Axe and G. H. Dieke, *J. Chem. Phys.* **37**:2364 (1962).

† L. G. DeShazer and G. H. Dieke, *J. Chem. Phys.* **38**:2190 (1963).

‡ K. S. Thomas, S. Singh, and G. H. Dieke, *J. Chem. Phys.* **38**:2180 (1963).

§ G. H. Dieke and B. Pandey, *J. Chem. Phys.* **41**:1952 (1964).

** F. Varsanyi and G. H. Dieke, *J. Chem. Phys.* **36**:2951 (1962).

found. This is an especially simple procedure in the case of integral J manifolds in a cubic field because the makeup of the wave functions in almost every case is independent of the magnitude of the crystal field [33, 34]. In that case, if the potential energy term due to the externally applied magnetic field is

$$\mathcal{3C}_m = (\mathbf{L}+2\,\mathbf{S})\,\frac{eH_0}{2\,mc} \tag{19.16}$$

then since

$$\psi(\gamma SJ\Gamma) = \sum a_i^{\Gamma}\,\psi_J^{M}(\gamma) \tag{19.17}$$

$$\langle \gamma SJ\Gamma \mid \mathcal{3C}_m \mid \gamma SJ\Gamma \rangle = \frac{eH_0}{2\,mc} \sum a_i^{\Gamma}\,\langle \gamma SJM \mid L+2\,S \mid \gamma SJM \rangle \tag{19.18}$$

and

$$\langle \gamma SJM \mid L+2\,S \mid \gamma SJM \rangle = M\left[1 + \frac{J(J+1)+S(S+1)-L(L+1)}{2\,J(J+1)}\right] \tag{19.19}$$

An example of the use of this method can be found in the case of Sm^{2+} in SrF_2 [35]. There one finds three $^3\Gamma_4$ wave functions for $J = 1$, and only one term occurs in each sum [33]. The level is $^7F_1$, $J = 1$, $L = 3$, $S = 3$, and $E = (eH_0/2mc) \times \frac{3}{2}M = \beta_0 H_0 \frac{3}{2} M$, giving three levels

$$^3\Gamma_4(^7F_1) \sum a_i \psi_J^{M} = \begin{cases} \Psi_1^{1} & E = \frac{3}{2}\beta_0 H \\[4pt] \psi_1^{0} & E = 0 \\[4pt] \psi_1^{-1} & E = -\frac{3}{2}\beta_0 H \end{cases} \tag{19.20}$$

and $g = \frac{3}{2}$.

The observed [35] value of g is 1.4 and the fluorescence line is at 14,353 cm^{-1}. In the case of $(^7F_2)\,^3\Gamma_5$, the wave functions are

$$^3\Gamma_5(^7F_1) \sum a_i \psi_J^{M} = \begin{cases} \psi_2^{1} & E = \frac{3}{2}\beta_0 H \\[6pt] \dfrac{1}{\sqrt{2}}\,(\psi_2^{2}-\psi_2^{-2}) & E = 0 \\[6pt] \psi_2^{-1} & E = -\frac{3}{2}\beta_0 H \end{cases} \tag{19.21}$$

and again $g = \frac{3}{2}$.

The other method of computing the Zeeman effect requires solving the crystal field and magnetic perturbation problem simultaneously. This is important in cases of low symmetry, since it is likely that no degeneracy remains for integral J manifolds, and only a nonlinear Zeeman effect can be observed. Thus the secular equation becomes

$$|\langle \psi_i | V_c + \mathcal{H} | \psi_j \rangle - E_{ij}| = 0 \tag{19.22}$$

where

$$\langle \psi_i | V_c + \mathcal{H} | \psi_j \rangle = \langle \psi_i | V_c | \psi_j \rangle + \langle \psi_i | \mathcal{H} | \psi_j \rangle \tag{19.23}$$

The Zeeman terms can then be shown (a useful summary is given by Fick and Joos [16]) to have the following forms:

For the component of H along z,

$$\langle LSJM | \mathcal{H}_z | LSJM \rangle = g_{JJ} \beta_0 HM \cos \zeta \tag{19.24}$$

For the component of H along x and y,

$$\langle LSJM | \mathcal{H}_x | LSJM \pm 1 \rangle = \tfrac{1}{2} g_{JJ} \beta_0 H \sqrt{(J \pm M)(J \mp M + 1)} \cos \xi \tag{19.25}$$

$$\langle LSJM | \mathcal{H}_y | LSJM \pm 1 \rangle = \tfrac{1}{2} g_{JJ} \beta_0 H \sqrt{(J \pm M)(J \mp M + 1)} \cos \eta \tag{19.26}$$

where

$$g_{JJ} = 1 + \frac{J(J+1) + S(S+1) - L(L+1)}{2 J(J+1)} \tag{19.27}$$

and ξ, η, and ζ are angles between the applied magnetic field H_0 and the crystal field axis of quantization with z the highest symmetry axis. This matrix has dimensions $(2J+1) \times (2J+1)$ and can be diagonalized by machine for $J < 8$ with little effort.

This latter method, then, is useful in calculating the nonlinear Zeeman effect which is often observed, and it can be used in two interesting cases where very high magnetic fields are encountered. In the first case the width of the lines observed is great enough to make it impossible to resolve the Zeeman pattern with a DC magnet. For example, with rare earths in the 1% concentration range, lines 10 cm^{-1} wide are common; for $g = 2$, $\Delta v = 9 \text{ cm}^{-1}$ for 30,000 G and the Zeeman pattern would not be well-resolved. Thus the use of high field magnets is required. For either pulsed fields (200,000 G) or superconducting magnets ($< 100,000$ G), the nonlinear Zeeman effect can easily be produced in the process of splitting wide lines.

The second case involves the occurrence of rare earth ions in magnetic crystals where exchange fields of the order of 10^5 G or more may be present. In either case, this theory permits an approach to the explanation of the nonlinear Zeeman effect.

IV. REFERENCES

1. J. Becquerel, *Z. Physik* **58**:205 (1929).
2. H. Bethe, *Ann. Physik* 3(5):135 (1929).
3. H. A. Kramers, *Proc. Acad. Sci. Amsterdam* **33**:953 (1930).
3a. J. H. Van Vleck and A. Sherman, *Rev. Mod. Phys.* **7**:167 (1935); J. H. Van Vleck, *Theory of Magnetic and Electric Susceptibilities*, Oxford University Press (Oxford), 1932.
4. Y. Tanabe and S. Sugano, *J. Phys. Soc. Japan* **9**:733 (1954); *J. Phys. Soc. Japan* **9**:766 (1954).
5. L. Orgel, *Chemistry of Transition Metal Complexes*, Methuen (London), 1960.
6. J. S. Griffith, *Theory of Transition Metal Ions*, Cambridge University Press (London), 1961.
7. C. J. Ballhausen, *Introduction to Ligand Field Theory*, McGraw-Hill (New York), 1962.
8. C. K. Jørgensen, *Absorption Spectra and Chemical Bonding in Complexes*, Pergamon Press (London), 1962.
9. F. A. Cotton, *Chemical Application of Group Theory*, Interscience (New York), 1963.
10. B. N. Figgis, *Introduction to Ligand Fields*, Interscience (New York), 1966.
11. F. Hund, *Z. Physik* **33**:345 (1925).
12. A. Landé, *Z. Physik* **15**:189 (1923); **19**:112 (1923).
13. J. C. Slater, *Quantum Theory of Atomic Structure*, McGraw-Hill (New York), 1960.
14. K. W. H. Stevens, *Proc. Phys. Soc. (London)* A**65**:209 (1952).
15. E. P. Wigner, *Group Theory and its Applications to the Quantum Mechanics of Atomic Spectra*, Academic Press (New York), 1959.
16. E. Fick and G. Joos, in: S. Flügge (ed.), *Handbuch der Physik, Vol. 28, Spectroscopy II*, Springer (Berlin), 1957, p. 246 *et seq.*
17. J. L. Prather, "Atomic Energy Levels in Crystals," *Natl. Bur. Std. (U.S.) Monograph* 19 (1961). (Available from Superintendent of Documents, U.S. Government Printing Office, Washington 25, D.C.)
18. J. C. Eisenstein, *J. Chem. Phys.* **34**:1628 (1961); Errata, *J. Chem. Phys.* **35**:2246 (1961).
19. Y. Tanabe and H. Kamimura, *J. Phys. Soc. Japan* **13**:394 (1958).
20. J. S. Griffith, *Trans. Faraday Soc.* **54**:1109 (1958); **56**:193 (1960).
21. A. D. Liehr and C. J. Ballhausen, *Ann. Phys. (N. Y.)* **6**:134 (1959).
22. A. D. Liehr, *J. Phys. Chem.* **67**:1314 (1963).
23. R. N. Enwema, *J. Chem. Phys.* **42**:892 (1965).
24. G. F. Koster, J. O. Dimmock, R. G. Wheeler, and H. Statz, *Properties of the 32 Point Groups*, MIT Press (Cambridge, Massachusetts), 1963.
25. O. Laporte, *Z. Physik* **23**:135 (1924).
26. D. L. Wood, J. Ferguson, K. Knox, and J. F. Dillon, Jr., *J. Chem. Phys.* **39**:890 (1963).
27. C. E. Moore, *Natl. Bur. Std. (U.S.) Circ.* 467, Vol. II, 16 (1950).
28. J. Ferguson, K. Knox, and D. L. Wood, *J. Chem. Phys.* **35**:2236 (1961); Erratum, *J. Chem. Phys.* **37**:193 (1962).
29. D. L. Wood and J. P. Remeika, *J. Chem. Phys.* **46**:3595 (1967).
30. M. Rotenberg, R. Bivins, N. Metropolis, and J. K. Wooten, Jr., *The 3-j and 6-j Symbols*, The Technology Press (MIT, Cambridge, Massachusetts), 1959.
31. M. T. Hutchings, *Solid State Phys.* **16**:227 (1964).
32. D. M. Dodd and D. L. Wood, *Proc. Intern. Symp. on Mol. Structure and Spectroscopy*, Science Council of Japan, Tokyo, 1962, p. A406.
33. R. A. Satten and J. S. Margolis, *J. Chem. Phys.* **32**:573 (1960).
34. R. Pappalardo, *J. Chem. Phys.* **34**:1380 (1961).
35. D. L. Wood and W. Kaiser, *Phys. Rev.* **126**:2079 (1962).

CHAPTER 20

Coupling of Modified Modes to Electronic Transitions at Defects

C. W. McCombie

*University of Reading
Reading, England*

I. INTRODUCTION

In this chapter, we shall discuss some aspects of the effect of lattice vibrations on the optical absorption associated with an electronic transition at an imperfection; other aspects will be treated by Silsbee (Chapter 21) and by Markham (Chapter 22). Our main concern will be to see how to take account of the fact that the forms of the normal modes of vibration of the lattice will be modified by the presence of the imperfection; it is the interaction of these modified modes with the electronic transition which has to be considered. In Chapter 15, Genzel discusses the actual form of such modified modes. Our approach will be rather different. We shall show that in calculations on the optical absorption problem it is convenient to take account of the modification of the modes by an indirect method in which explicit consideration of the forms of the modified modes is avoided.

Perhaps the main interest of the modified modes problem in optical absorption arises in connection with complicated absorption bands, the basic theory of which is presented by Silsbee in Chapter 21. Here in this chapter rather different situations are treated, in which we get broad, smooth absorption bands. These situations are of considerable intrinsic interest and from our point of view have the merit that the basic theory can be given rather briefly. As a result, we can proceed almost immediately to the lattice vibration problem. The treatment of the modified modes which is appropriate for the smooth band problems is, moreover, just what is needed for the problems treated by Silsbee, and the connection between the results given here and the quantities which appear in his theory will be indicated later.

The importance of taking account of the modification of the normal modes by the imperfection in calculating the temperature dependence of an associated broad absorption band was discussed for the F-center by McCombie, Matthew, and Murray [1]. The method used here to treat the modification of the modes is, however, more akin to that used by Brout and Visscher [2] in discussing how such modifications might be investigated by neutron scattering. This is a response function or Green's function treatment. In the approach presented in this chapter the Green's function idea will be embodied in a preliminary physical argument rather than used as a mathematical device. In this way, a considerable amount of elaborate formalism is avoided. For more standard applications of the Green's function procedure, reference may be made to review articles by Maradudin [3] and by Elliott [4].

II. TRANSITIONS BETWEEN NONDEGENERATE LEVELS

Consider first the electronic transition between two nondegenerate electronic states of a substitutional impurity in a cubic crystal. We shall suppose that the width of the absorption band can be regarded as arising from fluctuations in the positions of neighboring ions. Such fluctuations cause variations in the difference in energy between the two electronic states and these variations lead in turn to the variations in frequency of the absorbed photon which account for the width of the absorption band. In other words, we suppose that the Franck–Condon principle applies. Under these circumstances, one expects to get smooth absorption bands of essentially Gaussian shape. Thus, the shape of the absorption band is specified by giving its mean square width. Accordingly we shall discuss how to calculate this mean square width as a function of temperature.

To keep the discussion of the first example as simple as possible, two further assumptions will be made. First, we shall suppose that if the energy difference between the two states is expanded as a function of the displacements of neighboring atoms, quadratic and higher powers of the displacements may be neglected. Second, we shall assume that the energy difference between the two states depends only on the mean radial outward displacement ΔR of the nearest neighbor atoms from their ground state equilibrium positions. We shall see later how to proceed if this second assumption is dropped. It is also true that, in treating the smooth band problems, the linearity assumption can be avoided with only a small increase in complication. However, this shall not be considered here. In the type of problem discussed by Silsbee, nonlinearities introduce essentially new features.

The mean radial displacement ΔR, which has a definite value for any given set of displacements $\mathbf{a}_s(s = 1, \ldots, n)$ of the n nearest neighbor ions, may be defined more explicitly as follows. If $\mathbf{e}_s$ is a unit vector directed radially outward from the equilibrium position of the sth nearest neighbor ion, then

$$\Delta R = \frac{1}{n} \sum_{s=1}^{n} \mathbf{a}_s \cdot \mathbf{e}_s \tag{20.1}$$

The assumption that the energy difference depends only on ΔR would be justified if both electronic wave functions were spherically symmetrical and did not extend appreciably beyond the nearest neighbor ions. Our first assumption requires that the dependence on ΔR be linear, and if the energies of the upper and the ground electronic states are denoted by E_u and E_g, respectively, then

$$E_u - E_g = A - B\Delta R \tag{20.2}$$

where A and B are constants, which are assumed to be known. They might be determined experimentally by observing the frequency of the absorption band peak and its shift with pressure.

It is now convenient to introduce the normal coordinates q'_r associated with the normal vibrations of the lattice. The prime reminds us that we are not, in general, dealing with perfect lattice normal coordinates. The normal modes considered are those of the crystal with the imperfection present and in its ground electronic state. The normal vibrations, and thus the normal coordinates, will be slightly different when the imperfection is in its excited electronic state: the excited state normal coordinates will not appear explicitly in the present discussion. We shall take it that the scale for the normal coordinate has been so chosen that the kinetic energy of the rth mode is $\frac{1}{2} \dot{q}'^2_r$. Suppose that when all the normal coordinates except q'_r are zero, the mean outward displacement of the nearest neighbor ions is $\beta'_r q'_r$. Then,

$$\Delta R = \sum_r \beta'_r q'_r \tag{20.3}$$

and thus

$$E_u - E_g = A - B \sum_r \beta'_r q'_r \tag{20.4}$$

We can now calculate the mean value of $E_u - E_g$, which will give the mean frequency of the absorbed photon (in energy units). The mean will, of course, be calculated by averaging over the positions of neighboring atoms or, what comes to the same thing, averaging over the q'_r. Thus,

$$\overline{E_u - E_g} = A - B \sum_r \beta'_r \overline{q'_r}$$
$$= A \tag{20.5}$$

The difference between the frequency of the photon absorbed in a particular absorption process and the mean frequency of the photons absorbed will be given (in energy units) by

$$\Delta E = E_u - E_g - \overline{E_u - E_g} \tag{20.6}$$

$$= -B \sum_r \beta'_r q'_r$$

The mean square width of the absorption band (in units of energy squared) will therefore be given by

$$\overline{\Delta E^2} = B^2 \sum_{r,s} \beta'_r \beta'_s \overline{q'_r q'_s} \tag{20.7}$$

Now, if $r \neq s$, $\overline{q'_r q'_s}$ is zero and it is a simple statistical-mechanical result that at temperature T

$$\overline{q'^2_r} = \frac{\hbar}{2\omega_r} \coth \tfrac{1}{2} \frac{\hbar\omega_r}{kT} \tag{20.8}$$

$$= \phi(\omega_r, T)$$

where the notation $\phi(\omega, T)$ is introduced for brevity. Thus,

$$\overline{\Delta E^2} = B^2 \sum_r \beta'^2_r \phi(\omega_r, T) \tag{20.9}$$

It will be important to rewrite this expression in a different form by making use of the fact that the frequencies of the modes constitute a quasi-continuum (for simplicity we shall exclude the possibility of there being localized modes). We define now a function $C(\omega)$ such that

$$C(\omega)\, d\omega = \sum_{\omega < \omega_r < \omega + d\omega} \beta'^2_r \tag{20.10}$$

and we rewrite the expression for $\overline{\Delta E^2}$ in terms of $C(\omega)$ so that

$$\overline{\Delta E^2} = B^2 \int_0^\infty C(\omega)\, \phi(\omega, T)\, d\omega \tag{20.11}$$

It turns out that not only the present problem, but also the more elaborate problems of absorption band structure to be treated by Silsbee (Chapter 21), reduce to the determination of $C(\omega)$ and certain related functions. In fact $B^2 C(\omega)$ is essentially the function $\alpha(\omega)^2 \, g(\hbar\omega)$ of Silsbee's equation (21.14). Our main task is therefore to show how $C(\omega)$ may be determined, account being taken of the fact that the presence of the imperfection will modify the modes of vibration.

Let us first note that if the modifications of the modes are neglected and the eigenfrequencies and eigenvectors for a large sample of perfect lattice modes are available, then the determination of $C(\omega)$—denoted in this case by $C_{perf}(\omega)$—is straightforward. A knowledge of the eigenvector for a mode allows one to compute the displacement of each atom in the crystal. It is therefore easy for a computer to determine the displacements of the nearest neighbor atoms in the rth mode and then to evaluate β_r. It can then evaluate the sum of the β_r^2's for all modes in each of a suitably chosen set of frequency intervals. This will lead (after suitable smoothing) to an estimate of the function $C_{perf}(\omega)$. The choice of the suitable set of frequency intervals will depend on how dense a sample of the normal modes is available. The denser the sample the smaller the intervals, and thus the greater the detail obtained.

We calculate $C_{imp}(\omega)$—the $C(\omega)$ obtained when we take account of the modifications of the modes—by relating $C(\omega)$ to a response function and making use of a simple procedure to be described later for getting from the perfect lattice response function to that for the imperfect lattice. The appropriate response function is defined as follows. We apply an equal radial force $(F/n)e^{i\omega t}$ to each of the n nearest neighbors and denote the resulting radial displacement after decay of transients by $\Delta R e^{i\omega t}$. The resulting displacements of the neighbors will be radial and equal in magnitude because of the cubic symmetry, provided there is only one symmetry-adapted displacement of the neighbors which belongs to the identity representation of the cubic group, and we may assume this to be the case. The required response function is then defined by

$$\Delta R = [P(\omega) - iQ(\omega)] F \qquad (20.12)$$

In order to express this response function in terms of $C(\omega)$, we first relate it to the β_r' defined in equation (20.3). The generalized force on the rth mode will be $\beta_r' F e^{i\omega t}$ and the resulting response of that mode after transients have died away will be (transients being dealt with by introducing a damping term which will subsequently be allowed to tend to zero)

$$\ddot{q}_r' + \kappa \dot{q}_r' + \omega_r^2 q_r' = \beta_r' F e^{i\omega t} \qquad (20.13)$$

so that

$$q_r' = \frac{\beta_r'}{\omega_r^2 - \omega^2 + i\kappa\omega} F e^{i\omega t} \qquad (20.14)$$

The resulting displacement ΔR will be given by

$$\Delta R e^{i\omega t} = \sum \beta_r' q_r'$$

$$= \sum_r \frac{\beta_r'^2}{(\omega_r^2 - \omega^2) + i\kappa\omega} F e^{i\omega t} \qquad (20.15)$$

Thus,

$$P(\omega) - iQ(\omega) = \sum_r \frac{\beta_r'^2}{(\omega_r^2 - \omega^2) + i\kappa\omega} \qquad (20.16)$$

and thus

$$Q(\omega) = \sum_r \frac{\kappa\omega\beta_r'^2}{(\omega_r^2 - \omega^2)^2 + \kappa^2\omega^2}$$

$$= \int_0^\infty \frac{C(\omega')\,\kappa\omega\,d\omega'}{(\omega'^2 - \omega^2)^2 + \kappa^2\omega^2} \qquad (20.17)$$

The function of ω' multiplying $C(\omega')$ in the integrand becomes very sharply resonant at ω as κ tends to zero, and its integral has the value $\pi/2\omega$. Thus, letting κ tend to zero, we obtain

$$Q(\omega) = \frac{\pi}{2\omega} C(\omega) \qquad (20.18)$$

This relation applies, of course, both to the perfect and imperfect lattices, so that

$$Q_{\text{perf}}(\omega) = \frac{\pi}{2\omega} C_{\text{perf}}(\omega) \qquad (20.19)$$

and

$$Q_{\text{imp}}(\omega) = \frac{\pi}{2\omega} C_{\text{imp}}(\omega) \qquad (20.20)$$

As already discussed, there is no difficulty in determining $C_{\text{perf}}(\omega)$ if there is sufficient information about the normal modes of the perfect lattice available. Equation (20.19) then leads to $Q_{\text{perf}}(\omega)$, and, if we can get from $Q_{\text{perf}}(\omega)$ to $Q_{\text{imp}}(\omega)$, equation (20.20) will lead us to $C_{\text{imp}}(\omega)$. This is what we require.

One gets from $Q_{\text{perf}}(\omega)$ to $Q_{\text{imp}}(\omega)$ by regarding the crystal with the defect as a perfect crystal with extra springs and masses attached to it; these attached springs and masses correspond to the changes in force constants and masses of atoms in the vicinity of the defect. The response of this system to an appropriate applied force determines $Q_{\text{imp}}(\omega)$. From explicit consideration of the force exerted on the perfect lattice by the attached springs and masses, one can obtain the response to the applied force (and thus Q_{imp}) in terms of Q_{perf}.

In the present case, we regard the imperfect lattice as a perfect lattice with a set of equal springs with spring constant $\Delta k/n$, attached between the central atom of the crystal (chosen to be the impurity atom) and its neighbors. Here we are, of course, making an assumption, namely, that the only effect of the impurity on the lattice dynamics is to change the restoring forces on nearest neighbor ions. An extra mass corresponding to the difference between the mass of the impurity and that of the atom which it replaces can also be regarded as attached to the central atom. However, this will not concern us since we can suppose the modes chosen to be either all even or all odd about the imperfection, and only the even modes will contribute to ΔR. Since the central atom will not move in these even modes, its changed mass has no effect on them.

As already stated, we have to consider explicitly the forces exerted on the perfect lattice by the attached masses or springs. In the present case the radial force exerted by the springs will be $-(\Delta k/n)\Delta R e^{i\omega t}$, where $\Delta R e^{i\omega t}$ is the radial displacement. This force has to be added to the applied force $(F/n)e^{i\omega t}$. The radial displacement $\Delta R e^{i\omega t}$ and the applied force $(F/n)e^{i\omega t}$ for the imperfect lattice are related therefore by

$$\Delta R = [P_{\text{perf}}(\omega) - iQ_{\text{perf}}(\omega)] \ [F - \Delta k \Delta R] \qquad (20.21)$$

which gives

$$\Delta R = \frac{P_{\text{perf}}(\omega) - iQ_{\text{perf}}(\omega)}{[1 + \Delta k P_{\text{perf}}(\omega)] - i\Delta k Q_{\text{perf}}(\omega)} \ F \qquad (20.22)$$

Comparison of this with

$$\Delta R = [P_{\text{imp}}(\omega) - iQ_{\text{imp}}(\omega)] \ F$$

gives

$$Q_{\text{imp}}(\omega) = \frac{Q_{\text{perf}}(\omega)}{[1 + \Delta k P_{\text{perf}}(\omega)]^2 + \Delta k^2 Q_{\text{perf}}(\omega)^2} \qquad (20.23)$$

We note that the right-hand side of equation (20.23) contains $P_{\text{perf}}(\omega)$ as well as $Q_{\text{perf}}(\omega)$, but $P_{\text{perf}}(\omega)$ can be obtained from $Q_{\text{perf}}(\omega)$ by using the Kramers–Kronig relations. Therefore, there is no difficulty in getting from $Q_{\text{perf}}(\omega)$ to $Q_{\text{imp}}(\omega)$, and so from $C_{\text{perf}}(\omega)$ to $C_{\text{imp}}(\omega)$. This means that, if we know the coupling constant B and the eigenvectors and eigenfrequencies for a sufficiently large sample of the modes of the perfect lattice, we can calculate the mean square width of the absorption band as a function of temperature, taking account of the modification of the modes.

We shall now consider briefly how to proceed after removing some of the simplifying assumptions of the model just considered. Suppose first that we still consider only interactions with nearest neighbor displacements, but drop the requirement that the interaction will be only with ΔR. The displacements of the nearest neighbors may be represented as a superposition of symmetry-adapted displacements, one of which will be the "spherically symmetric" displacement already considered. Assume that we do not get two or more independent symmetry-adapted displacements belonging to a given row of a given irreducible representation. If we number the symmetry-adapted displacements by γ, the magnitude of the γth can be denoted by D_γ and

$$E_u - E_g = A - \sum_\gamma B_\gamma D_\gamma \qquad (20.24)$$

where A and B_γ are constants. The normal coordinates of the system may also be chosen to be symmetry-adapted. Only those with the same symmetry as the γth symmetry-adapted displacement of the nearest neighbors will contribute to D_γ, and conversely this will be the only D_γ to which these modes contribute. It follows that the contributions to the mean square width from the various D_γ's can be calculated quite independently and simply added together. In order to calculate this contribution for a given γ, taking account of the modification of the modes, one will have to introduce a response function in which the sinusoidal force system applied to the nearest neighbors (and thus the resulting displacement of the neighbors) is symmetry-adapted with symmetry γ. The calculation goes through for each in exactly the same way as in the case already considered (which was for the γ corresponding to the identity representation).

As a second example of a more complicated case, let us suppose that the interaction is with only "spherically symmetrical" displacements, but that we have to consider this displacement for second nearest neighbors as well as for nearest neighbors. More precisely, suppose that $E_u - E_g$ depends on both these displacements. We may also, if desired, suppose that the presence of the imperfection changes the restoring forces on second nearest neighbors as well as on first nearest neighbors: we may also introduce extra restoring forces depending on the relative displacement of these two shells. If the radial displacements of the first and second neighbors are denoted by ΔR_1 and ΔR_2, respectively, then

$$E_u - E_g = A - B_1 \Delta R_1 - B_2 \Delta R_2 \qquad (20.25)$$

If $\beta'_{1,r} q'_r$ and $\beta'_{2,r} q'_r$ are the contributions of the rth normal mode to ΔR_1 and ΔR_2, respectively, then

$$\overline{\Delta E}^2 = B_1{}^2 \sum_r \beta'_{1,r}{}^2 \phi(\omega_r, T) + B_2{}^2 \sum_r \beta'_{2,r}{}^2 \phi(\omega_r, T) + 2 B_1 B_2 \sum_r \beta'_{1,r} \beta'_{2,r} \phi(\omega_r, T) \qquad (20.26)$$

The first two sums can be handled in exactly the same way as in the simplest example with which we started, except that in calculating the modified response functions account must be taken of the modifications to the restoring forces on both shells. To do this, it will be necessary to introduce the perfect lattice cross response function relating the radial displacement of second nearest neighbors to radial sinusoidal forces on the first neighbors. This cross response function for the perfect lattice can be calculated just as easily as the two direct response functions. Moreover, it is a straightforward matter to write down the equations which relate the modified response functions (including the modified cross response function to which we shall refer in a moment) to the unmodified response functions.

The third sum in the expression for $\overline{\Delta E^2}$ can be handled equally easily. Define a function $C_{12}(\omega)$ such that

$$C_{12}(\omega)\,d\omega = \sum_{\omega < \omega_r < \omega + d\omega} \beta'_{1,r}\beta'_{2,r} \qquad (20.27)$$

in terms of which we can rewite the third sum as an integral. $C_{12}(\omega)$ can be related to the modified cross response function which can be obtained as already described. Therefore, there is no difficulty in evaluating $\overline{\Delta E^2}$ (taking account of the modifications to the modes) for the case under consideration. Clearly, the methods will extend to more elaborate problems, the calculations becoming more complicated but remaining straightforward.

III. TRANSITIONS BETWEEN DEGENERATE LEVELS

New features enter if, instead of a nondegenerate excited electronic state, we have a set of excited electronic states which are degenerate when the lattice is in the ground state equilibrium configuration. This situation has been treated by Henry, Schnatterly, and Slichter [5]. The determination of $\overline{\Delta E^2}$ will be discussed with the assumption that the Franck–Condon principle still applies. For simplicity, suppose that only displacements of the nearest neighbor ions need be considered. Assume also that there is only one electron at the imperfection and its position vector will be denoted by **r**. As before, the quantities D_γ will give the magnitudes of the various symmetry-adapted displacements into which the displacements of the nearest neighbors can be resolved. The extra potential seen by the electron at the imperfection as a result of the displacement of the neighboring ions may be written as

$$\Delta V(r) = \sum_\gamma D_\gamma V_\gamma(\mathbf{r}) \qquad (20.28)$$

provided, of course, that we again make the assumption that the interaction is linear in the displacements of the nuclei.

Let us denote the nondegenerate electronic ground state by $|\alpha\rangle$ and the set of degenerate excited electronic states by $|\beta\rangle$ where β takes N values, N being the degeneracy of the excited level. In the case of the triply degenerate *p*-state of the F-center, for example, the states $|\beta\rangle$ might be assumed to have the forms $xf(r)$, $yf(r)$, $zf(r)$ or perhaps $(x \pm iy)f(r)$ and $zf(r)$. When the neighboring ions are displaced from their ground state equilibrium positions, the degeneracy of the states will be removed and, if we neglect mixing in of states from other levels, the electronic states will now be N definite linear combinations of the states $|\beta\rangle$. These N states which are no longer degenerate, will be denoted by $|b\rangle$. The energy of the state $|b\rangle$ differs from the energy of states when the atoms are in their ground state equilibrium configuration by $\langle b|\Delta V|b\rangle$, where ΔV has the form appropriate to the configuration of neighboring ions which is being considered. We note that

$$\langle b\,|\Delta V|\,b'\rangle = 0 \qquad b' \neq b \tag{20.29}$$

since the (approximate) eigenstates $|b\rangle$ will diagonalize the perturbation.

If O is the operator representing the perturbation by the radiation field which induces the transition, the probability that the final state will be the particular state $|b\rangle$ (when absorption takes place with the ions in the configuration considered) is given by $C|\langle\alpha\,|O|\,b\rangle|^2$ where

$$C = \left\{ \sum_b |\langle\alpha\,|O|\,b\rangle|^2 \right\}^{-1} \tag{20.30}$$

When the transition is to the final state $|b\rangle$, the square of the deviation from the mean of the energy of the absorbed photon will be $\langle b|\Delta V|b\rangle^2$. [Here we are assuming for simplicity that $\langle\alpha|\Delta V|\alpha\rangle$ is zero. In general, it will be nonzero and one can take account of this by replacing ΔV by $\Delta V - \langle\alpha|\Delta V|\alpha\rangle$ in equation (20.28) and subsequent equations]. It follows that when the atoms are in the given configuration, the mean square fluctuation from the mean energy of the absorbed photon (i.e., from the energy of the photon absorbed when the atoms are in their ground state equilibrium position) will be

$$C \sum_b |\langle\alpha\,|O|\,b\rangle|^2 \, \langle b\,|\Delta V|\,b\rangle^2$$

This may be written

$$C \sum_b \langle\alpha\,|O|\,b\rangle \, \langle b\,|\Delta V|\,b\rangle \, \langle b\,|\Delta V|\,b\rangle \, \langle b\,|O|\,\alpha\rangle$$

Because, as already noted, $\langle b|\Delta V|b'\rangle$ is zero unless $b = b'$, this may be written as

$$C \sum_{b,\,b',\,b''} \langle\alpha\,|O|\,b\rangle \, \langle b\,|\Delta V|\,b'\rangle \, \langle b'\,|\Delta V|\,b''\rangle \, \langle b''\,|O|\,\alpha\rangle$$

We may now make use of the fact that the states $|b\rangle$ are related to the states $|\beta\rangle$ by a unitary transformation, which allows us to write the above expression as

$$C \sum_{\beta, \beta', \beta''} \langle \alpha |O| \beta \rangle \langle \beta |\Delta V| \beta' \rangle \langle \beta' |\Delta V| \beta'' \rangle \langle \beta''|O| \alpha \rangle$$

This last step greatly simplifies the expression because the electronic states which now appear are the simple electronic states with which we started, and not the linear combinations of those which diagonalize ΔV. The quantity C may also be expressed in terms of the simple electronic states, by use of the same results, which give

$$C = \left\{ \sum_{\beta} |\langle \alpha |O| \beta \rangle|^2 \right\}^{-1} \tag{20.31}$$

Substitution for ΔV in the expression yields

$$C \sum_{\beta, \beta', \beta'', \gamma, \gamma'} \langle \alpha |O| \beta \rangle \langle \beta |V_\gamma| \beta' \rangle \langle \beta' |V_{\gamma'}| \beta'' \rangle \langle \beta'' |O| \alpha \rangle D_\gamma D_{\gamma'}$$

As a final step, we average over the D_γ and note that, since D_γ and $D_{\gamma'}$ will be contributed to by different normal modes, $\overline{D_\gamma D_{\gamma'}} = 0$ when $\gamma \neq \gamma'$. Therefore,

$$\overline{\Delta E^2} = C \sum_{\beta, \beta', \beta'', \gamma} \langle \alpha |O| \beta \rangle \langle \beta |V_\gamma| \beta' \rangle \langle \beta' |V_\gamma| \beta'' \rangle \langle \beta'' |O| \alpha \rangle \overline{D_\gamma^2}$$

$$= \sum_\gamma E_\gamma \overline{D_\gamma^2} \tag{20.32}$$

where

$$E_\gamma = C \sum_{\beta, \beta', \beta''} \langle \alpha |O| \beta \rangle \langle \beta |V_\gamma| \beta' \rangle \langle \beta' |V_\gamma| \beta'' \rangle \langle \beta'' |O| \alpha \rangle \tag{20.33}$$

The coefficients E_γ can be evaluated when the matrix elements of V_γ between the simple excited electronic states $|\beta\rangle$ are known, either from theoretical calculations or from stress experiments. (The matrix elements of O between the initial and final states may be supposed known. In fact if we suppose the states $|\beta\rangle$ so chosen that $\langle \alpha|O|\beta\rangle$ is nonzero only for one of them, this matrix element cancels out.) The quantities $\overline{D_\gamma^2}$ are treated as follows. Suppose that the contribution of the rth normal mode to D_γ is $\beta'_{\gamma, r} q'_r$ (these β's are, of course, quite different from those specifying the electronic states). Then,

$$D_\gamma = \sum_r \beta'_{\gamma, r} q'_r \tag{20.34}$$

and it follows that

$$\overline{D_\gamma^2} = \sum_r \beta'^2_{\gamma, r} \phi(\omega_r, T) \tag{20.35}$$

This can be written in the now familiar way as an integral in terms of a function $C_\gamma(\omega)$ such that

$$C_\gamma(\omega)\, d\omega = \sum_{\omega < \omega_r < \omega + d\omega} \beta_{\gamma,r}'^2 \tag{20.36}$$

The quantity $C_\gamma(\omega)$ can be related to a response function in exactly the same way as the $C(\omega)$ considered earlier, and by this device one can determine all the $C_\gamma(\omega)$, and thus the $\overline{D_\gamma^2}$, for the imperfect lattice. This means that we have a procedure for calculating $\overline{\Delta E^2}$ as a function of temperature even when the excited states are degenerate.

In this section we have incorporated the procedure for handling modified modes into the treatment of the degenerate state problem given by Henry, Schnatterly, and Slichter [5]. These authors avoid the semiclassical treatment of the lattice vibrations which has been used here for simplicity (their more complete treatment is necessary if one is to consider moments higher than the second) and they treat more elaborate problems, taking account, for example, of spin-orbit coupling.

IV. CONCLUSION

A number of groups are currently engaged in applying response function or Green's function methods to treating the effect of modified lattice vibrations on electronic processes at defects. The actual lattice dynamics part of such problems is, as shown in this volume, relatively easy to handle. The difficulty comes in estimating force constant changes and coupling constants either from experimental data or from theoretical analysis. At the moment, only a few investigations have led to any critical comparison between theory and experiment, and we shall not try to survey them. The subject is, however, growing rapidly and it seems likely that extensive results will begin to emerge in the near future.

V. REFERENCES

1. C. W. McCombie, J. A. D. Matthew, and A. M. Murray, *J. Appl. Phys. Suppl.* **33**:359 (1962).
2. R. Brout and W. Visscher, *Phys. Rev. Letters* **9**:54 (1962).
3. A. A. Maradudin, *Repts. Progr. Phys.* **28**:331 (1965).
4. R. J. Elliott in: R. W. H. Stevenson (ed.), *Phonons,* Oliver & Boyd (Edinburgh), 1966.
5. C. H. Henry, S. E. Schnatterly, and C. P. Slichter, *Phys. Rev.* **137**:583 (1965).

CHAPTER 21

Optical Analog of the Mössbauer Effect

R. H. Silsbee

Laboratory of Atomic and Solid State Physics
Cornell University
Ithaca, New York

I. INTRODUCTION

This chapter consists of a discussion of certain aspects of the influence of the lattice motion in crystals upon the optical absorption spectrum associated with localized defects in the crystal. The central theme might be considered the phonon broadening of optical spectra. The relevance of these remarks to the title will become apparent shortly.

The most important source of thermal broadening of optical transitions of defects in solids is the dependence of the energy of the electronic states upon the location of the atoms in the immediate environment of the defect. Clearly for the F-center, where the potential field in which the electron moves is determined predominantly by the six neighboring alkali ions, the transition energies should depend sensitively upon the relative position of these six neighbors and one expects a strong coupling to the lattice motion. For a transition element or rare earth impurity in a crystal, the potential field seen by the electron is predominantly that of the impurity, but nonetheless one expects at least some small influence of the surrounding atoms on the transition frequencies.

The starting point for this discussion is then a model in which the electronic transition energy is expanded in a Taylor series in the displacements ξ_k, the normal coordinates for the system when the electron is in the ground state,

$$E_e = E_0 + \frac{1}{N} \sum_k \alpha_k \xi_k + \frac{1}{N} \sum_{kk'} \beta_{kk'} \xi_k \xi_{k'} + \ldots \quad (21.1)$$

The normal coordinates ξ_k are the real normal coordinates in the presence of the defect and need not be assumed to be plane waves. An assumption of convenience, not necessity, is that there are no localized modes, i.e., the amplitude of every mode at the defect goes to zero as the crystal size becomes infinite. N is three times the number of atoms, and normalizations are chosen such that the ξ_k's will have amplitudes of the order of atomic displacements and the α_k's, for high-frequency modes, the order of the energy shift due to displacements of the nearest neighbor atoms by one lattice constant. Furthermore, it is important to remember that this model will neglect the possible dependence of the electronic oscillator strength upon the nuclear displacements, and matrix elements between the electronic states of the potential associated with the nuclear displacements. Although the model may seem overly simplified, most of the features of thermal line broadening are illustrated by this system.

Consider first the linear coupling term of equation (21.1), the term involving the α_k's; the discussion of the quadratic coupling will be deferred until later. Qualitatively one argues that the optical line is broadened through the linear coupling since both thermal motion and zero point motion of the ions introduce variations in the normal coordinates ξ_k, giving in turn an uncertainty in the transition energy E_e or a width to the optical line. The mean square width is easily calculated as

$$\langle (E - E_0)^2 \rangle = \frac{1}{N} \sum_{kk'} \alpha_k \alpha_{k'} \langle \xi_k \xi_{k'} \rangle$$

$$= \frac{1}{N} \sum_k \alpha_k^2 \langle \xi_k^2 \rangle$$

$$\xrightarrow[T \to 0]{} \frac{1}{Nm} \sum_k \frac{\alpha_k^2}{2\omega_k}$$

$$\xrightarrow[T \to \infty]{} \frac{1}{Nm} \sum_k \frac{\alpha_k^2}{\omega_k} \frac{kT}{\hbar\omega_k} \qquad (21.2)$$

The low-temperature limit gives the broadening due to the zero point motion of the ions; the high-temperature limit, the thermal broadening, implies a line width proportional to $T^{\frac{1}{2}}$. The success of these simple arguments is indicated in Fig. 1. For the F-center in KCl, and many other "broad absorption bands", the qualitative prediction of a large zero point width and a weak $T^{\frac{1}{2}}$ high-temperature variation of the width is observed and the magnitudes are also reasonable. For the R line of the Cr^{3+} ion in MgO and many other "narrow absorption lines", the "theory" is a dismal failure. The coupling strengths α_k for MgO–Cr^{3+} are only one order of magnitude smaller than for the F-center, yet the observed widths are orders of magnitude different at low temperature. More serious is the apparent absence of broadening by the zero point motion and the wrong qualitative temperature

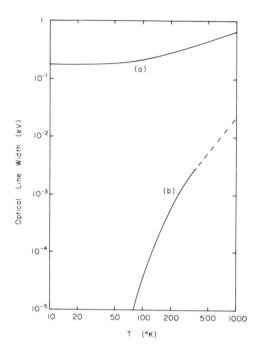

Fig. 1. Optical line width *versus* temperature for (a) F-centers in KCl and (b) Cr^{3+} ion (R line) in MgO.

dependence at high temperature. The discrepancy is particularly embarrassing because the observed width is much less than the predicted one. The problem is not to find another broadening mechanism but to find why the apparently general arguments above are inappropriate for discussing the Cr^{3+} problem.

II. THE LINEAR COUPLING SPECTRUM

To develop an understanding of this qualitative distinction between broad and narrow lines, consider first a classical argument concerning the qualitative nature of the spectrum and simplify even further by considering an Einstein solid or a defect with strong interaction with only a single mode. Classically, in terms of emission, the electron is considered to emit a frequency-modulated signal with instantaneous frequency

$$v_e(t) = v_0 + C \cos \omega_D t \qquad (21.3)$$

where C is the strength of the coupling to the normal mode with frequency ω_D of the order of the Debye frequency of a solid. The classical spectrum

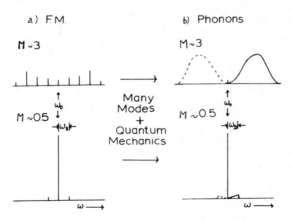

Fig. 2. Frequency modulation spectra for large and small modulation index. (a) A single modulating frequency. (b) A continuum of modulating frequencies.

associated with such a signal, illustrated in Fig. 2(a) (see any electrical engineering text), depends qualitatively upon the modulation index M, defined as

$$M = C/\omega_D \qquad (21.4)$$

or the ratio of the maximum frequency deviation to the modulating frequency. For small modulation index, most of the spectral intensity resides in the carrier at ν_0 and very little in the sidebands at $\nu = \nu_0 \pm n\omega_D$. For large M the situation is reversed. Most of the intensity resides in the side bands and the carrier is very weak.

In the case of a real solid (or a Debye solid) there is a continuum, not a single modulating frequency. Hence, there is no clearly resolved side-band structure as in Fig. 2(a) but rather a continuous structure as in Fig. 2(b). On the other hand, even with a continuum of modulating frequencies, the carrier signal still contributes a δ function spike to the spectral intensity. The spectrum, then, has two distinct features, a δ function-like carrier signal and a continuous side-band structure. The relative intensity of these two parts of the spectrum, as shown later, depends very strongly upon the effective modulation index or coupling strength and there are many examples in which only the carrier, "narrow line", or only the side band, "broad band", are observed.

The mean square width calculated by equation (21.2) is correct for the full spectrum. For the F-center, where the narrow component is too weak to be observed, the calculated width does correspond to the observed width. But for MgO–Cr^{3+}, it is only the narrow component which is normally observed experimentally. Since the calculated width is for the full spectrum, and contributed in a sense entirely by the side-band structure, whereas the observed width is that of the narrow component alone, one should not be offended by the lack of agreement. The calculated width is not incorrect; it simply bears no relevance to the experimentally observed width.

With use of the simple model described by equation (21.1) with the quadratic terms omitted, it is easy to apply perturbation theory to calculate the spectrum quantum mechanically. Establishment of the quantum state of the solid requires the specification of the electronic state of the defect and the degree of excitation of the vibrational motion. The normal coordinates ξ_k and normal mode frequencies ω_k are assumed known when the defect is in the ground electronic state, and the vibrational state of the crystal is specified by the set of occupation numbers $\{n_k\}^0$, where the superscript zero is a reminder that these are the ground state normal modes. The full crystal wave function will be denoted by $|\{n_k\}^0\,0\rangle$, where the zero indicates that the object is in the ground electronic state.

The electron phonon coupling terms of equation (21.1) imply that the potential energy function for the nuclear motion, in a Born–Oppenheimer treatment, is different depending upon whether the electron is in the ground or excited electronic state. Thus, in general, the normal coordinates with the electron excited will not be the same as the normal coordinates when the electron is in the ground state. In the linear approximation, however, there is no mixing of normal coordinates, but only a displacement of the equilibrium positions of the various oscillators. Thus, the full crystal states for the excited electron will be denoted by $|\{n_k'\}^1\,1\rangle$, where the superscript 1 indicates that the vibrational wave functions specified by the set of occupation numbers n_k' are centered about new equilibrium positions. The equilibrium displacement of each mode $\xi_k{}^0$ is calculated in terms of the coupling constant α_k and the force constant $m\omega_k{}^2$ for the kth mode as

$$\xi_k{}^0 = -\frac{\alpha_k}{\sqrt{N}\,m\omega_k{}^2} \tag{21.5}$$

The difference between the electronic energies for the two equilibrium configurations is then

$$E_0' = E_0 + \frac{1}{\sqrt{N}} \sum_k \alpha_k \xi_k{}^0$$

$$= E_0 - \frac{1}{N} \sum_k \frac{\alpha_k{}^2}{m\omega_k{}^2} \tag{21.6}$$

Figure 3 illustrates these relationships for a single coordinate model.

If the system is exposed to an optical radiation field of frequency ω which interacts with the electron via a term $\mathcal{H}_{\text{dip}}$, perturbation theory gives for the transition probability out of the assumed initial state, with vibra-

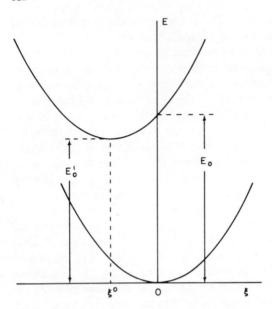

Fig. 3. Single coordinate model showing displacement of upper state minimum by linear coupling.

tional quantum numbers $\{n_k\}^0$

$$W(\hbar\omega) \sim \frac{1}{\hbar} \sum_{\substack{\text{final} \\ \text{states}}} |\langle\{n_k'\}^1 1| \mathscr{H}_{\text{dip}} |\{n_k\}^0 0\rangle|^2 \times \delta(\hbar\omega + E_{\{n_k\}} - E_0' - E_{\{n'_{k'}\}}) \quad (21.7)$$

$E_{\{n_k\}}$ is the vibrational energy associated with the set of occupation numbers $\{n_k\}$. The transition matrix element, in this simple model, factors into a product of an electronic matrix element and a vibrational overlap giving

$$W(\hbar\omega) \sim \frac{1}{\hbar} |\langle 1| \mathscr{H}_{\text{dip}} |0\rangle|^2 \sum_{\substack{\text{final} \\ \text{states}}} |\langle\{n_k'\}^1|\{n_k\}^0\rangle|^2 \times \delta(\hbar\omega + E_{\{n_{k'}\}} - E_0 - E_{\{n'_{k'}\}})$$
$$(21.8)$$

Remembering that the oscillator function $|\{n_k'\}^1\rangle$ differs from the function $|\{n_k'\}^0\rangle$ simply by a displacement of the center of the functions by an amount $\xi_k{}^0$, one can use the formal Taylor series expansion

$$|\{n_k'\}^1\rangle = \prod_k \left[\exp\left(-\xi_k{}^0 \frac{\partial}{\partial \xi_k}\right) \right] |\{n_k'\}^0\rangle \qquad (21.9)$$

to generate an excited state wave function with quantum numbers $\{n_k\}$ from the ground state vibrational function with the same set of quantum numbers.

Expanding the exponential, remembering that $\xi_k{}^0 \sim 0(1/\sqrt{N})$, and making the replacement $\partial/\partial\xi_k \to iP_k/\hbar$, with P_k the momentum operator for the kth oscillator, we get

$$W(\hbar\omega) \sim \frac{1}{\hbar} |\langle 1| \mathcal{H}_{\text{dip}} |0\rangle|^2 \sum_{\substack{\text{final} \\ \text{states}}} |\langle \{n_k\}^0| \prod_k \left[1 - \frac{i}{\hbar} \xi_k^0 P_k \right.$$

$$\left. - \frac{1}{2\hbar^2} (\xi_k^0)^2 P_k^2 + ... \right] |\{n'_k\}^0\rangle|^2 \times \delta(\hbar\omega + E_{\{n_k\}} - E'_0 - E_{\{n'_k\}}) \quad (21.10)$$

To simplify the discussion, consider the absorption for the case of the initial state being the ground state of the crystal, with all $n_k = 0$. It is convenient to classify the final states according to the number of oscillators which are found excited after the absorption has occurred. A particularly important final state is that in which no vibrational excitation takes place, the state in which the set $\{n'_k\}$ are also all zero. Since the term in equation (21.10) linear in P_k couples only states in which the occupation number changes, the "zero-phonon" transition probability is given by

$$W_0(\hbar\omega) \sim \frac{1}{\hbar} |\langle 1| \mathcal{H}_{\text{dip}} |0\rangle|^2 |\langle 0| \prod_k \left[1 - \frac{1}{2\hbar^2} (\xi_k^0)^2 P_k^2 + ... \right] |0\rangle|^2$$

$$\times \delta(\hbar\omega - E'_0)$$

$$\equiv I_0 \delta(\hbar\omega - E'_0) \quad (21.11)$$

There is no sum of final states since there is only a single final state in which no phonons are excited and the argument of the delta function (which governs the line shape for this part of the spectrum), simplifies because there are no vibrational contributions to the initial and final state energies. This sharp line is, of course, the quantum-mechanical analog of the classical carrier signal and is normally referred to as the zero-phonon line. Equation (21.11) may be rewritten for later use by evaluating the matrix elements of P_k^2, replacing $(\xi_k^0)^2$ with use of equation (21.5), and rewriting the product as the exponential of a sum of terms. Remembering the completeness of the set of oscillator functions $|\{n_k\}^1\rangle$ appearing in equation (21.8), one can argue that the intensity of the zero-phonon line relative to the total intensity of the spectrum is given by

$$I_0 / \int W(\hbar\omega) \, d(\hbar\omega) = (I_0/I_{\text{tot}}) = \exp \left[-\frac{1}{N} \sum_k (\alpha_k^2 / 2\hbar m\omega_k^3) \right] \quad (21.11a)$$

At finite temperatures, the result is similar except that each term in the exponent is multiplied by $(2 n_k + 1)$, where n_k is the thermal expectation value of the occupation number of the kth mode. This "Debye–Waller factor," a generalization of equation (21.11a), can, of course, lead to a very strong temperature dependence of the relative intensity of the zero-phonon line.

Next, consider the final states in equation (21.10) in which only one oscillator changes occupation number. These will be coupled to the initial state by that term in the expansion of the $\prod_k$ which involves the term linear in P_k for one mode and the term 1 for all the other modes. The resultant one-phonon spectrum is related to the transition probability

$$W_1(\hbar\omega) \sim \frac{1}{\hbar} |\langle 1| \mathscr{H}_{\text{dip}} |0\rangle|^2 \sum_k \frac{|\langle n_k = 0| - \frac{i}{\hbar} \xi_k^{\,0} P_k |n_k = 1\rangle|^2}{|\langle n_k = 0| \left(1 - \frac{1}{2\hbar^2} \xi_k^{02} P_k^{\,2} + ... \right) |n_k = 0\rangle|^2}$$

$$\times |\langle \{0\}| \prod_k \left(1 - \frac{1}{2\hbar^2} \xi_k^{02} P_k^{\,2} + ... \right) |\{0\}\rangle|^2 \, \delta(\hbar\omega - E_0' - \hbar\omega_k) \quad (21.12)$$

The term in the denominator is 1 to order $(1/N)$ and is replaced by 1, while the first and last squared matrix elements are related through equation (21.11) to the zero-phonon line intensity I_0. The matrix element of P_k is $(\hbar m\omega_k/2)^{\frac{1}{2}}$ and equation (21.5) is used to give for the one-phonon processes

$$W_1(\hbar\omega) = I_0 \frac{1}{2N} \sum_k \frac{\alpha_k^{\,2}}{\hbar\omega_k} \delta(\hbar\omega - E_0' - \hbar\omega_k) \quad (21.13)$$

If the sum is replaced by an integral over normal mode frequencies, introducing a density of modes $Ng(\hbar\omega)$ with $g(\hbar\omega)$ normalized to unity, and an effective coupling $\alpha(\omega)^2$ which is an average of α^2_k over modes of frequency ω, equation (21.13) becomes

$$W_1(\hbar\omega) = I_0 \frac{\alpha(\omega)^2}{2\hbar\omega} g(\hbar\omega) \quad (21.14)$$

The resulting one-phonon contribution to the spectrum is then a continuum, not a discrete line, and corresponds to the first-order side bands of the classical picture. Note that the one-phonon spectrum gives information about the density of normal modes in the crystal, $g(\omega)$, and the frequency dependence of the coupling strength, $\alpha(\omega)$.

Similarly, one can develop expressions for the intensities of the processes involving the emission of two, three, or more phonons. Furthermore, the generalization to finite temperatures introduces no intrinsic difficulties, requiring only a thermal averaging of expressions similar to the above over the possible initial states of the crystal. The resultant spectrum is, of course,

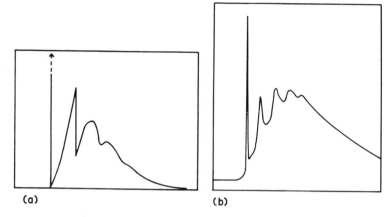

Fig. 4.   Spectra showing both the zero-phonon line and the phonon continuum (see Silsbee and Fitchen [2]). (a) A Debye model calculation by Visscher; $R/K\theta = 1.3$. (b) Experimental curve for the R-center in LiF; $R/K\theta = 2.6$.

similar to the classical system, the most striking difference being the existence of side bands on only one side of the zero-phonon line at low temperature, corresponding to the possibility of vibrational excitation but no de-excitation simultaneous with the electronic excitation. Figure 4 shows both an observed optical spectrum and a Debye model calculation illustrating the case where both the zero-phonon line and phonon continuum are observable. The several peaks in the continua correspond to the one-, two-, and higher multiphonon emission processes, the peaking being associated with the peak in the Debye density of states in Fig. 4(a) and presumably with a strong peaking of $g(\omega)$ for the real LiF crystal in Fig. 4(b).

III. EXAMPLES

After this brief sketch of the theory of optical spectra in the linear coupling approximation, it is appropriate to emphasize several points in terms of specific examples of spectra which may or may not exhibit the zero-phonon line.

A. Example 1

Consider the examples discussed in the introduction to these lectures, the F-center in KCl and the Cr^{3+} ion in MgO, and ask whether one would expect to observe the zero-phonon line for these systems. The experimental measurement of the shift of the optical absorption induced by applied mechanical stress allows a rough estimate of the coupling parameters, α_k, provided the unperturbed plane wave normal modes of the perfect crystal are assumed to be the normal modes in the presence of the impurity. Suppose a macroscopic strain ε produces an experimentally measured line shift $\Delta E \simeq \alpha\varepsilon$. Then, noting that the local strain due to a long-wavelength plane wave of wave vector k and amplitude ξ_k is roughly $\nabla\xi_k e^{ikx} \sim k\xi_k e^{ikx}$, one estimates the coupling strength for long waves to be

$$\alpha_k \sim \alpha k \sim \frac{\alpha\omega_k}{v} \tag{21.15}$$

where v is the sound velocity. These relationships are trivially made more precise by noting that the strain ε and strain coupling coefficients should be treated as tensor quantities—an unnecessary complication for the simple arguments to be made here.

Not so easily corrected is the "long-wave approximation." Although equation (21.15) gives a valid procedure for determining the coupling to the long-wavelength phonons, one has no simple direct way of determining the coupling to the bulk of the normal modes, namely, those with wavelengths of the order of interatomic distances. In some instances, with well-localized defects, such as transition element or rare earth impurity ions, equation (21.15) should be useful for order-of-magnitude estimates of the coupling to all modes. On the other hand, for impurities with weakly bound electrons, such as the shallow donor and acceptor states in semiconductors, equation (21.15) drastically overestimates the coupling to those modes with wavelength small compared with the electron orbit radius and is useless even for order-of-magnitude arguments.

Substitution of equation (21.15) into equation (21.11) gives for the strength of the zero-phonon line at absolute zero

$$I_0/I_{\text{tot}} \sim \exp\left(-\int_0^{\omega_D}\frac{\omega^2\,d\omega}{\omega_D{}^3}\frac{1}{2\hbar m\omega^3}\frac{\alpha^2\omega^2}{v^2}\right) \sim \exp\left[-\frac{\alpha^2}{(\hbar\omega_D)(mv^2)}\right] \tag{21.16}$$

where a Debye model is used to evaluate the sum over modes of equation (21.11a). The argument of the exponential is simply the square of a generalized modulation index and the main point to recognize in equation (21.16) is the strong dependence of the zero-phonon line intensity on the modulation index or coupling strength α. For our examples, α for the F-center is about 2 eV per unit strain and for the MgO–Cr^{3+} R lines about 0.2 eV, one order of magnitude smaller. Yet the appearance of the square of this coupling in the exponent implies that for the F-center the zero-phonon line intensity is of the order of e^{-30} or completely unobservable, while for the Cr^{3+} lines it is about $e^{-0.3} \sim 0.7$. Because of the narrowness of the line, the only commonly observed feature is the zero-phonon line; thus an order-of-magnitude change in coupling easily takes one from one extreme limit to the other. The intermediate cases do, of course, occur as illustrated by Fig. 4, but it is now not surprising that the more common behavior is one of the two limiting extremes.

B. Example 2

In solid state physics many problems are considered first with a one- or two-dimensional model in order to illustrate, for the simpler model, some physical effect. It is amusing to evaluate equation (21.11a) for a one-dimensional system. For such a system, the Debye model density of states is constant rather than varying as ω^2 and the integral appearing in equation (21.16), with two less powers of ω, diverges logarithmically at the low-frequency limit. A similar divergence occurs for the two-dimensional crystal at any finite temperature. Thus, the simple theory as outlined above is inadequate to describe the spectrum for these models and the zero-phonon line, in the sense discussed above as a truly discrete line, does not exist. There may, however, be a strong peaking at the position of the zero-phonon line; the detailed theory is discussed by Duke and Mahan [1]. This is one of a variety of examples that give warning that one should be skeptical of arguments based on one- or two-dimensional models; such arguments can lead to qualitatively false results.

C. Example 3

A divergence similar to that noted above also occurs in the case of three-dimensional piezoelectric crystals. In such materials the electric field associated with an acoustic mode is directly proportional to the normal mode displacement, not to the gradient of that displacement; the coupling in the long-wave approximation is independent of frequency [cf. equation (21.15)] and the integral in equation (21.16) again diverges. Again, no zero-phonon line exists and a more sophisticated approach [1] is required to understand the shape of the spectrum.

D. Example 4

To justify the title of this chapter it is, of course, important to remark that the recoil-free fraction of the γ-ray spectrum of a nucleus bound in a solid is a zero-phonon line. In the example of the Mössbauer effect, the mechanism of the coupling of the transition to the lattice vibrations is different but otherwise the physics is essentially the same. Under the assumption of plane wave normal modes and noting that for a signal of frequency ω_0 emitted by a nucleus moving with velocity $\dot{x}$, the frequency observed in a stationary frame is

$$\omega_N = \omega_0 \left(1 + \frac{1}{c} \dot{x} - \frac{1}{2c^2} \dot{x}^2 + \ldots \right)$$

$$= \omega_0 + \frac{\omega_0}{c} \frac{1}{\sqrt{N}} \sum_k \omega_k \xi_k - \frac{\omega_0}{2c^2} \frac{1}{N} \sum_{kk'} \omega_k \omega_{k'} \xi_k \xi_{k'} + \ldots \qquad (21.17)$$

one sees immediately the analog with equation (21.1) where

$$\alpha_k = \frac{\hbar \omega_0}{c} \omega_k \qquad \beta_{kk'} = \frac{\hbar \omega_0}{c^2} \omega_k \omega_{k'} \qquad (21.18)$$

Note that $\alpha_k \sim \omega_k$ as in the long-wave approximation [equation (21.15)] for the optical problem. The classical analysis is identical for this problem while the quantum-mechanical one is somewhat altered but essentially the same. The most important feature is that the oscillator displacement, the $\xi_k{}^0$ of equation (21.5), becomes for the Mössbauer problem a displacement in momentum rather than coordinate space. This is discussed in more detail by Silsbee and Fitchen [2] and by Maradudin [3].

An important feature of the Mössbauer problem is that the coupling parameters α_k for plane wave normal modes are known without ambiguity, in contrast to the optical problem. Equation (21.14) shows that a knowledge of the one-phonon side-band spectrum of a zero-phonon line, if the coupling coefficient is known, gives quite directly the phonon density of states. Thus, for a Mössbauer source or absorber which is pure (no modification of the normal modes by the γ-ray nucleus), the one-phonon spectrum would be most useful in learning about the lattice modes. Unfortunately, the side-band continuum seems to be inaccessible experimentally for the Mössbauer effect and only that information obtainable from the temperature variation and anisotropy of the recoil-free fraction or zero-phonon line is of use in learning about the phonons.

In the case of optical spectra, the one-phonon spectrum is sometimes easily deduced from the full spectrum and one is tempted to associate structure in this spectrum with structure in the phonon density of modes. Although these one-phonon spectra certainly are a powerful tool in studying lattice vibration problems, perhaps a word of caution is appropriate. There are three essentially different effects which importantly influence the one-phonon spectrum [equation (21.14)]. First, the density of modes does enter and structure in the density of modes may be reflected in the one-phonon spectrum. Second, the coupling coefficients may develop a very strong frequency dependence if an in-band resonance develops at the defect, becoming relatively large at the frequency of this resonance. Of course, the development of a true localized mode is also expected to give a sharp peak in the one-phonon spectrum which bears no direct relationship to peaks of the host crystal density of modes. Finally, the $\alpha(\omega)$'s may have important departures from the ω dependence of equation (21.15) because of the details of the coupling. As already noted, in semiconductors the coupling falls rapidly with increasing k for phonons of wavelength less than the Bohr radius of the orbits of the electrons associated with the defect. The observed peak in the one-phonon spectrum then is unrelated to details or modifications of the vibrational modes of the host. Thus, the analysis of such side-band spectra must be considered carefully to sort out properly the various possible sources of structure.

E. Example 5

The zero-phonon line appears as a feature not only of electronic and nuclear transitions but also of vibrational transitions. The U-center, as discussed in chapter 15 by Genzel, develops a localized vibrational mode with a corresponding absorption in the infrared at frequencies well above that of the host lattice. The anharmonic coupling within the crystal then gives a dependence of the local mode frequency upon the distortions of the local environment, and hence the possibility of frequency modulation of the local mode by the host crystal modes. The situation is entirely analogous to the electronic problem with the local mode transition playing the role of the electronic transition; the spectrum shows the familiar features of a zero-phonon line, corresponding to excitation of only the local mode, and a side-band spectrum, in which both the local mode and one or more lattice modes are excited. The concept is much less clear in the case of an in-band resonance but even here, as noted by Takeno and Sievers [4], there are features of the experimental results which are simply interpreted in terms of the properties of the zero-phonon lines of electronic transitions.

F. Example 6

A final example is in a rather unexpected area, namely electron spin resonance. Consider the ESR of the system MgO–Fe^{2+} in a field of a few thousand gauss. The corresponding Zeeman splitting is about $\frac{1}{3}$ cm^{-1} and the observed line width, due to random static strains in the crystal, is about $\frac{1}{10}$ cm^{-1}. The stress coupling of these transitions is measured to be about 0.05 eV per unit strain. This is less than the MgO–Cr^{3+} system but still sufficient that the rms zero point width estimated on the basis of equation (21.2) is about 10 cm^{-1}. Thus, the dynamic modulation of the energy levels by the zero point motion is large compared with the separation of the levels. Nonetheless one sees a relatively sharp absorption, the reason being, of course, that there is again a zero-phonon line which carries most of the line intensity. This example emphasizes the point that the important parameter determining the intensity of the zero-phonon line is the modulation index and hence the depth of modulation (10 cm^{-1} in this example) and the modulating frequency (hundreds of cm^{-1}) or a typical lattice frequency. The frequency of the transition itself, whether it be $\frac{1}{3}$ cm^{-1} as in this example or 10^8 cm^{-1} as in the Fe^{57} γ-ray, is irrelevant to the nature of the spectrum.

IV. LINE WIDTH OF THE ZERO-PHONON LINE

In the introductory section a rough argument for the line width of optical transitions in solids was outlined. Figure 1 indicated that for some examples (the broad band spectra) this was adequate, at least as a first approximation. One of the main points of the second section was that this argument was irrelevant to the line width of the zero-phonon line and that in the linear coupling approximation (with the further implicit restriction of no matrix elements linear in the phonon field and off diagonal in the electronic states) the zero-phonon line component of the spectrum would have zero line width. There are, of course, a wide variety of mechanisms which do contribute to the width of this line and this section will be devoted to a brief discussion of some though not all of these mechanisms: "Not all" both because the list would become tediously long and because it is doubtful that all have yet been discovered.

A. Mechanism 1

In principle the width of all optical lines is limited by the radiative lifetime of the excited state and for many systems this is ideally the only intrinsic limiting factor at low temperature. Note that a 30-msec lifetime for an optical transition implies a resonance Q of 10^{14}, a very attractive number for a variety of high-precision experiments. Unfortunately, this enticing Q does not seem to be even approximately achievable in practice in real systems with residual strains.

B. Mechanism 2

The least interesting from the point of view of the physics involved, but unfortunately also the most important in fact for many systems at low temperature, is the broadening by the random residual static strains resulting from dislocations or point defects in the crystals. For many systems this broadening limits the optical line width at low temperatures to values typically in the range 0.01 to 10 cm^{-1} depending upon the degree of perfection of the crystal and the magnitude of the coupling of the transition to the mechanical strain.

C. Mechanism 3

Many narrow lines show a temperature variation of the width similar to that of the Cr^{3+} ion, as illustrated in Fig. 1. This temperature dependence may be understood by considering the quadratic coupling terms of equation (21.1).

A simple argument shows that these terms may have a very important influence upon the zero-phonon line. Remember first that in the linear approximation the zero-phonon line appears only because in a three-dimensional crystal there are so very few modes of low frequency. Thus, the effective modulation index, related to an appropriate sum over the modes of the frequency deviation divided by the mode frequency, may remain small. The bilinear term in equation (21.1), $\beta_{kk'} \xi_k \xi_{k'}$, classically has time variations at the frequencies $\omega_k \pm \omega_{k'}$. If the two modes k and k' have nearly the same frequency, then this term gives modulation at a low frequency and hence with a relatively large modulation index. However, in a Debye spectrum there are many pairs of modes whose difference frequency is small and thus these terms may lead to a large modulation index and hence a dramatic effect upon the zero-phonon line. The effect turns out to be not the complete washing out of the line but a small broadening of the line which is strongly temperature-dependent.

To obtain this temperature dependence, consider for simplicity a model in which the linear coupling coefficients are zero and there is only the bilinear coupling. Suppose the system is initially in the state $|\{n_k\}^0\, 0\rangle$ and further let the excited states be described in a representation using the oscillator states of the ground state normal modes as basis states. Because of the quadratic coupling the excited state normal coordinates are not the same as the ground state and thus the states $|\{n_k'\}^0\, 1\rangle$ chosen to represent the excited states of the crystal are not eigenstates of the model Hamiltonian. The bilinear coupling term $\beta_{kk'} \xi_k \xi_{k'}$ can couple $|\{n_k\}^0\, 1\rangle$, for instance, to the state $|\{n_k'\}^0\, 1\rangle$ in which the set $\{n_k\}$ differs from the set $\{n_k'\}$ in that $n_k \to n_k \pm 1$ and $n_k' \to n_k' \mp 1$. If $\omega_k \approx \omega_k'$, energy conservation allows the state $|\{n_k\}\, 1\rangle$ to decay into the state $|\{n_k'\}\, 1\rangle$, and similar states involving the approximately elastic scattering of phonons, at a rate to be calculated below.

In the calculation of the optical absorption, the neglect of the linear coupling implies that, in the representation defined above, the initial state $|\{n_k\}\, 0\rangle$ is coupled only to the single final state $|\{n_k\}\, 1\rangle$. Without the quadratic coupling, this is simply the observation that all of the intensity appears in the zero-phonon line. The inclusion of the quadratic coupling does not alter this selection rule but does limit the lifetime of the final state, $|\{n_k\}\, 1\rangle$, and leads to a width of the optical line. The width is just the inverse of the lifetime of the state $|\{n_k\}\, 1\rangle$ against the approximately elastic scattering of phonons by the quadratic coupling term. Perturbation theory gives for this lifetime

$$\frac{1}{\tau} = \frac{2\pi}{\hbar} \sum_f |\langle\{n_k'\}| \frac{1}{N} \sum_{kk'} \beta_{kk'} \xi_k \xi_{k'} |\{n_k\}\rangle|^2 \, \delta(E_f - E_i) \qquad (21.19)$$

in which the final states $\{n_k'\}$ of interest are those in which two of the quantum numbers n_k change, one increasing by unity and the other decreasing by unity. To obtain the temperature dependence, replace the coupling coefficient $\beta_{kk'}$ by a suitable average over all pairs of modes of the same frequencies $\beta_{\omega\omega'}$, replace the sum over final states by a double integral over frequencies, the frequencies of the absorbed and emitted phonons in the scattering, and remember that the energy conservation condition $\delta(E_f - E_i)$ makes one of the integrations trivial. The formal result is

$$\frac{1}{\tau} = \frac{2}{\hbar^2} \int \frac{g(\omega)^2}{N^2} |\beta_{\omega\omega}|^2 \, n_\omega(n_\omega + 1) \left(\frac{\hbar}{2m\omega}\right)^2 d\omega \qquad (21.20)$$

where $g(\omega)$ is the mode density and the factors $n_\omega \hbar/2\,m\omega$ and $(n_\omega + 1)\hbar/2\,m\omega$ are the squares of the matrix elements of the coordinate operators.

In the long-wavelength approximation which led to equation (21.15), one has $\beta_{\omega\omega} \sim \omega^2$ and since the mode density $g(\omega)$ is proportional to ω^2

$$\frac{1}{\tau} \sim \int_0^{\omega_D} \omega^6 \, n_\omega(n_\omega + 1) \, d\omega \qquad (21.21)$$

where ω_D is the cutoff frequency of the Debye model. At high temperatures, $n_\omega \sim kT/\hbar\omega$ for all ω giving immediately

$$\frac{1}{\tau} \sim T^2 \qquad \text{as } T \to \infty \qquad (21.22a)$$

At temperatures low compared with the Debye temperature, the standard technique of introducing the dimensionless variable $x = \hbar\omega/kT$ in the integrand of equation (21.21) gives

$$\frac{1}{\tau} \sim T^7 \int_0^{\hbar\omega_D/kT} x^6 \, e^x \, dx/(e^x - 1)^2 \sim T^7 \qquad \text{if } T \ll \frac{\hbar\omega_D}{k} \qquad (21.22b)$$

The results are in qualitative accord with Fig. 1 and the more detailed result, equation (21.20), gives a satisfactory description of the thermal line broadening of a number of zero-phonon lines [5-7].

D. Mechanism 4

This thermal broadening via the quadratic coupling is not the only source of temperature variation of the line width. The model, equation (21.1), has neglected terms in which local distortions couple the two electronic states. Such terms will lead to temperature-dependent processes in which the electronic state changes and the energy is conserved by the emission or absorption of one or more phonons. Such nonradiative processes are very important for electronic states whose splitting is within the phonon spectrum such that energy is conserved with the absorption or emission of a single phonon [6] and become relatively unimportant as the electronic splitting becomes large compared with typical vibrational energies.

E. Mechanism 5

These are some of the broadening mechanisms for the single defect in a crystal. There is another class of mechanisms involving the interaction between defects, not in the sense of the uninteresting static strain broadening of equation (21.2) above, but in the sense of dynamic interactions involving resonant transfer of excitation from center to center via exchange interaction for close pairs [8] or electromagnetic and perhaps phonon interactions for more distant pairs.

V. CONCLUSION

The aim of this chapter was to develop some feeling for the effect of phonon modulation of electronic energy levels upon the optical line width of transitions involving these levels. This understanding requires first the recognition of the qualitative distinction between "narrow line" and "broad band" spectra and then the separate treatment of the broadening of these two components of the spectrum. The treatment has necessarily been a sketchy one and there are many qualifications and details which have been omitted with the justification that this was intended as an introduction to ideas, not as an exhaustive review of the topic.

It is hoped that this chapter has suggested that in inventing new broadening mechanisms and in elucidating those mechanisms already proposed there remains much interesting work in physics for the theorist. For the experimentalist it is no trivial task to find systems which clearly demonstrate the existence of these mechanisms or which serve to test the proposed theories, particularly with the ever-present strain broadening. This is an area with many challenging problems yet to be solved.

VI. REFERENCES

1. C. B. Duke and G. D. Mahan, *Phys. Rev.* **139**:A1965 (1965).
2. R. H. Silsbee and D. B. Fitchen, *Rev. Mod. Phys.* **36**:423 (1964).
3. A. A. Maradudin, *Solid State Physics, Vol. 18*, Academic Press (New York), 1966.
4. S. Takeno and A. J. Sievers, *Phys. Rev. Letters* **15**:1020 (1965).
5. W. M. Yen, W. C. Scott, and A. L. Schawlow, *Phys. Rev.* **136**:A271 (1964).
6. W. M. Yen, W. C. Scott, and P. L. Scott, *Phys. Rev.* **137**:A1109 (1965).
7. D. E. McCumber and W. D. Sturge, *J. Appl. Phys.* **34**:1682 (1963).
8. W. M. Yen, R. L. Greene, W. C. Scott, and D. L. Huber, *Phys. Rev.* **140**:A1188 (1965).
9. Hans Frauenfelder, *The Mössbauer Effect*, Benjamin, Inc. (New York), 1962.
10. E. O. Kane, *Phys. Rev.* **119**:40 (1960).
11. E. D. Trifonov, *Dokl. Akad. Nauk SSSR* **147**:826 (1962); *Soviet Phys. Dokl. (English Transl.)* **7**:1105 (1963).
12. A. F. Lubchenko and S. I. Dudkin, *Phys. Status Solidi* **14**:227 (1966).
13. R. Barrie and R. G. Rystephanick, *Can. J. Phys.* **44**:109 (1966).
14. M. A. Krivoglaz, *Zh. Eksperim. i Teor. Fiz.* **48**:310 (1965); *Soviet Phys. JETP (English Transl.)* **21**:204 (1965).
15. Yu Kagan, *Soviet Phys. JETP (English Transl.)* **20**:243 (1965).

CHAPTER 22

Configurational Coordinates

Jordan J. Markham

Physics Department
Illinois Institute of Technology
Chicago, Illinois

In these chapters, attempts were made to develop the basic concepts of configurational coordinates, using simple models which give most of the features found in more realistic cases. A configurational coordinate diagram is a plot of the system's energy as a function of the displacements of the nuclei (atoms or ions) which make up the system. The displacements are described by normal modes. For a diatomic molecule, the displacement is the change in the distance between the two nuclei.

In recent years, a great deal of work has been done on the configurational coordinates to be associated with F-centers. An F-center is an imperfection which forms in an alkali halide, when the site of a missing negative ion captures an electron. Here, in the simplest case, one plots the system's energy against the radial displacements of the six nearest neighbor ions. We used the developments on the F-center as an illustration, since they are a good starting point for the complete theory of configurational coordinates.

We developed the important concepts by examining the total Hamiltonian for the system made of ions and electrons. The approximations made were suggested by Born and Oppenheimer's treatment of diatomic molecules (for details see [1]). This approach is more rigorous than many and gives a deep insight into the basic assumptions and approximations that one must make.

The problem of particular interest was an imperfection with one or more electrons which have two bound states. In principle, every electronic state has a separate set of normal coordinates which describes the displacements of the surrounding atoms or ions. If the interaction between the trapped electron (or electrons) and the ions is not too large, one may obtain relations between the normal modes associated with the two trapped states. The phrase "not too large" has to be defined, and this is done by expanding the electronic eigenvalues which occur in the Born–Oppenheimer technique. In the simplest situation, one may show that the electronic transition causes a displacement of the equilibrium position of the ions, i.e., the position where $q = 0$ is altered.* The next approximation results in the change of the frequencies associated with the modes. The changes in the equilibrium positions during an electronic transition are primarily responsible for the width of many optical absorption and emission bands which occur in solids.

The view adopted by the author is that this problem can be treated in a completely quantum-mechanical fashion provided one is limited to relatively simple models. The author believes that the quantum-mechanical approach gives a depth of insight which is absent in the classical or semiclassical approach. In some complex situations the present quantum-mechanical approach cannot be employed and "simpler" classical methods are of value. Caution, however, must be used in employing these approaches.

The theory can be developed so that it will apply to two situations: (1) when an absorption (or emission) has some fine structure (this usually appears in low-temperature measurements), and (2) in the case where the band has only as overall shape. The difference depends on the size of the Huang-Rhys factor (this factor is defined in [1]). The theory presented here applies generally to the second case, although it can be modified to apply to the first. Several interesting modifications to the elementary approaches have appeared in the last several years.

A considerable portion of these chapters was devoted to an examination of the experimental data obtained from the absorption band associated with the F-center in KCl, since it has been studied with care. The major absorption due to this imperfection is known as the F-band, and it occurs in the visible spectra regions for most of the alkali halides. The shape of this band and its temperature dependence give an insight into the electron–phonon interaction and the true meaning of a configurational coordinate diagram.

Markham has just completed an introduction to this subject [1] and has also written a more advanced treatment of this field which reviews the development in the decade before 1960 [2]. Actually, the field of strong electron–phonon interaction is one of major interest in solid state physics at present, and many new ideas are appearing all the time. Several reviews of this field have recently appeared, in particular, one by Maradudin [3].

REFERENCES

1. J. J. Markham, *F-Centers in Alkali Halides*, Academic Press (New York), 1966, Chapter X.
2. J. J. Markham, *Rev. Mod. Phys.* **31**:956 (1959).
3. A. A. Maradudin, *Solid State Phys.* **18**:274 (1966).

---

* If the equilibrium position changes with the electronic state, we are studying the case of "strong electron–phonon interaction."

Author Index

Subject Index